Surface Hardening of Steels

Understanding the Basics

Edited by

J.R. Davis
Davis & Associates

ASM International
Materials Park, OH 44073-0002

www.asminternational.org

First printing, November 2002
Digital printing, September 2009
Third printing, July 2019

ASM International staff who worked on this project included; Bonnie Sanders, Manager of Production; Nancy Hrivnak and Jill Kinson, Production Editors; and Scott Henry, Assistant Director of Reference Publications.

Library of Congress Cataloging-in-Publication Data

Surface hardening of steels : understanding the basics / edited by J.R. Davis
p. c.m.
Includes bibliographical references and index.
1. Steel—Metallurgy. I. Title: Surface treatment of steels. II. Davis, J.R. (Joseph R.)

TN731.S85 2002
672.3′6—dc21

2002034248

ISBN-13: 978-0-87170-764-2 (print)
ISBN-10: 0-87170-764-0 (print)
ISBN: 978-1-61503-250-1 (pdf)
SAN: 204-7586

ASM International®
Materials Park, OH 44073-0002
www.asminternational.org

Printed in the United States of America

Contents

Introduction

Surface Hardening by Thermochemical Diffusion Treatments

Surface Hardening by Non-Diffusion Methods

Appendix

Preface

Surface hardening is a generic term covering a number of processes used to improve the wear resistance of ferrous parts without affecting the more soft, tough interior of the part. The combination of a hard surface and softer interior, made possible, for example, by case hardening methods, is of inestimable value in modern engineering practice. By the use of high-quality alloy steels, great strength and toughness in the core can be combined with extreme surface hardness, resulting in a composite structure capable of withstanding certain kinds of stress to a high degree. For less exacting requirements, there are many applications where low or moderate core properties, together with a high degree of surface hardness, can be obtained with cheaply fabricated, low-priced carbon steel.

Three of the major types of treatments that are conventionally used to enhance surface wear resistance are classified as thermochemical, thermal, and coating or plating. Examples of thermochemical treatments include carburizing, nitriding, and carbonitriding (case hardening methods). Examples of thermal or applied energy treatments include flame and induction hardening. Coating or plating processes include hard chromium plating, electroless nickel plating, and various hardfacing methods. These long established surface hardening technologies are continually being improved and remain among the most widely used. However, today completely different surface hardening technologies are being applied to and developed for steels. The objective remains the same, that is, enhanced surface performance, but technologies that incorporate high-energy beams, plasmas, magnetic and electrical fields, and vacuums are being applied. Some of these technologies have been used for some time in the electronics industry to fabricate thin film devices and circuits and therefore cannot be considered new technologies. Their application to steel, however, is relatively new, and they elevate the level of sophistication and control of surface hardening.

Surface Hardening of Steels: Understanding the Basics is a practical selection guide to help engineers and technicians choose the optimum surface hardening treatment for a given application from an ever increasing number of options. Emphasis is placed on characteristics such as processing temperature, case/coating thickness, bond strength, and hardness level obtained. The advantages and limitations of the various surface modification technologies are compared. Recent developments in the understanding of the relationships between microstructure and fatigue and wear performance are reviewed, as are more recently introduced surface hardening processes such as vacuum/plasma-related technologies, laser processing, chemical vapor deposition/physical vapor deposition, and ion implantation. Methods for evaluating hardness patterns and depths of hardness for quality control and failure analysis are described. Metallurgical comparisons are made between those processes that offer rapid heating and rapid cooling (self-quenching) characteristics—for example, induction hardening—and conventional furnace hardening. Metallurgical characteristics and properties obtained by atmosphere and vacuum carburizing are also compared. Wear and corrosion data are also provided to demonstrate the benefits of each process.

The successful completion of this book would not have been possible without the generous assistance of the ASM staff. In particular I would like to thank Scott Henry, Assistant Director, Technical Publications; Don Baxter, Managing Editor, *Advanced Materials & Processes* and *Heat Treating Progress*; and Eleanor Baldwin from the ASM Library. I called on them often for help, perhaps too often, but they always came through. Thanks again my fellow colleagues and friends.

Joseph R. Davis
Davis & Associates
Chagrin Falls, Ohio

CHAPTER 1

Process Selection Guide

SURFACE HARDENING, a process that includes a wide variety of techniques (Table 1), is used to improve the wear resistance of parts without affecting the more soft, tough interior of the part. This combination of hard surface and resistance to breakage on impact is useful in parts such as a cam or ring gear that must have a very hard surface to resist wear, along with a tough interior to resist the impact that occurs during operation. Further, the surface hardening of steel has an advantage over through hardening, because less expensive low- and medium-carbon steels can be surface hardened without the problems of distortion and cracking associated with the through hardening of thick sections.

There are three distinctly different approaches to the various methods for surface hardening (Table 1):

- Thermochemical diffusion methods, which modify the chemical composition of the surface with hardening species such as carbon, nitrogen, and boron. Diffusion methods allow effective hardening of the entire surface of a part and are generally used when a large number of parts are to be surface hardened.
- Applied energy or thermal methods, which do not modify the chemical composition of the surface but rather improve properties by altering the surface metallurgy; that is, they produce a hard quenched surface without additional alloying species.
- Surface coating or surface-modification methods, which involve the intentional buildup of a new layer on the steel substrate or, in the case of ion implantation, alter the subsurface chemical composition

Each of these approaches for surface hardening is briefly reviewed in this chapter, with emphasis placed on process comparisons to facilitate process selection. More detailed information on the various methods described can be found in subsequent chapters.

Table 1 Engineering methods for surface hardening of steels

Diffusion methods
Carburizing
Nitriding
Carbonitriding
Nitrocarburizing
Boriding
Thermal diffusion process
Applied energy methods
Flame hardening
Induction hardening
Laser beam hardening
Electron beam hardening
Coating and surface modification
Hard chromium plating
Electroless nickel plating
Thermal spraying
Weld hardfacing
Chemical vapor deposition
Physical vapor deposition
Ion implantation
Laser surface processing

Diffusion Methods of Surface Hardening

As previously mentioned, surface hardening by diffusion involves the chemical modification of a surface. The basic process used is thermochemical, because some heat is needed to enhance the diffusion of hardening species into the surface and subsurface regions of a part. The depth of diffusion exhibits a time-temperature dependence such that:

$$\text{Case depth} \propto K\sqrt{\text{Time}} \qquad \text{(Eq 1)}$$

where the diffusivity constant, *K*, depends on temperature, the chemical composition of the steel, and the concentration gradient of a given hardening species. In terms of temperature, the diffusivity constant increases exponentially as a function of absolute temperature. Concentration gradients depend on the surface kinetics and reactions of a particular process.

Methods of hardening by diffusion include several variations of hardening species (such as carbon, nitrogen, or boron) and of the process method used to handle and transport the hardening species to the surface of the part. Process methods for exposure involve the handling of hardening species in forms such as gas, liquid, or ions. These process variations naturally produce differences in typical case depth and hardness (Table 2). Factors influencing the suitability of a particular diffusion method include the type of steel, the desired case hardness, and the case depth.

It is also important to distinguish between total case depth and effective case depth. The effective case depth is typically approximately two-thirds to three-fourths the total case depth. The required effective depth must be specified so that the heat treater can process the parts for the correct time at the proper temperature.

Carburizing

Carburizing is the addition of carbon to the surface of low-carbon steels at temperatures (generally between 850 and 950 °C, or 1560 and 1740 °F) at which austenite, with its high solubility for carbon, is the stable crystal structure. Hardening of the component is accomplished by removing the part and quenching or allowing the part to slowly cool and then reheating to the austenitizing temperature to maintain the very hard surface property. On quenching, a good wear- and fatigue-resistant high-carbon martensitic case is superimposed on a tough, low-carbon steel core. Carburized steels used in case hardening usually have base carbon contents of approximately 0.2 wt%, with the carbon content of the carburized layer being fixed between 0.8 and 1.0 wt%. Carburizing methods include gas carburizing, vacuum carburizing, plasma (ion) carburizing, salt bath carburizing, and pack carburizing. These methods introduce carbon by use of an atmosphere (atmospheric gas, plasma, and vacuum), liquids (salt bath), or solid compounds (pack). The vast majority of carburized parts are processed by gas carburizing, using natural gas, propane, or butane. Vacuum and plasma carburizing are useful because of the absence of oxygen in the furnace atmosphere. Salt bath and pack carburizing have little commercial importance but are still done occasionally.

Gas carburizing can be run as a batch or a continuous process. Furnace atmospheres consist of a carrier gas and an enriching gas. The carrier gas is supplied at a high flow rate to ensure a positive furnace pressure, minimizing air entry into the furnace. The type of carrier gas affects the rate of carburization. Carburization by methane is slower than by the decomposition of carbon monoxide (CO). The enriching gas provides the source of carbon and is supplied at a rate necessary to satisfy the carbon demand of the work load.

Most gas carburizing is done under conditions of controlled carbon potential by measurement of the CO and carbon dioxide (CO_2) content. The objective of the control is to maintain a constant carbon potential by matching the loss in carbon to the workpiece with the supply of enriching gas. The carburization process is complex, and a comprehensive model of carburization requires algorithms that describe the various steps in the process, including carbon diffusion, kinetics of the surface reaction, kinetics of the reaction between the endogas and enriching gas, purging (for batch processes), and the atmospheric control system.

Vacuum carburizing is a nonequilibrium, boost-diffusion-type carburizing process in which austenitizing takes place in a rough vacuum, followed by carburization in a partial pressure of hydrocarbon gas, diffusion in a rough vacuum, and then quenching in either oil or gas. Vacuum carburizing offers the advantages of excellent uniformity and reproducibility because of the improved process control with vacuum furnaces, improved mechanical properties due to the lack of intergranular oxidation, and reduced cycle time. The disadvantages of vacuum carburizing are predominantly related to equipment costs and throughput.

Plasma (ion) carburizing is basically a vacuum process using glow-discharge technology to introduce carbon-bearing ions to the steel surface for subsequent diffusion. This process is effective in increasing carburization rates, because the process bypasses several dissociation steps that produce active soluble carbon. For example, because of the ionizing effect of the plasmas, active carbon for adsorption can be formed directly from methane (CH_4) gas. High temperatures can be used in plasma carburizing,

Table 2 Typical characteristics of diffusion treatments

Process	Nature of case	Process temperature, °C (°F)	Typical case depth	Case hardness, HRC	Typical base metals	Process characteristics
Carburizing						
Pack	Diffused carbon	815–1090 (1500–2000)	125 μm–1.5 mm (5–60 mils)	50–63(a)	Low-carbon steels, low-carbon alloy steels	Low equipment costs, difficult to control case depth accurately
Gas	Diffused carbon	815–980 (1500–1800)	75 μm–1.5 mm (3–60 mils)	50–63(a)	Low-carbon steels, low-carbon alloy steels	Good control of case depth, suitable for continuous operation, good gas controls required, can be dangerous
Liquid	Diffused carbon and possibly nitrogen	815–980 (1500–1800)	50 μm–1.5 mm (2–60 mils)	50–65(a)	Low-carbon steels, low-carbon alloy steels	Faster than pack and gas processes, can pose salt disposal problem, salt baths require frequent maintenance
Vacuum	Diffused carbon	815–1090 (1500–2000)	75 μm–1.5 mm (3–60 mils)	50–63(a)	Low-carbon steels, low-carbon alloy steels	Excellent process control, bright parts, faster than gas carburizing, high equipment costs
Nitriding						
Gas	Diffused nitrogen, nitrogen compounds	480–590 (900–1100)	125 μm–0.75 mm (5–30 mils)	50–70	Alloy steels, nitriding steels, stainless steels	Hardest cases from nitriding steels, quenching not required, low distortion, process is slow, is usually a batch process
Salt	Diffused nitrogen, nitrogen compounds	510–565 (950–1050)	2.5 μm–0.75 mm (0.1–30 mils)	50–70	Most ferrous metals including cast irons	Usually used for thin hard cases <25 μm (1 mil), no white layer, most are proprietary processes
Ion	Diffused nitrogen, nitrogen compounds	340–565 (650–1050)	75 μm–0.75 mm (3–30 mils)	50–70	Alloy steels, nitriding steels, stainless steels	Faster than gas nitriding, no white layer, high equipment costs, close case control
Carbonitriding						
Gas	Diffused carbon and nitrogen	760–870 (1400–1600)	75 μm–0.75 mm (3–30 mils)	50–65(a)	Low-carbon steels, low-carbon alloy steels, stainless steel	Lower temperature than carburizing (less distortion), slightly harder case than carburizing, gas control critical
Liquid (cyaniding)	Diffused carbon and nitrogen	760–870 (1400–1600)	2.5–125 μm (0.1–5 mils)	50–65(a)	Low-carbon steels	Good for thin cases on noncritical parts, batch process, salt disposal problems
Ferritic nitrocarburizing	Diffused carbon and nitrogen	565–675 (1050–1250)	2.5–25 μm (0.1–1 mil)	40–60(a)	Low-carbon steels	Low-distortion process for thin case on low-carbon steel, most processes are proprietary
Other						
Boriding	Diffused boron, boron compounds	400–1150 (750–2100)	12.5–50 μm (0.5–2 mils)	40–>70	Alloy steels, tool steels, cobalt and nickel alloys	Produces a hard compound layer, mostly applied over hardened tool steels, high process temperature can cause distortion
Thermal diffusion process	Diffused carbide layers via salt bath processing	800–1250°C (1475–2285 °F)	2–20 μm (0.08–0.8 mil)	>70	Tool steels, alloy steels, medium-carbon steels	Produces a hard compound layer, mostly applied over hardened tool steels, high process temperature can cause distortion

(a) Requires quench from austenitizing temperature

because the process takes place in an oxygen-free vacuum, thus producing a greater carburized case depth than both atmospheric gas and vacuum carburizing.

Salt bath or liquid carburizing is a method of case hardening steel in a molten salt bath that contains the chemicals required to produce a case comparable with one resulting from gas or pack carburizing. Carburizing in liquid salt baths provides a convenient method of case hardening, with low distortion and considerable flexibility and uniformity of control of the case. However, the expense and environmental problems associated with disposing of salt baths, particularly those containing cyanide, have limited the use of this process, although non-cyanide-containing salts have been developed.

Pack carburizing is the oldest carburizing process. In this case-hardening method, parts are packed in a blend of coke and charcoal with "activators" and then heated in a closed container. Although a labor-intensive process, pack carburizing is still practiced in some tool rooms, because facility requirements are minimal.

Nitriding

Nitriding is a process similar to carburizing, in which nitrogen is diffused into the surface of a ferrous product to produce a hard case. Unlike carburizing, nitrogen is introduced between 500 and 550 °C (930 and 1020 °F), which is below the austenite formation temperature (Ac_1) for ferritic steels, and quenching is not required. As a result of not austenitizing and quenching to form martensite, nitriding results in minimum distortion and excellent control. The various nitriding processes (Table 2) include gas nitriding, liquid nitriding, and plasma (ion) nitriding.

Gas nitriding is a case-hardening process that takes place in the presence of ammonia gas. Either a single-stage or a double-stage process can be used when nitriding with anhydrous ammonia. The single-stage process, in which a temperature of 495 to 525 °C (925 to 975 °F) is used, produces the brittle nitrogen-rich compound zone known as the white nitride layer at the surface of the nitrided case. The double-stage process, or Floe process, has the advantage of reducing the white nitrided layer thickness. After the first stage, a second stage is added either by continuing at the first-stage temperature or increasing the temperature to 550 to 565 °C (1025 to 1050 °F). The use of the higher-temperature second stage lowers the case hardness and increases the case depth.

Liquid nitriding (nitriding in a molten salt bath) uses temperatures similar to those used in gas nitriding and a case-hardening medium of molten, nitrogen-bearing, fused salt bath containing either cyanides or cyanates. Similar to salt bath carburizing, liquid nitriding has the advantage of processing finished parts because dimensional stability can be maintained due to the subcritical temperatures used in the process. Furthermore, at the lower nitriding temperatures, liquid nitriding adds more nitrogen and less carbon to ferrous materials than that obtained with high-temperature treatments because ferrite has a much greater solubility for nitrogen (0.4% max) than carbon (0.02% max).

Plasma (ion) nitriding is a method of surface hardening using glow-discharge technology to introduce nascent (elemental) nitrogen to the surface of a metal part for subsequent diffusion into the material. The process is similar to plasma carburizing in that a plasma is formed in a vacuum using high-voltage electrical energy, and the nitrogen ions are accelerated toward the workpiece. The ion bombardment heats the part, cleans the surface, and provides active nitrogen. The process provides better control of case chemistry, case uniformity, and lower part distortion than gas nitriding.

Carbonitriding and Ferritic Nitrocarburizing

Carbonitriding introduces both carbon and nitrogen into the austenite of the steel. The process is similar to carburizing in that the austenite composition is enhanced and the high surface hardness is produced by quenching to form martensite. This process is a modified form of gas carburizing in which ammonia is introduced into the gas-carburizing atmosphere. As in gas nitriding, elemental nitrogen forms at the work-piece surface and diffuses along with carbon into the steel. Typically, carbonitriding takes place at a lower temperature and a shorter time than gas carburizing, producing a shallower case. Steels with carbon contents up to 0.2% are commonly carbonitrided.

Ferritic nitrocarburizing is a subcritical heat treatment process, carried out by liquid, gaseous, or plasma techniques, and involves the diffusion of carbon and nitrogen into the ferritic phase. The process results in the formation of a thin white layer or compound layer with an underlying diffusion zone of dissolved nitrogen in iron, or alloy nitrides. The white layer improves surface resistance to wear, and the diffusion zone increases the fatigue endurance limit, especially in carbon and low-alloy steels. Alloy steels, cast irons, and some stainless steels can be treated. The process is used to produce a thin, hard skin, usually less than 25 μm (1 mil) thick, on low-carbon steels in the form of sheet metal parts, powder metallurgy parts, small shaft sprockets, and so forth.

Boriding

Boriding, or boronizing, is a thermochemical surface-hardening process that can be applied to a wide variety of ferrous, nonferrous, and cermet materials. The boronizing pack process is similar to pack carburizing, with the parts to be coated being packed with a boron-containing compound such as boron powder or ferroboron. Activators such as chlorine and fluorine compounds are added to enhance the production of the boron-rich gas at the part surface. Processing of high-speed tool steels that were previously quench hardened is accomplished at 540 °C (1000 °F). Boronizing at higher temperatures up to 1090 °C (2000 °F) causes diffusion rates to increase, thus reducing the process time. The boron case does not have to be quenched to obtain its high hardness, but tool steels processed in the austenitizing temperature range need to be quenched from the coating temperature to harden the substrate.

Boronizing is most often applied to tool steels or other substrates that are already hardened by heat treatment. The thin (12 to 15 μm, or 0.48 to 0.6 mil) boride compound surfaces provide even greater hardness, improving wear service life. Distortion from the high processing temperatures is a major problem for boronized coatings. Finished parts that are able to tolerate a few thousandths of an inch (75 μm) distortion are better suited for this process sequence, because the thin coating cannot be finish ground.

Thermal Diffusion Process

The thermal diffusion (TD) process is a method of coating steels with a hard, wear-resistant layer of carbides, nitrides, or carbonitrides. In the TD process, the carbon and nitrogen in the steel substrate diffuse into a deposited layer with a carbide-forming or nitride-forming element, such as vanadium, niobium, tantalum, chromium, molybdenum, or tungsten. The diffused carbon or nitrogen reacts with the carbide- and nitride-forming elements in the deposited coating to form a dense and metallurgically bonded carbide or nitride coating at the substrate surface.

The hard alloy carbide, nitride, and carbonitride coatings in the TD method can be applied to steels by means of salt bath processing or fluidized beds. The salt bath method uses molten borax with additions of carbide-forming elements, such as vanadium, niobium, titanium, or chromium, which combine with carbon from the substrate steel to produce alloy carbide layers. Because the growth of the layers is dependent on carbon diffusion, the process requires a relatively high temperature, from 800 to 1250 °C (1470 to 2280 °F), to maintain adequate coating rates. Carbide coating thicknesses of 4 to 7 μm are produced in 10 min to 8 h, depending on bath temperature and type of steel. The coated steels may be cooled and reheated for hardening, or the bath temperature may be selected to correspond to the steel austenitizing temperature, permitting the steel to be quenched directly after coating.

Surface Hardening by Applied Energy

The surface methods described in this section include conventional thermal treatments, such as flame and induction hardening, and technologies that incorporate high-energy laser or electron beams. All of these methods may be classified as simply a thermal treatment without chemistry changes. They can be used to harden the entire surface or localized areas. When localized heating is carried out, the term *selective surface hardening* is used to describe these methods.

Flame hardening consists of austenitizing the surface of steel by heating with an oxyacetylene or oxyhydrogen torch and immediately quenching with water. After quenching, the microstructure of the surface layer consists of hard martensite over a lower-strength interior core of other steel morphologies, such as ferrite and pearlite. A prerequisite for proper flame hardening is that the steel must have adequate carbon and other alloy additions to produce the desired hardness, because there is no change in composition. Flame-hardening equipment uses direct impingement of a high-temperature flame or high-velocity combustion product gases to austenitize the component surface and quickly cool the surface faster than the critical cooling rate to produce martensite in the steel. This is necessary because the hardenability of the component is fixed by the original composition of the steel. Thus, equipment design is critical to the success of the operation. Flame-heating equipment may be a single torch with a specially designed head or an elaborate apparatus that automatically indexes, heats, and quenches parts. With improvements in gas-mixing equipment, infrared temperature measurement and control, and burner rig design, flame hardening has been accepted as a reliable heat treating process that is adaptable to general or localized surface hardening for small or medium-to-high production requirements.

Induction heating is an extremely versatile heating method that can perform uniform surface hardening, localized surface hardening, through hardening, and tempering of hardened pieces. Heating is accomplished by placing a steel part in the magnetic field generated by high-frequency alternating current passing through an inductor, usually a water-cooled copper coil. The depth of heating produced by induction is related to the frequency of the alternating current: the higher the frequency is, the thinner or more shallow the heating. Therefore, deeper case depths and even through hardening are produced by using lower frequencies. The electrical considerations involve the phenomena of hysteresis and eddy currents. Because secondary and radiant heat are eliminated, the process is suited for production line areas. Table 3 compares the flame- and induction-hardening processes.

Laser surface heat treatment is widely used to harden localized areas of steel and cast iron machine components. The heat generated by the absorption of the laser light is controlled to prevent melting and is therefore used in the selective austenitization of local surface regions, which transform to martensite as a result of rapid cooling (self-quenching) by the conduction of heat into the bulk of the workpiece. This process is sometimes referred to as laser transformation hardening to differentiate it from laser surface melting phenomena. There is no chemistry change produced by laser transformation hardening, and the process, similar to induction and flame hardening, provides an effective technique to harden ferrous materials selectively.

The process produces typical case depths for steel ranging from 0.75 to 1.3 mm (0.030 to 0.050 in.), depending on the laser power range, and hardness values as high as 60 HRC. Laser processing has advantages over electron beam hardening in that laser hardening does not require a vacuum, wider hardening profiles are possible, and there can be greater accessibility to hard-to-get areas with the flexibility of optical manipulation of light energy.

Electron Beam Hardening. In electron beam hardening, the surface of the hardenable steel is heated rapidly to the austenitizing temperature, usually with a defocused electron beam to prevent melting. The mass of the workpiece conducts the heat away from the treated surface at a rate that is rapid enough to produce hardening. Materials for application of electron beam hardening must contain sufficient carbon and alloy content to produce martensite. With the rapid heating associated with this process, the carbon and alloy content should be in a form that quickly allows complete solid solution in the austenite at the temperatures produced by the electron beam. In addition, the mass of the workpiece should be sufficient to allow proper quenching; for example, the part thickness must be at least ten times the depth of hardening, and hardened areas must be properly spaced to prevent tempering of previously hardened areas.

To produce an electron beam, a high vacuum of 10^{-3} Pa (10^{-5} torr) is required in the region where the electrons are emitted and accelerated. This vacuum environment protects the emitter

Table 3 Comparison of flame- and induction-hardening processes

Characteristics	Flame	Induction
Equipment	Oxyfuel torch, special head quench system	Power supply, inductor, quench system
Applicable material	Ferrous alloys, carbon steels, alloy steels, cast irons	Same
Speed of heating	Few seconds to few minutes	1–10 s
Depth of hardening	1.2–6.2 mm (0.050–0.250 in.)	0.4–1.5 mm (0.015–0.060 in.); 0.1 mm (0.004 in.) for impulse
Processing	One part at a time	Same
Part size	No limit	Must fit in coil
Tempering	Required	Same
Can be automated	Yes	Yes
Operator skills	Significant skill required	Little skill required after setup
Control of process	Attention required	Very precise
Operator comfort	Hot, eye protection required	Can be done in suit
Cost		
Equipment	Low	High
Per piece	Best for large work	Best for small work

from oxidizing and avoids scattering of the electrons while they are still traveling at a relatively low velocity. Electron beam hardening in hard vacuum units requires that the part be placed in a chamber that is sufficiently large to manipulate the electron beam gun or the workpiece. Out-of-vacuum units usually involve shrouding the workpiece; a partial vacuum (13 Pa, or 10^{-2} torr), is obtained in the work area by mechanical pumps.

Surface Hardening by Coating or Surface Modification

Plating or coating treatments deposit hard surface layers of completely different chemistry, structure, and properties on steel substrates and are applied by well-established technologies such as electrodeposition, electroless deposition, thermal spraying, and weld hardfacing. In more recent years, coating or surface-modification methods long used in the electronics industry to fabricate thin films and devices have been used to treat steels. These include vapor deposition techniques and ion implantation. Laser surface processing (melting, alloying, and cladding) has also been carried out on steels. These various surface-engineering treatments can deposit very thin films (e.g., 1 to 10 μm for physical vapor deposition) or thick coatings (e.g., 3 to 10 mm for weld hardfacing).

Hard Chromium Plating

Process Description. Hard chromium plating is produced by electrodeposition from a solution containing chromic acid (CrO_3) and a catalytic anion in proper proportion. The metal so produced is extremely hard (850 to 1000 HV) and corrosion resistant. Plating thickness ranges from 2.5 to 500 μm (0.1 to 20 mils).

Applications. Hard chromium plating is used for products such as piston rings, shock absorbers, struts, brake pistons, engine valve stems, cylinder liners, and hydraulic rods. Other applications are for aircraft landing gears, textile and gravure rolls, plastic rolls, and dies and molds. The rebuilding of mismachined or worn parts comprises large segments of the industry.

Advantages:

- Low-temperature treatment (60 °C, or 140 °F)
- High hardness and wear resistance
- Low coefficient of friction
- Thick layers possible

Disadvantages:

- Poor thickness uniformity on complex components
- Hydrogen embrittlement
- Environmental problems associated with plating bath disposal. Chromium replacement coatings, such as electroless nickel and thermal spray coatings, are being used increasingly.

Electroless Nickel Coating

Process Description. The coating is deposited by an autocatalytic chemical reduction of nickel ions by hydrophosphite, aminoborane, or borohydride compounds. Currently, hot acid hypophosphite-reduced baths are most frequently chosen to coat steel. Heat-treated deposit hardness exceeds 1000 HV.

Applications. Electroless nickel coatings have good resistance to corrosion and wear and are used to protect machinery found in the petroleum, chemicals, plastics, optics, printing, mining, aerospace, nuclear, automotive, electronics, computers, textiles, paper, and food industries.

Advantages:

- Low-temperature treatment (<100 °C, or 212 °F)
- More corrosion resistance than electroplated chromium
- Ability to coat complex shapes uniformly
- Incorporation of hard particles to increase hardness
- Good solderability and brazeability

Disadvantages:

- Higher costs than electroplating
- Poor welding characteristics
- Slower plating rate, as compared to rates for electrolytic methods
- Heat treatment needed to develop optimal properties

Thermal Spraying

Process Description. Thermal spraying is a generic term for a group of processes in which a metallic, ceramic, cermet, and some polymeric materials in the form of powder, wire, or rod are fed to a torch or gun with which they are heated to or slightly above their melting point. The resulting molten or nearly molten droplets of material are accelerated in a gas stream and projected against the substrate to form a coating. Commonly employed methods of deposited thermal spray coatings can be classified as wire flame spray, powder flame spray, electric arc, plasma spray, and high-velocity oxyfuel (HVOF) spray. Process characteristics are compared in Table 4.

Applications. Thermal spray coatings are used for prevention against wear, corrosion, or oxidation. Table 19 in Chapter 11 lists a wide variety of wear-resistant applications for thermal spraying. Thermal spray coatings, particularly HVOF coatings, are being used increasingly to replace chromium electrodeposits.

Advantages:

- Most metals, ceramics, and some polymers can be sprayed.
- Significant substrate heating does not occur with most thermal spray processes.
- Worn or damaged coatings can be stripped without changing the properties or dimensions of the part.
- Localized treatments are possible.

Disadvantages:

- Line-of-sight process employed.
- Most sprayed coatings contain some porosity.
- The adhesion of sprayed coatings is generally poor, compared to other processes.

Table 4 Comparison of major thermal spray coating processes

Property or characteristic	Coating type	Process: Wire flame spray	Powder flame spray	Electric arc	Plasma spray	High-velocity oxyfuel (HVOF)
Bond strength, MPa (10^3 psi)	Ferrous metals	14 (2)	28 (4)	41 (6)	34+ (5+)	62 (9)
	Nonferrous metals	21 (3)	21 (3)	41+ (6+)	34+ (5+)	70 (10.2)
	Self-fluxing alloys	. . .	69+ (10+)(a)	. . .	. . .	62 (9)(b)
	Ceramics	. . .	14–34 (2–5)	. . .	21+ (3+)	. . .
	Carbides	. . .	34–48 (5–7)	. . .	55–69 (8–10)	83+ (12+)
Density, % that of equivalent wrought material	Ferrous metals	90	90	90	95	98+
	Nonferrous metals	90	90	90	95	98+
	Self-fluxing alloys	. . .	100(a)	. . .	. . .	100(a)
	Ceramics	. . .	95	. . .	95+	. . .
	Carbides	. . .	90	. . .	95+	98+
Hardness	Ferrous metals	84 HRB–35 HRC	80 HRB–35 HRC	95 HRB–40 HRC	80 HRB–40 HRC	90 HRB–50 HRC
	Nonferrous metals	95 HRH–40 HRC	30 HRH–20 HRC	40 HRH–80 HRB	40 HRH–40 HRC	100 HRH–55 HRC
	Self-fluxing alloys	. . .	30–60 HRC	. . .	. . .	50–60 HRC
	Ceramics	. . .	50–65 HRC	. . .	50–70 HRC	. . .
	Carbides	. . .	50–60 HRC	. . .	50–60 HRC	55–65 HRC
Permeability	Ferrous metals	Medium	Medium	High	Low	Negligible
	Nonferrous metals	Medium	Medium	High	Low	Negligible
	Self-fluxing alloys	. . .	None(a)	. . .	. . .	None(a)
	Ceramics	. . .	Medium	. . .	Low	. . .
	Carbides	. . .	Low	. . .	Low	Negligible
Coating-thickness limitation, mm (in.)	Ferrous metals	1.25–2.5 (0.05–0.1)	1.25–2.5 (0.05–0.1)	1.25–2.5 (0.05–0.1)	1.25–2.5 (0.05–0.1)	1.25–2.5 (0.05–0.1)
	Nonferrous metals	1.25–5 (0.05–0.2)	1.25–5 (0.05–0.2)	1.25–5 (0.05–0.2)	1.25–5 (0.05–0.2)	2.5–5 (0.1–0.2)
	Self-fluxing alloys	. . .	0.4–2.5 (0.015–0.1)	. . .	. . . (0.05)	1.25
	Ceramics	. . .	0.4 (0.015)	. . .	0.4 (0.015) max	. . .
	Carbides	. . .	0.4 (0.015)	. . .	0.4 (0.015) max	0.6 (0.025)

(a) Fused coating. (b) Unfused coating

- High-quality coatings on reentrant surfaces produced with difficulty.

Weld Hardfacing

Process Description. Welding is a solidification method for applying coatings with corrosion, wear, and erosion resistance. Weld-overlay coatings, sometimes referred to as hardfacing, offer unique advantages over other coating systems in that the overlay/substrate weld provides a metallurgical bond that is not susceptible to spallation and can easily be applied free of porosity or other defects. Welded deposits of surface alloys can be applied in thicknesses greater than most other techniques, typically in the range of 3 to 10 mm. Most welding processes are used for application of surface coatings, and on-site deposition can be more easily carried out, particularly for repair purposes.

Applications. Hardfacing applications for wear control vary widely, ranging from very severe abrasive wear service, such as rock crushing and pulverizing, to applications to minimize metal-to-metal wear, such as control valves where a few thousandths of an inch of wear is intolerable. Hardfacing is used for controlling abrasive wear, such as encountered by mill hammers, digging tools, extrusion screws, cutting shears, parts of earthmoving equipment, ball mills, and crusher parts. It is also used to control the wear of unlubricated or poorly lubricated metal-to-metal sliding contacts, such as control valves, undercarriage parts of tractors and shovels, and high-performance bearings. Hardfacing also is used to control combinations of wear and corrosion.

Advantages:

- Inexpensive
- Applicable to large components
- Localized coating possible
- Excellent coating/substrate adhesion
- High deposition rates possible

Disadvantages:

- Residual stresses and distortion can cause serious problems.
- Weld defects can lead to joint failure.
- Minimum thickness limits (it is impractical to produce layers less than 2 to 3 mm thick)
- Limited number of coating materials available, compared to thermal spraying

Chemical Vapor Deposition (CVD)

Process Description. Chemical vapor deposition involves the formation of a coating on a heated surface by a chemical reaction from the vapor or gas phase. Deposition temperatures are generally in the range of 800 to 1000 °C (1470 to 1830 °F). The most widely deposited wear-resistant coatings are titanium carbide (TiC), titanium nitride (TiN), chromium carbide, and alumina. Thicknesses are restricted to approximately 10 μm due to thermal expansion mismatch stresses that develop on cooling.

Applications. The use of the CVD process for steels has been largely limited to the coating of tool steels for wear resistance.

Advantages:

- High coating hardness; for example, TiN coatings have a hardness of 2500 HV.
- Good adhesion (provided the coating is not too thick)
- Good throwing power (i.e., uniformity of coating)

Disadvantages:

- High-temperature process (distortion a problem)
- Shard edge coating is difficult due to thermal expansion mismatch stresses.
- Limited range of materials can be coated.
- Environmental concerns about process gases

Physical Vapor Deposition (PVD)

Process Description. Physical vapor deposition processes involve the formation of a coating on a substrate by physical deposition of atoms, ions, or molecules of the coating species. There are three main techniques for applying PVD coatings: thermal evaporation, sputtering, and ion plating. Thermal evaporation involves heating of the material until it forms a vapor that condenses on a substrate to form a coating. Sputtering involves the electrical generation of a plasma between the coating species and the substrate. Ion plating is essentially a combination of these two processes. A comparison of the process characteristics of PVD, CVD, and ion implantation is provided in Table 5.

Applications. Similar to CVD, the PVD process is used to increase the wear resistance of tool steels by the deposition of thin TiN or TiC coatings at temperatures ranging from 200 to

550 °C (400 to 1025 °F). This temperature range is much more suitable for the coating of tool steels than the temperatures required for CVD.

Advantages:

- Excellent process control
- Low deposition temperature
- Dense, adherent coatings
- Elemental, alloy, and compound coatings possible

Disadvantages:

- Vacuum process with high capital cost
- Limited component size treatable
- Relatively low coating rates
- Poor throwing power without manipulation of components

Ion Implantation

Process Description. Ion implantation involves the bombardment of a solid material with medium- to high-energy ionized atoms and offers the ability to alloy virtually any elemental species into the near-surface region of any substrate. The advantage of such a process is that it produces improved surface properties without the limitations of dimensional changes or delamination found in conventional coatings.

Applications. For steels, the most common application of ion implantation is nitrogen-implanted tool steels used for forming and cutting tools. Titanium plus carbon implantation has also proved beneficial for tool steels.

Advantages:

- Produces surface alloys independent of thermodynamic criteria
- No delamination concerns
- No significant dimensional changes
- Ambient-temperature processing possible
- Enhance surface properties while retaining bulk properties
- High degree of control and reproducibility

Disadvantages:

- Very thin treated layer (1 μm or less)
- High-vacuum process
- Line-of-sight process
- Alloy concentrations dependent on sputtering
- Relatively costly process; intensive training required, compared to other surface-treatment processes
- Limited commercial treatment facilities available

Laser Surface Processing

Process Description. Laser surface processing involves the melting of a surface with a

Table 5 Comparison of PVD, CVD, and ion implantation process characteristics

Process	Processing temperature, °C	Throwing power	Coating materials	Coating applications and special features
Vacuum evaporation	RT–700, usually <200	Line of sight	Chiefly metal, especially Al (a few simple alloys/a few simple compounds)	Electronic, optical, decorative, simple masking
Ion implantation	200–400, best <250 for N	Line of sight	Usually N (B, C)	Wear resistance for tools, dies, etc. Effect much deeper than original implantation depth. Precise area treatment, excellent process control
Ion plating, ARE	RT–0.7 T_m of coating. Best at elevated temperatures	Moderate to good	Ion plating: Al, other metals (few alloys). ARE: TiN and other compounds	Electronic, optical, decorative. Corrosion and wear resistance. Dry lubricants. Thicker engineering coatings
Sputtering	RT–0.7 T_m of metal coatings. Best >200 for nonmetals	Line of sight	Metals, alloys, glasses, oxides. TiN and other compounds(a)	Electronic, optical, wear resistance. Architectural (decorative). Generally thin coatings. Excellent process control
CVD	300–2000, usually 600–1200	Very good	Metals, especially refractory TiN and other compounds(a), pyrolytic BN	Thin, wear-resistant films on metal and carbide dies, tools, etc. Free-standing bodies of refractory metals and pyrolytic C or BN

PVD, physical vapor deposition; CVD, chemical vapor deposition; RT, room temperature; ARE, activated reactive evaporation; T_m, absolute melting temperature. (a) Compounds: oxides, nitrides, carbides, silicides, and borides of Al, B, Cr, Hf, Mo, Nb, Ni, Re, Si, Ta, Ti, V, W, and Zr laser, with or without surface additions. With laser surface melting, melting and controlled cooling are combined to refine the microstructure or to produce an amorphous (or nearly amorphous) structure. No external material is added during this process. The composition and properties of the surface can also be modified by adding external material via powder injection or wire feed. External material can also be placed on the surface by powder deposition, electroplating, vapor deposition, or thermal spray, then incorporated by laser scanning. The nature of incorporating material in the modified surface varies depending on laser processing parameters, such as energy density and traverse speed. Alloy, clad, and composite surface layers may be formed in this way.

Applications. Although laser surface processing has not reached commercial significance for steels, various carbon and low-alloy steels, tool steels, and stainless steels have been laser processed to varying degrees of success. See Chapter 11 for application examples.

Advantages:

- Rapid rates of processing produce novel structures in the surface region not possible with conventional processing.
- For laser cladding, low weld-metal dilution rates and distortion, compared with competing arc welding methods

Disadvantages:

- High capital equipment costs
- Some substrate materials are not compatible with the laser thermal conduction requirements.

Important Considerations for Process Selection

Performance Requirements. The key to proper selection of surface-hardening techniques is in the identification of the performance requirements for a given surface-modified material system in a given application. Not only must the properties of the surface be considered but also the properties of the substrate and the interface between the surface and substrate. In some systems there is a gradual change in properties between the surface and interior, as, for example, in nitrided and carburized components, while in others there is an abrupt change, as, for example, for parts where a coating of vapor-deposited titanium nitride has been deposited on steel. Such interface characteristics may significantly influence the performance of a surface-modified system.

The performance requirements of surface-engineered systems may vary widely. For example, heavily loaded systems, such as bearings and gears, require deep cases to resist rolling contact and bending stresses that result in fatigue damage. Other applications may require only very thin surface films to resist near-surface abrasion or scuffing or to reduce friction between moving surfaces. Many of these requirements are based on complex interactions between applied static and cyclic stress states and gradients in structures and properties of the surface-modified systems (see, for example, the discussion of bending fatigue strength of carburized steels in Chapter 2).

Design Constraints. The component design constraints include consideration of the size and shape of the component, because they may affect surface-treatment process capabilities. Will the component fit in the coating equipment? Can a line-of-sight process be used? Do small holes or channels require a process with high throwing power? What kind of masking will be required to prevent coating unwanted areas? Is the temperature required by the surface treatment compatible with the temperature limitations of the component material? What kind of postcoating treatment, including heat treatment and finishing, will be required?

Economic Analysis. The economic issue of fundamental importance is a cost/benefit analysis. This analysis should be based on the full life-cycle costs of the surface treatment, including the process costs (preparation, application, finishing, quality control, waste disposal), utilization costs and benefits (productivity of coated components), and added value of any products produced or treated. Other economic factors that must be considered include the availability of the process, number of components to be treated, quality-control requirements, delivery schedules, and so on. Note that the analysis should not be based on just the initial cost of the surface treatment. Life-cycle costs are as important when comparing the cost of various surface treatments as in comparing a surface treatment to an untreated surface.

Process Comparisons

Hardness versus Wear Resistance. The wear processes that are *usually* mitigated by the use of hard surfaces are low-stress abrasion, wear in systems involving relative sliding of conforming solids, fretting wear, galling, and, to some extent, solid-particle erosion. Unfortunately, there are many caveats to this statement, and substrate/coating selection should be carefully studied, with proper tests carried out if necessary. Coating suppliers should also be consulted. Chapter 3 provides additional information on wear processes and the means to prevent specific types of wear.

Figure 1 shows typical ranges in hardness for many of the surface-engineering processes used to control wear. All of the treatments shown in this figure have hardness values greater than ordinary constructional steel or low-carbon steel. The surface-hardening processes that rely on martensitic transformations all have comparable hardness, and the diffusion treatments that produce harder surfaces are nitriding, boronizing (boriding), and chromizing. The hardest metal coating is chromium plate, although hardened electroless nickel plate can attain values just under that of chromium. The surfaces that exceed the hardness of chromium are the cermets or ceramics or surfaces that are modified so that they are cermets or ceramics. These include nitrides, carbides, borides, and similar compounds. The popular solid ceramics used for wear applications—aluminum oxide, silicon carbide, and silicon nitride—generally have hardnesses in the range of 2000 to 3000 kg/mm^2. As shown in Fig. 1, when materials such as aluminum oxide are applied by plasma spraying or other thermal spray process, they have hardnesses that are less than the same material in solid pressed-and-sintered form. This is because the sprayed materials contain porosity and oxides that are not contained in the sintered solid form.

Cost must be weighed against the performance required for the surface-treatment system. A low-cost surface treatment that fails to perform its function is a wasted expense. Unfortu-

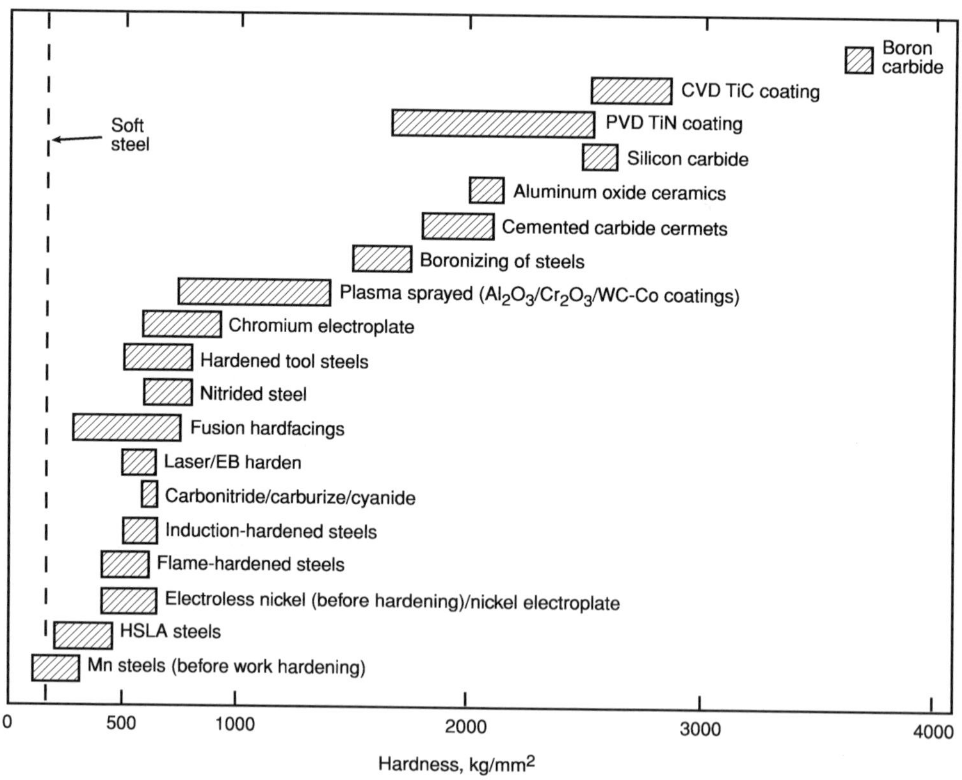

Fig. 1 Range of hardness levels for various materials and surface treatments. EB, electron beam; HSLA, high-strength, low-alloy

nately, it is nearly impossible to give absolute comparative costs for different surface-engineering options. Often, a range of prices is offered for a particular job from different, equally competent candidate suppliers. Probably the most important factor that relates to costs of producing a wear-resistant surface on a part is part quantity. Treating many parts usually allows economies in treatment and finishing.

Another consideration when assessing surface-treatment costs is part size. There are some critical sizes for each surface-treatment process above which the cost of obtaining the treatment may be high. A number of surface treatments require that the part fit into the work zone of a vacuum chamber. The cost of vacuum equipment goes up exponentially with chamber volume.

Other factors to be considered are:

- The time required for a given surface treatment
- Fixturing, masking, and inspection costs
- Final finishing costs
- Material costs
- Energy costs
- Labor costs
- Environmentally related costs, for example, disposal of spent plating solutions
- Expected service life of the coating

Because of these various factors, it is difficult to compare costs with a high degree of accuracy. Figure 2 provides some general guidelines for cost comparisons.

Distortion or Size Change Tendencies. Figure 3 shows the surface temperatures that are encountered in various surface-engineering processes. As indicated in the figure, the processes are categorized into two groups: one group produces negligible part distortion, and the other group contains processes that have varying potential for causing distortion. Obviously, if a part could benefit from a surface treatment but distortion cannot be tolerated, processes that require minimal heating should be considered.

Coating Thickness Attainable. Figure 4 shows the typical thickness/penetration capabilities of various coating and surface treatments. As indicated in the figure, some surface-engineering treatments penetrate into the surface,

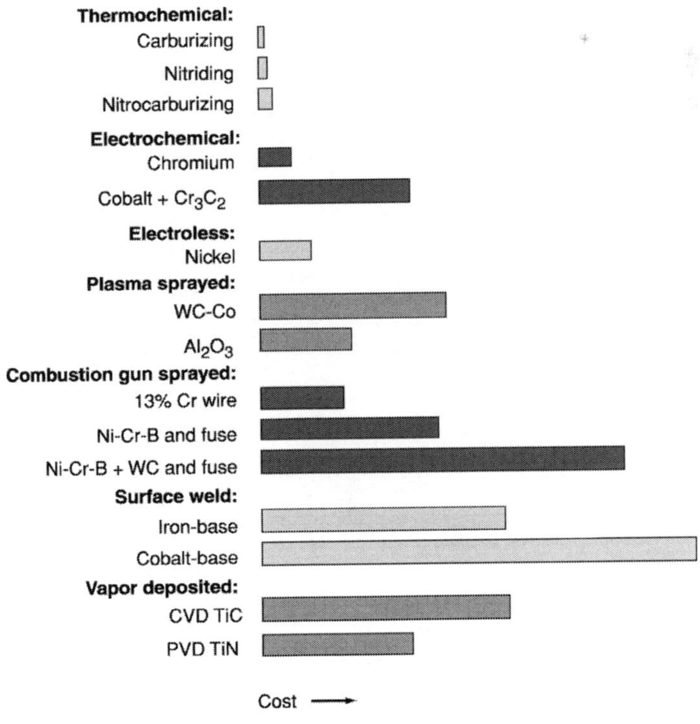

Fig. 2 Approximate relative costs of various surface treatments

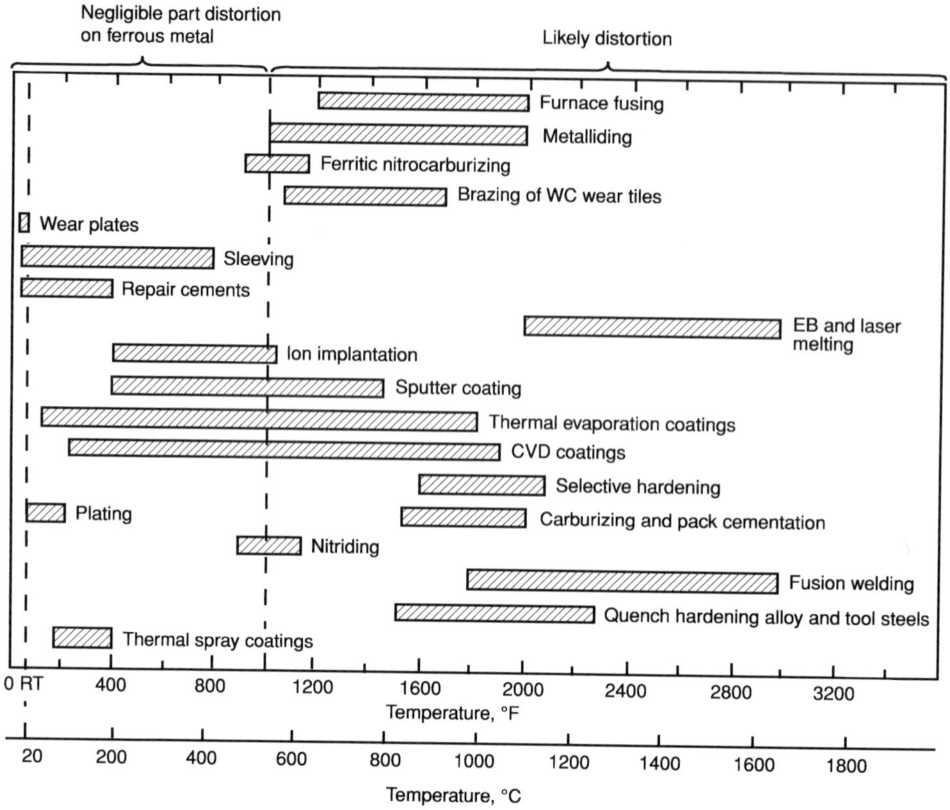

Fig. 3 Maximum surface temperatures that can be anticipated for various surface-engineering processes. The dashed vertical line at 540 °C (1000 °F) represents the temperature limit for distortion for ferrous metals. Obviously, a temperature of 540 °C (1000 °F) would melt a number of nonferrous metals, and it would cause distortion on metals such as aluminum or magnesium. However, this process temperature information can be used to compare the heating that will be required for a particular process. EB, electron beam

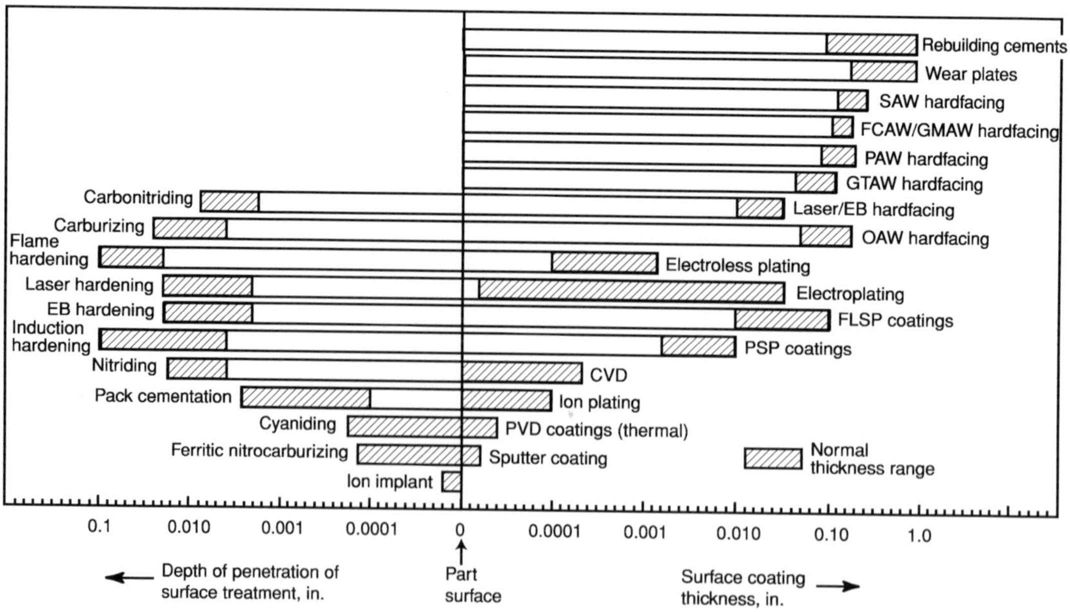

Fig. 4 Typical coating thickness/depth of penetration for various coating and surface-hardening processes. SAW, submerged arc welding; FCAW, flux cored arc welding; GMAW, gas metal arc welding; PAW, plasma arc welding; GTAW, gas tungsten arc welding; EB, electron beam; OAW, oxyacetylene welding; FLSP, flame spraying; PSP, plasma spraying

and there is no intentional buildup on the surface. Other surface treatments coat or intentionally build up the surface. This is a selection factor. Can a part tolerate a buildup on the surface? If not, the selection process is narrowed to the treatments that penetrate into the surface. Other factors affecting the thickness of a given surface treatment include dimensional requirements, the service conditions, the anticipated/allowable corrosion or wear depth, and anticipated loads on the surface. Questions or concerns related to coating thickness should be discussed with the contractor. Available specifications should also be reviewed.

SELECTED REFERENCES

- K.G. Budinski, *Surface Engineering for Wear Resistance,* Prentice-Hall, 1988
- J.R. Davis, Ed., *Surface Engineering for Corrosion and Wear Resistance,* ASM International and IOM Communications, 2001
- G. Krauss, Advanced Surface Modification of Steels, *J. Heat Treat.,* Vol 9 (No. 2), 1992, p 81–89
- S. Lampman, Introduction to Surface Hardening of Steels, *Heat Treating,* Vol 4, *ASM Handbook,* 1991, p 259–267

CHAPTER 2

Gas Carburizing

CARBURIZING is a case-hardening process in which carbon is dissolved in the surface layers of a steel part at a temperature sufficient to render the steel austenitic, followed by quenching and tempering to form a martensitic microstructure. The resulting gradient in carbon content below the surface of the part causes a gradient in hardness, producing a strong, wear-resistant surface layer on a material, usually low-carbon steel, which is readily fabricated into parts. In *gas carburizing,* the source of carbon is carbon-rich furnace atmosphere produced either from gaseous hydrocarbons, for example, methane (CH_4), propane (C_3H_3), and butane (C_4H_{10}), or from vaporized hydrocarbon liquids. High-production-rate gas carburizing is commonly carried out on highly stressed components such as gears, bearings, and shafts.

This chapter is divided into three major Sections. Section I, "Gas Carburizing Practices," deals with the furnaces, atmospheres, kinetics, and control of the carburizing process. Section II, "Microstructures of Carburized Steels," examines the various microstructural constituents that may be present and which significantly influence the performance of carburized parts. Section III, "Properties of Carburized Steels," describes the relationships between microstructures and properties, with emphasis placed on bending fatigue, although other important properties such as rolling-contact fatigue, wear resistance, hot hardness, and toughness are also reviewed.

Gas Carburizing Practices

Gas carburizing is the most used modern production carburizing technique. Batch or continuous furnaces are used with parts that are either placed on fixtures or loaded in baskets. A carbon-rich furnace atmosphere is provided by an endothermic carrier gas enriched with a hydrocarbon gas, such as methane (natural gas) or propane. Furnaces are often fitted with "sealed oil quenches." Oxidation prior to quenching is prevented by maintaining parts in the carburizing atmosphere during transfer to the quenchant.

Although quenching in oils at temperatures from 50 to 75 °C (120 to 165 °F) is common, many carburized parts are martempered by quenching in oils or molten salts at temperatures from 175 to 200 °C (345 to 390 °F). Because the quench temperature for martempering is above the martensite start temperature (M_s) of the high-carbon case, the case transforms to martensite during subsequent air cooling.

Characteristic Features of Carburized Cases

Some of the characteristic features of carburized cases are due to the fact that they are created by the diffusion of carbon. First, there is a gradual transition in carbon content, as well as a transition in microstructure and mechanical properties, between case and core. As a rule, the deeper the case, the less steep is the slope of the carbon gradient (Fig. 1). The absence of any sharp transition in properties assures excellent adherence of the case.

Second, carburized cases are most frequently produced in thicknesses that range from 0.5 to 1.5 mm (20 to 60 mils). At 925 °C (1700 °F), which is a typical processing temperature, this thickness range can be produced in processing times from about 2 to 15 h. Cases as thin as 0.1 mm (4 mils), which require less than 10 min at 927 °C (1700 °F), are sometimes produced on small parts by salt bath carburizing. Cases as deep as 3 mm (120 mils), which require more than two days of carburizing at 927 °C (1700 °F), are occasionally produced on large parts. The ease of adjusting the case depth to resist the anticipated contact loading is one of the great advantages of carburizing.

Carbon Sources

Low-carbon steel parts exposed to carbon-rich atmospheres derived from a wide variety of sources will carburize at temperatures of 850 °C

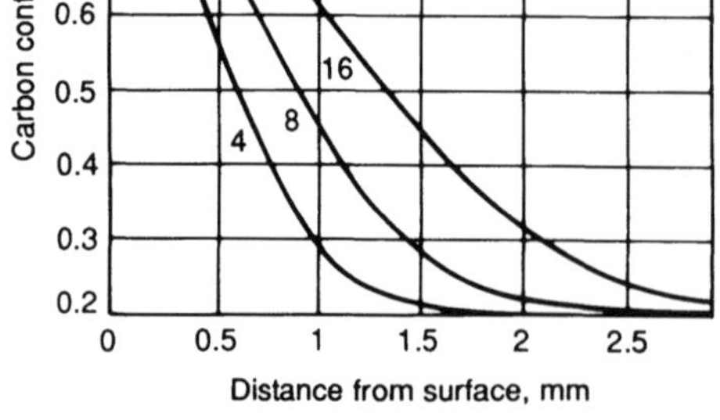

Fig. 1 Computed carbon concentration gradients resulting from gas carburization of SAE 8620 steel for 4, 8, and 16 h at 927 °C (1700 °F). Carbon potential of furnace atmosphere assumed to be 1 wt% C during process

(1560 °F) and above. In the most primitive form of this process, the carbon source is so rich that the solubility limit of carbon in austenite is reached at the surface of the steel and some carbides may form at the surface. (The carbon gradient produced by maintaining saturated austenite at the surface of the steel is referred to as the normal carbon gradient, such as in earlier editions of *ASM Handbook.*) Such atmospheres will also deposit soot on surfaces within the furnace, including the parts. While this mode of carburizing is still practiced in parts of the world in which resources are limited, the goal of current practice in modern manufacturing plants is to control the carbon content of furnace atmospheres so that:

- The final carbon concentration at the surface of the parts is below the solubility limit in austenite.
- Sooting of the furnace atmosphere is minimized.

Controlled carburizing atmospheres are produced by blending a carrier gas with an enriching gas, which serves as the source of carbon. The usual carrier, endothermic gas, is not merely a diluent but plays a role (described in the following paragraphs) in accelerating the carburizing reaction at the surface of the parts. The amount of enriching gas required by the process depends primarily on the carbon demand, that is, the rate at which carbon is absorbed by the work load.

Endothermic gas (Endogas, Seco-Warwick Corp.) is a blend of carbon monoxide, hydrogen, and nitrogen (with smaller amounts of carbon dioxide, water vapor, and methane) produced by reacting a hydrocarbon gas such as natural gas (primarily methane), propane, or butane with air. Endogas is usually produced in a separately fired retort furnace (endothermic atmosphere generator) using an air-to-hydrocarbon feed ratio that will produce an oxygen-to-carbon atom ratio of about 1.05 in the endothermic atmosphere. For endothermic gas produced from pure methane, the air-to-methane ratio is about 2.5; for endothermic atmosphere produced from pure propane, the air-to-propane ratio is about 7.5. These ratios will change depending on the composition of the hydrocarbon feed gases and the water vapor content of the ambient air. Table 1 lists typical compositions of natural gas. Propane for atmosphere generation should contain less than 5% propylene ($CH_3CH{=}CH_2$) and less than 2.5% butane or heavier hydrocarbons, satisfying the requirements in ASTM D 1835 for so-called special-duty propane or the Gas Producers Association specification 2140, grade HD 5.

A carrier gas similar in composition to endothermic atmosphere produced from methane can be formed from a nitrogen-methanol blend. The proportions of nitrogen and methanol (CH_3OH) are usually chosen to give the same nitrogen-to-oxygen ratio as that of air, that is, about 1.9 volumes of nitrogen for each volume of gaseous methanol. On entering the furnace, each volume of gaseous methanol cracks to form approximately one volume of carbon monoxide and two volumes of hydrogen.

A carrier gas can be generated in situ by the direct addition of air and a hydrocarbon gas to the carburizing furnace (Ref 1). Special precautions (low flow rates, high temperatures, and good mixing) must be taken in setting up and controlling such a process to ensure a thorough reaction of the feed gases and uniformity of carburizing. Similar precautions are needed if carrier gases high in either carbon dioxide or water vapor content (such as exothermic gas) are used.

Carburizing Equipment

Gas carburizing furnaces vary widely in physical construction, but they can be divided into two major categories: batch and continuous furnaces. In a batch-type furnace, the work load is charged and discharged as a single unit or batch. In a continuous furnace, the work enters and leaves the furnace in a continuous stream. Continuous furnaces are favored for the high-volume production of similar parts with total case depth requirements of less than 2 mm (0.08 in.)

Batch Furnaces. The most common types of batch furnaces are pit furnaces and horizontal batch furnaces, although fluidized bed furnaces are also used. Pit furnaces are usually placed in a pit with the cover or lid located just above floor level and are often loaded and unloaded with the aid of an overhead crane (Fig. 2). Pit furnaces are frequently used for large parts requiring long processing times. If the work is to be direct quenched, the load must be moved through air before quenching. As a result, parts will be covered by an adherent black scale, which, depending on the needs of the application, may have to be removed by shot blasting or acid pickling.

Horizontal batch furnaces are frequently used for carburizing and direct quenching. Many of these furnaces are so-called sealed quench, or integral quench, furnaces; that is, parts are discharged from the furnace into a vestibule that covers an oil quench tank (Fig. 3). Because the furnace atmosphere also flows through the vestibule, parts can be kept free of oxidation prior to quenching. Sealed-quench batch furnaces are capable of processing many different types of loads with widely varying case depth requirements. Like pit furnaces, they can be made quite gas tight, with the result that positive furnace pressures are easily achieved.

Continuous Furnaces. Types of continuous furnaces used for carburizing include mesh belt, shaker hearth, rotary retort, rotary hearth, roller hearth, and pusher designs. Many of these furnaces can be built with sealed oil quenching so that oxide-free parts can be produced. Most of these furnaces can be sealed well enough that positive furnace pressures can be maintained. Some continuous mesh belt furnaces, on the other hand, are open to the air at either end. Because air cannot be positively excluded, carburizing in these furnaces is often difficult to control.

Furnace Atmosphere Parameters. Certain principles of operation apply to all controlled-atmosphere furnaces regardless of

Table 1 Specific gravity and composition of natural gas in selected regions of the United States

State	Specific gravity	Composition(a), vol%			
		CH_4	CH_3CH_3	N_2	CO_2
New York	0.58–0.59	94.1–96.3	1.8–2.0	0.3–1.8	0.83–0.96
Illinois	0.57–0.61	89–97.5	1.6–4.4	0.31–5.7	0.39–0.75
California	0.60–0.63	92–98.8	3.9–5	1.2–1.24	0.76–3.0

(a) CH_4, methane; CH_3CH_3, ethane; N_2, nitrogen. Source: American Gas Association design. First, in order to ensure the uniformity of carburizing, the furnaces must be equipped with internal fans so that the atmosphere is well circulated through the work load. In addition, the individual parts within the work load must be well spaced to allow the atmosphere to penetrate the load. Critical parts, such as gears, are usually placed on fixtures to control not only their spacing, but also their orientation entering the quenchant. At times the weight of trays and fixtures in pusher furnaces is two to three times the weight of the parts processed.

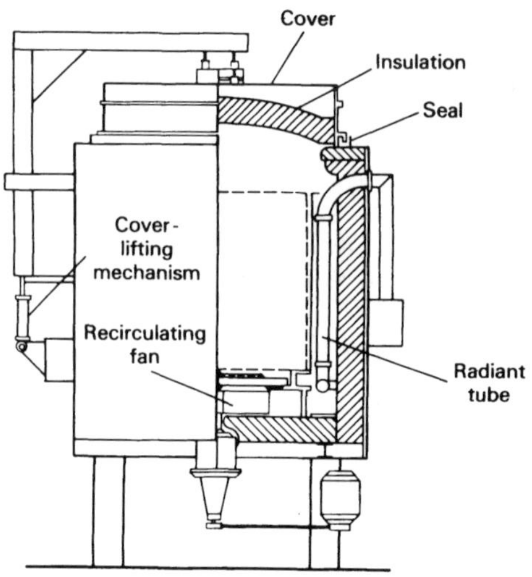

Fig. 2 A pit batch carburizing furnace. Dashed lines outline location of workload.

Second, the furnace should be operated at a positive pressure so that if the furnace has small leaks, air does not enter the furnace. Pressures of 12 to 37 Pa (0.09 to 0.28 torr, or 0.05 to 0.15 in. water column) are usually satisfactory for carburizing furnaces. The furnace pressure can be controlled by adjusting the orifice size in atmosphere vent lines and the carrier gas flow rate.

Because the hot gases inside a furnace are low in density, the pressure differential (furnace pressure minus ambient pressure, measured at the same height) will have its smallest value at the lowest point in the furnace. The minimum furnace pressure needed (at any height) to maintain a positive differential at all heights (P_{min}) can be computed from the relation:

$$P_{min} = H(D_A - D_F) \quad \text{(Eq 1)}$$

where H is the internal height of the furnace chamber, D_A is the density of ambient air outside the furnace, and D_F is the density of the

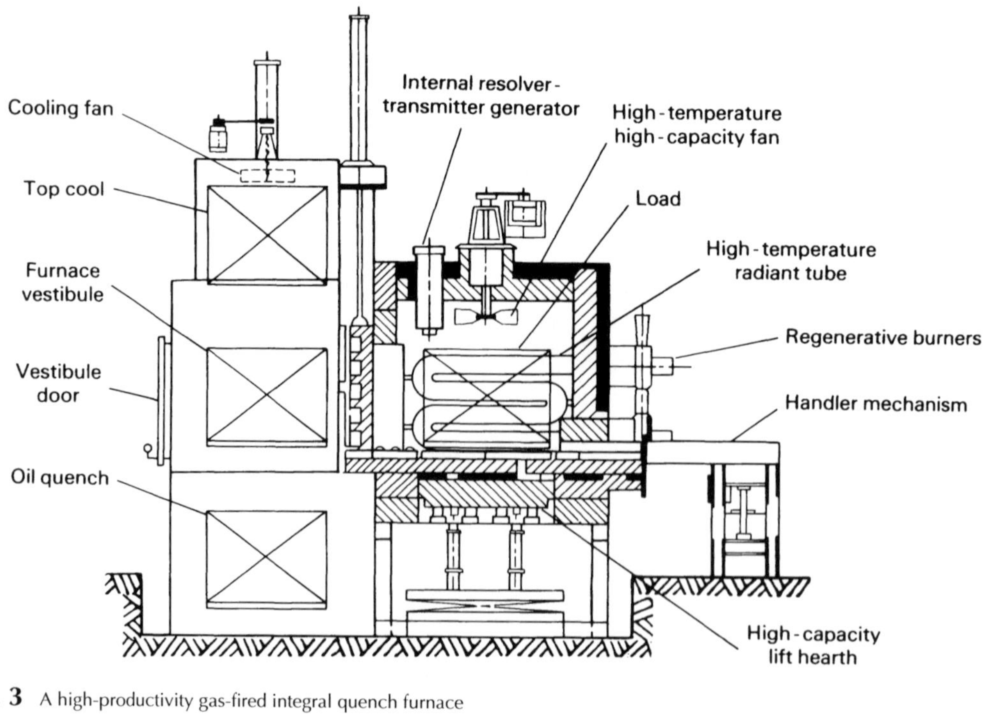

Fig. 3 A high-productivity gas-fired integral quench furnace

atmosphere inside the furnace. Because $D_A >> D_F$, a suitable minimum value for the furnace pressure in pascals is:

$$P_{min} = H(0.117) \quad \text{(Eq 2)}$$

where H is in centimeters and it is assumed that the ambient air is at 1 kPa (1 atm) pressure and 20 °C (70 °F). Even though the furnace pressure is nominally positive, air can still enter the furnace through small openings if there are local fluctuations in the ambient pressure. A large cooling fan blowing at the furnace might raise the ambient pressure locally by as much as 25 Pa (0.19 torr, or 0.1 in. water column).

Third, the rate at which the furnace atmosphere responds to changes in inlet gas composition depends on the mean residence time of the atmosphere gases in the furnace. The mean residence time (t_{res}) is approximately:

$$t_{res} = \frac{(V \cdot T_A)}{F \cdot T_F} \quad \text{(Eq 3)}$$

where V is the furnace volume; F is the carrier gas flow rate measured at T_A, the absolute ambient temperature; and T_F is the absolute furnace temperature. Residence times in carburizing furnaces vary from about 2 to 15 min. If the inlet gas composition is changed, it takes about three residence times for 95% of the effect of the change to be felt in the furnace. Therefore, batch furnaces, in which the atmosphere composition must be changed during the course of a processing cycle, are usually operated with shorter residence times than are continuous furnaces. It is often considered an advantage to use high flow rates of carrier gas to speed the purging of air that enters the furnace when parts are charged. However, the same result can usually be achieved more economically by using an automatic control system to regulate the flow of the hydrocarbon enriching gas.

Preparation of Parts for Carburizing

Parts, trays, and fixtures should be thoroughly cleaned before they are charged into a carburizing furnace. Often they are washed in a hot alkaline solution. Some users heat washed parts, trays, and fixtures in an oxidizing atmosphere at 400 °C (750 °F) before carburizing to remove traces of organic contaminants (Ref 2).

Very thin oxide layers on parts (such as those produced by oxidation below 500 °C, or 900 °F) are reduced by the carburizing atmosphere. Heavy oxide layers, such as forging scale, will be reduced to iron flakes, which are not adherent to the part.

Residues from alkaline washer solutions deposited on parts, particularly those with silicates, can cause spotty carburizing, as well as give the parts a blotchy appearance. In addition, alkaline residues can adversely affect the life of heat-resistant furnace alloys. Quenching salts remaining on trays and fixtures can also damage furnace hardware (for example, silicon carbide rails in pusher furnaces). Chlorine- or sulfur-containing residues on parts will release gases that can react with brickwork, the protective oxide films on heat-resistant alloy fixtures, or the work load.

Carburizing Process Variables

The successful operation of the gas carburizing process depends on the control of three principal variables:

- Temperature
- Time
- Atmosphere composition

Other variables that affect the amount of carbon transferred to parts include the degree of atmosphere circulation and the alloy content of the parts.

Temperature. The maximum rate at which carbon can be added to steel is limited by the rate of diffusion of carbon in austenite. This diffusion rate increases greatly with increasing temperature; the rate of carbon addition at 925 °C (1700 °F) is about 40% greater than at 870 °C (1600 °F).

The temperature most commonly used for carburizing is 925 °C (1700 °F). This temperature permits a reasonably rapid carburizing rate without excessively rapid deterioration of furnace equipment, particularly the alloy trays and fixtures. The carburizing temperature is sometimes raised to 955 °C (1750 °F) or 980 °C (1800 °F) to shorten the time of carburizing for parts requiring deep cases. Conversely, shallow case carburizing is frequently done at lower temperatures because case depth can be controlled more accurately with the slower rate of carburizing obtained at lower temperatures.

For consistent results in carburizing, the temperature must be uniform throughout the work load. Temperature gradients through the work load will persist for a substantial period of time while the work is being heated to the carburizing temperature. Because parts at the exterior of the load reach the furnace temperature first, they will begin carburizing well before parts at the interior of the load. The consequence is variability in case depth from part to part and within a single part. In addition, soot can be deposited on cold parts exposed to a carburizing atmosphere. Therefore, for best results, the work load should be heated to the carburizing temperature in a near-neutral furnace atmosphere. In batch furnaces, parts can be heated in endothermic atmosphere until they reach the furnace temperature; then carburizing can commence with the addition of the enriching gas. Many new continuous furnaces are being built with separate preheat chambers to ensure that the load is at a uniform temperature before entering the carburizing zone. In continuous furnaces that lack positive separation between heating and carburizing stages, the best that can be done is to:

- Add only endothermic gas to the front of the furnace.
- Establish a front-to-back internal flow of atmosphere gases by adjusting flow rates and orifice size in the effluent lines at either end of the furnace.

In batch furnaces, the thermocouple used for temperature control is usually positioned so that it reaches the set-point temperature before the work load does. In continuous furnaces that are not positively separated into zones, the thermocouple in the first zone (used for heating) should be placed near the end of that zone. This prevents overheating of the work load. The control thermocouple is usually positioned near the center of the carburizing zone. If the last zone is at a lower temperature than the carburizing zone, the control thermocouple is usually placed near the discharge end of the zone. However, the features of individual furnaces, such as the location of radiant tubes, must be considered when positioning control thermocouples.

Time. The effect of time and temperature on total case depth is shown in Fig. 4. The data given, originally published by Harris (Ref 3) in 1943 are computed assuming saturated austenite at the surface of the workpieces. When the surface carbon content is controlled so that it is less than the saturation value, case depths will be less than they otherwise would be. Figure 5 shows how the carburizing time decreases with increasing carburizing temperature for a case depth of 1.5 mm (0.06 in.). In addition to the time at the carburizing temperature, several hours may be required to bring large workpieces or heavy loads of smaller parts to operating temperature. For work quenched directly from the carburizing furnace, the cycle may be lengthened further by allowing time for the work to cool from the carburizing temperature to approximately 845 °C (1550 °F) prior to quenching. If the workload is exposed to the carburizing atmosphere during heating, some carburizing will occur before the nominal start of carburizing. Similarly, additional diffusion and interchange of carbon with the atmosphere will occur during cooling prior to quenching. Thus, the actual case depth achieved may differ significantly from the values listed in the table in Fig. 4. More complex mathematical models that allow for variations in temperature and atmosphere carbon potential with time can be constructed to allow a better prediction of case depth.

Carbon Potential. The carbon potential of a furnace atmosphere at a specified temperature is defined as the carbon content of pure iron that is in thermodynamic equilibrium with the atmosphere. The carbon potential of the furnace atmosphere must be greater than the carbon potential of the surface of the workpieces in order for carburizing to occur. It is the difference in carbon potential that provides the driving force for carbon transfer to the parts.

Carbon Diffusion. The combined effects of time, temperature, and carbon concentration on the diffusion of carbon in austenite can be expressed by Fick's laws of diffusion. Fick's first law states that the flux of the diffusing substance perpendicular to a plane of unit cross-sectional area is proportional to the local carbon gradient perpendicular to the plane. The constant of proportionality is the diffusion coefficient D, which has the units $(\text{distance})^2/\text{time}$. Fick's second law is a material balance within an elemental volume of the system; the flux of carbon into an elemental volume of iron minus the flux of carbon out of the elemental volume equals the rate of accumulation of carbon within the volume. Combining the two laws leads to a partial differential equation that describes the diffusion process. Reference 5 provides solutions to the diffusion equation for a variety of boundary conditions and part configurations, for example, plate, rod, sphere, and so on. With these solutions and values of the diffusion coefficient, it is possible to predict the carbon gradient and depth of penetration occurring for any combination of time, temperature, and surface carbon concentration.

The diffusion coefficient for carbon in austenite is a function of carbon content and temperature. The following expression, proposed by Tibbetts (Ref 6), summarizes the experimental data:

$$D = 0.47 \exp[-1.6\,C - (37{,}000 - 6600\,C)/RT] \quad \text{(Eq 4)}$$

where D is in cm²/s, C is the weight percent carbon, T is temperature in degrees Kelvin, and R is the gas constant. Exact solutions to the diffusion equation when the diffusion coefficient depends on composition are available for steady-state diffusion (Ref 5), but for time-dependent solutions, numerical methods must be used. Figure 6 lists a BASIC computer program for finding the carbon concentration gradient below a flat surface using Eq 4 for the diffusion coefficient. In this program, the surface boundary condition is:

$$\text{Carbon flux} = \beta\,\sigma\,(CP - C_0) \quad \text{(Eq 5)}$$

where carbon flux is in g/s · cm², β is an effective reaction rate constant in s^{-1}, σ is the density in g/cm³, CP is the atmosphere carbon potential, and C_0 is the surface carbon content. The value of β may vary depending on the degree of atmosphere circulation within the furnace. Taking β equal to 0.00002 s^{-1} seems to produce results that are in reasonable agreement with experience. However, values of β that are as much as a factor of two larger or smaller may be needed to model carburizing behavior in specific furnaces.

Alloy Effects. The various alloying elements found in carburizing steels have an influence on the activity of carbon dissolved in austenite. A definition of carbon activity (a_C) is:

$$a_C = (\text{wt\% } C)\,\Gamma \quad \text{(Eq 6)}$$

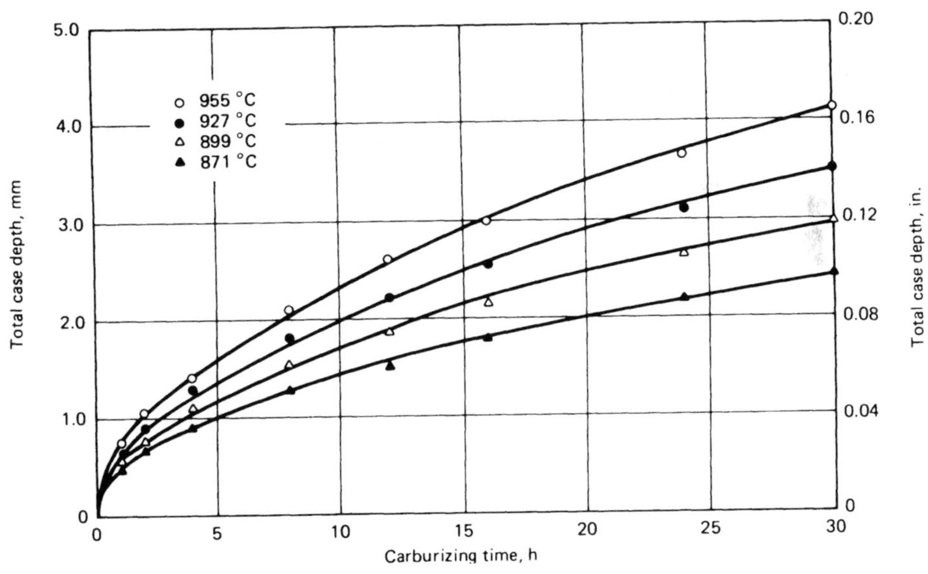

Time, h	871 °C (1600 °F) mm	in.	899 °C (1650 °F) mm	in.	927 °C (1700 °F) mm	in.	955 °C (1750 °F) mm	in.
1	0.46	0.018	0.53	0.021	0.64	0.025	0.74	0.029
2	0.64	0.025	0.76	0.030	0.89	0.035	1.04	0.041
4	0.89	0.035	1.07	0.042	1.27	0.050	1.30	0.051
8	1.27	0.050	1.52	0.060	1.80	0.071	2.11	0.083
12	1.55	0.061	1.85	0.073	2.21	0.087	2.59	0.102
16	1.80	0.071	2.13	0.084	2.54	0.100	2.97	0.117
24	2.18	0.086	2.62	0.103	3.10	0.122	3.66	0.144
30	2.46	0.097	2.95	0.116	3.48	0.137	4.09	0.161

Fig. 4 Plot of total case depth versus carburizing time at four selected temperatures. Graph based on data in table

where Γ, the activity coefficient, is chosen so that $a_C = 1$ for an amount of carbon in solution that is in equilibrium with graphite. Chromium tends to decrease the activity coefficient, and nickel tends to raise it. As a consequence, foils of a chromium-bearing steel equilibrated with a specific furnace atmosphere will take on more carbon than pure iron, and nickel-bearing steels will take on less carbon. It is also true that carbides are produced at lower carbon potentials in chromium-bearing steels than in carbon steels. The primary effect of alloying elements on the diffusion of carbon is due to their effect on the driving force for the surface reaction (Eq 5). To obtain the true driving force, the surface carbon content in an alloy must be converted into the equivalent carbon content in pure iron. Methods of correcting the activity coefficient of carbon for alloy content are available (Ref 7). However, the quantity of experimental data upon

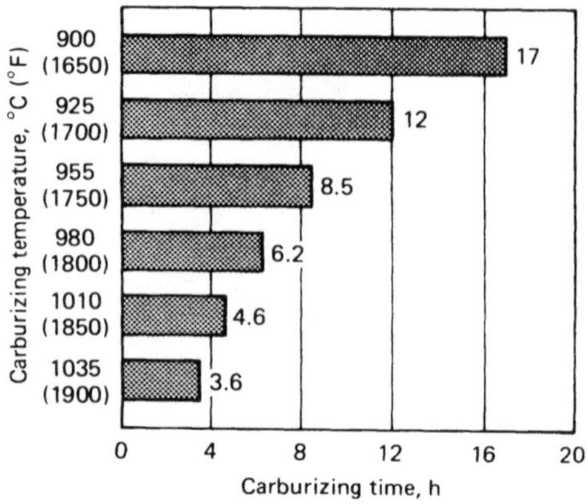

Fig. 5 Reducing effect of increased process temperature on carburizing time for 8620 steel. Case depth: 1.5 mm (0.060 in.). Source: Ref 4

```
1Ø 'Diffusion of Carbon in Austenite
12 'Tibbetts' Expression for the Diffusion Constant of Carbon
15 DEF FNDFC(T,WC) = .47*EXP(-1.6*WC - (37ØØØ! - 66ØØ*WC)/(1.987*T))
2Ø INPUT "Temperature in Degrees C  ",TC
3Ø INPUT "Time in Hours  ",TH
4Ø INPUT "Carbon Potential, wt. pct.  ",CP
5Ø INPUT "Surface Reaction Rate Constant  ",BETA
6Ø INPUT "Initial Carbon Content of Steel  ",CØ
7Ø T = TC + 273.15  :  TSEC = TH*36ØØ
8Ø DT = 15  :  DX = .Ø1
9Ø DIM C(31), CL(31)
1ØØ L1 = 2*DT/DX  :  L2 = DT/DX^2
11Ø NN% = INT(TSEC/DT)  :  V% = Ø
12Ø FOR I% = Ø TO 31
13Ø   CL(I%) = CØ
14Ø NEXT I%
15Ø FOR J% = 1 TO NN%
16Ø   WC = (CL(Ø) + CL(1))/2
17Ø   DB = FNDFC(T,WC)
18Ø   ETA = .75*CL(Ø) + .25*CL(1) + L1*(BETA*(CP - CL(Ø))
          + DB*(CL(1) - CL(Ø))/DX)
19Ø   FOR I% = 1 TO 3Ø
2ØØ     IF (V% = Ø) THEN GOTO 22Ø
21Ø     C(I%) = CØ  :  GOTO 27Ø
22Ø     DA = DB
23Ø     WC = (CL(I%) + CL(I%+1))/2
24Ø     DB = FNDFC(T,WC)
25Ø     C(I%) = CL(I%) + L2*(DB*CL(I%+1) - (DA+DB)*CL(I%)
          + DA*CL(I%-1))
26Ø     IF (ABS(C(I%) - CØ) < .ØØØØ5) THEN V% = 1
27Ø   NEXT I%
28Ø   C(Ø) = 4*(ETA - .25*C(1))/3  :  V% = Ø
29Ø   FOR I% = Ø TO 31
3ØØ     CL(I%) = C(I%)
31Ø NEXT I%,J%
32Ø PRINT "Depth, mm","Wt. Pct. C","Depth, mm.","Wt. Pct. C"
33Ø FOR I% = Ø TO 15
34Ø   PRINT USING "#.####        ";DX*I%*1Ø,C(I%),DX*(I%+15)*1Ø,C(I%+15)
35Ø NEXT I%
36Ø END
```

Fig. 6 Finite-difference computation of the diffusion of carbon in austenite using BASIC computer program which such correlations are based is rather limited. Therefore, predictions should be verified by experiments, particularly when an alloy contains substantial amounts of more than one alloying element.

Gas Carburizing Atmospheres

In the discussion of carburizing atmospheres in this section, it will be assumed that the atmosphere consists of an endothermic carrier gas (produced from methane) that is enriched by a methane addition, which serves as the source of the carbon being transported to the work load. The main constituents of the atmosphere are CO, N_2, H_2, CO_2, H_2O, and CH_4. Of these constituents, N_2 is inert, acting only as a diluent. The amounts of CO, CO_2, H_2, and H_2O present are very nearly the proportions expected at equilibrium from the reversible reaction:

$$CO + H_2O \leftrightarrow CO_2 + H_2 \quad \text{(Eq 7)}$$

given the particular ratios of carbon, oxygen, and hydrogen in the atmosphere. Methane is invariably present in amounts well in excess of the amount that would be expected if all the gaseous constituents were in equilibrium.

Although the sequence of reactions involved in carburizing is not known in detail, it is known that carbon can be added or removed rapidly from steel by the overall reversible reactions:

$$2CO \leftrightarrow C\text{ (in Fe)} + CO_2 \quad \text{(Eq 8)}$$

and

$$CO + H_2 \leftrightarrow C\text{ (in Fe)} + H_2O \quad \text{(Eq 9)}$$

A carburization process based solely on the decomposition of CO would require large flow rates of atmosphere gas to produce appreciable carburizing. As an example, the loss of just 0.47 g C from a cubic meter of endothermic gas at 925 °C (1700 °F) is sufficient to reduce the CO-to-CO_2 ratio from 249 to 132 and the carbon potential from 1.25 to 0.8%. The loss of 0.47 g C represents about the same amount present in a steel part of 100 cm^2 (15.5 $in.^2$) surface area carburized to a depth of 1 mm (0.040 in.).

The methane enrichment of endothermic gas provides carbon for the process by slow reactions such as:

$$CH_4 + CO_2 \rightarrow 2CO + 2H_2 \quad \text{(Eq 10)}$$

and

$$CH_4 + H_2O \rightarrow CO + 3H_2 \quad \text{(Eq 11)}$$

which reduce the concentrations of CO_2 and H_2O, respectively. These reactions regenerate CO and H_2, thereby directing the reactions of Eq 8 and 9 to the right. Because the methane content of carburizing atmospheres is usually far above the content that is expected at equilibrium, given the CO_2 and H_2O contents present, it is evident that the reactions in Eq 10 and 11 do not approach equilibrium. The sum of the reactions in Eq 8 and 10 and in Eq 9 and 11 is reduced to:

$$CH_4 \rightarrow C\text{ (in Fe)} + 2H_2 \quad \text{(Eq 12)}$$

Thus, with constant CO_2 content and constant dew point, the net atmosphere composition change during carburizing is a reduction in methane content and an increase in the hydrogen content. In most commercial operations, atmosphere flow rates are high enough and the rate of methane decomposition is low enough to prevent a large buildup of hydrogen during a carburizing cycle. However, with carburizing loads having high surface area, there is a drop in the CO content of 1 to 3% at the beginning of the cycle when the carbon demand is greatest. This is caused by the dilution of the furnace atmosphere with hydrogen.

Carbon potential control during carburizing is achieved by varying the flow rate of the hydrocarbon-enriching gas while maintaining a steady flow of endothermic carrier gas. As a basis for regulating the enrichment gas flow, the concentration of some constituent of the furnace atmosphere is monitored:

- Water vapor content by dew point measurement
- Carbon dioxide content by infrared gas analysis
- Oxygen potential using a zirconia oxygen sensor

The first two quantities provide measures of carbon potential according to the reactions of Eq 8 and 9. Oxygen potential is related to carbon potential by the reaction:

$$C\text{ (in Fe)} + 1/2\ O_2 \leftrightarrow CO \quad \text{(Eq 13)}$$

When the carbon monoxide content of the atmosphere remains relatively constant, both the carbon dioxide and the oxygen potential provide good measures of the carbon potential. For the dew point to be a valid measure of carbon potential, the product of the hydrogen and carbon monoxide contents must be stable. If the hydrogen content of the furnace atmosphere rises, as a result of either carburizing or sooting, the relationships between CO_2 content, oxygen potential, dew point, and the carbon potential will be altered. For this reason, some process control systems include infrared analysis of CO and the measurement of CO_2 or oxygen potential so that a true carbon potential may be computed for all operating conditions.

Calculation of Equilibrium Compositions. Gas carburizing is a nonequilibrium process; that is, the gaseous constituents of the atmosphere are not fully in equilibrium with one another, and the atmosphere is not in equilibrium with the steel being carburized. Nevertheless, several important reactions, for example, reactions in Eq 7 to 9, approach equilibrium rapidly enough to permit predictions of the rate of carburizing from the atmosphere composition. Thus, the same case carbon gradient can be expected from different furnaces with different atmosphere gas flow rates when certain factors are held constant:

- Carbon potential, as inferred from CO_2, H_2O, or oxygen potential measurements
- Carburizing time
- Carburizing temperature

To compute the equilibrium gas composition resulting from the reaction of a blend of a hydrocarbon gas and air:

- Six gaseous species are assumed to be present in the reacted gas: CO, CO_2, H_2, H_2O, CH_4, and N_2. The partial pressures of five of these are unknowns to be determined; the sixth can be found by computing the difference after the others are known.
- The carbon-to-hydrogen and oxygen-to-nitrogen ratios are fixed by the nature of the hydrocarbon and the composition of air. With the air-to-hydrocarbon ratio fixed, three equations can be found that relate the five unknown partial pressures.
- Two equilibrium relationships, for example, the reactions in Eq 7 and 10, provide two additional equations; thus, all the partial pressures are determined.

Because explicit expressions for the unknowns cannot be written, trial-and-error computation methods must be used with the aid of a computer (Ref 8). After the equilibrium composition of the gas is known, its carbon potential is found by writing the equilibrium relationship for Eq 8:

$$K_8 = (a_C \cdot P_{CO_2})/(P_{CO})^2 \quad \text{(Eq 14)}$$

where P_{CO} and P_{CO_2} are partial pressures of CO and CO_2, respectively; a_C is the activity of carbon ($a_C = 1$ when the atmosphere is in equilibrium with graphite); and K_8 is the equilibrium constant for the reaction in Eq 8.

The constant K_8 can be computed from the Gibbs free energy of formation of CO and CO_2 at the temperature of interest (free energy values are tabulated in Ref 9):

$$\Delta F^\circ_{CO_2} - 2\Delta F^\circ_{CO} = -R\,T \cdot \ln K_8 \quad \text{(Eq 15)}$$

where ΔF°_{CO} and $\Delta F^\circ_{CO_2}$ are free energies of CO and CO_2, respectively; T is the absolute temperature; and R is the gas constant.

The carbon activity is related to the carbon content of austenite by the expression (Ref 10):

$$\ln a_C = \ln y_C + (9167\, y_C + 5093)/T - 1.867 \quad \text{(Eq 16)}$$

where

$$y_C = (4.65\, w)/(100 - w)$$

and T is the temperature in degrees Kelvin, w is the weight percent carbon in austenite, and y_C is the atom ratio of carbon to iron. Combining Eq 14 and 16 gives a relation between carbon potential (that is, the equilibrium carbon content in austenite) and the carbon dioxide and carbon monoxide contents. As long as the assumptions of the calculation are satisfied, measuring just the CO_2 content suffices to define the carbon potential. However, when carbon is removed from the atmosphere by carburizing, there is a reduction in the carbon-to-hydrogen ratio characterizing the atmosphere. When the carbon-to-hydrogen ratio may vary, it is necessary to measure two constituents of the atmosphere. CO and CO_2 content, to define the carbon potential accurately. However, unless the changes in the atmosphere are large, it is seldom worth measuring both CO and CO_2 because of the effect of the added measuring error on the uncertainty of the computed carbon potential.

Endothermic gas derived from propane has a carbon monoxide content of about 23%, whereas that derived from methane has a carbon monoxide content of about 20%. Therefore, the carbon dioxide contents corresponding to a given carbon potential are higher for atmospheres derived from propane than for those derived from methane. Figures 7 and 8 show the relationship between carbon dioxide content and carbon potential for endothermic gas atmospheres derived from methane and propane, respectively.

The oxygen potential (or oxygen partial pressure), as measured by a zirconia oxygen sensor, is related to the carbon activity by the equilibrium equation for the reaction given in Eq 13:

$$a_C = \frac{P_{CO}}{K_{13}P_{O_2}} \qquad \text{(Eq 17)}$$

where P_{O_2} is the oxygen partial pressure and K_{13} is the equilibrium constant. In carburizing atmospheres, the oxygen partial pressure is approximately 10^{-14} to 10^{-20} Pa (10^{-19} to 10^{-25} atm).

The voltage output of a zirconia oxygen sensor, with air as a reference gas, is a function of the absolute temperature (T) and the oxygen partial pressure (P_{O_2}) according to the expression:

$$\text{emf} = 0.000049593\, T \log_{10}\left(\frac{P_{O_2}}{0.209}\right) \qquad \text{(Eq 18)}$$

where emf is in volts. Combining Eq 16, 17, and 18 yields a relation between carbon potential and emf. Because this relation also depends on the carbon monoxide content of the atmosphere (Eq 18), emf measurements corresponding to a certain carbon potential for endothermic gas atmospheres derived from methane (Fig. 9) differ from those for atmospheres derived from propane (Fig. 10).

The water vapor content of the atmosphere is related to the carbon potential by the reaction in Eq 11. The equilibrium relation is:

$$a_C = \frac{K_{11}P_{CO}P_{H_2}}{P_{H_2O}} \qquad \text{(Eq 19)}$$

where K_{11} is the equilibrium constant.

For atmospheres derived from a particular hydrocarbon, the product of the carbon monoxide and hydrogen contents varies little for large changes in water vapor content. The water vapor content of the atmosphere is usually measured by determining its dew point. An

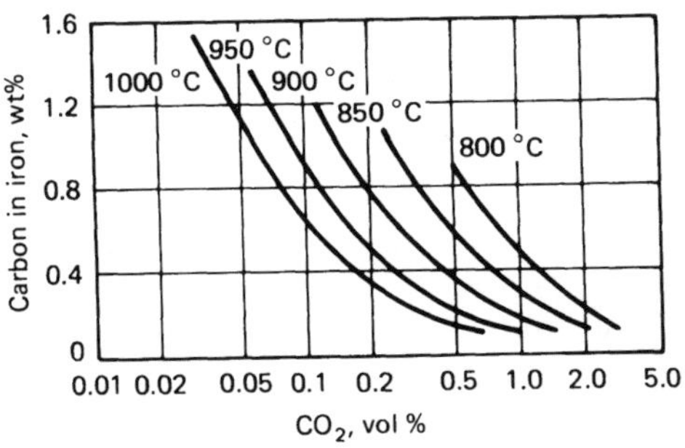

Fig. 7 Relationship between CO_2 content and carbon potential for endothermic gas from methane

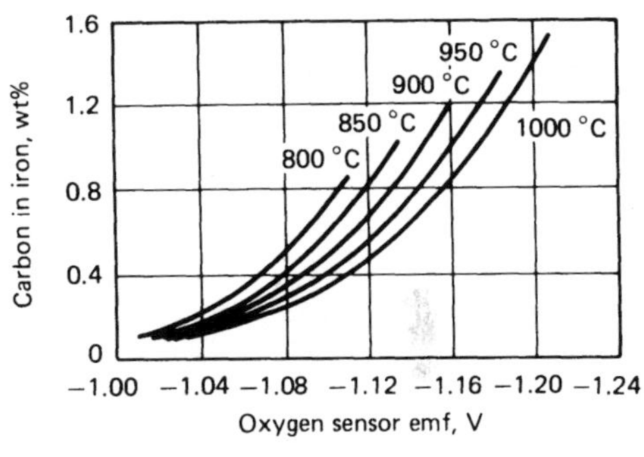

Fig. 9 Measurements of the emf of endothermic gas from methane

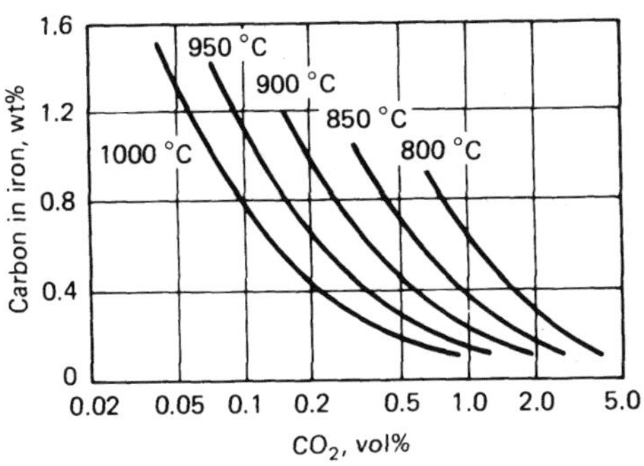

Fig. 8 Relationship between CO_2 content and carbon potential for endothermic gas from propane

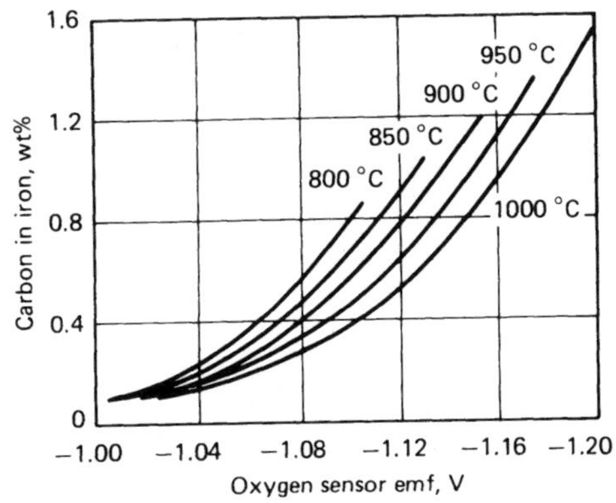

Fig. 10 Measurements of the emf of endothermic gas from propane

equation relating dew point in degrees Celsius and P_{H_2O} in atmospheres is:

$$\text{Dew point} = \frac{5422.18}{14.7316 - \ln P_{H_2O} - 273.16} \quad \text{(Eq 20)}$$

Equations 16, 19, and 20 can be combined to produce the relations between dew point and carbon potential plotted in Fig. 11 and 12 for endothermic gas atmospheres derived from methane and propane, respectively.

Sooting. If the carbon potential of a furnace atmosphere is allowed to become very high, sooting will occur. As carbon is lost from the atmosphere, hydrogen is produced. One symptom of a sooting condition is a pronounced drop in the CO content due to hydrogen dilution. When a furnace becomes sooted, it is found that the carbon potential of the furnace atmosphere is no longer controllable; that is, changes to the flow of enriching gas no longer produce the expected changes in atmosphere composition and carbon potential. The only practical solution is to empty the furnace of parts, introduce air, and burn out the soot.

In burning out accumulated soot, some care is necessary to prevent local overheating in the furnace. Typically, the furnace temperature will be set at about 815 °C (1500 °F) for burnout, and air will be admitted to the furnace by either opening doors or introducing a flow of air. The rise in temperature due to the combustion of soot should be monitored. The air supply can be reduced if the temperature rise is excessive.

Modern automatic atmosphere control systems minimize sooting by maintaining a constant atmosphere composition, thereby matching carbon supply to carbon demand. However, even with the best system, a furnace can be sooted up if the control carbon potential is set too high.

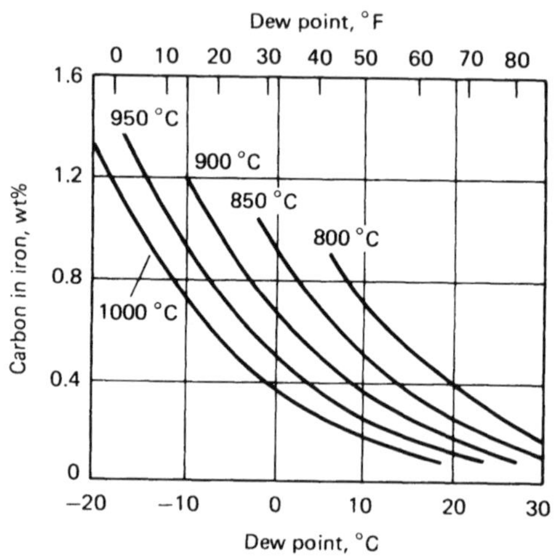

Fig. 11 Relationship between dew point and carbon potential for endothermic gas derived from methane

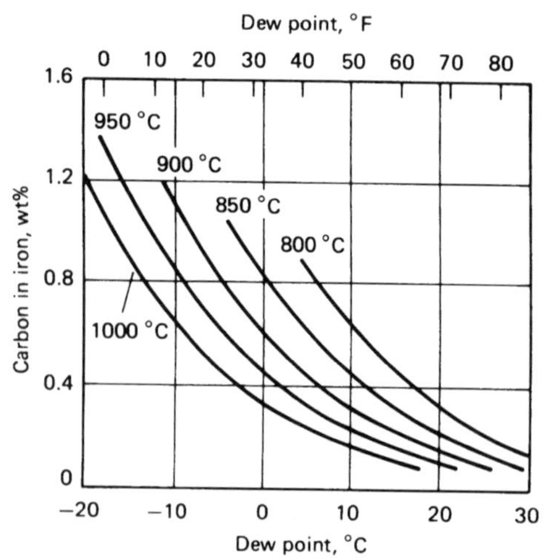

Fig. 12 Relationship between dew point and carbon potential for endothermic gas derived from propane

Furnace Conditioning. When a carburizing atmosphere is introduced into an empty furnace that has been idle for some time or that has just been burned out to remove soot, it is found that the amount of enriching gas needed to maintain a given carbon potential is much higher than would be expected. As time passes (typically 12 to 24 h), the amount of enriching gas needed to maintain the carbon potential decreases to a steady-state value. This process is called conditioning a furnace.

During the time in which conditioning is occurring, carbon is deposited in the crevices of the brickwork and at other locations where the temperature is low. This occurs because the carbon potential of a carburizing atmosphere increases as the temperature falls. An atmosphere with a carbon potential of 1% at 927 °C (1700 °F) is capable, thermodynamically, of depositing soot at or below 843 °C (1550 °F). Eventually, the crevices will be filled with carbon, and the exposed carbon will be at a high enough temperature that further carbon deposition cannot occur. However, there will be some locations, such as sight ports and gas sample lines where soot will continue to be deposited as the furnace is used. Soot may also form in unheated furnace vestibules as the furnace atmosphere gas cools on entering the vestibule.

Nitrogen-Methanol Atmospheres. The in situ generation of a carrier gas from a blend of nitrogen and methanol has become more common in recent years. The primary advantage of this approach is that separately fired endothermic gas generators are no longer required. The primary disadvantage is that operating costs are often higher, although this obviously depends on the local cost and availability of natural gas and propane.

Upon entering a heat-treating furnace, methanol cracks to form CO and H_2. The nitrogen-to-methanol ratio is usually chosen so that the nitrogen-to-oxygen ratio is the same as that for air:

$$CH_3OH + 1.89N_2 \rightarrow CO + 2H_2 + 1.89N_2$$

$$CH_4 + 2.39\frac{\text{Air}}{(0.2095\ O_2 + 0.7905\ N_2)} \rightarrow CO + 2H_2 + 1.89N_2 \quad \text{(Eq 21)}$$

These material balances show that when the nitrogen-to-methanol mole ratio is 1.89, the atom ratios of oxygen, hydrogen, carbon, and nitrogen will be the same as those produced from an air-to-methane mole ratio of 2.39. Because endothermic gas is usually produced with an air-to-methane ratio of about 2.5, it is clear that a nitrogen-methanol carrier gas is richer than conventional endothermic gas. This poses no special problems when carburizing but may lead to sooting if the nitrogen-methanol blend is run into an empty furnace for long periods of time. Air is often added to nitrogen-methanol to reduce its carbon potential for neutral hardening applications.

Heat is required to vaporize methanol, and the cracking of methanol to CO and hydrogen is quite endothermic. Therefore, successful cracking is favored by higher furnace temperatures and lower rates of carrier gas usage. The presence of lower than expected CO contents in furnace atmospheres is symptomatic of incomplete cracking. However, low CO contents can also arise if the proportions of nitrogen and methanol change with changes in the carrier gas flow.

Because of the similarity in atmosphere composition, a 1.89 mole ratio nitrogen-methanol carrier gas enriched with methane can be controlled using the same control points (Fig. 7, 9, and 11) as an endothermic gas atmosphere produced from methane. The only additional complication is the need for maintaining a constant mole ratio for the carrier gas, particularly when changing the carrier gas flow rate. Rather than invest in sophisticated flow metering, some users have chosen to control their processes by measuring two atmospheric constituents (CO and oxygen potential, for example), rather than just one, to determine carbon potential.

Nitrogen of high purity (about 10 ppm residual CO_2 and water vapor) is produced by liquifying air and then using the differences in boiling points of the various constituents to effect the separation of nitrogen. In most instances, liquid nitrogen is shipped from an air separation plant to the heat-treating plant and stored in vacuum-insulated vessels. Nitrogen of lesser purity can be produced on site by a variety of means:

- *Combustion processes.* Air is burned with natural gas, and the CO_2 and H_2O are stripped from the gas by absorption and condensation.
- *Pressure swing absorption, vacuum swing absorption.* Air separation using zeolite molecular sieves
- *Membrane air separation.* Differences in molecular diffusion rates through thin-walled fibrous tubes are used to separate oxygen and nitrogen.

Low-purity less expensive nitrogen can be used with methanol to form satisfactory furnace atmospheres. However, because the oxygen content of the nitrogen should be kept relatively constant, the nitrogen generation process must be designed with this requirement in mind.

Carbon Concentration Gradients and Surface Carbon Content

The carbon concentration gradients of carburized parts are influenced by carburizing temperature, carburizing time, the type of cycle (various combinations of carburizing and diffusion times), the carbon potential of the furnace atmosphere, and the original composition of the steel. Compositions of steels most often carburized are given in Table 2.

The term carbon gradient not only encompasses the rate of change of carbon content with depth below the surface, but also alludes to the absolute value of carbon content in any layer other than the surface layer. As will be described later in this chapter, carbon gradients, as well as hardness gradients, are associated with the microstructural gradients of gas carburized steels.

Carbon Gradients. Single-potential carburizing involves carburizing at a single temperature and constant atmosphere composition for the entire cycle. In some instances, single-potential carburizing is done to produce saturated austenite at the surface of the steel. Figure 13 shows the influence of carburizing temperature on the carbon gradient for single-potential carburizing of 1020 steel in a batch furnace. Comparable data are also given for 8620 steel carburized for 7.5 h in an atmosphere containing 12% methane. Carburizing temperatures for both steels were 870, 900, and 925 °C (1600, 1650, and 1700 °F).

Figure 14 presents carbon gradient curves obtained in a batch furnace at a carbon potential corresponding to saturated austenite for four steels after carburizing for 4 h at 870 and 925 °C (1600 and 1700 °F). Typical influences of methane content, carbon potential, and time on the carbon gradient after single-potential carburizing of 1022 steel are shown in Fig. 15 and 16.

To develop optimum mechanical properties in the case, it is common practice to use carburizing cycles that consist of two or more combinations of temperature, time, and atmosphere composition (carbon potential). The main objective of using multiple-potential carburizing is to decrease the total cycle time.

The most widely used carburizing cycles are constant-temperature cycles starting with an atmosphere carbon potential that approaches the value for carbon saturation in austenite. Part way through the cycle the flow of the enriching gas is reduced, which lowers the carbon poten-

Table 2 Composition of commonly used carburizing steels

Steel	Composition, %					
	C	Mn	Ni	Cr	Mo	Other
Carbon steels						
1010	0.08–0.13	0.30–0.60	. . .	. . .	. . .	(a), (b)
1018	0.15–0.20	0.60–0.90	. . .	. . .	. . .	(a), (b)
1019	0.15–0.20	0.70–1.00	. . .	. . .	. . .	(a), (b)
1020	0.18–0.23	0.30–0.60	. . .	. . .	. . .	(a), (b)
1021	0.18–0.23	0.60–0.90	. . .	. . .	. . .	(a), (b)
1022	0.18–0.23	0.70–1.00	. . .	. . .	. . .	(a), (b)
1524	0.19–0.25	1.35–1.65	. . .	. . .	. . .	(a), (b)
1527	0.22–0.29	1.20–1.50	. . .	. . .	. . .	(a), (b)
Resulfurized steels						
1117	0.14–0.20	1.00–1.30	. . .	. . .	. . .	0.08–0.13 S
Alloy steels						
3310	0.08–0.13	0.45–0.60	3.25–3.75	1.40–1.75	. . .	(b), (c)
4023	0.20–0.25	0.70–0.90	. . .	. . .	0.20–0.30	(b), (c)
4027	0.25–0.30	0.70–0.90	. . .	. . .	0.20–0.30	(b), (c)
4118	0.18–0.23	0.70–0.90	. . .	0.40–0.60	0.08–0.15	(b), (c)
4320	0.17–0.22	0.45–0.65	1.65–2.00	0.40–0.60	0.20–0.30	(b), (c)
4615	0.13–0.18	0.45–0.65	1.65–2.00	. . .	0.20–0.30	(b), (c)
4620	0.17–0.22	0.45–0.65	1.65–2.00	. . .	0.20–0.30	(b), (c)
4815	0.13–0.18	0.40–0.60	3.25–3.75	. . .	0.20–0.30	(b), (c)
4820	0.18–0.23	0.50–0.70	3.25–3.75	. . .	0.20–0.30	(b), (c)
5120	0.17–0.22	0.70–0.90	. . .	0.70–0.90	. . .	(b), (c)
5130	0.28–0.33	0.70–0.90	. . .	0.80–1.10	. . .	(b), (c)
8617	0.15–0.20	0.70–0.90	0.40–0.70	0.40–0.60	0.15–0.25	(b), (c)
8620	0.18–0.23	0.70–0.90	0.40–0.70	0.40–0.60	0.15–0.25	(b), (c)
8720	0.18–0.23	0.70–0.90	0.40–0.70	0.40–0.60	0.20–0.30	(b), (c)
8822	0.20–0.25	0.75–1.00	0.40–0.70	0.40–0.60	0.30–0.40	(b), (c)
9310	0.08–0.13	0.45–0.65	3.00–3.50	1.00–1.40	0.08–0.15	(b), (c)
Special alloys						
CBS-600	0.16–0.22	0.40–0.70	. . .	1.25–1.65	0.90–1.10	0.90–1.25 Si
CBS-1000M	0.10–0.16	0.40–0.60	2.75–3.25	0.90–1.20	4.00–5.00	0.40–0.60 Si 0.15–0.25 V
Pyrowear Alloy 53	0.10	0.35	2.00	1.00	3.25	1.00 Si, 2.00 Cu, 0.10 V

(a) 0.04 P max, 0.05 S max. (b) 0.15–0.35 Si. (c) 0.035 P max, 0.04 S max tial of the atmosphere. During the remainder of the time at temperature, known as the diffusion cycle, the atmosphere is maintained at a carbon potential equal to the final desired surface carbon content. During the diffusion period, the surface layer is partly decarburized because it is in an environment of lower carbon activity. It must not be assumed that all of the carbon added during the carburizing cycle will be diffused inward during the diffusion cycle; most of it does diffuse inward, but some returns to the atmosphere.

Carbon gradients from multiple-potential carburizing are used for the twofold purpose of reducing the amount of retained austenite that would normally occur on quenching and increasing the depth of the effective case and making its properties more uniform. The two examples that follow describe carbon gradients that are typical of those produced in continuous furnaces with zone control.

Example 1: Comparison of Carbon Gradients Produced with and without Zone Control. Two types of carburized cases were produced on 4027 steel by gas carburizing in the same continuous furnace using identical time-temperature cycles but different conditions of atmosphere control. For both cycles, total time in the furnace was 10.3 h, and the zone temperatures were 900, 925, 925, and 845 °C (1650, 1700, 1700, and 1550 °F). The carbon gradients produced by the two cycles are shown in Fig. 17. One case was produced without zone control, that is, without regard to the fact that there is a high demand for carbon early in the carburizing cycle and a low demand during the final

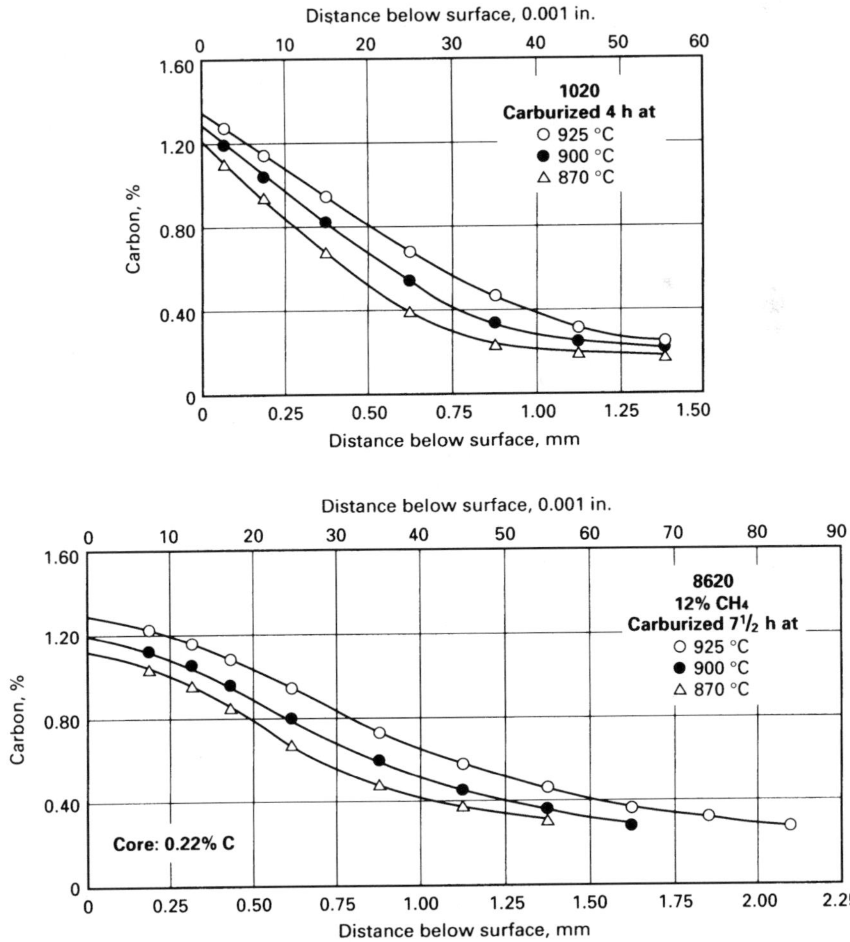

Fig. 13 Carbon gradients for 1020 and 8620 steels carburized at three temperatures. The 1020 steel was carburized in a batch furnace; the 8620 steel, in a recirculating pit furnace.

part of the cycle. A uniformly distributed mixture of carrier gas plus hydrocarbon gas was admitted continuously at various points along the length of the furnace. The ratio of carrier gas to hydrocarbon gas was high enough to ensure that only a minimum amount of soot would be deposited during the carburizing and cooling portions of the cycle. As can be seen from the curve labeled "Without zone control" in Fig. 17, the case had a high carbon content at the surface and a slightly concave curvature to the carbon gradient.

The other case (see curve labeled "With zone control" in Fig. 17) was produced under identical conditions, except for the distribution of the atmosphere within the furnace. Hydrocarbon gas was admitted only within the carburizing zones, with the greatest amount added in the first of these zones. The total amount of hydrocarbon gas was equal to that for the case produced without zone control. Carrier gas without enrichment was admitted to the furnace in the diffusion and cooling zones. The case had a lower surface carbon content and a greater effective depth than the case produced without zone control, and the carbon gradient curve was convex in the near-surface region of the case.

Example 2: Variation of Carbon Gradient with Processing Conditions in a Three-Zone Continuous Furnace under Manual Control. The carbon gradients shown in Fig. 18 are for 4815 steel that was carburized in a three-zone, continuous pusher furnace. Carbon potential was manually controlled. For all three cycles, the temperature in zones 1 and 2 was 920 °C (1690 °F), and the atmosphere was carrier gas

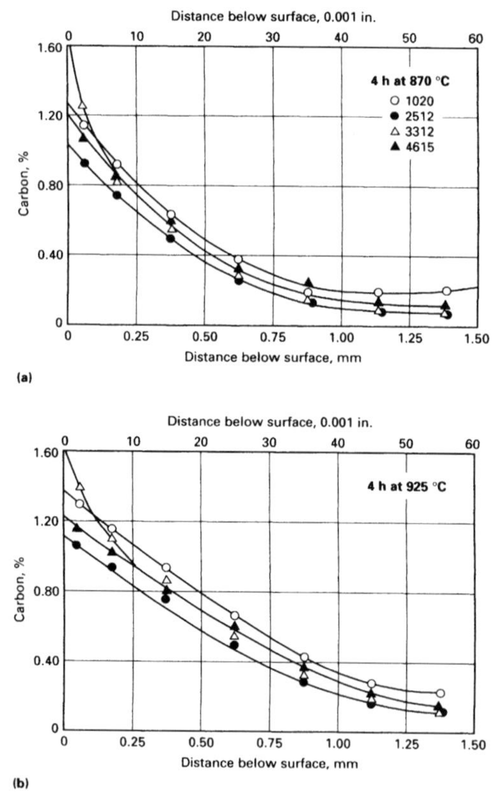

Fig. 14 Carbon gradients for four steels. (a) After carburizing for 4 h at 870 °C (1600 °F). (b) After carburizing for 4 h at 925 °C (1700 °F)

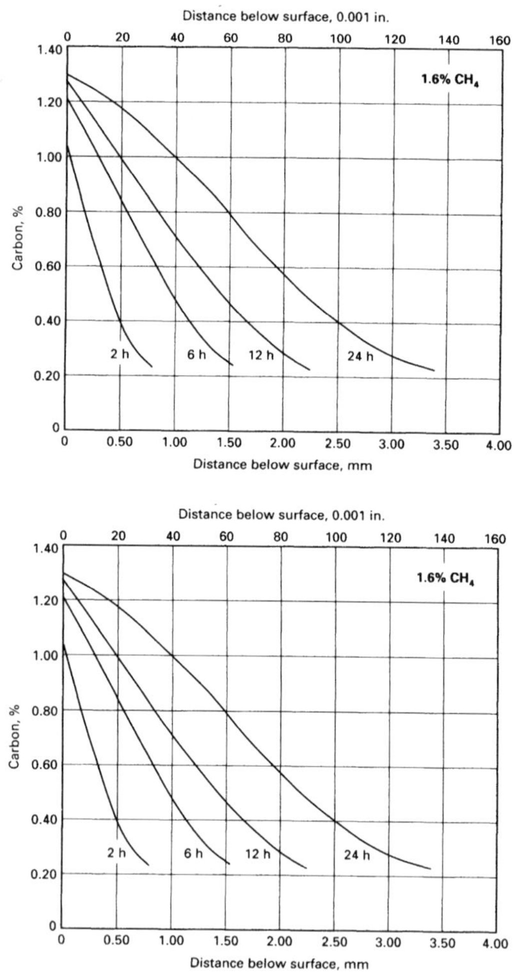

Fig. 15 Carbon gradients for carburized 1022 steel test bars carburized at 918 °C (1685 °F) in 20% CO-40% H_2 gas with 1.6 and 3.8% CH_4 added

enriched with methane; in zone 3, the temperature was 910 °C (1670 °F), and the atmosphere was carrier gas slightly diluted with air. The three cycles differed in the carbon potentials established in the third zone and in the time that parts were exposed in each of the three zones.

Cycle 1 was 35 h long, consisting of 8 h in zone 1 at 1.25% carbon potential, 14 h in zone 2 at 1.25% carbon potential, and 13 h in zone 3, where the carbon potential varied from 0.80% at the center of the zone to 0.55% at the discharge end. This cycle was designed to produce a total case depth of 2.8 mm (0.110 in.) at 0.25% C.

Cycle 2 was 25 h long, consisting of 6 h in zone 1 at 1.25% carbon potential, 10 h in zone 2 at 1.25% carbon potential, and 9 h in zone 3, where the carbon potential was 0.80% at the center and 0.60% at the discharge end. This

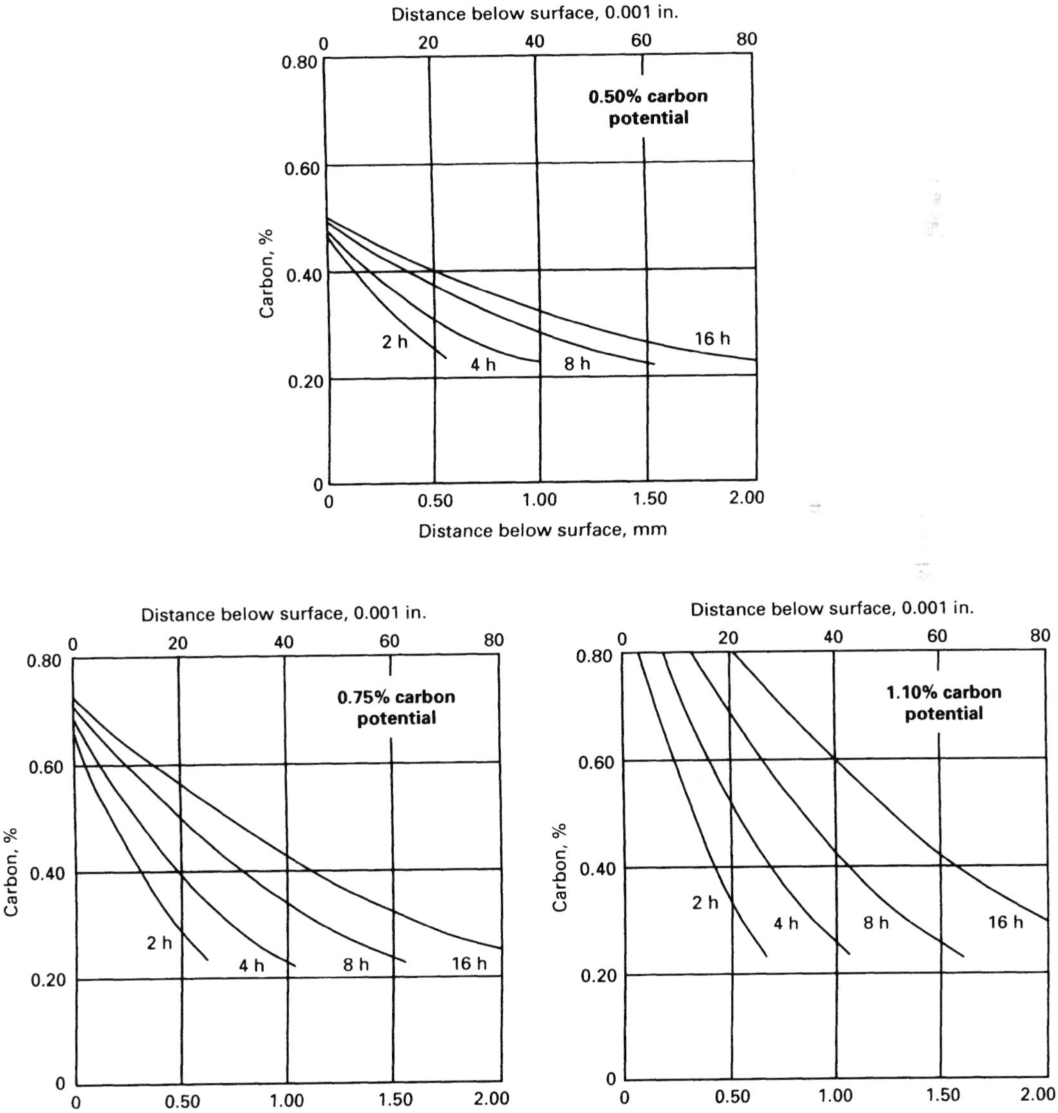

Fig. 16 Carbon gradients for carburized 1022 steel carburized at 918 °C (1685 °F) in 20% CO-40% H_2 gas containing enough H_2O to produce the carbon potentials shown

cycle produced a total case depth of 2.3 mm (0.090 in.) at 0.25% C. The carbon gradient exhibited a slightly higher surface carbon content and a narrower zone of constant carbon content in the near-surface region than did the gradient resulting from cycle 1.

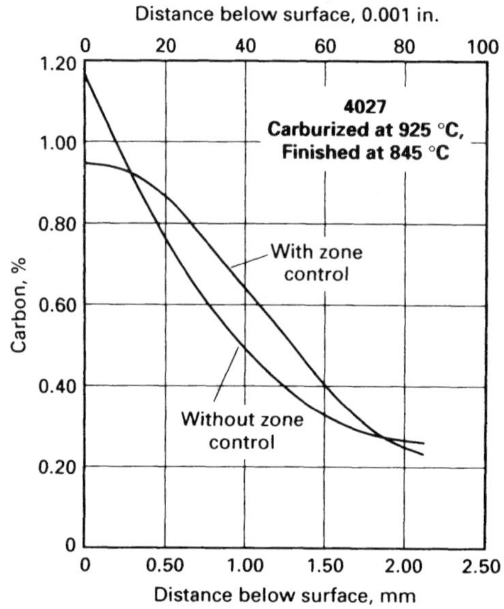

Fig. 17 Effect of zone control on carbon gradient for 4027 steel carburized in a continuous furnace (see Example 1)

Cycle 3 consisted of 4 h in zone 1 and 7 h in zone 2, both at 1.25% carbon potential, plus 6 h in zone 3, where the carbon potential was 0.8% at the center and 0.75% at the discharge end, for a total carburizing-plus-diffusion time of 17 h. This cycle produced a total case depth to 0.25% C of 1.8 mm (0.070 in.). The carbon gradient showed a slightly higher surface carbon content than did either cycle 1 or 2 (Fig. 18) and a slightly shallower plateau of constant carbon content in the near-surface region of the case.

Surface Carbon Content. As indicated in Fig. 17, carbon potential control also affects surface carbon concentration. Surface carbon content has a pronounced effect on the properties of the steel and must be controlled for optimum results.

The amount of carbon in solution affects the amount of austenite that is retained after quenching, and the amount of retained austenite affects the hardness of the quenched case. The amount of austenite retained after quenching is determined by carbon potential, carburizing cycle, diffusion cycle, quenching temperature and rate, and steel composition.

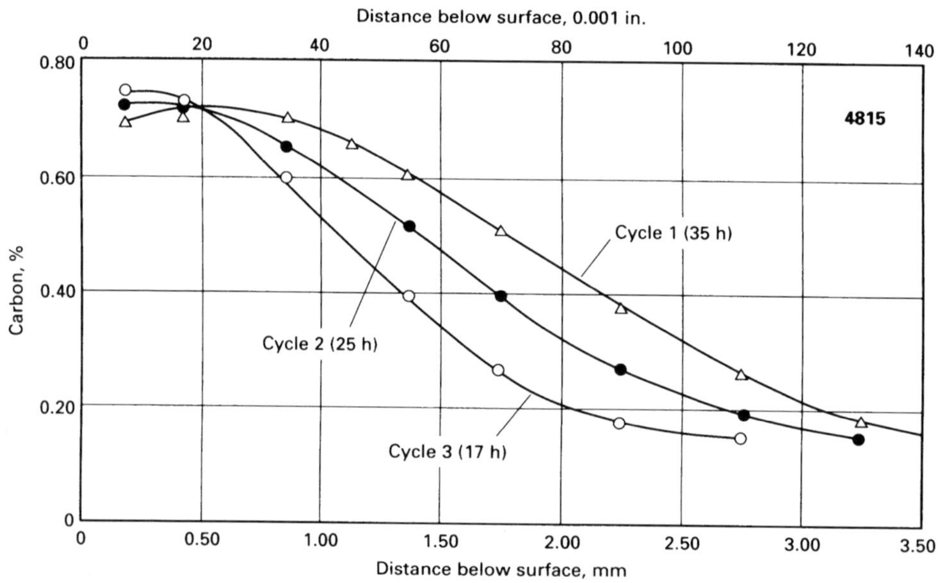

Fig. 18 Carbon gradients from three-zone, continuous pusher-type carburizing furnace with manual control of carbon potential (see Example 2)

Effect of Carbon Gradient on Case Properties. Carbon gradient is related to case hardness, as shown by the data in Fig. 19 for specimens of 1024 steel. The specimens were carburized for 2.25 h at 900 °C (1650 °F) in an atmosphere containing 10% methane. They were oil quenched from 845 °C (1550 °F) and achieved a maximum surface hardness of about 62 HRC. However, a hardness of about 66 HRC was obtained at a depth of 0.25 mm (0.010 in.) below the surface, indicating the presence of some retained austenite at the surface. Retained austenite is frequently considered a normal constituent in the microstructure, corresponding to the hypereutectoid portion of the carbon gradient. Alloy steels are more likely to exhibit retained austenite than are carbon steels.

In many instances, finely dispersed retained austenite in amounts up to 30% is not detrimental to pitting-fatigue strength, whereas much lower amounts are often harmful if the austenite is not finely dispersed. However, small amounts of finely dispersed retained austenite apparently allow mating surfaces to conform at a slightly faster rate and to spread the load more evenly, thereby reducing local areas of high stress.

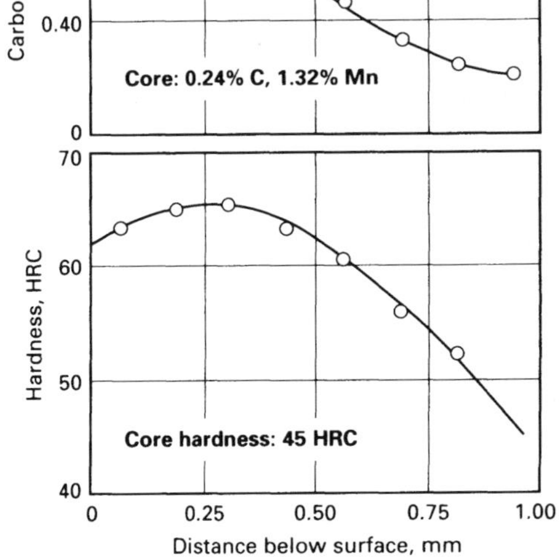

Fig. 19 Carbon and hardness gradients for 1024 steel carburized in a recirculating batch furnace under conditions that produced saturated austenite at the surface. The effect of retained austenite on surface hardness can be seen. Specimens were 25 mm (1 in.) in diameter by 150 mm (6 in.) long.

The effect of carbon content on pitting-fatigue strength has been the subject of controversy for many years. Undoubtedly other microstructural changes have clouded the issue. Rolling-contact fatigue tests have indicated that pitting-fatigue strength increases with increasing hardness in medium-carbon and high-carbon steels.

The effect of carbon content on resistance to wear and scuffing is too complicated to warrant a general conclusion. Excessive retained austenite permits surfaces to deform plastically under heavy loads, resulting in ripples, or "orange peel."

Low Surface Carbon. Most steels can be carburized by the carbon saturation-carbon diffusion type of cycle to produce surface carbon content well below saturation (for example, from 0.90 to 1.00% C). There is a strong preference in present carburizing practice for surface carbon concentrations of eutectoid composition or slightly higher. With the lean-alloy steels that are most often used, it is especially important to utilize the full hardenability of the steel. Maximum hardenability of the lean-alloy carburizing steels is obtained at carbon concentrations near eutectoid composition. The excess carbides formed at high carbon concentrations promote transformation to products other than martensite and may remove part of the carbide-forming elements from the austenite, thereby decreasing the effective hardenability.

Low carbon concentrations at the surface also permit direct quenching of work from the carburizing temperature, which is more economical than reheating before quenching. Direct quenching of carburized parts having high surface carbon concentrations favors the retention of austenite. Such austenite is undesirable because it lowers the indentation hardness of the case and promotes secondary hardening with the formation of untempered martensite, which may change the dimensions of the finished parts as well as embrittle them.

When surface carbon concentrations near eutectoid composition are desired, a multiple-manifold arrangement may be used by which part of the carrier gas and all of the hydrocarbon gas are introduced through ports at the front of the carburizing zone, where carbon demand is high. Only carrier gas is supplied to the other zones, and the carbon potential is adjusted to give the desired final surface carbon concentration.

Because, for a given carbon content of the core, the diffusion rate of carbon in austenite decreases with a lowering of surface carbon concentration, any "starvation" method of carburizing in which the workpiece surfaces are never carburized above the desired final value requires a longer cycle. Also, good recirculation of gases is essential. The atmosphere adjacent to the work quickly becomes depleted of carbon, and, if stagnant areas are allowed, excessively low surface carbon concentrations and shallow case depths will result on the work in these areas. This potential detriment to part quality is the chief reason why most continuous carburizing furnaces in use today are equipped with effective recirculating fans.

Variability in Surface Carbon. Statistical data relating to the control of surface carbon content are given in Fig. 20 to 23 for several plain carbon and low-alloy steels. These data represent practice in eight different plants and therefore involved several different types of furnaces, steel compositions, and "aim" carbon contents. Control varied from good to poor, as the graphs indicate.

The data summarized in Fig. 20 compare carbon concentrations 0.25 mm (0.010 in.) below the surface obtained in gas carburizing 25 batches in each of three similar batch furnaces.

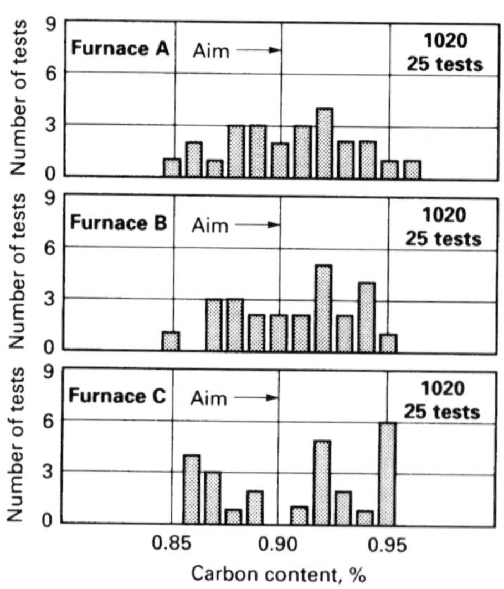

Fig. 20 Variation in carbon content 0.25 mm (0.010 in.) below the surface for 1020 steel carburized in three similar batch furnaces

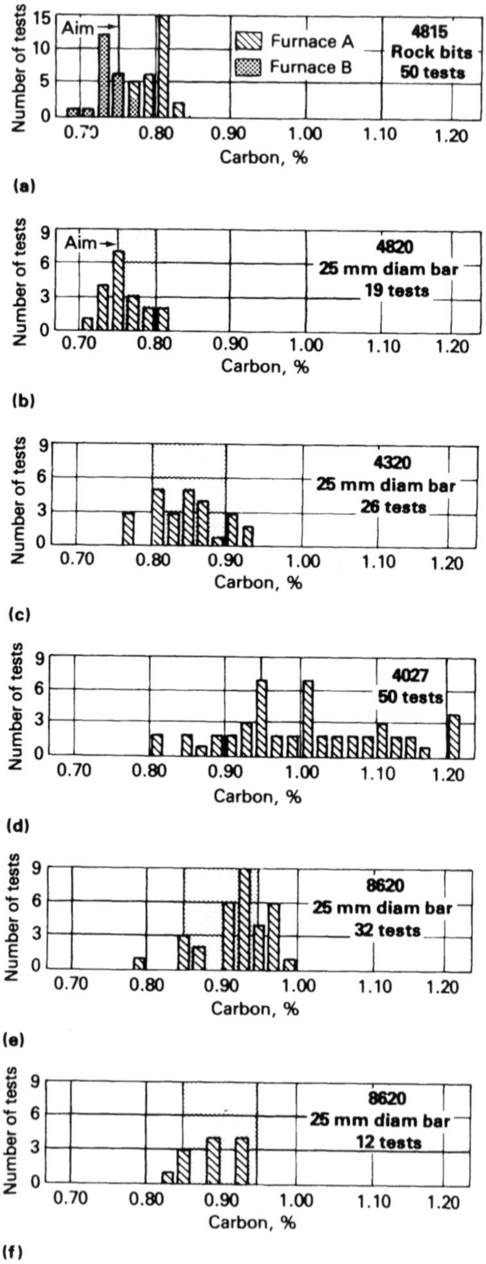

Fig. 21 Variation in surface (or near-surface) carbon content for alloy steels carburized under different conditions (see text)

The 75 tests show a large majority of the concentrations to be within ±0.05% of the 0.90% C aim. Parts were carburized at 925 °C (1700 °F). The carbon potential was controlled through the dew point. A carrier gas with a dew point of 1.5 °C (35 °F) was enriched with natural gas to maintain a dew point of –5.5 °C (22 °F) in the furnace.

The data presented in Fig. 21(a) compare surface carbon content in rock bit cutters carburized in two pit furnaces to a desired carbon content of 0.75%. Furnaces were operated simultaneously, using the same carrier gas generator. Each furnace was about 1 m (3.3 ft) in diameter by 2 m (6.6 ft) deep and contained a load of about 1360 kg (3000 lb). A carburize-

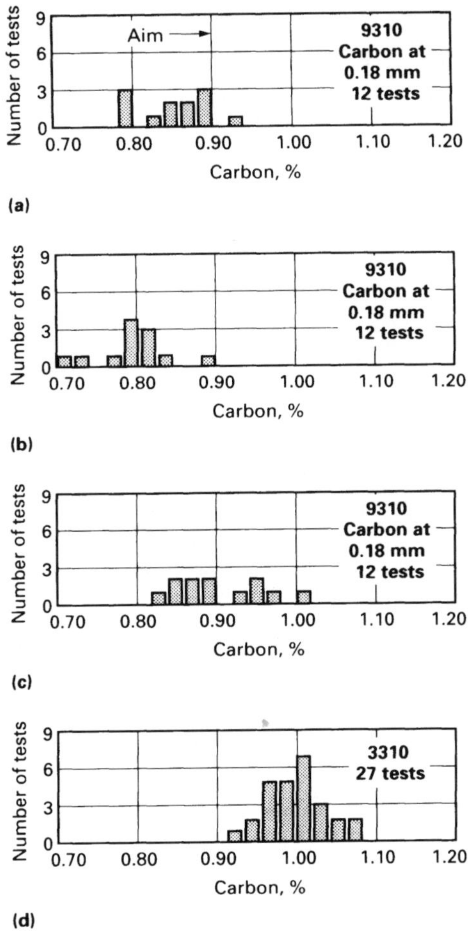

Fig. 22 Variation in surface (or near-surface) carbon content for 9310 and 3310 alloy steels carburized under different conditions (see text)

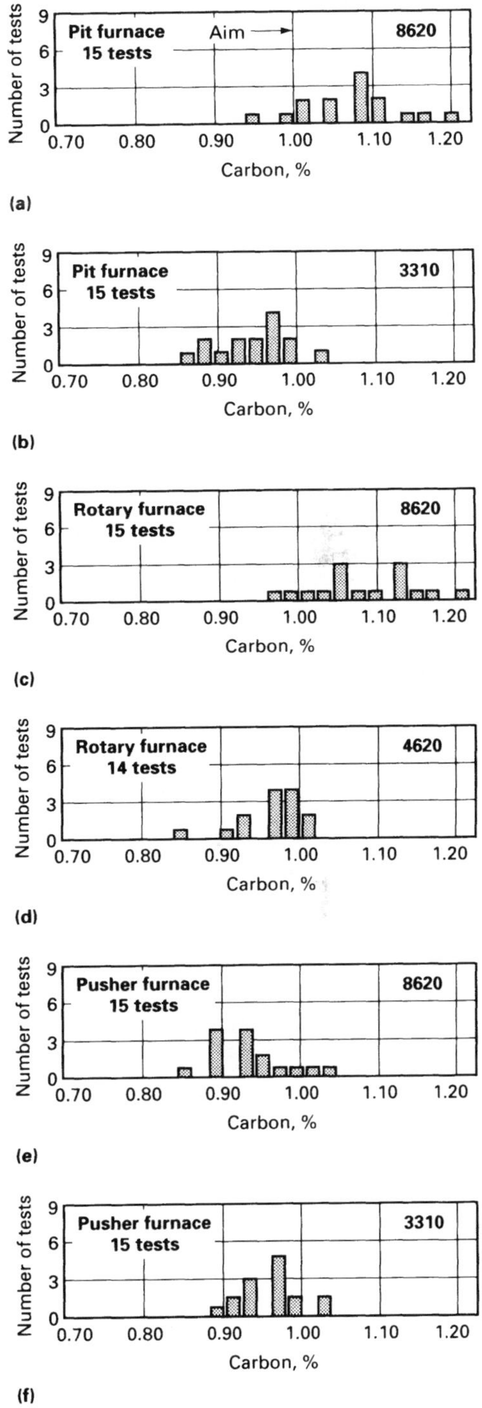

Fig. 23 Variation in near-surface (0.13 to 0.25 mm, or 0.005 to 0.010 in.) carbon content for alloy steels carburized in different types of furnaces (see text)

diffuse cycle was used. An automatic dew point controller was employed on the carrier gas generator but not on the furnace.

The results plotted in Fig. 21(b) show surface carbon contents (carbon content of the first 0.075 mm, or 0.003 in., cut) for 19 specimens of 4820 steel carburized along with loads of production parts in a two-row, pusher-type continuous furnace. The desired carbon content was 0.75 to 0.80%, with a specification of 0.70 to 0.90% carbon. All results were within specification, and most were within desired limits. Specimens were carburized at 925 °C (1700 °F) using a diffusion cycle, quenched from 815 °C (1500 °F) in oil at 60 °C (140 °F), tempered in lead at 620 °C (1150 °F) for 5 min, and cleaned with a wire brush and liquid abrasive. The atmosphere at the charge end of the furnace was automatically controlled at a dew point of –15 °C (5 °F) and at the discharge end was controlled at a dew point of 3 °C (37 °F). Endothermic gas enriched with straight natural gas was used as the carburizing medium; air was added at the discharge end.

The data presented in Fig. 21(c) for 4320 steel bar specimens were obtained under the same carburizing and test conditions used for Fig. 21(b) for lengths of 4820 steel bar, except that the dew point was –1 °C (30 °F) at the discharge end and the specimens were tempered at 650 °C (1200 °F). The desired surface carbon content was 0.85 ± 0.05%.

The results of 50 tests that represent variations in surface carbon content for continuous carburizing of 4027 steel are given in Fig. 21(d). These results were obtained over a period of 3 to 4 years. An automatic dew point controller was used.

The data given in Fig. 21(e) for 8620 steel bar specimens were obtained under the same carburizing and test conditions used for carburizing 4320 steel (Fig. 21c). The desired surface carbon content was 0.90 ± 0.05%.

The data shown in Fig. 21(f) were based on the surface carbon content (the first 0.075 mm, or 0.003 in., depth of cut) of 25 mm (1 in.) rounds. The rounds were carburized with production parts in a horizontal batch furnace, quenched in oil from 845 °C (1550 °F), tempered in lead at 650 °C (1200 °F), and cleaned with a wire brush and liquid abrasive. The desired carbon content was 0.90 ± 0.05%. The carburizing medium was endothermic gas enriched with natural gas. No air was added. A dew point of –15 °C (5 °F) was used throughout the 925 °C (1700 °F) carburizing cycle, –12 °C (10 °F) for the diffusion cycle (also at 925 °C, or 1700 °F), and –3 °C (27 °F) for the 845 °C (1550 °F) cycle that preceded quenching.

The data in Fig. 22(a) represent carbon content at 0.18 mm (0.007 in.) below the surface. Actual surface carbon ranged from 0.85 to 1.03%. The steel was carburized in a batch-type furnace using a hot-wire analyzer for automatic atmosphere control. Actual case depth to 50 HRC was 1.10 to 1.32 mm (0.043 to 0.052 in.); required case depth was 1.15 to 1.25 mm (0.045 to 0.050 in.); total case depth ranged from 1.93 to 2.05 mm (0.076 to 0.081 in.).

The data summarized in Fig. 22(b) also represent carbon content at 0.18 mm below the surface. Actual surface carbon ranged from 0.84 to 1.19%. The steel was carburized in a batch-type furnace with manual atmosphere control. Actual case depth to 50 HRC was 0.38 to 0.58 mm (0.015 to 0.023 in.); required case depth was 0.38 to 0.64 mm (0.015 to 0.025 in.); total case depth ranged from 0.64 to 0.80 mm (0.025 to 0.031 in.).

Carburizing and test conditions for the data shown in Fig. 22(c) were the same as those for the data in Fig. 22(b). In this instance, surface carbon ranged from 0.89 to 1.48%, and case depth to 50 HRC was 0.64 to 0.94 mm (0.025 to 0.037 in.). Specified case depth was 0.75 to 0.90 mm (0.030 to 0.035 in.); total depth ranged from 0.86 to 1.35 mm (0.034 to 0.053 in.). The same batch-type furnace was used to obtain the data for Fig. 22(a), (b), and (c).

The data shown in Fig. 22(d) represent surface carbon contents of 27 samples taken at 8 h intervals. Parts were carburized in trays in a continuous pusher furnace with dew point control of the atmosphere. Desired surface carbon was 1.0%.

The carbon concentration data for 8620 and 3310 steels that are shown in Fig. 23(a) and (b) were obtained on round-bar test specimens placed at several locations throughout each charge of production parts of the same steel composition. Specimens were carburized in a vertical pit furnace having a single circulating fan. The atmosphere was endothermic gas enriched with methane. The 8620 steel was carburized at 955 °C (1750 °F) without automatic atmosphere control, using a diffusion cycle in which only endothermic gas was introduced during the last third of the carburizing cycle.

The 3310 steel was carburized at 925 °C (1700 °F) with automatic atmosphere control but no diffusion cycle.

Round bars were used to obtain the data shown for 8620 and 4620 steels in Fig. 23(c) and (d). Both steels were carburized at 955 °C (1750 °F) in a rotary retort furnace under an atmosphere of endothermic gas plus methane. A diffusion cycle, in which the methane was shut off for the last third of the carburizing cycle, was used for the 8620 steel. The 4620 steel was carburized without a diffusion cycle, and the carbon content ranged predominantly below the desired 1.0%. Results for the 8620 steel were largely above the desired carbon content.

The data in Fig. 23(e) and (f) for 8620 and 3310 steels were obtained from round test specimens distributed throughout loads in continuous pusher furnaces. The atmospheres were endothermic gas enriched with methane. The 8620 specimens were carburized at 925 °C (1700 °F) in a furnace with three fans, without automatic atmosphere control or a diffusion cycle. The 3310 specimens were carburized at 925 °C in a furnace equipped with two fans, with automatic atmosphere control but no diffusion cycle.

Surface carbon content may vary significantly as a function of location in the furnace. Figure 24 indicates a variation in surface carbon on 4620 steel bearing races from the tip to the bottom of the pit furnace used. The furnace was about 0.75 m (30 in.) in diameter by 0.90 m (36 in.) deep. An open load of races was carburized for 7 h in a natural gas atmosphere, and this step was followed by a 3.5 h diffusion cycle. Desired surface carbon was 1.00%. Each bar on the chart represents 14 heats of steel. It is apparent that in this instance surface carbon content was consistently higher in specimens carburized in the bottom portion of the furnace, possibly indicating a slightly higher operating temperature there.

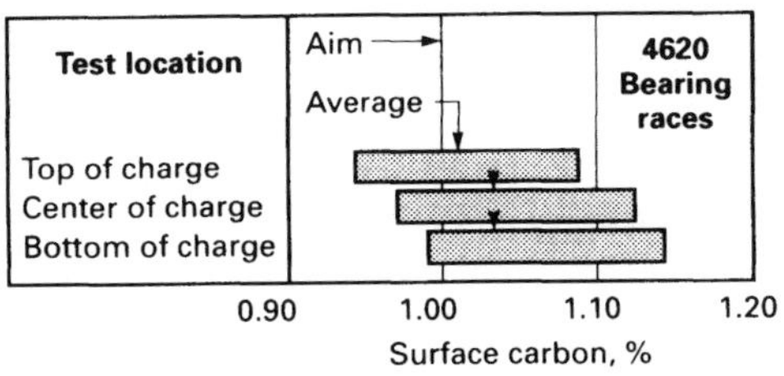

Fig. 24 Variation in surface carbon content with position in a pit furnace for 4620 steel bearing races

Process Planning

Designers usually specify the case hardness, case depth, and core hardness required to meet the service loads they anticipate for a particular part. It is the task of the process engineer to develop the carburizing treatment that will produce the properties desired. Some of the considerations involved in setting up processes include:

- Case microstructure
- Residual stress
- Alloy selection
- Operating schedules
- Quenchants
- Reheating for quenching
- Tempering

Case Microstructure. As described later in this chapter in the section "Microstructures of Carburized Steels," a carburized case is usually a mixture of tempered martensite and retained austenite. Other microconstituents, such as primary carbides, bainite, and pearlite, are generally avoided. For a particular alloy, the amount of retained austenite in the case increases as the case carbon content increases. An appreciable decrease in case hardness is usually found when the amount of retained austenite exceeds about 15%, but for applications involving contact loading, such as rolling element bearings, the best service life is found when the retained austenite content is quite high, for example, 30 to 40%. In other applications, especially when dimensional stability is critical, the retained austenite content should be low.

Figure 25 shows the dependence of retained austenite content on carbon content for iron-carbon alloys (Ref 11). For a given carbon content, more retained austenite is usually present in alloy steels. Therefore, by controlling the case carbon content, the amount of retained austenite can be regulated. Tempering above 260 °C (500 °F) will also eliminate retained austenite, but at the expense of case hardness.

Residual Stress. Carburized parts have a compressive residual stress in the case that is balanced by a tensile residual stress in the core.

The presence of this stress distribution is an advantage for applications such as bending or torsional loading, in which maximum tensile stresses are experienced at the surface of the part. When other factors are the same, the magnitude of the compressive residual stress at the surface depends on the ratio of case and core thicknesses. When the core is much thicker than the case, compressive stresses at the surface will be high. When the reverse is true, surface compressive stress will be low and core tensile residual stress will be high. Because retained austenite is more dense than is martensite, the former will tend to reduce the magnitude of the compressive residual stress in the surface. The influence of residual stresses on the microstructures and properties of carburized steels are discussed in subsequent sections of this chapter.

Alloy Selection. Frequently used carburizing steels are listed in Table 2. Carburizing steels are usually selected on the basis of case and core hardenability; comparative hardenability data for these alloys can be found in the SAE specifications J1268 and J1868 (*SAE Handbook,* Volume 1 or the article "Hardenability Curves" in Volume 1 of the *ASM Handbook*). As a rule, carburized plain carbon steels with less than 1% Mn must be water quenched to form a martensitic case. If an oil quench must be used, adequate case hardenability can usually be obtained by carbonitriding a coarse-grained steel.

Alloy steels are used for most heavily loaded parts, not only because of their increased hardenability, but also because the standards of steel cleanliness are more restrictive. Many of the alloy steels are similar in hardenability but differ in other important respects. For example, nickel-molybdenum alloys provide the most trouble-free processing because the principal alloying elements are neither strong oxide nor strong carbide formers. However, economic considerations often dictate the use of less expensive steels, such as chromium-manganese alloys, which are prone to alloy depletion (and loss of hardenability) because of the formation of grain-boundary oxides near the surface during carburizing. Chromium-nickel-molybdenum steels, such as SAE 8620 and 8720, provide a balance between cost, hardenability, and ease of processing that leads to their specification for many parts. More expensive steels, such as SAE 9310 and 3310, are used for critical gearing applications. Some special carburizing alloys, CBS 1000M, for example, have secondary hardening characteristics that provide resistance to softening for temperatures up to about 550 °C (1025 °F).

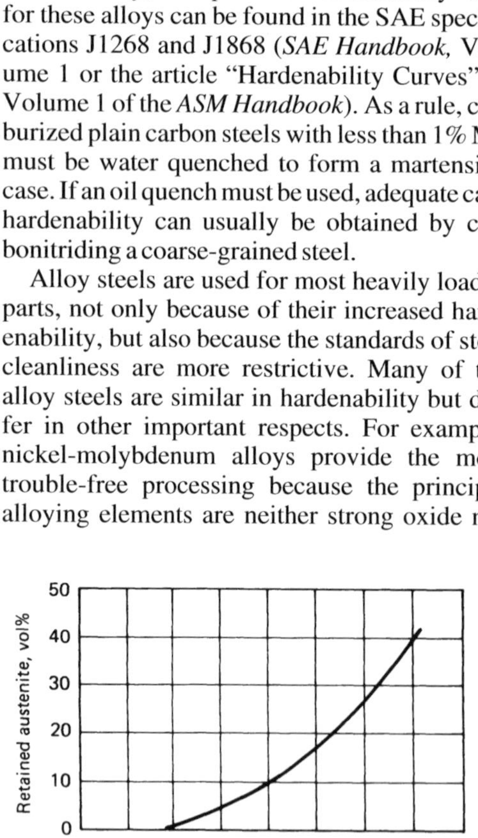

Fig. 25 Effect of retained austenite content on carbon content for iron-carbon alloys

Operating Schedules. To minimize the total amount of time required for carburizing to a given case depth, most processes are set up in a boost-diffuse mode. The boost step is at a relatively high temperature and high carbon potential to facilitate the rapid development of a deep case; the diffuse step is at a lower carbon potential, allowing the case carbon content to decrease to the level desired. Mathematical models for diffusion, described previously, are very helpful in choosing the times and temperatures required for each step.

In practical situations, many considerations, in addition to minimizing processing time, enter into the choice of processing parameters. In continuous furnaces that cannot be separated into distinct zones by means of internal doors, there is a limit to the magnitude of the differences in temperature and carbon potential that can be sustained over the length of the furnace. Similarly, in batch furnaces, the rate at which the temperature can be lowered depends on the thermal inertia of the furnace and load and the magnitude of the heat losses; the rate at which the atmosphere gas composition can be changed depends on the mean residence time of the gas in the furnace. In the absence of a detailed mathematical model incorporating furnace operating characteristics, some trial-and-error experimentation may be required to find the operating set points that produce the desired results. The maximum operating temperature can be limited by the accelerated degradation of fixtures experienced at higher temperatures. Lower processing temperatures are generally favored when the distortion of parts must be minimized.

Thin-cased parts (0.5 mm, or 0.02 in., effective case or less) tend to be processed at 870 °C

(1600 °F) or below simply because the processing time at higher temperatures is so short that it is difficult to control the results. For the same reason, thin-cased parts are carburized using a constant carbon potential rather than a boost-diffuse process. Sometimes the available equipment, for example, a continuous pusher furnace, is larger than the process requires. In this situation the furnace may be run at a lower-than-customary temperature to avoid the frequent door openings, which tend to upset the furnace atmosphere, that would be necessary at higher temperatures.

Quenchants. Carburized parts can be quenched in brine or caustic solutions, water-based polymer quenchants, oils, or molten salt. The more rapid the quench, the lower the requirements for hardenability, but the greater the likelihood of distortion. If a part is used as-heat treated (no finishing operations to control dimensions), mar-quenching into molten salt or hot oil may be used. Rings and shafts are often press quenched, that is, clamped in a fixture while hot and sprayed with oil, to reduce distortion. The choice of alloy and quenchant and the manufacturing-process design are inter-related and must be compatible.

Reheating for Quenching. It has been the practice in some industries to carburize parts at a relatively high temperature (925 °C, or 1700 °F, or above), cool slowly to ambient temperature, then reheat to a lower temperature (845 °C, or 1550 °F, for example), and quench. The advantages of this approach are:

- The final austenite grain size, which controls the size of the martensite plates and laths, is smaller, and therefore the microstructure is more refined.
- The low reheat temperature places an upper limit on the amount of carbon dissolved in austenite, thereby limiting the amount of retained austenite present in the case.

The main disadvantage of this processing, aside from added cost, is increased distortion of the parts. As better controls have become available for regulating furnace atmospheres and as the use of fine-grain steels has become the norm, the need for this practice has diminished.

Tempering. As high-carbon martensite is tempered, carbides precipitate, increasing the density and reducing the hardness of this constituent. The retained austenite, which accompanies the high-carbon martensite, transforms in the temperature range of 220 to 260 °C (425 to 500 °F), reducing the density and increasing the hardness of this constituent. Thus, the response of case hardness to tempering can be complex, affected by the fraction of retained austenite present initially.

Density changes during tempering affect the relief of residual stresses produced by carburizing. Figure 26 (Ref 12) shows the effect on stress relief of tempering for 1 h at various temperatures. Stress relief occurs at lower tempering temperatures, as the amount of carbon dissolved in austenite increases. The dependence of stress relief on tempering temperature in carburized parts is qualitatively similar to these results, although more complex behavior can be expected in steels with high levels of retained austenite.

The decrease in hardness with tempering is shown in Fig. 27 for four carburized steels. The HRC hardness values shown were converted from Rockwell A for surface hardness and from Vickers microhardness for the subsurface values. It should be noted that the hardness changes very little for tempering temperatures up to 205 °C (400 °F). However, a substantial degree of stress relief will have been experienced as indicated in Fig. 26.

Parts that are to be ground after heat treatment should always be tempered. The tempering temperature should be greater than the sur-

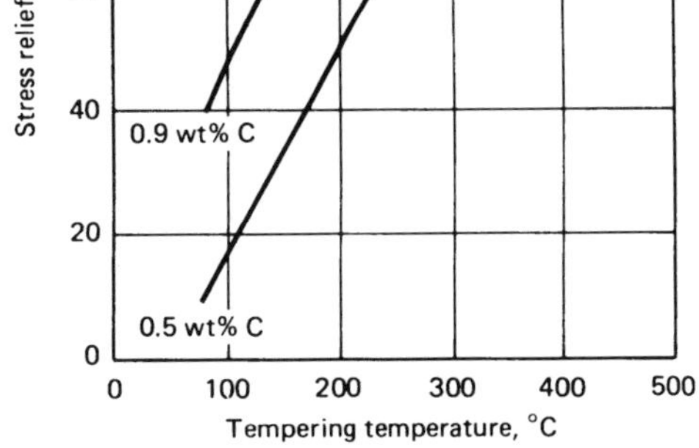

Fig. 26 Plot of stress relief versus tempering temperatures held for 1 h for two carbon concentrations in austenite. Source: Ref 12

face temperature experienced during grinding. If it is not, grinding will produce surface tensile residual stresses. Tempering in the range of 150 to 205 °C (300 to 400 °F) improves subsequent dimensional stability. If the best possible stability is required, as in some instrument bearings, the treatment should also be designed to eliminate retained austenite. On the other hand, because tempering reduces the compressive residual surface stresses, resistance to bending and torsional fatigue may be best if the parts are not tempered at all. Tempering is omitted on many thin-cased parts (0.5 mm, or 0.02 in., case depth or less) that are not finished after heat treatment.

Selective Carburizing

To function properly, some parts must be selectively carburized, that is, carburized only on certain surfaces. Some gears, for example, are carburized only on teeth, splines, and bearing surfaces. In addition to satisfying the performance requirements of a part, selective carburizing may facilitate the machining or welding of noncarburized surfaces in the hardened condition.

Surfaces that are not to be carburized must be protected by a coating or shield that is impervious to the carburizing atmosphere. Various means are employed to protect or stop off selected surfaces from the atmosphere.

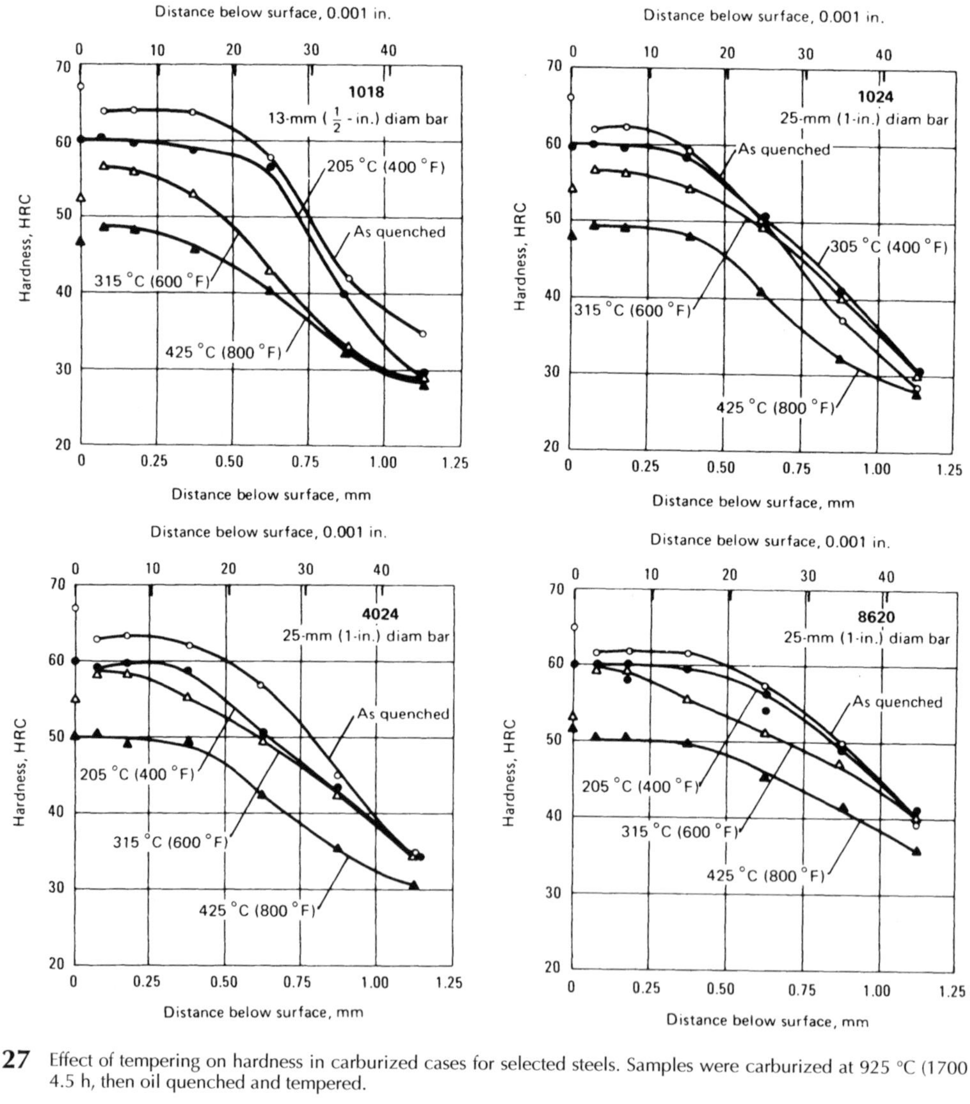

Fig. 27 Effect of tempering on hardness in carburized cases for selected steels. Samples were carburized at 925 °C (1700 °F) for 4.5 h, then oil quenched and tempered.

Copper plating. to a minimum thickness of 13 µm (0.5 mil) is widely used for this purpose because it is relatively easy to apply, is machinable, and does not contaminate furnace atmospheres. Prior to carburizing a large, 915 mm (36 in.) diameter, 4620 steel ring gear, for example, the gear is copper plated on the inside diameter flange area only to permit finish machining, after hardening, of the bore and bolt holes located in the flange. Surfaces that are not to be copper plated can be coated with a chemical-resistant lacquer, which is removed prior to carburizing. After carburizing, the copper can be chemically stripped from the part or removed in subsequent machining operations.

Ceramic coatings. in the form of paint can also protect selected surfaces from carburizing. Surfaces must be thoroughly cleaned before ceramic paint is applied, and the first coat is allowed to dry before a second coat is applied. Ceramic paint coatings must adhere tightly in order to be impervious to the carburizing atmosphere. The application of ceramic paint to the bushing and button recesses of a rock bit cutter, for example, has been used in production.

Stopoffs. Blind holes can be stopped off by inserting copper plugs or by filling them with clay. If air is entrapped by the plug, a means of venting must be provided to relieve the pressure buildup during heating. Through holes may be plugged at either end to limit access by the atmosphere, thereby minimizing carburization. Internal threads can be protected by the insertion of a copper screw and external threads, by capping with a copper nut. If a steel screw or nut is used, the threads should be coated with a stopoff material to facilitate removal. The success of all stopoff methods depends largely on the care used in their applications. Seldom are mechanical stopoffs completely effective in production.

Case Removal. If parts are cooled slowly after carburizing, they will be soft enough to permit the removal of the case in selected areas by machining. After subsequent reheating and quenching, these areas will remain lower in carbon and relatively soft. Operations can also be planned so that the case on a hardened part can be removed selectively by grinding. This practice is usually confined to small areas and generally to cases less than 1.3 mm (0.05 in.) in depth.

Localized Softening. A part that has been carburized may be locally softened by induction heating. The heat may just temper the part, or, for steels with low hardenability, may actually normalize the area to be softened. This practice is used to soften threads on carburized shafts in the automotive industry.

Dimensional Control

Parts should be as near to final dimensions as possible before carburization so that heat treating times may be kept short. Nevertheless, some growth and distortion will be encountered in all carburized parts. Part size and shape strongly influence distortion, but a host of other factors can play a role as well, such as:

- Residual stresses existing in parts prior to heat treatment
- Shape changes induced by heating too rapidly
- Method of stacking or fixturing parts during carburizing and quenching
- Increasing growth as the case depth increases
- Severity of quenching (including variations caused by changing quench temperatures)
- Hardenability, as influenced by variations in steel composition

A variety of methods, to be described, are used to minimize distortion. All of them add to heat treating cost, but they may be cost effective when total production costs are considered.

In the high-volume production of precision components, such as automotive transmission gears, a general objective is to hold constant all of the processing variables affecting growth and distortion and then to compensate for the changes that do take place by adjusting the shape of the green machined (unhardened) part. For example, the lead angle of a helical gear becomes smaller after it has been carburized because the gear becomes longer. Once the change has been documented, the machining operation prior to heat treatment can be adjusted to compensate for the growth. Trial lots of parts are processed using the intended production process and then measured to determine shape changes. After production has begun, parts may be checked periodically to ensure that processing conditions that affect distortion have not changed.

Fixturing can be used to separate parts, allowing more uniform heating and access to the carburizing atmosphere, and to orient parts so that each part enters the quench bath the same way. Fixturing adds labor cost (fixtures are often loaded and unloaded by hand) and adds operating cost because the fixtures may weigh as much as, or more than, the work load. However, proper fixturing is often effective in reducing variations in growth and lessening distortion, thereby allowing compensation for the changes in the shape of the green part.

Marquenching or martempering, involves quenching in molten salt or in hot oil; the quenchant temperature is above the M_s temperature (the temperature at which martensite starts to form from austenite upon cooling) of the case, but below the M_s temperature of the core. Consequently, the core transforms, while parts are in the quenchant, to a mixed microstructure of martensite, bainite, and ferrite. The case transforms to martensite while the parts are cooling in air after leaving the quenchant. Distortion is reduced because the transformation strains in the core are reduced, and because thermal gradients are much lower during transformation in the case and core.

Press quenching and other means of quenching in fixtures, such as plug or cold die, are effective for reducing distortion. While quench presses may have automated loading systems, they are often loaded manually, one part at a time. Quenching in fixtures adds significantly to the cost of heat treating, but there may be no practical alternative for limiting distortion of certain parts, for example, thin-walled rings and slender shafts.

Straightening is employed to reduce the distortion of heat-treated shafts. Most straightening is done by flexing the part. There is always a risk of cracking the case when straightening carburized parts, so it is the usual practice to straighten after tempering. Some manufacturers find it best to straighten while parts are still hot from the tempering furnace. In one instance, shafts that cracked upon straightening after tempering could be straightened without cracking immediately after quenching. Many straighteners are equipped with acoustic emission sensors to detect cracking. Selective peening has also been used as a method for straightening parts.

Case Depth Measurement

Case depth is usually specified as the depth below the surface at which a defined value of some property occurs. A case depth to a hardness of 50 HRC and a case depth to a carbon content of 0.4 wt% are examples of specifications for an effective case depth. Also used is the term total case depth, which is too vague for general use as a specification because of the gradual transition between case and core properties in most carburized parts. However, some carbonitrided parts and induction case hardened parts will have a sharp transition in microstructure clearly separating case and core. Only in these instances is total case depth defined well enough to be useful as a specification. Methods of measuring case depth are described in detail in SAE Recommended Practice J423.

Many parts, such as gears, consist of some surfaces that are convex, some that are relatively flat, and some that are concave. It is usually found that the case depth will be least at the concave surface (the root) and greatest at the convex surface (the tooth tip). The effect of surface curvature on effective case depth is shown in Fig. 28. These data are computed for a particular case:

- Surface carbon, 1%
- Core carbon, 0.2%
- Diffusion time temperature sufficient to produce a case depth to 0.4% of 1 mm (0.04 in.)

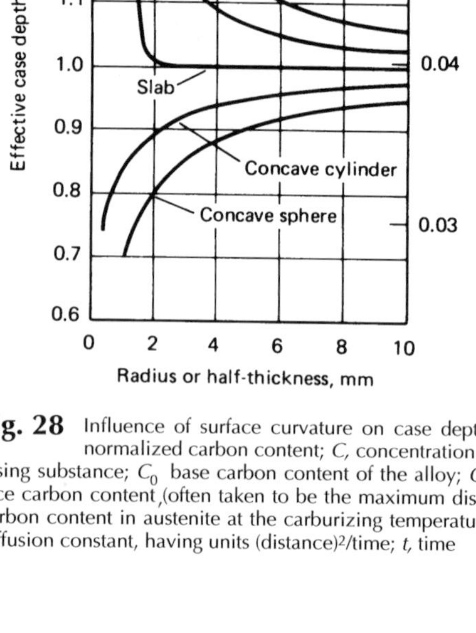

Fig. 28 Influence of surface curvature on case depth. C^*, normalized carbon content; C, concentration of diffusing substance; C_0 base carbon content of the alloy; C_s, surface carbon content,(often taken to be the maximum dissolved carbon content in austenite at the carburizing temperature); D, diffusion constant, having units (distance)^2/time; t, time

beneath the flat surface of a slab of infinite thickness

From Fig. 28, the same treatment that will produce a case depth of 1 mm (0.04 in.) in a plane slab of 3 mm (0.12 in.) half-thickness will produce a case depth of 1.13 mm (0.0445 in.) in a rod of 3 mm (0.12 in.) radius, a case depth of 1.28 mm (0.0504 in.) in a sphere of 3 mm (0.12 in.) radius, and a case depth of 0.93 mm (0.037 in.) in a circular hole of 3 mm (0.12 in.) radius. The effect of surface curvature is quite pronounced if the radius of curvature of the surface is less than about five times the effective case depth beneath a flat surface. The effect of curvature is small when the radius of curvature of the surface is more than ten times the effective case depth. Surface curvature also affects cooling so that in parts with marginal hardenability there may be a greater difference between convex and concave surfaces when effective case depth is defined in terms of depth to a specified hardness than when it is defined as depth to a specified carbon content. These facts must be kept in mind when comparing case depth at various locations on a part and when comparing case depths on parts with case depths on testpieces.

Effective case depth to a specified hardness value is best measured by means of a microhardness traverse on a polished metallographic section cut normal to the surface. Vickers (diamond pyramid hardness) or Knoop or Rockwell microficial indenters may be used with loads of 0.5 to 2 kg. Specifications can be written in terms of a microhardness value or in terms of HRC hardness, where it is understood that microhardness (HV, HK, or HRMF) is converted to HRC using standard tables (50 HRC = 513 HV = 542 HK = 870 HRMF). Surface hardness, which is often a part of carburizing specifications, can be obtained by extrapolating subsurface microhardness values to the surface. Alternatively, the microhardness at a depth of 0.1 mm (0.004 in.) beneath the surface can be defined as the surface hardness. It is difficult to get reliable microhardness readings closer than 0.1 mm (0.004 in.) to the edge of a polished section because of edge rounding, which occurs during polishing.

Effective case depth to a specified carbon content is most frequently measured on testpieces of the same alloy and shape as the workpieces and processed simultaneously with the work load. Typically, a ground bar about 25 mm (1 in.) in diameter is used. If it is quenched with the work load, it must be tempered at about 650 to 700 °C (1200 to 1300 °F) prior to machining. Bars are cut dry, at very slow speeds to avoid burning, and washed between cuts with a nonflammable fluorinated hydrocarbon. Turnings are collected on clean paper, rewashed, and analyzed for carbon. To detect the effect of the final atmosphere exposure on the near surface carbon gradient, three or four initial cuts of 0.025 mm (0.001 in.) thickness should be made. Thereafter, cuts of 0.075 to 0.125 mm (0.003 to 0.005 in.) thickness can be made through the case. The accuracy of the method depends on good machining practice in addition to well-standardized analytical procedures. One user reports repeatability of 0 to 0.03% C at depths greater than 0.075 mm (0.003 in.) and 0 to 0.05% C for samples taken closer to the surface.

As a plant floor test, test bars about 10 to 12.5 mm (0.4 to 0.5 in.) can be processed with the work load, quenched, and fractured. The depth of the case, evident on the fracture surface, can be read with a 10 × Brinell glass. Even workpieces can be fractured if their shape (and cost) allow easy fracture. A variant of this test involves reheating a test bar to 780 °C (1440 °F) and then quenching and fracturing. In plain carbon steels, only material with a carbon content of 0.4 wt% or greater will become austenitic at this temperature. Therefore, the line of demarcation between case and core on the fracture surface should occur at a carbon content of 0.4 wt%.

Another test for case depth makes use of the principle that the M_s temperature depends primarily on the dissolved carbon content of austenite. A carburized testpiece or part is austenitized, quenched in a bath at a precisely controlled temperature, held for a few minutes to allow tempering of the martensite formed, and then quenched to room temperature. A polished, etched section will show a line of demarcation between the martensite tempered in the salt bath (dark) and the untempered martensite formed during the final quench (white). The location of the line of demarcation corresponds to a specific carbon content. The appropriate quenching bath temperature can be found from a correlation between M_s temperature (in degrees Fahrenheit) and composition (Ref 13):

$$M_s = 930 - 600(\%C) - 60(\%Mn) - 50(\%Cr) - 30(\%Ni) - 20(\%Si + \%Mo + \%W) \quad \text{(Eq 22)}$$

If the quenching bath temperature is chosen to be the M_s temperature for 0.4 wt% C in a given alloy, then the austenitizing temperature must be high enough to dissolve at least this much carbon in austenite, typically 830 °C (1525 °F).

Surface hardness is often measured on the plant floor on workpieces using a superficial hardness test, such as Rockwell 15N. Because of the rather light load, the necessity for supporting pieces well, and the importance of a smooth surface, values tend to be unreliable. However, such checks of surface hardness are often necessary to detect the presence of a shallow, decarburized layer such as might occur if parts experience a delay when transferred from the carburizing furnace to the quench bath.

Microstructures of Carburized Steels

At first glance, the microstructures of carburized steels appear to be quite straightforward: high-carbon martensite is gradually replaced by martensite of lower carbon content with increasing distance from the surface. This view of the microstructures of carburized steel is essentially correct. Lightly tempered martensite is the dominant microstructural constituent of properly carburized steel. However, the martensite changes in morphology, amount, and properties as a function of distance from the surface. Other case microstructural constituents may also be present and may significantly affect the performance of carburized parts. These other microstructural features include retained austenite; carbides of various origins, sizes, and morphologies; inclusions; prior austenite grain boundaries embrittled by phosphorus segregation; microcracks; and processing-induced surface oxides. Superimposed on the case microstructures are compressive residual stresses produced during quenching. Core microstructures, depending on hardenability, may consist of tempered martensite, bainite, or ferrite and pearlite.

Martensite (Ref 14)

How Martensite Forms. Martensite tempered at low temperatures—150 to 200 °C (300 to 400 °F)—is the primary component of carburized microstructures that imparts the desired properties to carbur-carbon has been introduced into austenite, typically at temperatures around 930 °C (1700 °F), low-temperature-tempered (LTT) martensite is produced by quenching and subsequent low-temperature reheating or tempering. Martensite is formed from austenite by a diffusionless shear transformation, and quenching has the important function of providing cooling rates rapid enough to suppress competitive diffusion-controlled transformations of austenite. The latter transformations involve the diffusion of carbon and the formation of lower-strength mixtures of the phases ferrite and cementite. The martensitic transformation, on the other hand, traps carbon atoms between iron atoms within the crystal structure of martensite. In carbon-containing steels, the martensitic crystal structure is body-centered-tetragonal, and the carbon atoms are located in octahedral interstitial sites between iron atoms.

Carburizing is applied to low-carbon steels containing typically 0.2 wt% C. The process introduces carbon to the surface of the steel, and the carbon diffuses into the low-carbon interior. Depending on atmosphere control and carburizing temperature and time, surface carbon contents of 0.8% or more and carbon gradients over a range of distances, or case depths, into the steel are produced. The carbon gradients have a profound influence on martensitic transformation, morphology, and properties. As described earlier in this chapter, the temperature at which martensite begins to form during the quench is designated the martensite start temperature or M_s, and as carbon content increases, M_s decreases.

Martensite Morphologies. The martensite crystals that form at low M_s temperatures have a three-dimensional plate geometry, and hence are called plate martensite. Figure 29 is a micrograph of martensite in the high-carbon case of a carburized steel. Adjacent plates of martensite are nonparallel, and the appearance of the microstructure under the light microscope is that of a zigzag array of needles or acicular shapes. The needlelike shapes are in fact cross sections of martensite plates. The plates etch dark after tempering and are surrounded by white-appearing retained austenite. The amount of martensite that forms is determined only by the amount of cooling below M_s; therefore, the lower the M_s, the smaller the amount of martensite and the larger the amount of retained austenite after cooling to room temperature.

In contrast to high-carbon regions, the martensite that forms in austenite of medium or low carbon content assumes a completely different morphology. The martensite crystals appear to have a lath or board-shaped geometry—the crystals are relatively thin and flat with one long dimension, and adjacent crystals form parallel to each other in stacks or packets. Figure 30 is a micrograph of martensite in the low-carbon core of a carburized steel. The parallel arrangement of the martensite crystals is discernible, but most of the crystals are too thin to be resolved in the light microscope.

The high M_s temperature of low-carbon austenite provides a much wider temperature range over which martensite can form during quenching, and therefore there is little or no retained austenite in low-carbon lath martensites cooled to room temperature. With increasing carbon content, the amount of retained austenite in lath martensitic microstructures increases, but the austenite is retained as thin films between the laths and is not resolvable in the light microscope.

The carbon gradients introduced into austenite by carburizing result, after quenching, in a gradient of martensitic microstructures, ranging from plate martensitic morphologies with large amounts of retained austenite to lath martensitic microstructures with essentially no retained austenite (Fig. 31). These gradients in microstructure are related directly to the hardness and strength gradients produced by carburizing. Although features resolvable in the light microscope influence hardness (especially retained austenite, which coexists with plate martensite in the case region), it is the carbon-dependent fine structure of LTT martensite crystals that primarily determines the strength gradients in carburized steels.

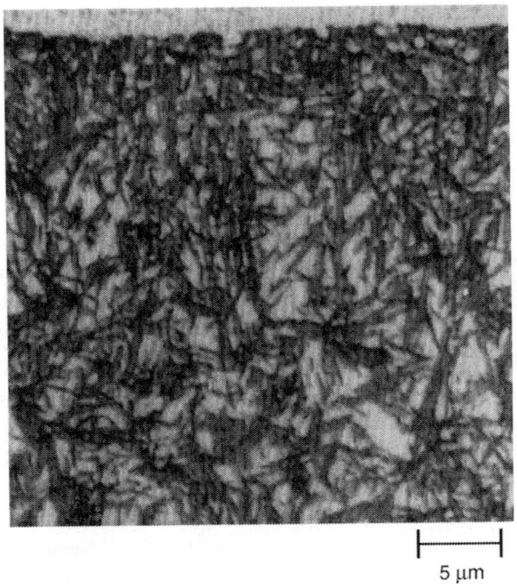

Fig. 29 Plate martensite microstructure in the case of a gas-carburized AISI 8719 alloy steel (1.06Mn, 0.52Cr, 0.5Ni, 0.17Mo). The "needles" are cross sections of plates. Retained austenite appears white. Light photomicrograph, nital, 2000×. Source: Ref 15

Fig. 30 Lath martensite microstructure in the core of a gas-carburized AISI 8719 alloy steel (the same as in Fig. 29). Light photomicrograph, nital, 1000×. Source: Ref 15

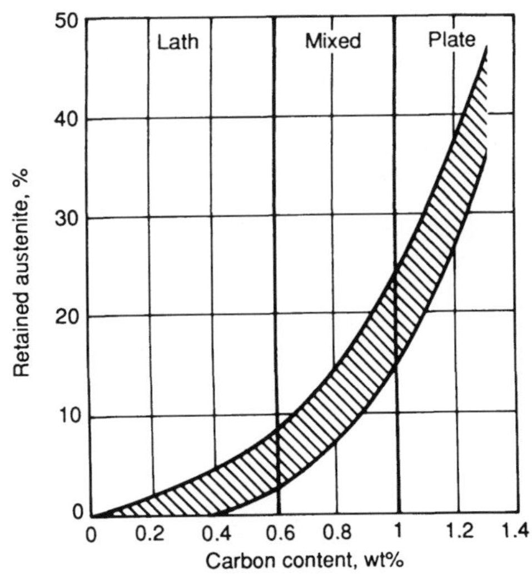

Fig. 31 Percent retained austenite as function of carbon content. Vertical lines show carbon ranges in which lath and plate martensites are found in Fe-C alloys. Source: Ref 16

Effect of Tempering. As previously noted, the diffusionless formation of martensite traps carbon atoms in the crystal structure of martensite. Under cooling conditions slower than the quenching that forms martensite, equilibrium requires that austenite first transform to body-centered-cubic ferrite having a very low solubility for carbon. Any carbon rejected from the ferrite forms carbide crystals in pearlitic or bainitic microstructures. Therefore, the trapped carbon atoms in as-quenched martensite represent a nonequilibrium, highly supersaturated state for a body-centered arrangement of iron atoms. Also, the martensitic transformation, by virtue of the constrained shears and volume changes that accompany the diffusionless transformation, results in a work-hardened or highly strained dislocation or twinned substructure within martensite crystals. When the as-quenched martensite is tempered at low temperatures, very fine carbides precipitate to relieve the carbon supersaturation. Low tempering temperatures of 150 to 200 °C (300 to 400 °F) ensure that the precipitated carbides remain fine. In addition, retained austenite remains stable at these low temperatures. Consequently, the only microstructural changes produced by tempering occur on a very fine scale within the martensite plates.

The transmission electron micrographs in Fig. 32 and 33 show the fine structure typical of LTT high-carbon martensite crystals. Figure 32, a bright-field micrograph, shows the regions with fine carbides as short dark bands. The carbides, however, are masked by strain contrast. When a diffracted electron beam from the carbides is used to create the micrograph (produce a dark-field image), the carbides appear bright, as shown in Fig. 33. The individual carbide particles are very fine, ~2 nm, and are arrayed in rows, which correspond to the dark bands in Fig. 32. The carbides are not cementite, which forms on tempering at higher temperatures, but are transition carbides that have been designated either epsilon carbides or eta carbides, depending on small differences in their crystal structures and diffraction patterns. Also associated with the transition carbides are high densities of dislocations, which were produced by the deformations that accompany the martensitic and subsequent tempering transformations.

Role of Transition Carbides. The higher the carbon content of the martensite, the higher the density of the clusters of transition carbides after tempering. These carbide arrays, and the associated dislocation substructures, make deformation by slip or dislocation motion difficult. As a result, the hardness and strength of LTT martensite increases with increasing carbon content. This is illustrated in Fig. 34, which shows the hardness of as-quenched martensitic

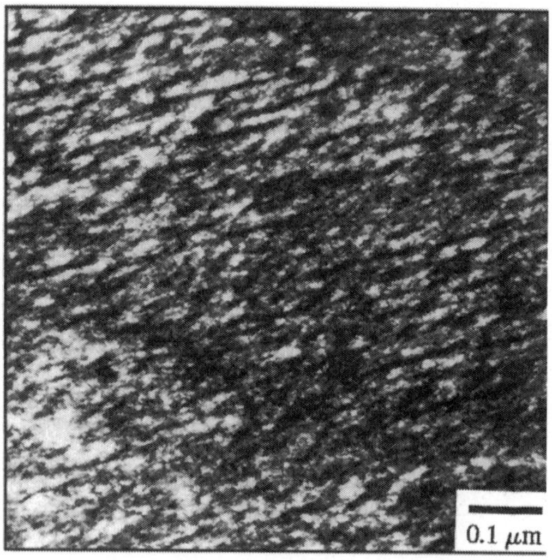

Fig. 32 Bright-field transmission electron micrograph of the fine structure within a plate of martensite in an Fe-1.22% C alloy tempered at 150 °C (300 °F). Short dark bands are associated with transition carbides, which are masked by strain contrast. 100,000×. Source: Ref 17

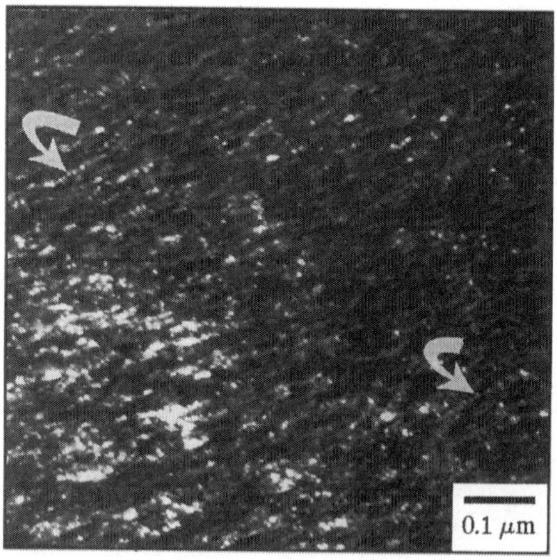

Fig. 33 Dark-field transmission electron photomicrograph of the fine structure within a plate of martensite in an Fe-1.22% C alloy tempered at 150 °C (300 °F). Arrows point to linear arrays of very fine transition carbides, which are located in the dark bands shown in Fig. 32. The image was created using a diffracted electron beam from transition carbide particles. 100,000×. Source: Ref 17

microstructures, and LTT martensitic microstructures tempered between 150 and 200 °C (300 and 400 °F). The hardness increase is almost linear up to 0.8% C. At higher carbon contents, the rate of hardness increase falls because of larger amounts of retained austenite. Low-temperature tempering lowers the hardness of as-quenched martensite because of the relief of carbon supersaturation and stress relief, but increases toughness.

The strengthening of LTT martensite is associated with very high rates of strain hardening due to dynamic dislocation interactions under stress with the LTT martensitic substructure. Thus, the dislocations move under stress, but their collective motion, or the plastic strain that they produce, is limited, and the ductility and toughness values for LTT martensitic structures are quite low. However, LTT martensite is not brittle, since it is able to sustain some dislocation motion, and overload fracture occurs by ductile mechanisms associated with microvoid nucleation and growth around second-phase particles. Brittle, intergranular fracture does occur in the case regions of carburized steels, but this brittle fracture is attributed to microstructural features other than LTT martensite, for example, phosphorus segregation at austenite grain boundaries as described subsequently in "Intergranular Fracture at Austenite Grain Boundaries."

Austenite (Ref 19)

Although LTT martensite dominates strengthening in the case regions of carburized steels, case microstructures are actually composites of LTT martensite and retained austenite. Depending on the amount of retained austenite and the loading conditions in service, the austenite may either degrade or enhance performance of the part. The effects of retained austenite on carburized components and the factors that affect the amount of austenite retained in carburized steels are discussed in this section.

Characteristic Features of Austenite

Austenite is essential to the manufacture of carburized steels. It has a high solubility for the carbon introduced into steel by high-temperature exposure to carburizing atmospheres.

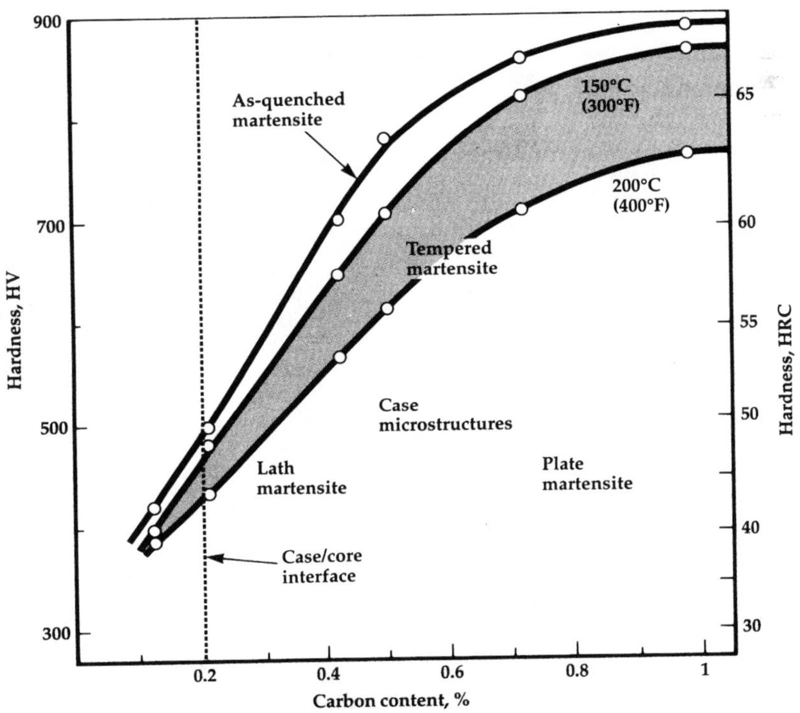

Fig. 34 Hardness as a function of carbon content for as-quenched martensite, and LTT martensite tempered between 150 and 200 °C (300 and 400 °F). Case martensitic microstructures have similar hardness gradients, with maximum and minimum values determined by surface and core carbon contents, respectively. Source: Ref 18

When the austenite transforms on quenching, the carbon is "inherited" by the martensite to produce, after tempering, the high-strength tempered martensite that is the backbone of carburized steels. The parent austenite phase influences the microstructure of carburized steels in several ways. Most important, the chemical composition of the austenite, as described in the next section, directly determines M_s temperatures and the amount of austenite retained after quenching to room temperature.

Also, the grain structure of the parent austenite affects the martensitic microstructure after quenching, because austenite grain size limits the size of martensite crystals. Thus, fine-grained austenite results in fine distributions of martensite plates and improved mechanical properties. For this reason, most carburizing grades are aluminum-killed, fine-grain steels. Small amounts of aluminum are added during melting and dissolve in austenite during subsequent high-temperature bar rolling or forging. At lower temperatures (about 900 °C, or 1650 °F), where carburizing is typically performed, the solubility of aluminum in austenite is low, and the aluminum combines with nitrogen to form aluminum nitrides. These fine aluminum nitride particles restrict grain growth during carburizing. At higher carburizing temperatures the aluminum nitride particles are coarser and, because of higher solubility, are present in lower densities, resulting in coarser austenite grain sizes.

Figure 35 shows the austenite grain structure from the core of a carburized SAE 8719 steel, a low-alloy carburizing grade containing small amounts of chromium, nickel, and molybdenum for hardenability. (Note: Special polishing and etching techniques must be used to reveal the austenite grain structure and suppress the etching of martensite, which would mask prior austenite grain boundaries.) Generally the core and case austenite grain sizes of steels direct quenched after carburizing are the same, but reheating of carburized steels may result in significant refinement of the case austenite grain size.

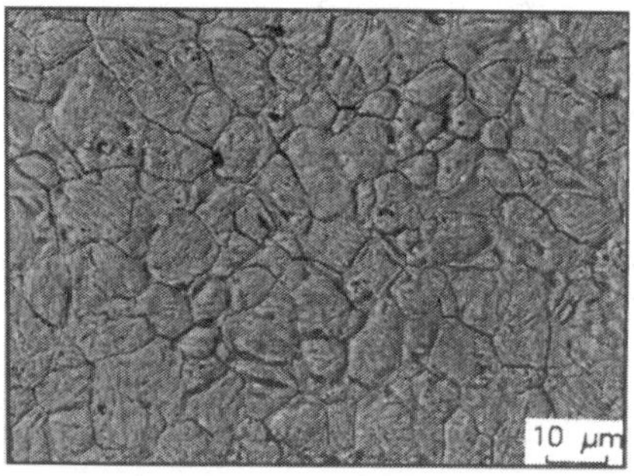

Fig. 35 Prior austenite grain structure in the core of a carburized SAE 8719 steel. The microstructure has been etched to emphasize austenite grain boundaries and suppress the appearance of the martensitic structure. Light photomicrograph, sodium tridecylbenzene sulfonate, 1000×

Effect of Austenite Composition. The temperature at which austenite starts to form martensite on quenching (the M_s temperature) is a function of austenite composition. Typical of a number of empirical equations that have been determined to relate M_s temperature to composition is the one published by Andrews (Ref 20):

$$M_s\ (°C) = 539 - 423\ C - 30.4\ Mn - 12.1\ Cr - 17.7\ Ni - 7.5\ Mo \quad \text{(Eq 23)}$$

where the element amounts are in weight percent. Although this equation was established for medium-carbon steels having a maximum of 0.6 wt% C, it demonstrates the very strong dependence of M_s on carbon content and the significant but lesser dependence of M_s on the substitutional alloying elements manganese, chromium, nickel, and molybdenum, which are often added to improve hardenability. When the M_s temperature is calculated from the composition of a steel of uniform carbon content, or at a point in a carburized case, the amount of martensite that forms on cooling to a given quench temperature, T_q, can be estimated from (Ref 21):

$$f_m = 1 - \exp - (0.011\ \Delta T), \quad \text{(Eq 24)}$$

where f_m is the volume fraction of martensite, and ΔT is the undercooling below M_s ($\Delta T = M_s - T_q$). Note that the amount of martensite that forms is not a function of time, and assuming no diffusion-controlled transformation takes place, the amount of austenite retained at a given quench temperature can be estimated by subtracting f_m from 1.

Figure 36 shows martensite and retained austenite in the case of a carburized SAE 8620 steel, a carburizing grade that contains small amounts of chromium, nickel, and molybdenum. The martensite plates appear dark, and the retained austenite appears white between the martensite crystals. Typically, the retained austenite content of the near-surface case microstructure of direct-quenched carburized steels is between 20 and 30 vol%. At this level, the retained austenite can be readily observed by light microscopy. However, if the retained austenite content is much less than 20 vol%, it becomes very difficult or impossible to resolve with the light microscope, and its amount is much more reliably determined by x-ray diffraction techniques.

Figure 37 shows retained austenite gradients, measured by x-ray diffraction, as a function of distance from the surface of a carburized SAE 8620 steel. The steel has been direct quenched from the carburizing temperature, reheated to the austenite phase field (single reheat), and reheated to the austenite-cementite two-phase field (double reheat or intercritical-temperature reheat). The retained austenite profiles for the direct quenched and single reheat specimens are almost the same because they were both quenched from completely austenitic parent microstructures of identical compositions. In accord with Eq 23 and 24, the amount of retained austenite decreases with increasing distance into the specimen as carbon content decreases.

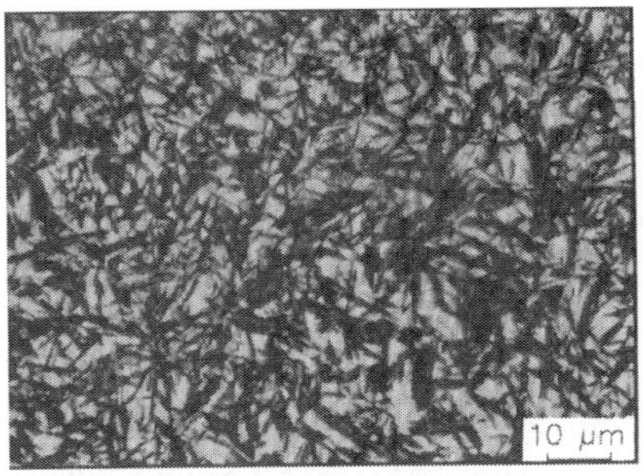

Fig. 36 Tempered martensite (black) and retained austenite (white) in the case of a carburized SAE 8620 steel. Light photomicrograph, nital, 1000×

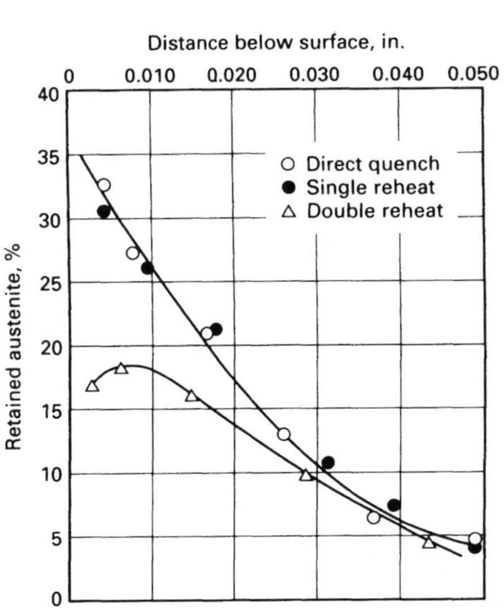

Fig. 37 Retained austenite, measured by x-ray diffraction, as a function of distance from the surface of an 8620 steel carburized at 925 °C (1700 °F). The single and double reheats were accomplished by heating to 845 and 790 °C (1550 and 1450 °F), respectively. Source: Ref 22

The retained austenite profile for the specimen reheated into the austenite-cementite two-phase field is significantly different from that of the other specimens. Overall, the austenite content is lower because some of the carbon is combined with iron to form cementite or iron carbide. Thus, the carbon content of the austenite is reduced, M_s increases, and less austenite is retained on quenching to room temperature. The decrease in retained austenite at the surface of the double reheated specimen may be due to a small amount of decarburization that occurred during reheating.

Excessive Retained Austenite

Effect on Properties. Too much retained austenite, ~40 vol% or more, may be detrimental to the performance of carburized steels in several ways. Most important, because austenite is of much lower strength than tempered martensite, the hardness, wear resistance, and fatigue resistance of a carburized part may be reduced. Also, high retained austenite content may reduce surface compressive residual stresses and, accordingly, contribute to a decrease in fatigue resistance. These stresses are in part determined by the transformation of austenite in the near-surface regions in the carburized case during the final stages of cooling, and constraint of the martensite volume expansion by transformed structures in the core of carburized parts. Consequently, stresses may decrease if less surface martensite is formed.

Excessive retained austenite, again by virtue of the volume expansion that accompanies austenite transformation to martensite, also may cause unacceptable dimensional changes for precision parts in service. In this case, the retained austenite transformation is mechanically induced by applied service loads, rather than thermally induced by cooling.

The major cause of excessive retained austenite is a case carbon content that is too high. Equations 23 and 24 show that M_s and the amounts of martensite formed on cooling are very sensitive to carbon content. If carburizing temperatures are high, and if the austenite is allowed to become saturated with carbon, then large amounts of retained austenite can be expected. This situation generally is avoided in gas carburizing by adopting a two-step process: the first step introduces a high surface carbon content (>1 wt%), while the second is performed at a lower carbon potential, which lowers the surface carbon content by permitting carbon introduced in the first stage to diffuse deeper into the steel. This process is often referred to as boost-diffuse carburizing, and if properly controlled, results in reasonable levels of retained austenite.

However, excessive austenite may be retained even in boost-diffuse carburizing in regions of a part where surfaces meet at included angles of 90° or less. At such corners, carbon diffuses in from the carburizing atmosphere from two surfaces, but the high carbon content that results has only a limited area through which to diffuse into the interior of the part. The result is that carbon concentration at the corner remains high throughout the diffusion step. Excessive retained austenite (~60 vol%) at the corner of a vacuum-carburized SAE 8620 specimen is shown in Fig. 38(a). Figure 38(b) shows that surface hardness at the corner was considerably lower than that of the plane surface with its more moderate amount of retained austenite. Massive carbides may also be present at corners because of the high carbon concentration. In this example, the combination of high retained austenite content and coarse carbide particles at corners led to reduced resistance to fatigue crack initiation.

Reducing Retained Austenite

In steels directly quenched after carburizing, excessive retained austenite—present because of high case carbon content, high substitutional alloy content, high carburizing temperature, or geometry—may be reduced by several methods, including tempering, refrigeration or "subzero" treating, intercritical-temperature reheating, and shot peening. Process selection depends on the desired performance level of the carburized part.

Tempering. As noted earlier in this chapter, typical tempering temperatures for high-performance carburized steels are between 150 and 200 °C (300 and 400 °F). Retained austenite is stable during tempering in this range. Tempering at higher temperatures, however, causes the retained austenite to transform to a mixture of ferrite and relatively coarse carbides, and causes the dislocation transition-carbide substructure of the low-temperature-tempered martensite to coarsen. These changes increase microstructural stability, but reduce toughness, hardness, and strength.

Cold Treating. Refrigeration treatments effectively reduce retained austenite levels of direct-quenched carburized steels, and thereby increase hardness. Equation 24 shows that the greater the undercooling below M_s, the larger the amount of austenite that will transform to martensite. However, subzero treatments should be used with caution, and, according to Parrish and Harper, "considered as a last resort" (Ref 24). As discussed in the section "Properties of Carburized Steels," reduced fracture and

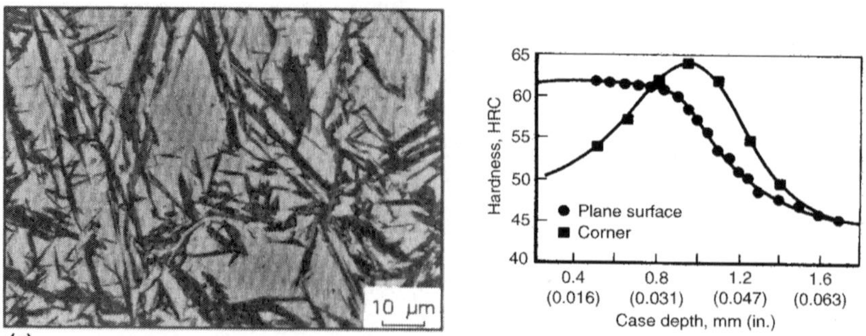

Fig. 38 (a) Excessive retained austenite at the corner of a specimen of vacuum-carburized SAE 8620 steel. Light photomicrograph, nital, 1000×. (b) Hardness as a function of distance from a specimen corner or a specimen plane surface in the same steel. Source: Ref 23

bending fatigue resistance have been documented in refrigeration-treated carburized steels. The performance reduction is attributed to tensile microstresses that lead to increased sensitivity to cracking. Tempering both before and after refrigeration is recommended to reduce detrimental microstresses.

Intercritical-temperature reheating is a very effective method of reducing high retained austenite content in the case of direct-quenched carburized steels. Intercritical heating is accomplished in the two-phase austenite-carbide field between the upper critical temperature (A_{cm}), above which only austenite exists, and the lower critical temperature (A_1) below which only ferrite and carbides are stable. Because some of the carbon is tied up in carbides, the austenite carbon content is reduced and, per Eq 23 and 24, more martensite and less austenite will be present at room temperature. The effect of intercritical reheating is shown in Fig. 37, while the case microstructure of an intercritically reheated carburized specimen is shown in Fig. 39. The white circular features are fine, spheroidized retained carbide particles, and the dark-etching matrix consists of a very fine mixture of martensite and retained austenite that cannot be resolved by the light microscope.

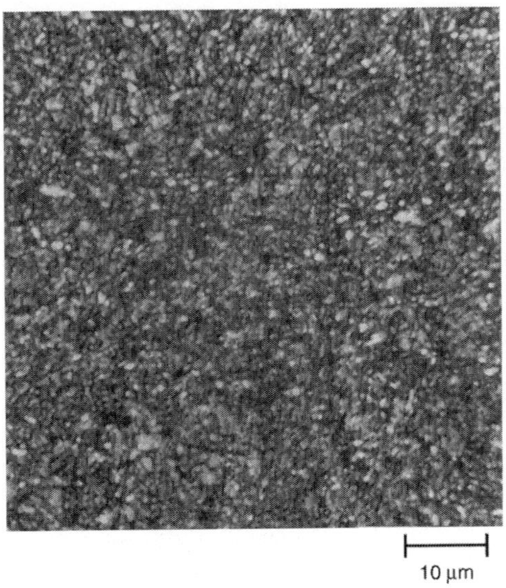

Fig. 39 Typical near-surface case microstructure of carburized steel (SAE 8620: 0.81% Mn, 0.19% Mo, 0.48% Ni) reheated between A_1 and A_{cm}. Retained carbides are small, white spherical particles, and matrix consists of a dark etching of mixture of martensite and austenite too fine for resolution in the light microscope. Light micrograph, nital etch. Source: Ref 23

When intercritical reheating is applied, the higher the case carbon content, the greater will be the density of retained carbides. The carbon content of the austenite will be fixed by the steel alloy content, which sets the carbon dependence of the A_{cm}. The presence of strong carbide-forming elements such as chromium and molybdenum shifts the A_{cm} to lower carbon contents.

Peening. Retained austenite in the case of carburized steels also can be effectively reduced by shot peening. The surface deformation introduced by the impact of the shot causes the retained austenite to mechanically transform to martensite. And, as in the final stages of quenching, the restraint of the volume expansion caused by the formation of martensite introduces favorable residual compressive stresses. Figures 40 and 41 illustrate the effect of shot peening on retained austenite and compressive residual stresses in carburized SAE 4320 steel. In this instance, the very high surface compressive stresses introduced by peening substantially increased bending fatigue performance. More detailed information on the effect of retained austenite and residual stresses on bending fatigue can be found in the section "Properties of Carburized Steels" in this chapter.

Carbides (Ref 16)

Composite structures that contain primary carbides can also be produced by carburizing. Three common morphologies are coarse primary carbides, carbide networks, and fine primary carbides.

Coarse primary carbides (from 1 to 10 μm, or 40 to 400 μin.) (Fig. 42) are produced by carburizing in an atmosphere with a carbon potential that is high enough to exceed the solubility limit for carbon in austenite. Coarse, or massive, carbides are often found at corners and edges of parts made of alloys that are rich in strong carbide formers, such as chromium. Plate martensite and high levels of retained austenite can be found with coarse carbides in parts quenched directly from the carburizing temperature. Processing is sometimes designed to produce large primary carbides as a means of enhancing wear resistance. More often, large carbides are avoided, because they deplete the matrix in alloying elements such as chromium, thereby reducing hardenability.

Carbide networks (Fig. 43) form in austenite grain boundaries when parts are carburized at an elevated temperature and then slowly cooled. The solubility limit in austenite is exceeded as cooling occurs, and carbon is rejected to austenite grain boundaries. Because this structure tends to embrittle the case, it is usually avoided.

Fine primary carbides (from 0.1 to 0.5 μm, or 4 to 20 μin. diam) (Fig. 44) result when a part is carburized at a high temperature, such as 950 °C (1740 °F), cooled to form pearlite or bainite,

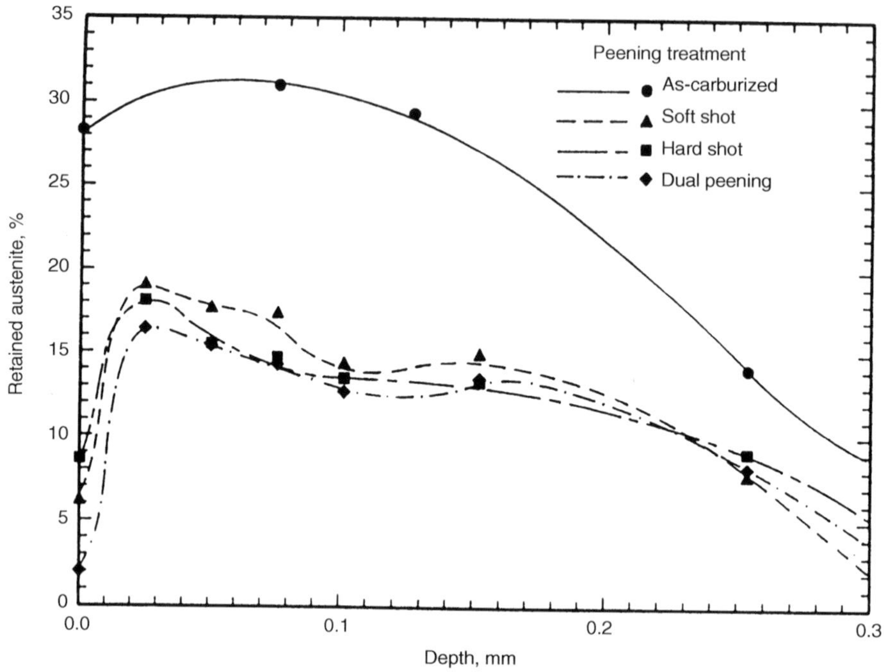

Fig. 40 Retained austenite as a function of depth below the carburized surface for gas-carburized 4320 specimens in the as-carburized, direct-quenched, and various shot-peened conditions after direct quenching. Source: Ref 25

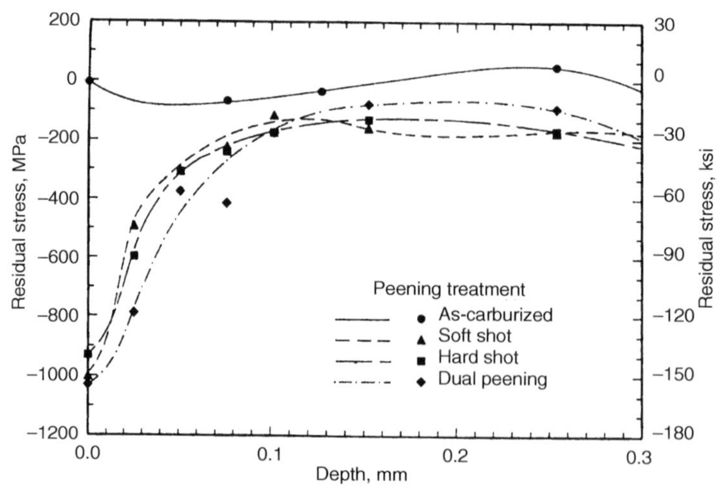

Fig. 41 Residual stress profiles for the specimens described in Fig. 40. Source: Ref 25

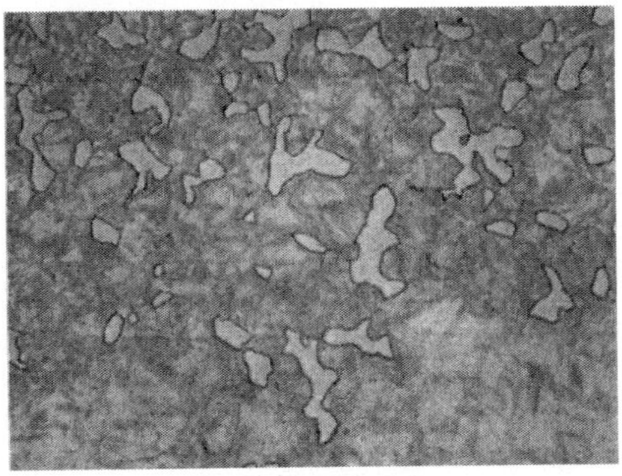

Fig. 42 Coarse primary carbides produced by carburizing SAE 4130 steel at 950 °C (1740 °F), and then quenching. Matrix microstructure is plate martensite and retained austenite. Picral etch. 600×

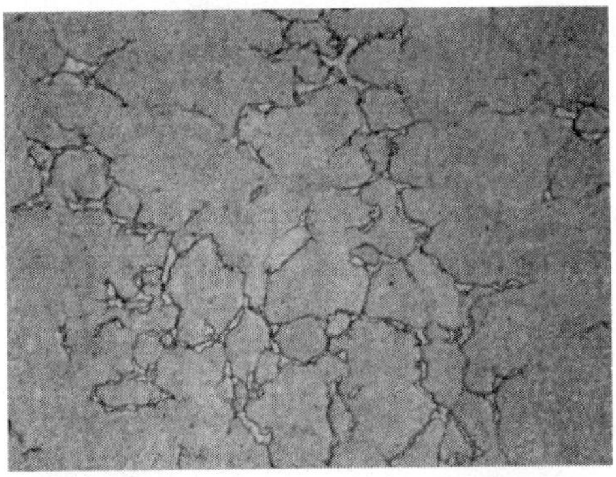

Fig. 43 Carbide networks in prior austenite grain boundaries. Produced by carburizing 4130 steel at 950 °C (1740 °F), furnace cooling to 800 °C (1470 °F), and then quenching. Picral etch. 600×

Fig. 44 Fine primary carbides in lath martensite produced by carburizing at 950 °C (1740 °F), air cooling to room temperature, then reheating to 820 °C (1510 °F) for 20 min and quenching. Picral etch. 600×

and then reheated to a lower temperature, such as 830 °C (1525 °F), for a brief time and quenched. Because the carbon solubility at 950 °C (1740 °F) is approximately 1.5 times the solubility at 830 °C (1525 °F), substantial quantities of fine carbides can be produced. The carbides will not coarsen significantly if the time at the lower austenitizing temperature is approximately 30 min or less. Because the retained austenite content is a function of the carbon dissolved in austenite, the hardness will be near the maximum for the alloy. Finely dispersed primary carbides can incrementally increase the hardness. Their main contribution is in restricting austenite grain growth, thereby assuring fine martensite plates and finely dispersed retained austenite. Many gears and bearings are heat treated in this manner.

Alloying Effects

The primary concern in alloy development and the selection of carburizing steels is hardenability. In carburizing steels, a given composition must provide adequate hardenability over a range of carbon contents because hardenability is important for both the case region and the core. The objective is to produce a high-carbon martensitic case (for wear and fatigue resistance) and a low-carbon martensitic core to provide sufficient strength to resist case-core failures.

The goal of hardenability is the formation of hard martensite in preference to microstructures of lower hardness (Ref 26, 27). The controlling factors may be metallurgical, such as the effects of substitutional alloying elements that retard solid-state, diffusion-controlled transformation of austenite to bainite, pearlite, or ferrite; or they may be technological, such as the selection of quenchants, or compensating for slow cooling rates in heavy sections, which provide time for diffusion-controlled transformation at the expense of martensitic transformation. The alloying elements traditionally used for improving hardenability in carburized steels are manganese, chromium, molybdenum, and nickel. Combinations of moderate amounts of several elements have been found to be more effective than large amounts of a single element. Boron is most effective in improving the hardenability of low-carbon steels but loses its effectiveness as carbon content increases. Therefore, it is not expected to improve case hardenability. However, a German carburizing steel, 20MnCr5B, uses boron to remove nitrogen from solution and thereby improve toughness (Ref 28).

Most carburizing steels are deoxidized with aluminum for grain size control. Aluminum combines with nitrogen to form aluminum nitride particles, which limit austenite grain growth during carburizing. Fine grain size reduces hardenability, and Cook has shown that the case hardenability of plain carbon steels is reduced because of the grain-refining effect of aluminum additions (Ref 29). This effect of aluminum on hardenability is not noted in alloy carburizing steels.

Alloy Effects on Hardenability. Hardenability is important for both the case and core regions of carburized steels, and a given steel must have adequate hardenability over a range of carbon contents. Figure 45 shows Jominy end-quench data for the core and for all carbon levels up to 0.9% for an SAE 4620 carburizing steel. The powerful, beneficial effect of increasing carbon on hardenability is shown. Nevertheless, in heavy sections, where cooling rates are low, even case regions may transform to microstructures other than martensite. Jatczak (Ref 31) has evaluated the effects of various alloying elements on the hardenability of high-carbon steels and has shown that higher austenitizing temperatures increase hardenability by dissolving alloy carbides and increasing the amount of carbon and alloying elements in solution in the austenite. Other investigators have also evaluated various aspects of the hardenability of carburized steels (Ref 32–35).

Other Alloy Effects. Although hardenability is a major concern in alloying and the selection of carburizing steels, alloy elements also affect other aspects of microstructure. Many of the alloying elements, in particular chromium and molybdenum, are strong carbide and ferrite formers. Those elements shift A_{cm} temperatures (Fig. 46) and raise Ae_1 temperatures, the lowest temperatures at which austenite is stable under equilibrium conditions. The shift in A_{cm} by carbide-forming elements limits the amount of carbon that can be dissolved in austenite and increases the possibility of carbide formation in carburized steels.

Many alloying elements also lower M_s temperatures and the transformation temperature ranges for martensite formation (Ref 20, 37). Therefore increased alloying increases the amounts of austenite that are retained in carburized and hardened steels.

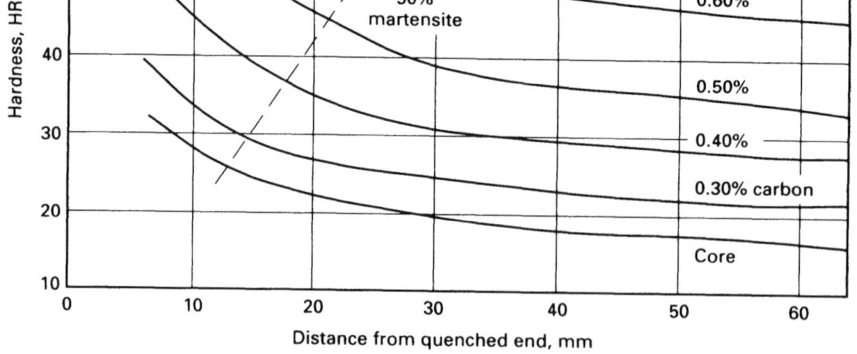

Fig. 45 Jominy end-quench curves showing hardenability differences as a function of carbon content in direct-quenched SAE 4620 steel. Source: Ref 30

Intergranular Fracture at Austenite Grain Boundaries

Figure 47 shows an example of intergranular fracture at prior austenite grain boundaries in the case overload fracture zone of a carburized 8620 steel. Such intergranular cracking is a major fracture mode of high-carbon hardened steels that have been quenched from temperatures at which the microstructure consists only of polycrystalline austenite. Thus cracking can occur in the case regions of steel directly quenched after carburizing, but rarely in carburized steels that have been reheated to produce finer-grained structures with retained carbides (Ref 38, 39). Fracture in the latter microstructures is low-toughness ductile fracture characterized by closely spaced microvoids that form around the dispersed carbide particles. Intergranular fracture occurs even in carburized steels tempered between 150 and 200 °C (300 and 400 °F), tempering temperatures that are too low to cause tempered martensite embrittlement.

The reasons for the intergranular fracture of high-carbon case microstructures have been difficult to establish because no associated grain-boundary features are discernible in the light microscope. However, several studies suggest that the sensitivity to grain-boundary fracture is due to a two-step process (Ref 38, 40, 41): first, the segregation of phosphorus to austenite grain boundaries during carburizing or austenitizing for hardening, and second, the nucleation and growth of very thin cementite particles on austenite grain boundaries during

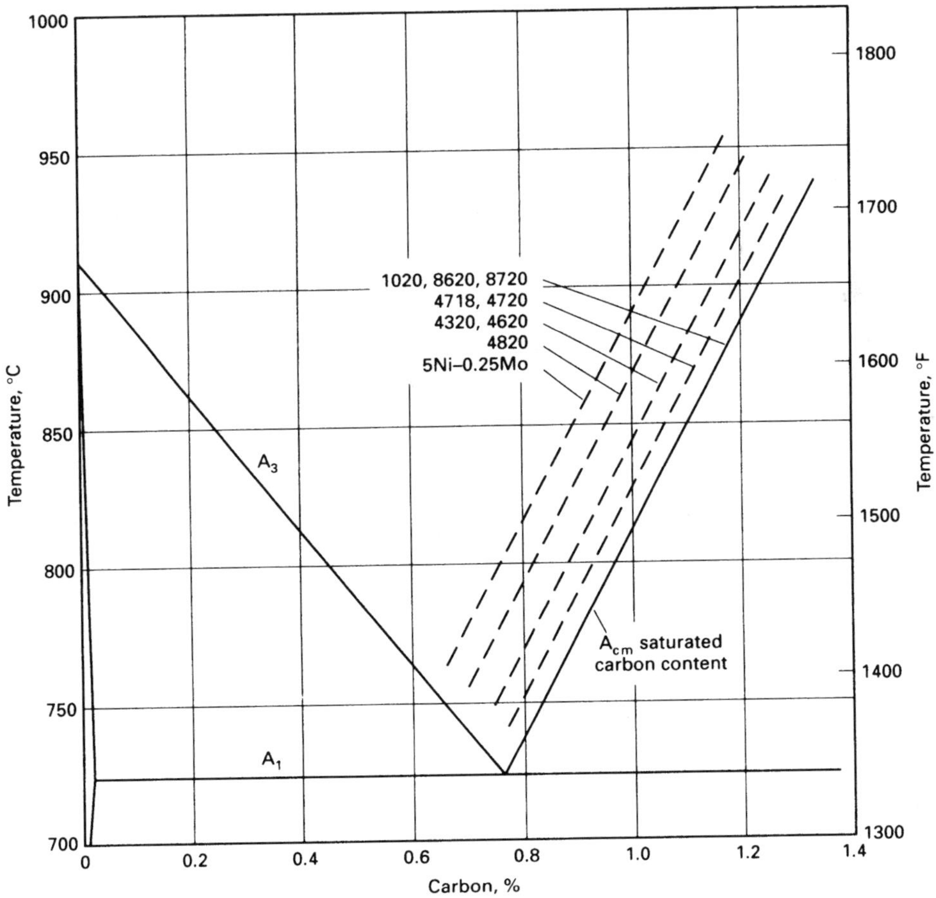

Fig. 46 The shift in A_{cm} temperatures with alloying in various carburizing steels. Source: Ref 36

quenching. The phosphorus segregation and carbide formation are largely on an atomic scale, but their combined effect is sufficient to produce interfaces that fracture at lower stresses than do the matrix martensite and austenite.

The key to the identification of microstructural features leading to intergranular fracture has been the application of Auger electron spectroscopy (AES), an analytical technique that has very high depth resolution. Auger electrons of specific energies are emitted from specific atoms on a surface that is irradiated with an electron beam in a high vacuum chamber. The Auger electrons have very low energy and originate from a depth of less than 1 nm from the surface of a specimen (Ref 42).

Figure 48 shows Auger spectra from case fracture surfaces of a carburized 8620 steel. The spectra in Fig. 48(a) is from a transgranular fracture surface, while that in Fig. 48(b) is from an intergranular fracture surface. No phosphorus peak is detectable in the spectrum produced from the transgranular fracture, and a small phosphorus peak, clearly shown by the 10× magnification, is produced from the intergranular fracture surface. These observations are consistent with other investigations that show that phosphorus segregates to austenite grain boundaries during austenitizing (Ref 41, 43).

A major difference between the two spectra is the significantly larger carbon peak in the Auger spectrum from the intergranular fracture surface. The carbon peak shape, characterized by a major peak and several auxiliary peaks, is identical to that produced by AES of cementite (Ref 41). Thus AES analysis provides evidence for cementite formation on austenite grain boundaries.

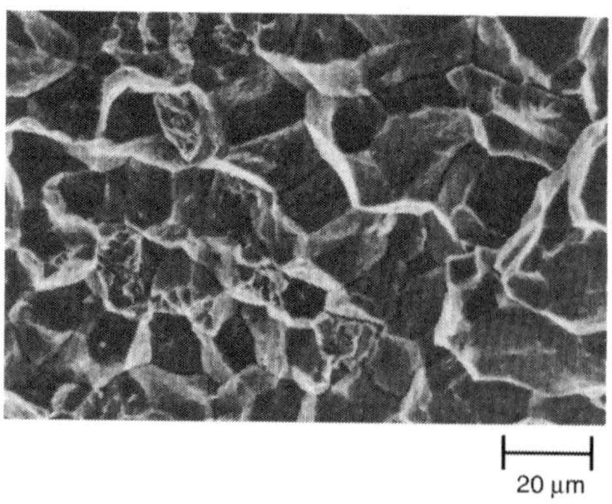

Fig. 47 Intergranular fracture from the overload fracture zone in the case of a carburized SAE 8620 steel. Scanning electron microscope (SEM) micrograph

A fracture toughness study on a set of EX24-type steels (or SAE 4121) containing 0.85% C with 0.044 and 0.002% P verified the above observations (Ref 40). The high carbon content was designed to simulate the high-carbon case of carburized specimens. High phosphorus content greatly increased the amount of intergranular fracture. Intergranular fracture was significantly reduced, but not completely eliminated, in the low-phosphorus steel. Oil-quenched specimens developed more intergranular fracture than brine-quenched specimens, a result explained by increased coverage of austenite grain boundaries by cementite due to more time for diffusion during quenching at slower rates.

Ando (Ref 44, 45) has modeled the growth kinetics of cementite allotriomorphs in high-carbon, iron-chromium-carbon alloys. Figure 49 shows that growth takes place in several stages. At first, rapid thickening occurs in a stage controlled only by carbon diffusion, and no partitioning of chromium takes place. However, equilibrium considerations eventually require the diffusion of chromium to the carbide particles, and at that stage the growth of grain-boundary cementite slows significantly. Therefore, the formation of very thin cementite particles, even during the oil quenching of carburized steels, is explained by the very rapid first-stage growth, shown in Fig. 49. High phosphorus content has also been shown to accelerate the formation of cementite grain-boundary allotriomorphs (Ref 41).

Prevention of Intergranular Fracture. Intergranular fracture frequently initiates fatigue cracks in carburized steels. Lower phosphorus contents would reduce intergranular cracking, but the reduction of phosphorus to extremely low levels that might completely eliminate intergranular cracking is dependent on the economics of steelmaking. Reheating carburized specimens with nominal phosphorus contents to produce very fine austenite grain sizes eliminates intergranular cracking (Ref 15, 22), perhaps because of the dilution of phosphorus segregation by a high grain-boundary area. Finally, alloying might be used to eliminate intergranular fracture. Carburizing steels with high nickel contents have high toughness and do not seem to be sensitive to intergranular fracture (Ref 46, 47).

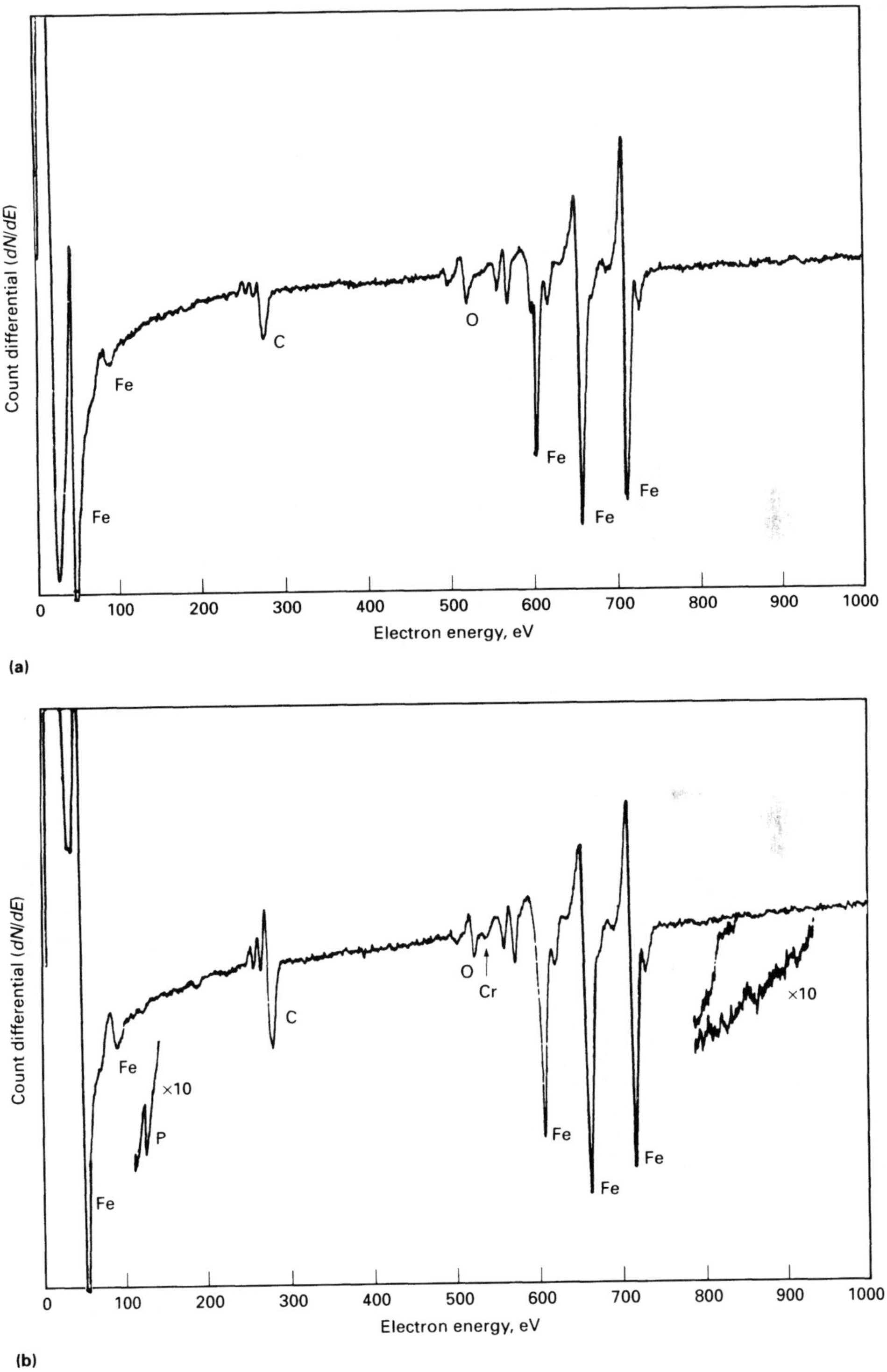

Fig. 48 Auger electron spectra from case fracture surfaces of carburized 8620 steel. (a) From transgranular fracture surface. (b) From intergranular fracture surface. Source: Ref 38

Microcracking in Carburized Steels

Microcracks frequently form in martensite plates of high-carbon steels. Examples of martensite microcracks are shown in Fig. 50. Marder and Benscoter have shown by serial metallographic sectioning that the cracks form at points of contact between impinging martensite plates (Ref 49). Because the microcracks are formed by the impingement of nonparallel plates of martensite, microcracking density decreases with the transition from plate to lath martensite (Ref 50). Fine austenite grain size also limits microcracking (Ref 48, 51), apparently because smaller martensite plates do not create sufficient stresses to produce cracks.

Microcracks have long been known to be present in the case microstructures of carburized steels (Ref 22, 52, 53), especially in coarse-grained microstructures with large martensite plates. The presence of microcracks may contribute to impaired fatigue performance of carburized steels. However, microcracks may have only a very secondary effect on fatigue, in light of the fact that many fatigue cracks initiate at embrittled austenite grain boundaries, as discussed in the previous section. Such grain-boundary cracks effectively bypass the microcracked martensite plates within austenite grains, and therefore the presence or absence of microcracks is immaterial to fatigue crack initiation. On the other hand, the influence of microcracks in martensite plates on transgranular crack propagation would be expected.

A study of grain size effects on microcracking in an Fe-1.22% C alloy provides a link between microcracking and grain-boundary fracture (Ref 51). In that study microcracks were found both in martensite plates and at prior austenite grain boundaries. As grain size decreased, both types of microcracks decreased, but the number of grain-boundary microcracks became a higher fraction of the total. The intergranular microcracks may have formed partly because of martensite plate impingement on embrittled grain boundaries. If such grain-boundary microcracks are present in carburized steels, intergranular fatigue crack initiation will occur at lower stresses.

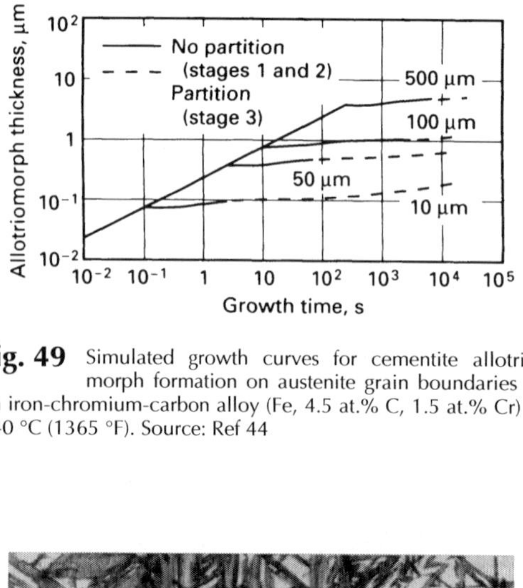

Fig. 49 Simulated growth curves for cementite allotriomorph formation on austenite grain boundaries in an iron-chromium-carbon alloy (Fe, 4.5 at.% C, 1.5 at.% Cr) at 740 °C (1365 °F). Source: Ref 44

Fig. 50 Microcracks in martensite plates of an Fe-1.86C alloy. Light micrograph. Source: Ref 48

Excessive Retained Austenite and Massive Carbides

Moderate amounts of retained austenite are proper and unavoidable in the high-carbon case microstructure of carburized steels. However, excessive amounts of retained austenite, that is, greater than 50%, lower hardness significantly and reduce bending fatigue resistance. The most important cause of excessive amounts of retained austenite is too high a surface carbon content. This condition drives M_s temperatures down and shifts the balance of the temperature range for martensite transformation to well below room temperature. High alloy content also lowers M_s temperatures.

Common locations of excessive surface carbon concentration are specimen corners at which the austenite is saturated with carbon during the first part of a carburizing cycle (Ref

23, 54). The carbon has access to both surfaces of the corner during carburizing but has little physical access to the interior of the specimen during the diffusion part of a cycle. As a result, although carbon content falls to desired levels on the flat or gradually curved surfaces of a part, the carbon content at a corner remains much higher than desired. Figure 51 shows carbon contours determined at the corners of an 8620 steel specimen carburized at 1050 °C (1920 °F). Carbon contents as high as 1.20% were measured at the corner. Excessive retained austenite in the corner microstructure of a 4121 carburized specimen is shown in Fig. 52. Figure 53 shows hardness profiles from corner and plane surface regions of 8620 steel carburized at 930 °C (1700 °F). The corner surface hardness is much lower than that of the plane surface because of high retained austenite content.

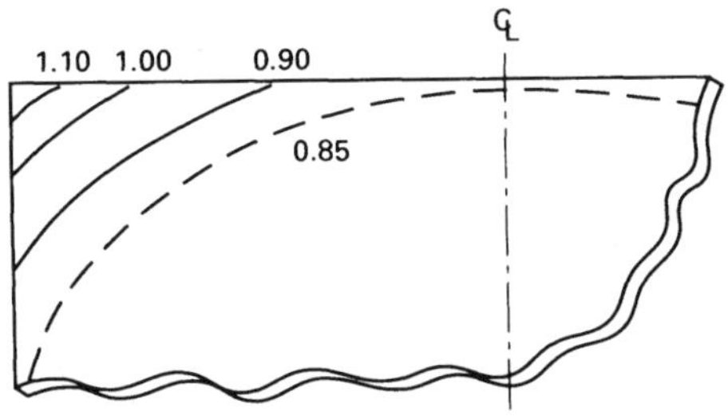

Fig. 51 Schematic of carbon concentrations at the corners of an 8620 steel specimen subjected to carburizing and diffusion at 1050 °C (1920 °F). Based on chemical analysis of chips milled from various locations of the specimen. Source: Ref 55

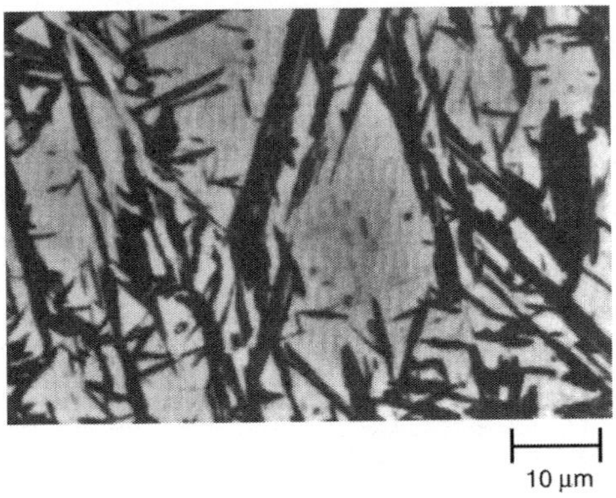

Fig. 52 High retained austenite content in corner of SAE 4121 steel (formerly EX24) specimen carburized at 1050 °C (1920 °F). Source: Ref 50

Reheating of direct-quenched specimens eliminates excessive retained austenite and raises surface hardness (Ref 23, 54).

Another consequence of a surface carbon content that is too high is the formation of massive carbides. The carbides form at austenite grain boundaries and may have different morphologies, depending on alloy content. As discussed relative to Fig. 49, large carbide grain-boundary allotriomorphs require considerable diffusion to grow, and therefore they form during the high-temperature stages of carburizing or when the temperature of the part is lowered to about 845 °C (1550 °F) just prior to quenching. Figure 54 shows two morphologies of massive carbides that have formed in the corners of carburized specimens. Figure 54(a) shows blocky, angular particles formed in an 8620 steel containing nominally 0.5% Cr, 0.5% Ni, and 0.2% Mo. Figure 54(b) shows long, thin carbides formed in an SAE 4121 steel containing 0.55% Cr and 0.24% Mo but no nickel.

Effect on Fatigue Cracking. The combination of excessive retained austenite and massive carbides, together with stress concentration at sharp changes in section, causes fatigue crack initiation at specimen corners. Figure 55(a) shows fatigue initiation at the corner of a carburized specimen of SAE 4121 steel. Details of the corner fracture along the massive carbides are shown in Fig. 55(b). Carbide grain-boundary allotriomorphs grow by ledges, which make up the interface between the carbides and the martensite-austenite matrix (Ref 45). These

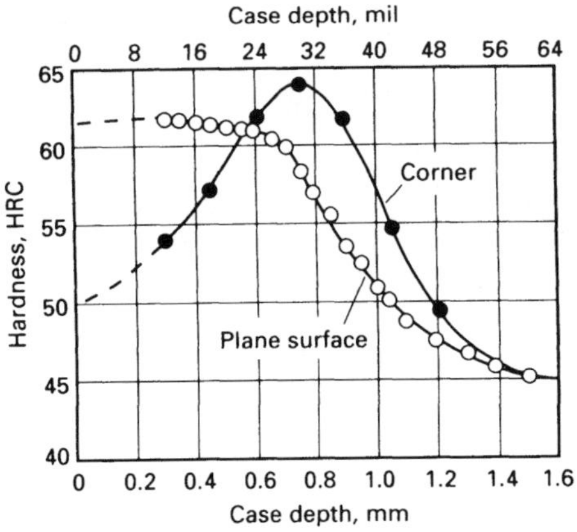

Fig. 53 Corner and plane surface microhardness profiles from 8620 steel specimen carburized at 930 °C (1700 °F). Source: Ref 54

ledges provide preferred fracture paths, as shown in Fig. 55(b).

Although massive network carbides are detrimental to the bending fatigue and fracture performance of carburized steels in many applications, a process referred to as super carburizing is occasionally used for special applications (Ref 55). This process supersaturates a part surface with carbon and results in the formation of large volumes of massive carbide particles. Surface carbon contents of 1.80% to greater than 3.0% are produced, and steels with large amounts of carbide-forming elements such as chromium and molybdenum respond most effectively. High volume fractions of hard alloy carbide particles significantly increase resistance to abrasive wear and also may create problems in the grinding of the very hard surfaces.

Residual Stresses

Effects of Residual Stresses. As discussed earlier in this chapter, a major benefit of carburizing is the introduction of compressive residual stresses into the surfaces of carburized parts. These stresses counteract applied tensile stresses and therefore improve bending fatigue performance. Because of the importance of residual stresses to the performance of carburized parts, considerable effort has been devoted to modeling, measuring, and understanding their effects (Ref 56–59).

Figure 56 shows schematically the residual stress profiles that develop in properly carburized and hardened steels. The compressive stresses reach a maximum at some distance from the surface, gradually decrease, and are eventually balanced by tensile residual stresses in the core of the carburized part. A survey of a number of carburized parts showed that measured peak compressive stresses ranged from –200 to –450 MPa (–29 to –65 ksi) (Ref 24).

Surface compressive residual stresses in carburized steels arise from transformation and temperature gradients induced during cooling and the volume expansion that accompanies the

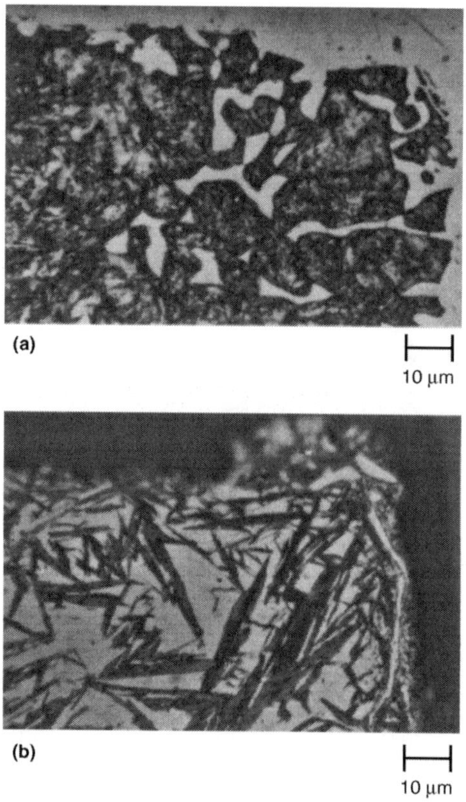

Fig. 54 Examples of massive carbides formed at the corners of carburized specimens. (a) Blocky carbides in 8620 steel. (b) Thin, continuous grain-boundary carbides in SAE 4121 steel. Light micrographs. Source: Ref 23

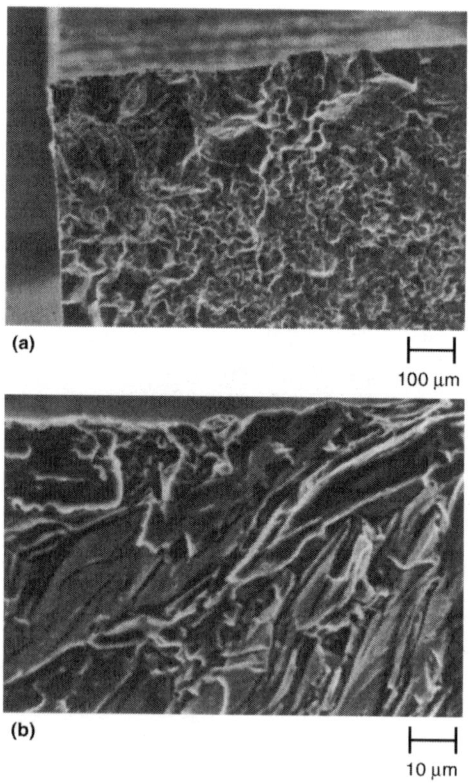

Fig. 55 Fracture surfaces of carburized SAE 4121 steel. (a) Low-magnification view of corner initiation. (b) Detail of fracture at carbide-matrix interface. SEM micrographs. Source: Ref 54

transformation of austenite to martensite (Ref 60). The carbon profiles produced by carburizing introduce the M_s temperature and transformation gradients: the M_s temperature is lowest at the surface, where carbon content is the highest, and increases with increasing distance from the surface as carbon content approaches that of the core. Temperature gradients are due to heat flow and thermal-conductivity factors; at any given time during quenching, the surface temperature is lower than temperatures in the part interior.

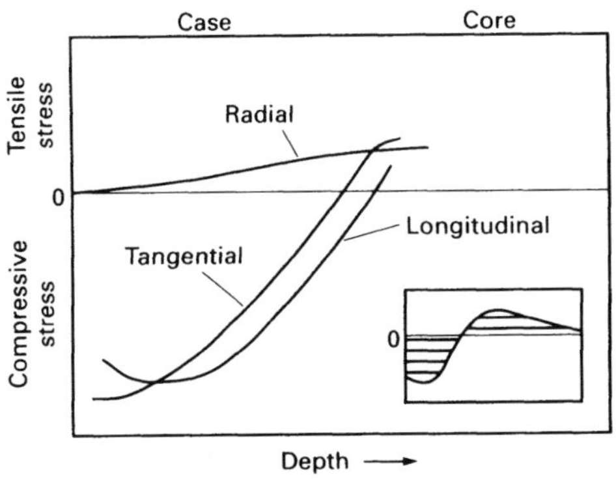

Fig. 56 Schematic diagram of residual stresses in carburized steels. Insert shows that surface compressive residual stresses are balanced by interior tensile stresses. Source: Ref 24

In the early stages of cooling, martensite first forms at some distance from the surface, where the part temperature has fallen below the higher, interior M_s temperatures. The volume changes at this stage are readily accommodated by the surrounding austenite because of its low flow stresses and the high temperatures. The surface austenite does not transform because of its low M_s. The temperature continues to fall and eventually drops below the M_s in the surface regions. The expansion at this point is constrained by the interior martensite that has formed earlier, and as a result the surface microstructure is placed in compression. Many factors affect this process, including alloy and carbon levels, which set hardenability and M_s temperatures; case depths; temperature at the start of quenching; quenchant temperature; and the temperature-dependent plastic flow behavior of martensite and austenite. Despite the complexity of the interactions that affect the formation of residual stresses, hardened carburized parts with the martensite-austenite microstructures described earlier generally develop favorable compressive stresses.

Effects of Shot Peening on Residual Stresses. Surface compressive residual stresses can be increased by shot peening. Figure 57 shows the dramatic effect of shot peening at different velocities on the compressive residual stresses of carburized steel. These improve-

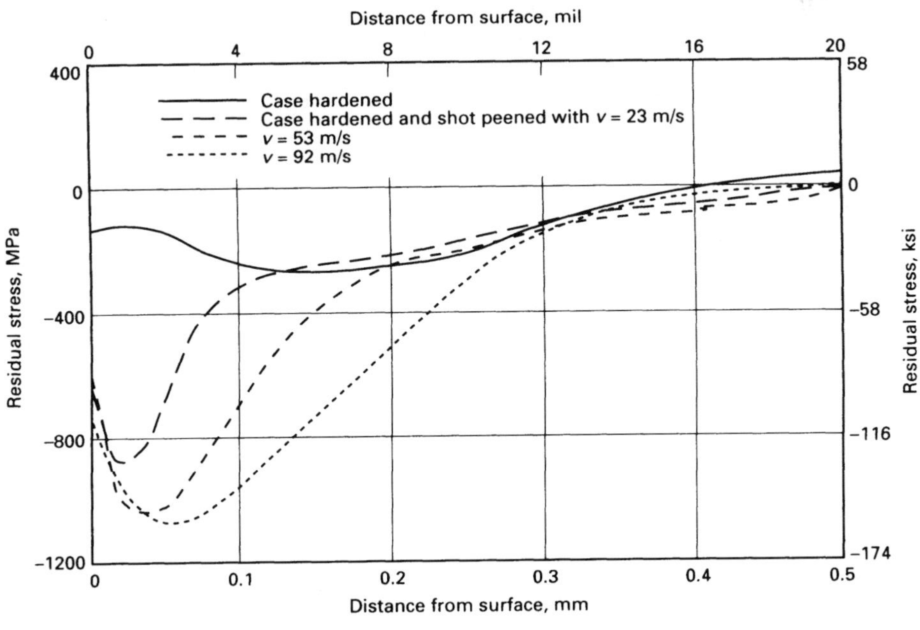

Fig. 57 Effect of shot peening at different velocities on compressive residual stresses in carburized 16MnCr5 steel (1.23% Mn, 1.08% Cr). Source: Ref 57

ments in stresses translate into improved bending fatigue performance.

Effects of Surface Oxidation on Residual Stresses. Residual stresses can be adversely affected by surface oxidation during gas carburizing. As discussed in the next section, certain alloying elements are preferentially oxidized and removed from solid solution in the austenite. As a result, hardenability decreases, and, in severe cases, pearlite instead of martensite forms at the surface (Ref 61). Thus the surface transformation occurs at high temperatures, and the beneficial effect of austenite-to-martensite transformation late in the quenching process is lost. Even if the oxidation is not severe enough to cause pearlite formation, surface M_s temperatures may be raised by the removal of some of the alloying element, resulting in a thin surface zone with lower compressive stresses.

Effects of Refrigeration Treatments on Residual Stresses. Subzero, or cryogenic, refrigeration is sometimes used to lower retained austenite contents. As a result, surface hardness increases. Also, dimensional stability in service is increased because there is less austenite available to transform to martensite by stress- or strain-controlled mechanisms. However, a number of investigations have shown that the refrigeration treatment of carburized parts lowers fatigue performance (Ref 54, 62, 63). The transformation of additional surface retained austenite would be expected to continue the process established during quenching to room temperature; that is, the volume expansion associated with the formation of new martensite would be constrained, and compressive stresses would be increased. Increased compressive stresses are in fact measured in the martensite of refrigerated specimens (Ref 62, 63). However, Kim et al. (Ref 62) have shown that the stresses in the remaining retained austenite are tensile. Such localized tensile stresses would lower the applied stresses required to initiate and propagate fatigue cracks.

Surface and Internal Oxidation

The H_2O/H_2 and CO_2/CO equilibria in gas carburizing atmospheres cause the internal oxidation of certain alloying elements in carburizing steels (Ref 64). Figure 58 shows the oxidiz-

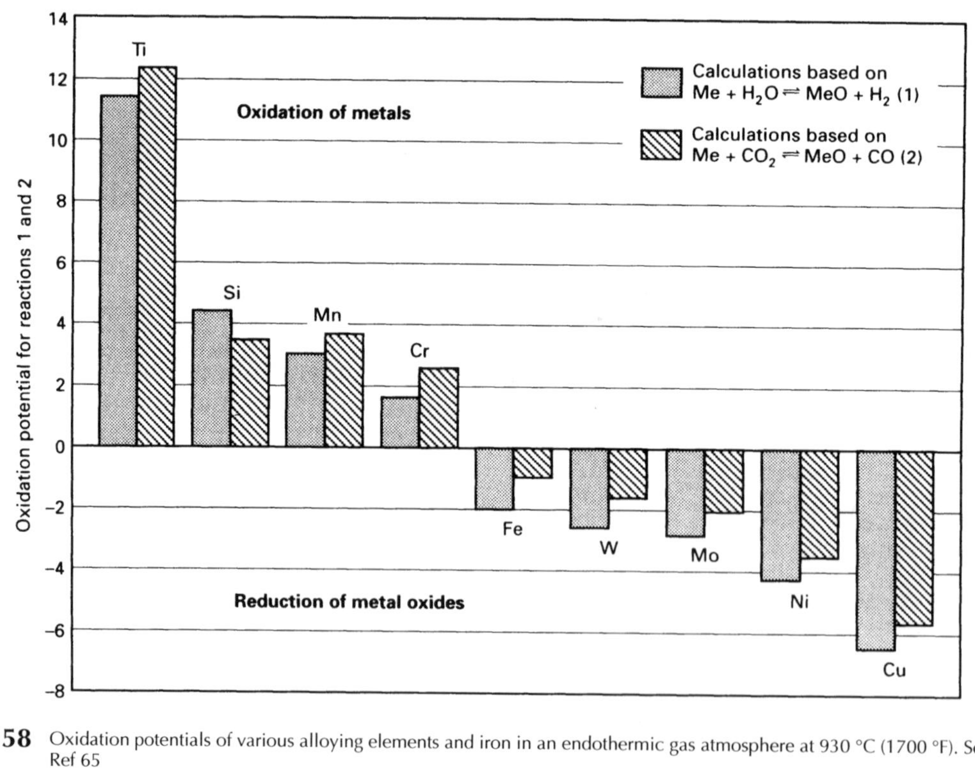

Fig. 58 Oxidation potentials of various alloying elements and iron in an endothermic gas atmosphere at 930 °C (1700 °F). Source: Ref 65

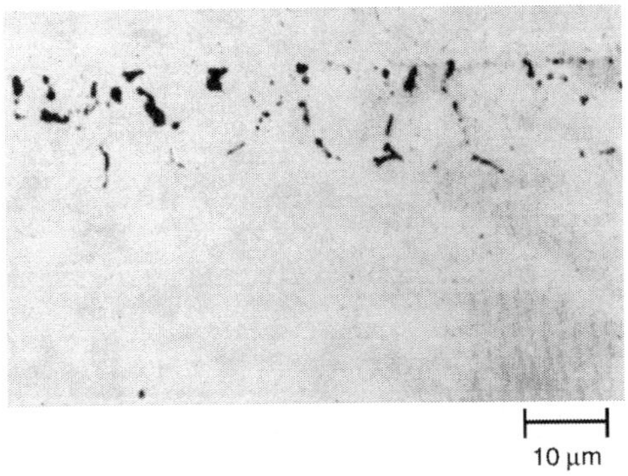

Fig. 59 Internal oxidation (dark features) at surface of gas-carburized steel containing 1.06% Mn, 0.21% Si, 0.52% Cr, 0.50% Ni, and 0.17% Mo. Light micrograph. Source: Ref 15

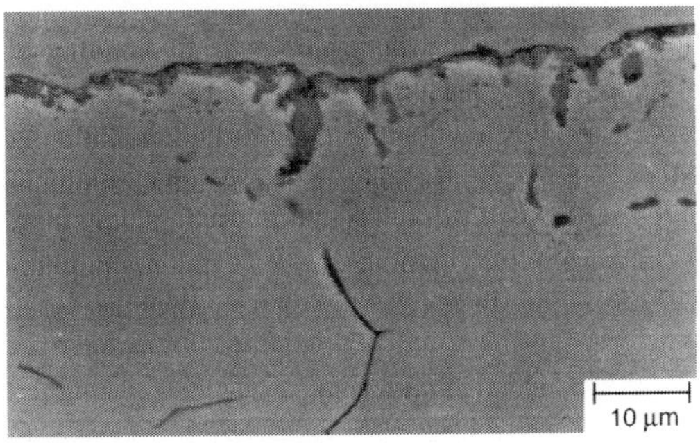

Fig. 60 Internal oxidation of gas-carburized 20MnCr5 steel containing 1.29% Mn, 0.44% Si, 1.25% Cr, 0.25% Ni, and 0.0015% B. SEM micrograph. Source: Ref 66

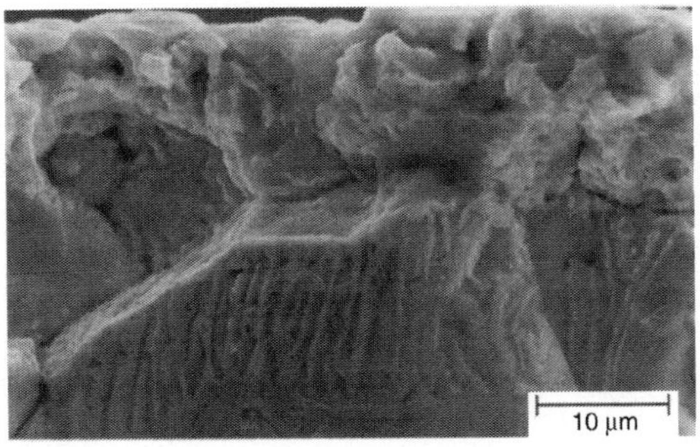

Fig. 61 Lamellar internal grain-boundary oxides on fracture surface of carburized 20MnCr5 steel containing boron. SEM micrograph. Source: Ref 66

ing potentials for various elements in endothermic gas at 930 °C (1700 °F). Chromium, silicon, and manganese, all commonly found in carburizing steels, oxidize readily, while molybdenum, nickel, and iron are not oxidized. The oxidation is diffusion dependent, and therefore the depth and extent of oxide formation is a function of carburizing time and temperature. The oxides may form on austenite grain boundaries or within austenite grains.

Figure 59 shows internal oxidation at the surface of a carburized specimen of a steel containing 1.06% Mn, 0.21% Si, 0.52% Cr, 0.50% Ni, and 0.17% Mo. The oxidation has followed the austenite grain boundaries to a depth on the order of an austenite grain diameter, about 10 μm (0.4 mil). This depth of penetration is typical for steels carburized to a case depth of about 1 mm (40 mil).

Figure 60 shows the internal oxidation of a carburized 20MnCr5 steel containing 1.29% Mn, 0.44% Si, 1.25% Cr, 0.25% Ni, and 0.0015% B. There are two zones of oxidation. The outer zone, about 5 μm (0.2 mil) deep, consists of chromium-rich oxides penetrating into the austenite grains. The other zone, about 30 μm (1.2 mils) deep, consists of manganese-rich and silicon-rich oxides along austenite grain boundaries. These oxide chemistries and morphologies agree with those presented by Chatterjee-Fischer (Ref 64). In addition, silicon appears to form intergranular dispersed oxide particles.

The grain-boundary oxides shown in Fig. 60 appear to be discontinuous. Examination of fractured specimens of the same carburized steel shown in Fig. 60 showed that the intergranular oxides grew as lamellae (Fig. 61). Thus the oxide structure appears to develop by a discontinuous or cellular transformation in which grain-boundary austenite initially containing nominal amounts of silicon and manganese decomposes to alloy oxides and austenite depleted in silicon and manganese. The discontinuous appearance of the oxides in Fig. 60 is therefore due to a sectioning effect through the oxide and austenite lamellae.

Properties of Carburized Steels

Because most carburized parts are subjected to cyclic loading, by far the most important property or measure of their performance is fatigue resistance. As such, emphasis in this section is placed on bending fatigue. Other properties of interest that are briefly reviewed include rolling contact fatigue, bend ductility, hot hardness, wear resistance, and toughness. Additional property data on carburized steels can be found in Ref 36.

Bending Fatigue Strength

Bending fatigue of carburized steel components is a result of cyclic mechanical loading. The bending produces stresses, which are tensile at the surface, decrease with increasing distance into the component, and at some point become compressive. Such loading is a characteristic of rotating shafts and the roots of gear teeth. Carburizing produces a high-carbon, high-strength surface layer, or high-strength case, on a low-carbon, low-strength interior or core and is therefore an ideal approach to offset the high surface tensile stresses associated with bending. Thus when the design of a component maintains operating stress gradients below the fatigue strength of the case and core microstructures, excellent bending fatigue resistance is established.

There are, however, many alloying and processing factors that produce various microstructures, and therefore variable strength and fracture resistance, of the case regions of carburized steels. When applied surface stresses exceed the surface strength, surface fatigue crack initiation and eventual failure will develop. When the surface strength is adequate, depending on the steepness of the applied stress gradients in relationship to the case/core strength gradient, subsurface fatigue cracking may develop.

Bending fatigue performance of carburized steels can vary significantly. One study reported values of experimentally measured endurance limits ranging from 200 to 1930 MPa (29 to 280 ksi), with most values between 700 and 1050 MPa (100 and 152 ksi) (Ref 67). As will be described subsequently, this wide variation in fatigue performance is a result of variations in specimen design and testing, alloying, and processing interactions that produce large variations in carburized microstructures and the response of the microstructures to cyclic loading.

Bending Fatigue Testing

Data Presentation and Analysis. Most bending fatigue data for carburized steels are presented as plots of maximum stress, *S*, versus the number of cycles, *N*, to fracture for a specified stress ratio (*R*), which is the ratio of minimum (or compressive) stress to maximum tensile stress (*R* = min-*S*/max-*S*). Figure 62 shows an example of typical *S*-*N* plots for a series of carburized alloy steels (Ref 30). The *S*-*N* curve consists of two parts: a straight section with negative slope at low cycles and a horizontal section at high cycles. The horizontal line defines the fatigue limit or endurance limit, which is taken to be the maximum applied stress below which a material is assumed to be able to withstand an infinite number of stress cycles without failure. Pragmatically, the endurance limit is taken as the stress at which no failure occurs after a set number of cycles, typically on the order of 10 million cycles. The low-cycle portion of the *S*-*N* plot defines various fatigue strengths or the stresses to which the material can be subjected for a given number of cycles. The more cycles at a given strength, the better the low-cycle fatigue resistance of a material.

Analysis of bending fatigue behavior of carburized steels based on *S*-*N* curves represents a stress-based approach to fatigue and assumes that the carburized specimens deform nominally only in an elastic manner (Ref 68). This assumption is most valid at stresses up to the endurance limit and is useful when machine components are designed for high-cycle fatigue. However, as maximum applied stresses increase above the endurance limit, plastic strain becomes increasingly important during cyclic loading, and fatigue is more appropriately analyzed by a strain-based approach. In this approach, the total strain range is the sum of the applied elastic and plastic strains, and the strain amplitude is plotted

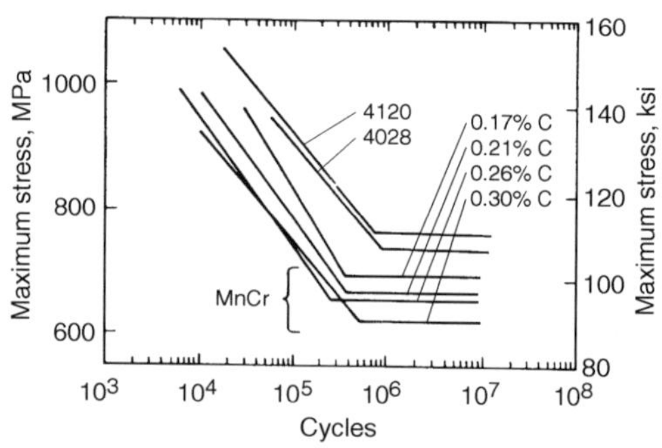

Fig. 62 Typical maximum stress (*S*) vs. number of cycles (*N*) bending fatigue plots for 6 carburized steels. *R* = –1. Source: Ref 30

as a function of the strain reversals required for failure at the various levels of strain.

According to the strain-based approach, low-cycle fatigue behavior is determined by plastic strains, while high-cycle fatigue behavior is determined by elastic strains. In particular, ductile materials with microstructures capable of sustaining large amounts of plastic deformation have better low-cycle fatigue resistance, while high-strength materials with high elastic limits and high yield strengths have better high-cycle fatigue resistance. Figure 63 shows the results of strain-based bending fatigue testing of uncarburized and carburized 4027 steel (Ref 69). The more ductile, low-strength uncarburized specimens show better fatigue resistance at low cycles than do the carburized specimens. The performance is reversed at high cycles, where the carburized specimens with their high-strength surfaces show better fatigue resistance, especially those specimens with the deeper cases.

Specimen Design. Many types of specimens have been used to evaluate bending fatigue in carburized steels. Rotating beam, unnotched four-point bend, notched four-point bend, and cantilever beam specimens have all been used, and they have in common a maximum applied surface tensile stress and decreasing tensile stress with increasing distance into the specimen. Axial fatigue testing of carburized specimens, which applies the maximum tensile stresses uniformly over the cross section of a specimen, invariably results in subsurface initiation and propagation of fatigue cracks in the core of carburized specimens, and therefore it does not permit evaluation of the resistance of case microstructures to fatigue.

Brugger was the first to use cantilever bend specimens to evaluate the fracture and fatigue of carburized steels (Ref 70). Figure 64 shows a cantilever bend specimen that has evolved from the Brugger specimen. The radius between the change in section simulates the geometry at the root of gear teeth and results in maximum applied surface stresses just where the cross section begins to increase, as shown in Fig. 65 (Ref 71). An important feature of this specimen is the rounding of the corners of the beam section. If the corners are square, the carbon introduced into the corner surfaces cannot readily diffuse into the interior of the specimen. As a result, the corner microstructures may have significantly elevated levels of retained austenite and coarse

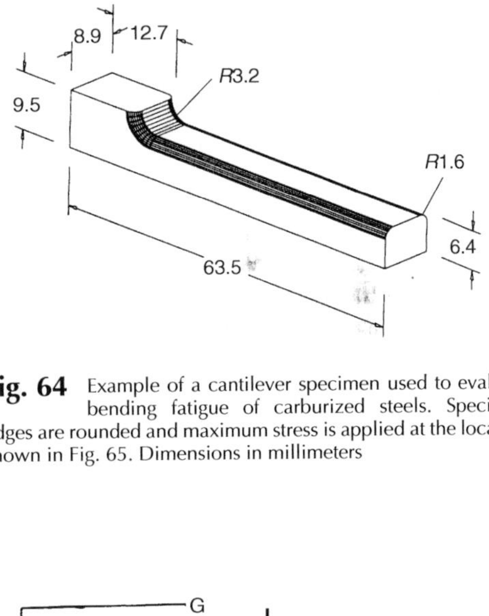

Fig. 64 Example of a cantilever specimen used to evaluate bending fatigue of carburized steels. Specimen edges are rounded and maximum stress is applied at the location shown in Fig. 65. Dimensions in millimeters

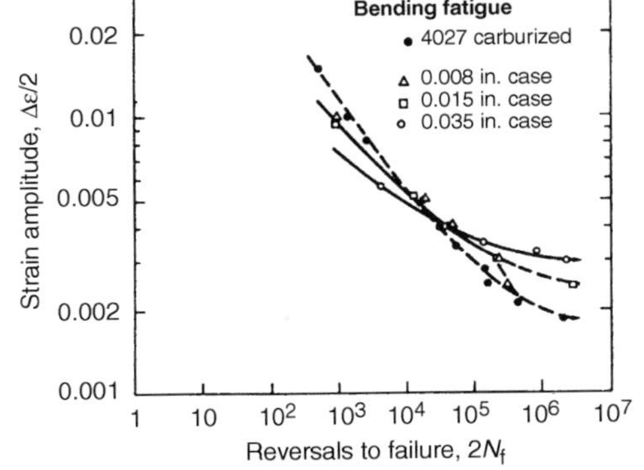

Fig. 63 Strain amplitude vs. reversals to failure for uncarburized (solid symbols) and carburized (open symbols) 4027 steel (0.80% Mn, 0.28% Si, 0.27% Mo). Source: Ref 69

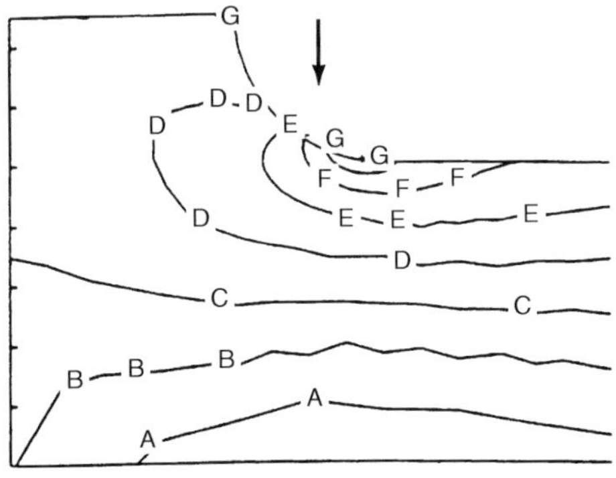

Fig. 65 Location of maximum stress on the cantilever bend specimen shown in Fig. 64 as determined by finite element modeling

carbide structures—both microstructural features that influence bending fatigue resistance (Ref 23, 71).

Mean Stress and Stress Ratio. The maximum applied surface tensile stress is the testing parameter plotted in *S-N* curves that characterize bending fatigue. However, the applied stress ranges between maximum and minimum values during a fatigue cycle, and two other parameters, mean stress and stress ratio, are important for the characterization of fatigue. The mean stress is the algebraic average of the maximum and minimum stresses in a cycle, and as discussed, the stress ratio, *R*, is the ratio of the minimum stress to the maximum stress in a cycle. Thus *R* values may range from –1, for fully reversed loading that ranges between equal maximum tensile and compressive stresses, to positive values where the stress is cycled between two tensile values (Ref 68). Much of the bend testing of cantilever specimens described subsequently is performed with *R* values of 0.1 in order to preserve details of the fracture surface.

Figure 66 shows a typical allowable-stress diagram that plots fatigue strength versus mean stress for a given material (Ref 72). The diagram shows that the most severe condition for fatigue is for fully reversed testing with $R = -1.0$. As the mean stress increases, the fatigue strength in terms of maximum applied stress increases, but the allowable stress range decreases. Zurn and Razim (Ref 73) have examined the effect of notch severity and retained austenite on allowable-stress/mean-stress diagrams of carburized steels, and they conclude that carburizing, relative to the use of through-hardened steels, is especially effective for parts with sharp notches. In the absence of notches, carburizing is most suitable for parts subjected to fatigue loading at low values of mean stress.

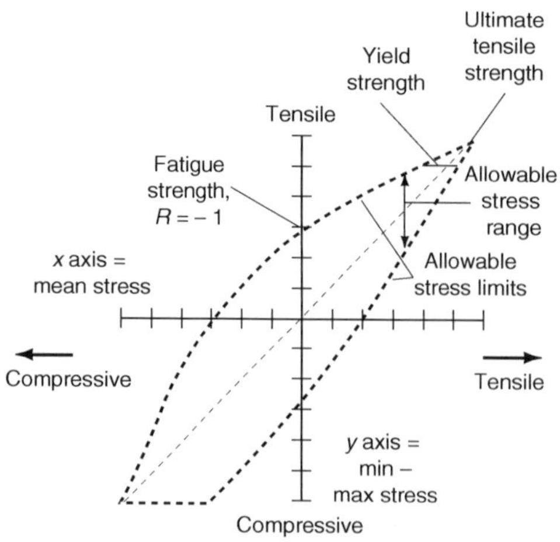

Fig. 66 Typical allowable-stress-range diagram. Source: Ref 72

The testing of actual machine components is another important approach to the fatigue evaluation of carburized steels. An example of component testing is the bending fatigue testing of single teeth in gears (Ref 74). Gears are fabricated, carburized, and mounted in a fixture so that one tooth at a time is subjected to cyclic loading. More recently, identically carburized specimens of the same steel were subjected to cantilever bend and single tooth bending fatigue testing (Ref 75). The mechanisms of fatigue failure, based on fracture surface examination, were found to be the same, but the single tooth testing showed higher levels of fatigue resistance than did the cantilever testing, a result attributed to the higher surface compressive stresses that were measured in the gear tooth specimens.

Stages of Fatigue and Fracture

Bending fatigue fractures of carburized steels consist of well-defined stages of crack initiation, stable crack propagation, and unstable crack propagation. The fracture sequence is strongly influenced by the gradients in strength, microstructure, and residual stress that develop in carburized steels. Figure 67 shows a series of SEM fractographs that characterize the typical fracture sequence of a direct-quenched carburized steel with a nearsurface case microstructure similar to that shown earlier in Fig. 29. The cantilever bend specimen from which the fractographs of Fig. 67 were taken was a 4320 steel carburized to a 1 mm case depth at 927 °C (1700 °F), quenched from 850 °C (1560 °F) into oil at 65 °C (150 °F), and tempered at 150 °C (300 °F) for 1 h. The specimen was tested in bending fatigue with an *R* value of 0.1 (Ref 76). Figures 67(a) and (b) show a low-magnification overview of the initiation, stable propagation, and unstable fracture surfaces, and Fig. 67(c) shows the intergranular initiation and transgranular stable crack propagation zones of the fracture at a higher magnification.

Intergranular cracking at prior-austenite grain boundaries is an almost universal fracture mode in the high-carbon case of direct-quenched carburized steels (Ref 22, 38, 76). Not only do the fatigue cracks initiate by intergranular cracking, but also the unstable crack propagates largely by intergranular fracture until it reaches the lower-carbon portion of the case, where ductile fracture becomes the dominant fracture mode. In fact, sensitivity of the case microstructures to intergranular fracture makes possible the quantitative characterization of the size and shape of the stable fatigue crack, as shown in Fig. 67(b). The transition from the transgranular fracture of the stable crack to the largely intergranular fracture of the unstable fracture is identified by the dashed line.

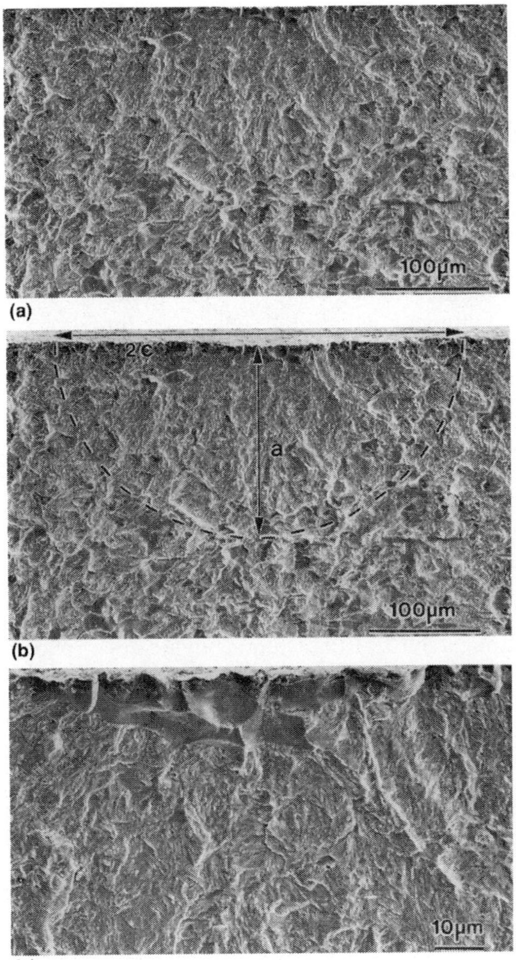

Fig. 67 Fatigue fracture in gas-carburized and modified 4320 steel. (a) Overview of initiation, stable crack propagation, and unstable crack propagation. (b) Same area as shown in (a), but with extent of stable crack indicated by dashed line. (c) Higher magnification of intergranular initiation and transgranular stable crack propagation areas. SEM Fractographs. Source: Ref 76

A replica study of carburized specimens subjected to incrementally increasing stresses showed that surface intergranular cracks initiated when the applied stresses exceeded the endurance limits (Ref 76). Thus, it appears that in direct-quenched carburized specimens, intergranular cracks are initiated as soon as the applied surface bending stress reaches a level sufficient to exceed the surface compressive residual stress and the cohesive strength of the prior-austenite grain boundary structures. The surface intergranular cracks are shallow, typically approximately two to four austenite grains, and are arrested, perhaps because of a plastic zone smaller than the grain size at the tip of the sharp intergranular cracks, and the fact that strain-induced transformation of retained austenite in the plastic zone ahead of the crack introduces compressive stresses (Ref 77, 78). The fatigue crack then propagates in a transgranular mode, and when the stable crack reaches critical size, as defined by the fracture toughness, unstable fracture occurs.

The initiation and stable crack zones of carburized steels are quite small and are often difficult to identify. Figure 68, based on the measurement of critical crack sizes in a number of direct-quenched carburized 4320 steels, shows that the size of the unstable cracks ranges from 0.170 to 0.230 mm, and that the cracks therefore become unstable well within the high-carbon portion of the carburized specimens. The small critical crack sizes are consistent with the low fracture toughness of high-carbon steel LTT microstructures susceptible to intergranular

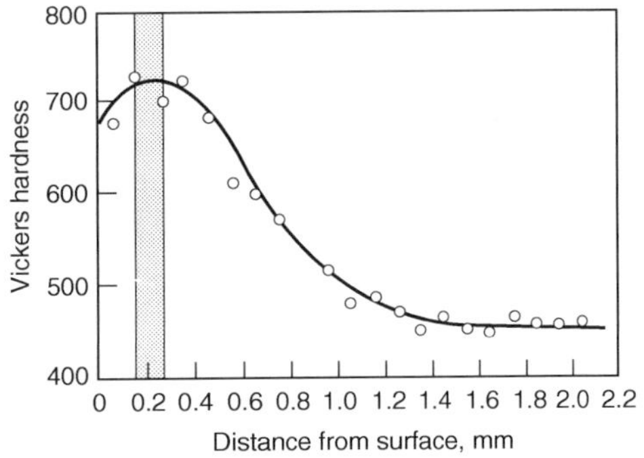

Fig. 68 Hardness vs. distance from the surface of direct-cooled gas-carburized SAE 4320 steel. Superimposed on the hardness profile is the range of critical depths (vertical dashed band) at which stable fatigue cracks became unstable in bending fatigue of similarly processed steels. Source: Ref 76

fracture (Ref 39). When the critical crack sizes and the stresses at which the cracks become unstable are used to calculate the fracture toughness of the case microstructures of carburized steels (Ref 76), the results show good agreement with the range of fracture toughness, 15 to 25 MPa $\sqrt{m}$, that has been measured from through-hardened specimens with high-carbon LTT martensitic microstructures (Ref 39). Table 3 shows the data used to calculate the various case fracture toughness values in gas-carburized 4320 steel and the fracture toughness values calculated according to three different fracture toughness equations (Ref 76):

$$K_{IC} = \frac{1.2\sigma_a \sqrt{a\pi}}{Q} \quad \text{(Eq 25)}$$

$$K_{IC} = \frac{\left[1.2\sigma_a + 0.683\left(\frac{\partial\sigma}{\partial_x}\right)a\right]\sqrt{a\pi}}{Q} \quad \text{(Eq 26)}$$

and

$$K_{IC} = \frac{M\sigma_a \sqrt{a\pi}}{Q} \quad \text{(Eq 27)}$$

where

$$Q = \phi^2 - 0.212\left(\frac{\sigma_a}{\sigma_{ys}}\right)^2$$

with ϕ the aspect ratio of crack depth (a) and crack length (c) such that:

$$\phi^2 = 1 + 1.464\left(\frac{a}{c}\right)^{1.65}$$

The unstable crack that proceeds through the high- and medium-carbon martensitic portions of the case may be arrested when the sensitivity to intergranular fracture decreases at a case carbon content between 0.5 and 0.6% (Ref 79). The continued application of cyclic loading at this point then may cause a secondary stage of stable fatigue crack propagation, characterized by transgranular fracture, resolvable fatigue striations, and secondary cracking (Ref 63, 75). This stage of low-cycle, high-strain fatigue is short and gives way to ductile overload fracture of the core.

Intergranular Fracture and Fatigue

As noted earlier, intergranular fracture at the prior-austenite grain boundaries of high-carbon case microstructures dominates bending fatigue crack initiation and unstable crack propagation of direct-quenched carburized steels. The intergranular cracking may be associated with other microstructural features, such as the surface oxides generated by gas carburizing, but it generally extends much deeper into a carburized case than the oxide layer. Several studies have documented bending fatigue crack initiation by intergranular fracture even in the absence of surface oxidation, where, for example, the oxidized surface has been removed by chemical or electropolishing (Ref 22, 80) or no oxidation is present because the specimens were vacuum or plasma carburized (Ref 15).

Figure 69 shows an example of intergranular fatigue crack initiation in a direct-quenched specimen of gas-carburized type 8719 steel (Ref 81). There is a shallow zone of surface oxidation, about 10 μm in depth, but the intergranular cracking extends much deeper into the specimen. Figure 70 shows extensive intergranular

Table 3 Fracture toughness results for carburized SAE 4320 bending fatigue specimens

Max stress, MPa	Cycles to failure	Depth, a, μm	Width, c, μm	a/c	M	K_{Ic}, MPa$\sqrt{mw}$, Eq 25	K_{Ic}, MPa$\sqrt{m}$, Eq 26	K_{Ic}, MPa$\sqrt{m}$, Eq 27
1370	6400	175	388	0.45	0.87	30	29	22
	16,900	170	355	0.48	0.86	29	28	21
	15,300	200	210	0.95	0.78	18	18	12
1285	17,400	230	300	0.77	0.81	22	22	15
	18,700	230	295	0.78	0.81	22	22	15
1235	34,100	210	300	0.70	0.81	22	22	15
1160	21,700	230	295	0.78	0.81	19	19	13
	32,300	220	295	0.75	0.81	19	19	14

Equations used to determine the information in this table are defined in text. Source: Ref 76 cracking in the unstable crack propagation zone in the case of a direct-quenched, gas-carburized 4320 steel. Auger electron spectroscopy (see Fig. 48 and corresponding text) shows that such intergranular fracture surfaces have higher concentrations of phosphorus and carbon, in the form of cementite, than do transgranular fracture surfaces removed from prior-austenite grain boundaries (Ref 38, 41, 82). Thus, the brittle intergranular fracture that occurs in stressed high-carbon case microstructures of carburized steels is associated with the combined presence of segregated phosphorus and cementite at prior-austenite grain boundaries. These grain boundary structures are present in as-quenched specimens and do not require tempering for cementite formation, as is typical in the intergranular mode of tempered martensite embrittlement in medium-carbon steels (Ref 37). This embrittlement, termed *quench embrittlement,* is found in quenched steels with carbon contents as low as 0.6% (Ref 83). There is evidence that phosphorus segregation stimulates the formation of the grain boundary cementite (Ref 41, 82).

The higher the phosphorus content of a carburized steel, the lower its bending fatigue resistance and case fracture toughness. Figure 71 shows *S-N* curves for a series of gas-carburized and direct quenched modified 4320 steels with systematic variations in phosphorus content from 0.031 to 0.005% (Ref 82). Endurance limits and low-cycle fatigue resistance increase with decreasing phosphorus content, but little difference is noted between the performance of the 0.005 and 0.017% phosphorus specimens. All of the specimens, even those with the lowest phosphorus content, failed by intergranular initiation of fatigue cracks.

The bending endurance limits of gas-carburized specimens in which fatigue is initiated by intergranular fracture typically range between 1050 and 1260 MPa (Ref 84, 85). This range is based on studies of cantilever bend specimens with good surface finish, rounded specimen corners, nominal amounts of surface oxidation, and loading at $R = 0.1$. Variations within this range may be due to variations in austenitic grain size, inclusion contents, retained austenite content, or residual stresses, as discussed subsequently. Nevertheless, the common mechanism of bending fatigue crack initiation of direct-quenched specimens is intergranular fracture at embrittled

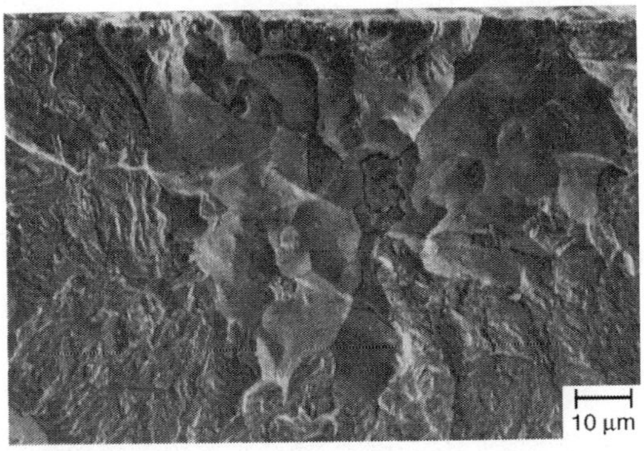

Fig. 69 Intergranular bending fatigue crack initiation at the surface of a gas-carburized and direct-cooled SAE 8719 steel specimen. Source: Ref 81

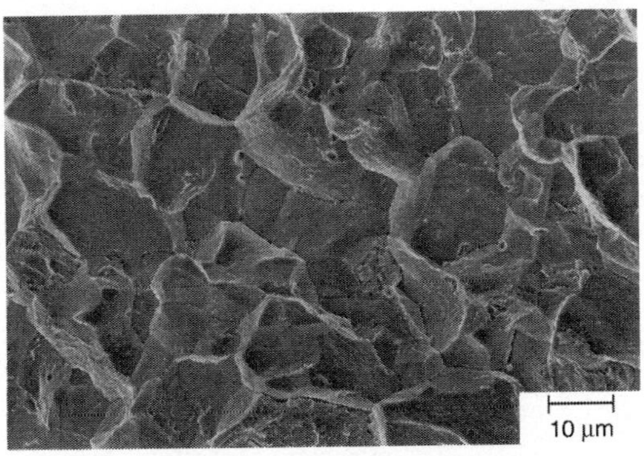

Fig. 70 Intergranular fracture in case unstable crack propagation zone in gas-carburized and direct-cooled SAE 4320 steel

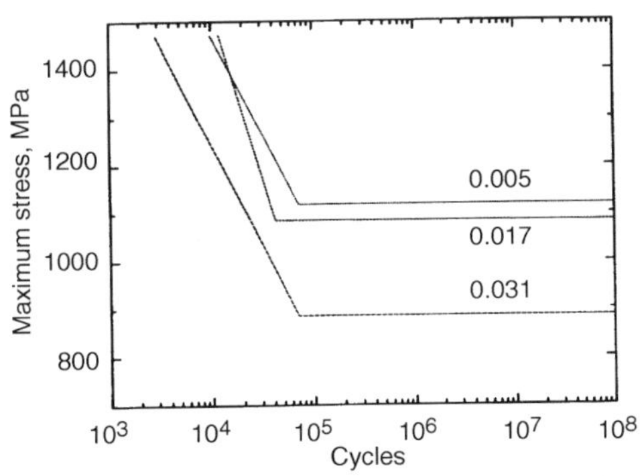

Fig. 71 Effect of phosphorus content on the bending fatigue of direct-quenched, gas-carburized modified 4320 steel with 0.005, 0.017, and 0.031 wt% phosphorus, as marked. Source: Ref 82

grain boundaries in a microstructure of LTT martensite and retained austenite, as shown in Fig. 29.

Carburized steels with high nickel content do not appear to be as susceptible to intergranular cracking as steels with low nickel content (Ref 46, 47). Also, major changes in the case microstructures of carburized steel, such as those produced by reheating and described relative to Fig. 39, result in bending fatigue crack initiation sites other than embrittled prior-austenite grain boundaries. Microstructural conditions that produce fracture initiation other than by intergranular cracking are also described herein.

Inclusions and Fatigue

Inclusions—phases formed between metallic elements and nonmetallic elements such as sulfur and oxygen—are an important microstructural component of steels. Coarse or high densities of inclusions initiate fracture and lower the toughness of steels. As a result, modern steelmaking practices, which incorporate improved deoxidation, shrouding of liquid steel to prevent reoxidation, vacuum degassing, argon blowing, and desulfurization, are designed to substantially increase the "cleanliness" of steels by lowering the number and/or modifying the morphology of inclusions.

In carburized steels, the very high strength of the carburized case makes the plastic zone ahead of surface discontinuities (such as machining marks), flaws, or cracks very small, and therefore, in clean steels, distributed inclusions play a smaller role in fracture than, for example, uniformly distributed and closely spaced grain boundary embrittling structures or surface oxides. In other words, the high stresses in the plastic process zone have a much higher probability of acting on grain boundaries or surface oxides than on widely spaced inclusions.

Although inclusions often play a secondary role in the fatigue of gas-carburized steels, especially when there are other microstructural causes of fatigue crack initiation, they may be involved in the fatigue process in several ways. Inclusions in the carburized steel may either combine with other features that initiate fatigue cracks, or in the absence of such features, serve as the sole source of fatigue crack initiation.

An example of the first type of effect was demonstrated in a study that examined the effects of systematic variations in sulfur content on the bending fatigue resistance of a gas-carburized low alloy steel (Ref 81). Sulfur combines with manganese to form manganese sulfide (MnS) inclusions in steel. The MnS particles are plastic during hot rolling, and as a result, are elongated in the rolling direction. This elongation imparts an anisotropy to the mechanical properties of the steel, which makes the effect of inclusions a function of the orientation of the particles relative to the direction of the applied load.

S-N (stress vs. life) curves for specimens of a gas-carburized SAE 8219-type steel with three levels of sulfur are plotted in Fig. 72. The endurance limit decreases with increasing sulfur content. A number of the specimens with the higher sulfur contents showed runouts at 10 million cycles at stress levels higher than the endurance limits shown, but the specimens that failed at the lower stresses were used to establish the endurance limits. In these specimens, elongated MnS inclusion particles that happened to be close to the specimen surfaces were associated with the fatigue fracture initiation. Fatigue fracture initiation of the direct-quenched, gas-carburized specimens was still dominated by intergranular fracture, but if sulfides were present at the highly stressed surfaces of the bending fatigue specimens, they apparently provided an extra source of stress concentration and reduced fatigue performance.

An example of MnS particles associated with fatigue crack initiation in a carburized 8219-type steel is shown in Fig. 73. Although fatigue resistance is lowered somewhat by the presence of increased densities of MnS particles, the decrease may be outweighed by a gain in machinability associated with higher levels of sulfur.

Austenitic Grain Size and Fatigue

Effect of Fine Grain Size on Microstructures and Properties. Prior-austenite grain size of carburized steels correlates strongly with bending fatigue resistance. Generally, the finer the prior-austenite grain size, the better the fatigue performance. For example, Fig. 74 shows a direct relationship of bending fatigue endurance limit on prior-austenitic grain size, plotted as the inverse square root of the grain size, for several sets of carburized 4320 steels (Ref 86).

The refinement of austenite grain size has several effects on the case microstructure of carburized steels. A finer prior-austenite grain size produces a finer martensitic microstructure on quenching and therefore raises the strength of the carburized case. Increases in strength are beneficial to high-cycle fatigue resistance, as discussed previously in the section on bending fatigue testing.

Another very important consequence of fine austenitic grain size is the dilution of the grain boundary segregation of phosphorus. In fact, very fine austenitic grain sizes can eliminate the sensitivity of high-carbon case microstructures to intergranular fracture. As a result, other mechanisms of fatigue crack initiation replace intergranular cracking, generally to the benefit of fatigue performance. The high values of endurance limits shown for the very fine grain specimens in Fig. 74 were associated with fatigue crack nucleation at surface oxidation, not at embrittled prior-austenite grain boundaries. Although, as discussed below, surface oxides form on austenite grain boundaries and fatigue cracks may nucleate on the oxide-covered austenite boundaries, fine-grain specimens show no intergranular fracture below the oxidized surface layers.

Reheat Treatments to Achieve Fine Grain Size. The most effective way to produce very fine grains in carburized steels is by slow cooling and reheating of carburized parts at temperatures below the A_{cm} where austenite and cementite are stable. The cementite particles effectively retard austenite grain growth and reduce the carbon content of the austenite, causing the type of microstructure shown in Fig. 39 to form on quenching. Specimens reheated to above the A_{cm} may show grain refinement, depending on the temperature of heating, but because all carbides are dissolved, grain size refinement is not as effective as in specimens heated below the A_{cm}, and the type of microstructure shown in Fig. 29 develops upon quenching (Ref 22). Figure 75 shows austenitic grain size as a function of distance from the carburized surface of gas-carburized 4320 steel specimens in the direct-quenched condition and after one and three reheating treatments (Ref 86). The reheat treatments very effectively reduce the near-surface case grain size where carbon content is the highest, and therefore the greatest density of carbide particles is retained during intercritical reheating.

Intercritical-temperature reheat treatments of carburized steels produce very fine austenitic grain sizes and high endurance limits. Typically the endurance limits are above 1400 MPa (Ref 15, 22, 86). However, the beneficial effects of the reheating on bending fatigue resistance are not due to grain size refinement alone. The reduced carbon content of the austenite when carbides are retained raises M_s temperatures and reduces the amount of retained austenite in the as-quenched case microstructures. Reduced levels of retained austenite raise the strength of case microstructures and therefore also may

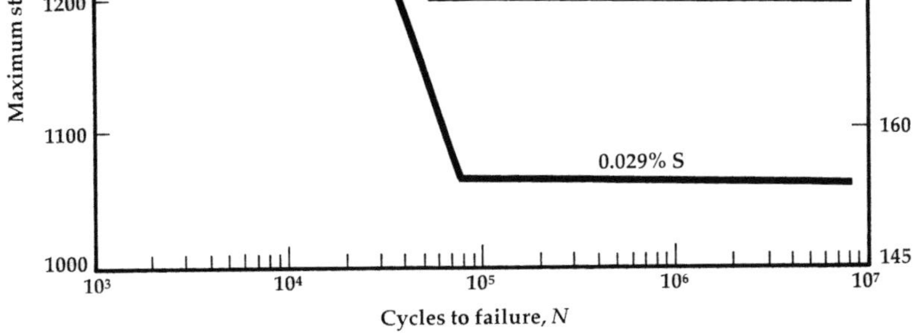

Fig. 72 *S-N* curves determined by bending fatigue of a gas-carburized SAE 8219-type steel containing 1.40 Mn, 0.61 Cr, 0.30 Ni, 0.20 Mo, and three levels of sulfur. Source: Ref 81

contribute significantly to the improved high-cycle fatigue performance of fine-grain, intercritically reheated carburized steels.

Surface Oxidation and Fatigue

The surface oxidation produced during gas carburizing may or may not significantly reduce bending fatigue resistance. The most severe effects of such oxidation are associated with a reduction in near-surface case hardenability, which results from the removal of chromium, manganese, and silicon from solution in the austenite by the oxide formation (Ref 61, 87, 88). The reduced case hardenability can cause nonmartensitic microstructures, such as ferrite, bainite, and pearlite, to form at the surface of the carburized steel. Not only is the surface hardness reduced, but the residual surface stresses may become less compressive or even tensile.

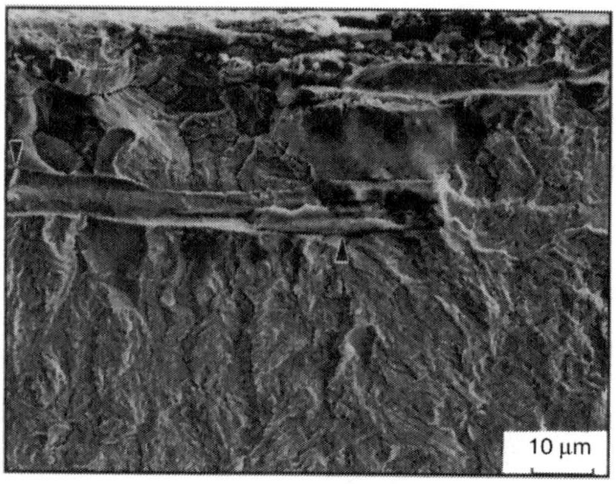

Fig. 73 Elongated manganese sulfide particles associated with bending fatigue crack initiation at the surface of a gas-carburized SAE 8219-type steel. SEM photomicrograph. Source: Ref 81

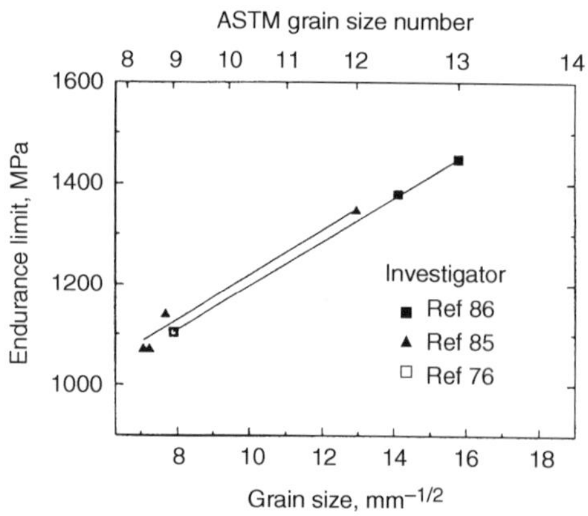

Fig. 74 Endurance limits as a function of prior-austenite grain size from various studies of bending fatigue of gas-carburized 4320 steels. Source: Ref 86

Figures 76 and 77 show the effects of surface oxidation with reduced hardenability on the bending fatigue and residual stresses of 8620 and 4615 gas-carburized specimens (Ref 89). The 4615 steel has higher hardenability by virtue of higher nickel and molybdenum contents and a lower sensitivity to surface oxidation by virtue of reduced manganese and chromium contents (Ref 89). As a result of the different chemistries, the 8620 steel formed pearlite in the near-surface regions of the case, while the microstructure of the 4615 steel, despite some oxidation, consisted only of plate martensite and retained austenite at the surface (Ref 89). These differences in microstructures due to surface oxidation and reduced case hardenability are consistent with the differences in bending fatigue performance and residual stresses shown between the two steels in Fig. 76 and 77. This study illustrates the importance of steel chemistry on controlling surface oxidation and the associated formation of nonmartensitic microstructures in gas-carburized steels. Another approach used to reduce surface oxide formation in steels with low hardenability is to use more severe quenching with higher cooling rates.

If the hardenability of a steel is sufficient to prevent the formation of nonmartensitic microstructures for a given gas carburizing and quenching schedule, surface oxidation has a much reduced effect on bending fatigue performance. In direct-quenched specimens, as discussed previously and demonstrated in Fig. 69, intergranular fracture to depths much deeper than the oxidized layers dominates fatigue crack initiation. However, when the conditions for intergranular crack initiation are minimized, as, for example, by reheating (Ref 22, 86) or shot peening (Ref 25), the surface oxide layers become a major location for bending fatigue crack initiation. Figure 78 shows crack initiation in the oxidized zone of a gas-carburized and reheated specimen of 4320 steel. The initiation is confined to the oxidized zone, and stable transgranular fatigue propagation proceeds directly below the oxidized zone with no evidence of intergranular fracture.

Retained Austenite and Fatigue

As described earlier in the section "Microstructures of Carburized Steels," next to LTT martensite, retained austenite is the most important microstructural component in the case of carburized steels. The amounts of retained austenite vary widely, depending on carbon and alloy content, heat-treating conditions, and special processing steps such as shot peening and subzero cooling. Generally, the higher the carbon and alloy content, the lower the M_s temperature and the higher the retained austenite content in the microstructure. The low-temperature tempering applied to carburized steels, generally performed at temperatures below 200 °C (400 °F), is not high enough to cause the retained austenite to transform, and

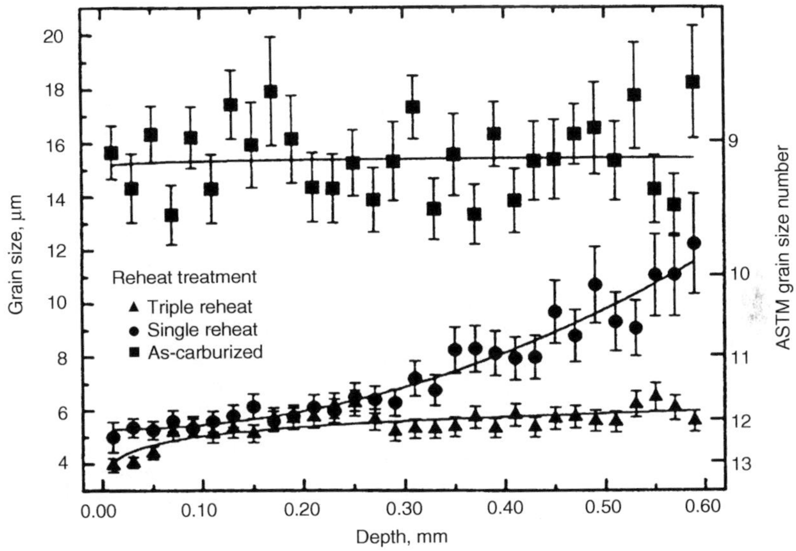

Fig. 75 Prior-austenite grain size as a function of depth from the surface of gas-carburized 4320 specimens in the as-carburized, direct-quenched condition and reheated conditions. Source: Ref 86

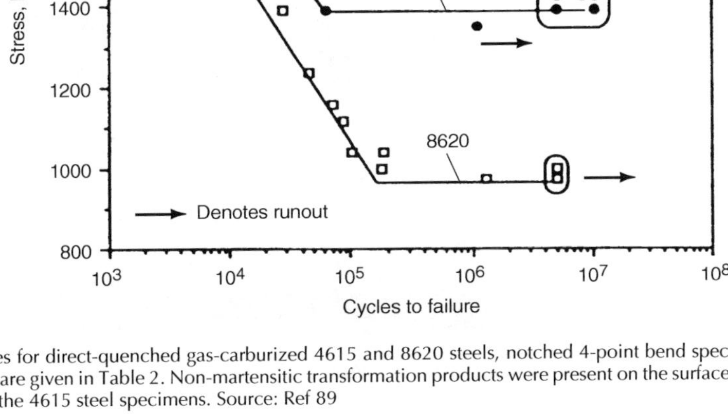

Fig. 76 *S-N* curves for direct-quenched gas-carburized 4615 and 8620 steels, notched 4-point bend specimens. Compositions of the steels are given in Table 2. Non-martensitic transformation products were present on the surfaces of the 8620 steel specimens and absent on the 4615 steel specimens. Source: Ref 89

therefore retained austenite remains an important component of the microstructure. At higher tempering temperatures, retained austenite transforms to cementite and ferrite with attendant decreases in hardness and strength as the martensitic microstructure coarsens.

The role that retained austenite plays in the bending fatigue performance of carburized steels has been difficult to identify because of the variable loading conditions that may be applied to carburized machine components and the complicating effects of other factors, such as

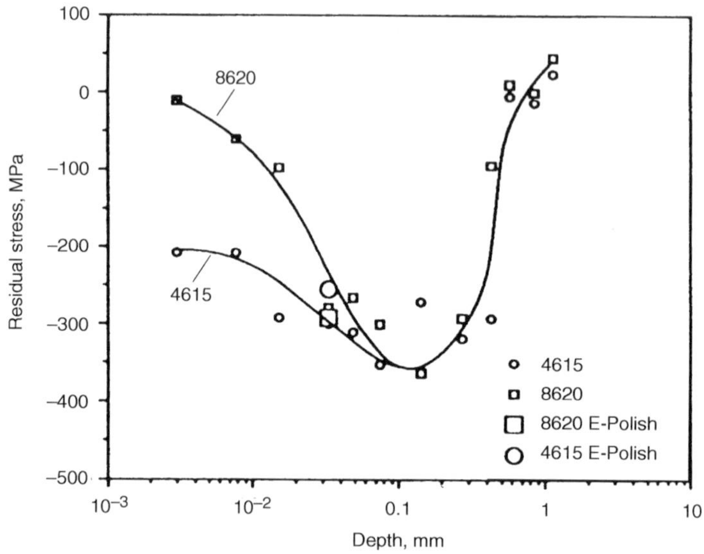

Fig. 77 Residual stress as a function of depth below the surface of the direct-quenched gas-carburized 4615 and 8620 steel specimens described in Fig. 76. Source: Ref 89

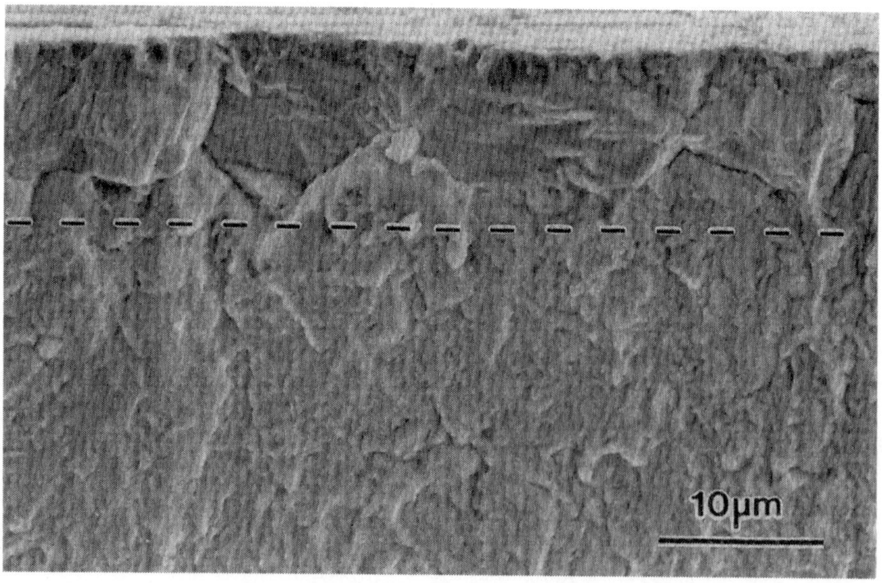

Fig. 78 Bending fatigue crack initiation in gas-carburized and reheated 4320 steel. The dashed line corresponds to maximum depth of surface oxidation, and all fracture below dashed line is transgranular. Source: Ref 86

residual stresses, grain boundary embrittling structures, and surface oxidation. With respect to loading conditions, it appears that higher amounts of retained austenite are detrimental to high-cycle fatigue and reduce endurance limits (Ref 15, 73, 77), while higher amounts of retained austenite are beneficial for low-cycle, high-strain fatigue (Ref 77, 90, 91).

Reduced retained austenite contents of LTT martensite/austenite composite microstructures increase elastic limits and yield strengths (Ref 92) and therefore benefit stress-controlled, high-cycle fatigue. One of the approaches to reducing the retained austenite content in the case microstructures of carburized steels, as noted previously, is to reheat carburized specimens to temperatures below the A_{cm} and quench to produce the type of microstructure shown in Fig. 39. Invariably such reheating and quenching significantly increases bending fatigue endurance limits compared to direct-quenched specimens of identically carburized specimens (Ref 22, 86). The reheating not only reduces retained austenite but also refines the austenitic grain size, refines the martensitic structure, and reduces susceptibility to intergranular fracture—all features that are known to improve fatigue resistance. Therefore, improved high-cycle fatigue resistance of reheated and quenched specimens is related to a combination of microstructural changes, including low retained austenite contents.

The benefit of retained austenite to strain-controlled, low-cycle bending fatigue is related to the improved ductility and reduced strength and hardness that retained austenite contributes to a composite LTT martensite/austenite case microstructure. In addition, retained austenite, at sufficiently high applied strains and stresses, undergoes deformation-induced transformation to martensite (Ref 93). The volume expansion associated with the strain-induced formation of martensite creates compressive stresses (Ref 38) that lead to reduced rates of fatigue crack growth, accounting for the enhanced low-cycle fatigue performance that is observed in carburized steels with high amounts of retained austenite in the case (Ref 91).

Subzero Cooling and Fatigue

Cooling carburized steel below room temperature is a processing approach sometimes used to reduce the retained austenite content in the case regions of carburized steels. The transformation of austenite to martensite is driven by temperature changes, and the low M_s temperatures of high-carbon case regions of carburized alloy steels limit the temperature range between M_s and room temperature over which martensite forms. Therefore, the temperature range for martensite formation and the reduction of retained austenite is extended by cooling below room temperature. The cooling treatments are variously referred to as subzero cooling, refrigeration treatments, or deep cooling.

In addition to the effects of retained austenite on bending fatigue, as discussed, any deformation-induced transformation of retained austenite during cyclic loading in service, because of the volume expansion that accompanies the transformation of austenite to martensite, may change the dimensions of a carburized component. Therefore, subzero cooling is one approach to reduce retained austenite in parts that require high precision and stable dimensions throughout their service life. However, several studies show that subzero cooling lowers the bending fatigue resistance of carburized steels. Nevertheless, high-quality, high-performance aircraft and helicopter gears are routinely subjected to subzero cooling without apparent detrimental effects (Ref 94). For example, a commonly used carburizing steel for aircraft gears is 9310, which contains about 3 wt% nickel (see Table 2). The high nickel content lowers the M_s temperature and increases the amount of austenite at room temperature. The austenite content can be reduced by subzero cooling, probably with adverse effects on localized residual stress, as discussed subsequently. However, the latter adverse effect of subzero cooling may be offset by fine austenite grain size, and high nickel content may improve the fracture toughness and fatigue resistance of carburized steels (Ref 46, 47, 95) to a level where fatigue resistance is not adversely affected by subzero cooling.

In the low-alloy steels commonly used for carburizing, subzero cooling used to reduce retained austenite may reduce the bending fatigue resistance of carburized components. If refrigeration treatments are applied, parts should be tempered both before and after. Figure 79 shows an example of the detrimental effects of subzero cooling on the bending fatigue resistance of carburized specimens. The data were produced in an experimental study of vacuum-carburized specimens of 8620 and EX 24 steel that were deep cooled to –196 °C (–321 °F) in liquid nitrogen (Ref 23). The overall fatigue performance in this study was complicated by high retained austenite contents and coarse carbide particles at square specimen corners, but the detrimental effect of subzero cooling on bending fatigue performance is clearly demonstrated in Fig. 79.

The detrimental effect of subzero cooling on the bending fatigue of carburized specimens has been related to changes in residual stress by several investigations (Ref 62, 63, 96). The overall surface residual stresses become increasingly compressive, as measured from the martensite in the case and as expected from the constraint of the expansion that accompanies the transformation of austenite to martensite as temperature is decreased. However, the residual stresses in the austenite phase are measured to be tensile, especially at the surface of the carburized specimens. These tensile stresses would then be expected to lower the surface tensile stresses applied in bending to initiate fatigue cracks. Microcrack formation within martensite plates and at plate/austenite interfaces (Ref 48) may be enhanced by the localized residual stresses induced by subzero cooling (Ref 96), but they could be minimized by maintaining a fine prior-austenite grain size and applying reheating treatments (Ref 50, 51).

Residual Stresses, Shot Peening, and Fatigue

Compressive residual stresses are formed in the case microstructures of carburized steels as a result of transformation and temperature gradients induced by quenching (Ref 59, 60). The magnitude and distribution of the residual stresses therefore are complex functions of the temperature gradients induced by quenching (Ref 97), which in turn are dependent on specimen size and geometry, the hardenability of the steel, the carbon gradient, and the case depth. The residual stresses as a function of case depth are routinely measured by x-ray diffraction, and considerable effort has been applied to modeling residual stress profiles in carburized steels as a function of cooling and hardenability (Ref 56, 98, 99).

Figure 80 shows the range and pattern of compressive residual stresses typically formed in the case regions of direct-cooled carburized steels (Ref 24). The compressive residual stresses offset the adverse effects of factors such as quench embrittlement and intergranular fracture to which high-carbon microstructures are susceptible (Ref 83), and they increase the fracture and fatigue resistance of direct-quenched parts to levels that provide good engineering performance. As discussed earlier, case residual stresses in carburized steels are adversely modified by subzero cooling, and they are positively modified (made locally more compressive) by the strain-induced transformation of austenite to martensite. Tempering lowers residual compressive stresses because of dimensional changes that accompany the recovery and coarsening of the martensitic microstructure during tempering (Ref 100).

Shot peening is an effective way to increase the case compressive residual stresses in carburized steels (Ref 57, 101, 102) and as a result improve the bending fatigue performance. Shot peening causes deformation-induced transformation of case retained austenite, and the constraint of the associated volume expansion causes the development of additional compressive stresses. Figures 40, 41, and 81 show, rela-

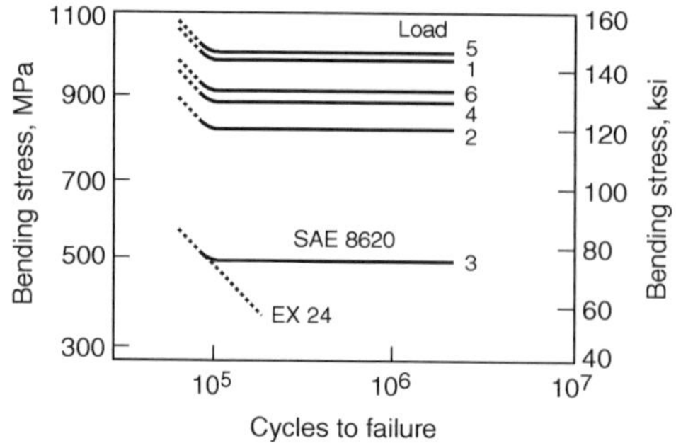

Fig. 79 *S-N* curves of vacuum-carburized 8620 and EX 24 (0.89% Mn, 0.24% Mo, 0.55% Cr) steels. The lower curves were obtained from specimens subzero cooled to −196 °C, and the upper curves were obtained from specimens not subjected to subzero cooling. Source: Ref 23

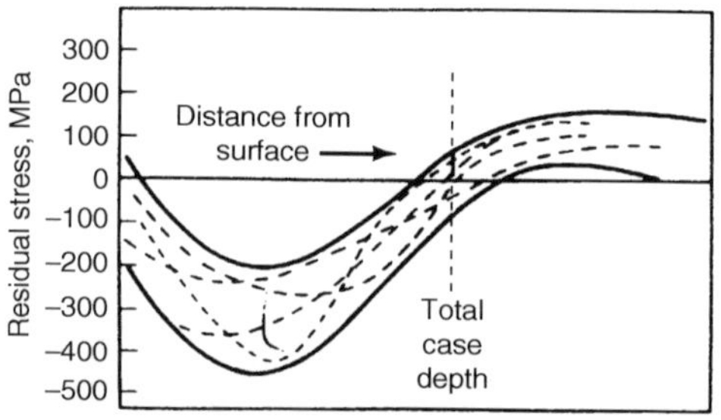

Fig. 80 Ranges and patterns of residual stresses as a function of depth for 70 carburized steels. Source: Ref 24

tive to unpeened specimens, the decrease in retained austenite, the increase in the case compressive stresses, and the increased bending fatigue performance, respectively, that are associated with shot peening of direct-quenched carburized 4320 specimens (Ref 25).

Other Properties of Interest

Although much of the recent research on properties of carburized steels has centered around bending fatigue (see previous section), there are other properties that affect the service life of carburized components. As will be described forthwith, these mechanical properties are strongly influenced by core and case microstructure, case depth, residual stresses, and alloy chemistry. Additional property data for carburized steels may be found in Ref 36.

Rolling Contact Fatigue

Rolling contact fatigue is a surface-pitting-type failure commonly found in ball or roller bearings and gears (Ref 103, 104). Rolling contact fatigue differs from classic structural fatigue (bending or torsional) in that it results from contact or Hertzian stress state. This localized stress state results when curved surfaces are in contact under a normal load. Generally, one surface moves over the other in a rolling motion as in a ball rolling over a race in a ball bearing. The contact geometry and the motion of the rolling elements produces an alternating subsurface shear stress. Subsurface plastic strain builds up with increasing cycles until a crack is generated. The crack then propagates until a pit is formed. Once surface pitting has initiated, the bearing becomes noisy and rough running. If allowed to continue, fracture of the rolling element and catastrophic failure occurs. Fractured races can result from fatigue spalling and high hoop stresses.

Extreme cases of spalling are associated with case crushing or cracking initiated at the case-core interface. Figure 82 shows an example of a spall on a carburized SAE 4118 interface. If sliding is coupled with contact loading, surface pits develop. Very high contact loads cause microstructural changes within high-carbon martensite that are revealed by various types of etching (Ref 105–107). Generally retained austenite is regarded as a microstructural constituent that is beneficial for rolling-contact fatigue resistance (Ref 108, 109).

Wear Resistance

Another important property afforded by the carburizing process is wear resistance resulting from the high hardness of the case. As with bending fatigue, there are a number of microstructure/property relationships that must be considered for wear-resistant applications as well as factors associated with steel selection.

Microstructural Features. The independent variables available for controlling the microstructure/properties of carburized cases

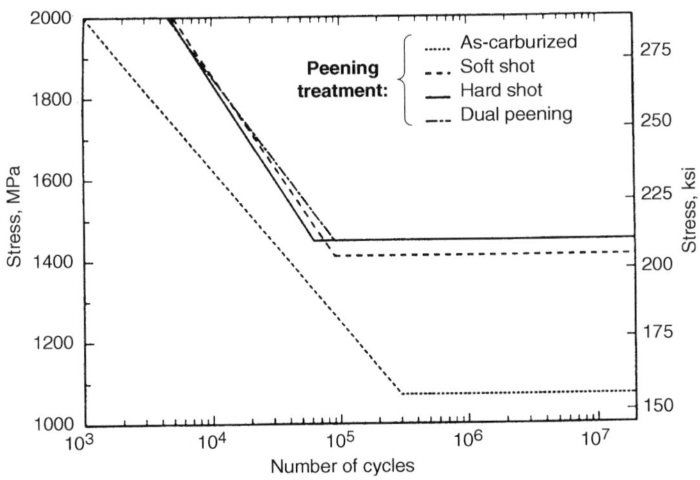

Fig. 81 *S-N* curves for gas-carburized 4320 specimens in the as-carburized, direct-quenched and various shot peened conditions after direct quenching. Source: Ref 25

are those that define the carburizing alloy (composition, cleanliness) and those that define the carburizing process (time/temperature/carbon-potential carburizing history, time/temperature quenching history, time/temperature tempering history). These tools provide a considerable degree of control over these microstructural features:

- Martensite
  a. Carbon content of source austenite
  b. Plate size (austenite grain size)
  c. Strength
  d. Secondary hardening
- Primary carbides
  a. Size
  b. Volume fraction
- Retained austenite
  a. Volume fraction
  b. Carbon content
- Nonmetallic inclusions

and these global features:

- Case depth
- Residual stress distribution

which determine the tribological properties of the case. The combination of properties that is best for each application must then be decided.

The necessary case depth and case hardness can be estimated from a Hertzian stress calculation, but other microstructural objectives can be specified only qualitatively. For many applications, the following "rules of thumb" apply:

- Sufficient case depth and case hardness must be provided to prevent indentation or case crushing under the anticipated contact loads. For gears and bearings loaded in "line contact," a minimum case hardness of 58 HRC frequently is specified. When high contact loads are accompanied by sliding, the near-surface hardness (to a depth of about 50 μm, or 2 mils) may have to be raised to prevent shearing of surface layers.
- The retained austenite content should be as high as possible, consistent with the requirements of the previous rule. The retained austenite content should be controlled by adjusting the case carbon content, not by subzero quenching after carburizing or by tempering at temperatures above 200 °C (390 °F).
- The tempering temperature chosen should be as low as possible, but above the surface temperatures anticipated in finishing operations and in service.
- The content of nonmetallic inclusions should be no higher than that needed for economical machining.
- Coarse primary carbides can be helpful in resisting abrasive wear. Fine primary carbides can permit more retained austenite at the same hardness level. Experiments should be conducted to verify any benefits presumed to be associated with primary carbides.

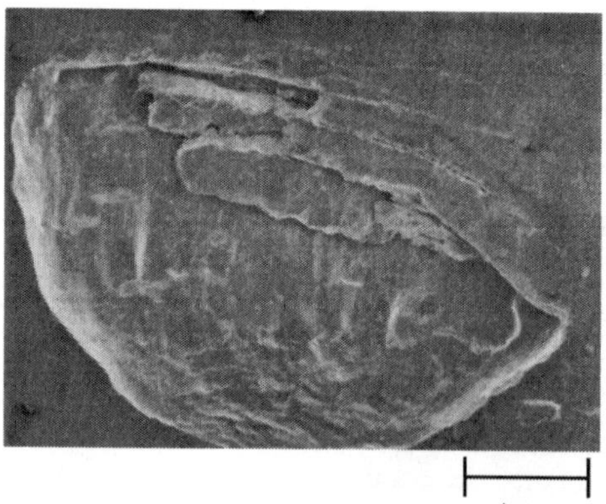

Fig. 82 An example of spalling in carburized SAE 4118 steel subjected to rolling contact loading

Steel Cleanliness. For the best resistance to rolling contact fatigue (spalling), the content of aluminate, silicate, and globular oxide inclusions (Types B, C, and D, respectively, in the Jernkontoret system, ASTM E 45) must be as low as possible. Manganese sulfide inclusions (Type A) are generally not regarded as detrimental to rolling contact fatigue life (Ref 16).

The inclusion standards specified in ASTM A 534, "Carburizing Steels for Anti-Friction Bearings," have been steadily tightened since the late 1960s, reflecting improvements in steelmaking. Some individual steel suppliers claim to be able to furnish premium-quality carburizing steels with oxygen contents below 15 ppm, titanium contents below 30 ppm, and inclusion ratings considerably better than ASTM A 534. Calcium treatment of bearing-quality steels to modify aluminates, thereby improving machinability, is usually avoided because of the possibility of forming large inclusions.

When sliding is combined with rolling contact, near-surface inclusions (including oxides formed in grain boundaries during heat treatment) promote pitting (Ref 16). The role of inclusions in most forms of sliding wear is not as well defined as their role in rolling contact fatigue, probably because the conditions that are possible at a sliding interface are more diverse and more difficult to characterize than at a rolling contact interface. Some insight into possible effects of inclusions on sliding wear comes from the machining literature (Ref 16). It is known that some inclusions in a steel workpiece can promote tool wear during machining, whereas others can reduce wear.

Steel Hardenability. The alloy content of carburizing steels is usually selected on the basis of hardenability. If the application involves high contact loads (roller bearings, for example), the uncarburized core must be martensitic to prevent the subcase from yielding. An alloy that allows the part to attain full hardness (through-harden) in whatever quenchant is employed will be selected. The selection of an alloy with sufficient core hardenability almost always assures sufficient case hardenability. When contact loads are well within the capability of the case to support them, it is often neither necessary nor desirable for the core microstructure to be martensitic. Shape distortion during quenching, for example, is usually reduced if the core transforms at a relatively high temperature to a nonmartensitic structure. For such parts, the alloying need only be sufficient to ensure case hardenability.

Special Alloy Considerations. Several secondary hardening carburizing alloys have been developed for applications that require resistance to elevated temperatures, such as helicopter gearing and rock drill bits (see the "Special alloys" listed in Table 2). These alloys make use of the precipitation of copper and/or M_2C and MC carbides to provide resistance to softening for temperatures up to 550 °C (1020 °F). Because they contain substantial amounts of Mo and V, these alloys resemble low-carbon versions of tool steels. Some of the alloys are difficult to carburize because of high Si and Cr contents; preoxidation prior to carburizing is necessary to permit carbon penetration (Ref 16). Secondary hardening alloys could also be useful in ambient-temperature applications in which lubrication is marginal, because the heat generated by intermittent metal-to-metal contact would not readily soften the underlying metal.

Hot Hardness

Retention of hardness at elevated temperatures (hot hardness) is vitally important in applications where high local temperature conditions can be encountered. Examples include helicopter gears, speed reduction gear sets, and turbine gearing. Figure 83 provides hot hardness data for several carburized steels. It is evident from these data that higher alloy content is needed to assure sufficient hardness at temperatures above 315 °C (600 °F) encountered in severe service.

Bending Strength and Bend Ductility

In service, carburized and hardened steels are subjected to bending loads and must be able to resist design loads and overloads without fracture. A variety of laboratory tests have been performed to define the resistance of carburized and hardened steels to failure under bending loads, and to provide information on the contribution of alloys in resisting failure.

Bend ductility of several carburized and hardened steels over a range of test temperatures from room temperature to –195 °C (–320 °F) is given in Fig. 84. A comparison of the data for carburized SAE 4817 steel with those for

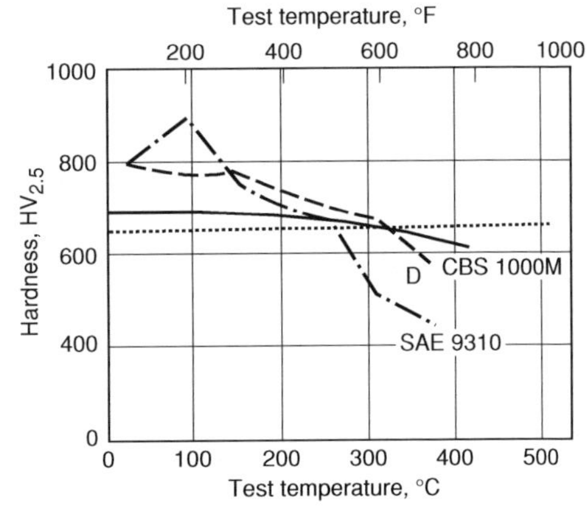

Fig. 83 Hot hardness of three carburized steels. The dashed line corresponds to a surface hardness of 58 HRC. Compositions for SAE 9310 and CBS 1000M are listed in Table 2. The nominal composition for steel D is 0.12%C, 0.5% Mn, 1.1% Si, 1.0% Cr, 2.0% Ni, 2.3% Mo, and 1.2% V. Source: Ref 36

SAE 4027 (both with about the same amount of retained austenite) shows that the steel alloyed with a substantial amount of nickel exhibited much greater ductility.

Razim (Ref 110) has observed that the static bend test is useful in evaluating the ability of a carburized and hardened surface zone to sustain plastic deformation without cracking. A suitable measurement can be the initial crack strength. His summary of such tests indicates that with low surface carbon contents of about 0.6% carbon, the initial crack strength increases with increasing core strength; while with high surface carbon contents (about 1.2% C), the initial crack strength becomes less dependent on core strength, but drops to a considerably lower level than the crack strength exhibited by the steels with 0.6% carbon at the surface.

Impact Toughness

Various studies of single-blow impact tests—whether with notched or unnotched Charpy specimens, or specimens designed to simulate gears or other actual components—indicate that carburized and hardened steels exhibit considerable resistance to impact, despite the high hardness of their cases. The data consistently demonstrate that steels containing substantial amounts of nickel and smaller but controlled amounts of molybdenum exhibit greater energy absorption in single-blow impact toughness tests.

The Charpy V-notch impact toughness of two carburized nickel steels is shown in Fig. 85. Test bars of SAE 8620 (0.40–0.70% Ni) and 9310 (3.0–3.5% Ni) steels were carburized to 0.38 mm (0.015 in.) case depth and heat treated to similar case and core hardness levels. Heat treatment conditions were:

- SAE 8620: Carburized at 860 °C (1580 °F) for 1.7 h, quenched in agitated oil at 65 °C (150 °F), then tempered at 205 °C (400 °F). Core hardness was 40 HRC; case hardness was 57 HRC.
- SAE 9310: Carburized at 870 °C (1600 °F) for 1.7 h, quenched in agitated oil at 65 °C (150 °F), refrigerated at –80 °C (–110 °F), then tempered at 205 °C (400 °F). Core hardness was 39 HRC; case hardness was 56 HRC.

The maximum absorbed energy of the carburized 9310 steel is higher than that of the carburized 8620 steel, and the impact transition temperature of the carburized 9310 steel is more than 97 °C (175 °F) below that of the carburized 8620 steel, despite the relative severity of the test.

Figure 86 shows the results of impact fracture tests of simulated gear specimens that were

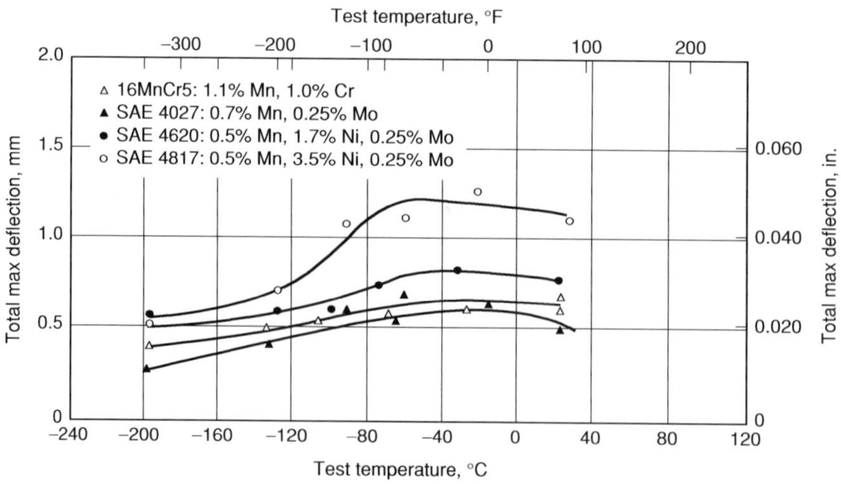

Fig. 84 Bend ductility transition curves for carburized and hardened steels. Nominal alloy contents of the steels are listed within the diagram. Source: Ref 36

loaded in a standard pendulum type testing machine, with the specimen held in a vertical position, similar to an Izod impact test. An instrumented tup permitted recording the energy absorbed by the specimen as a function of time during the test. For carburized and hardened steels, the investigators found that alloy content and core carbon content significantly influenced resistance to fracture under impact bending conditions. In Fig. 86, the Cr-Mo steels exhibited higher fracture strengths than the Mn-Cr steels, but the Ni-Cr-Mo steel, PS55, not only exhibited much higher fracture strength, but that fracture strength did not decrease with

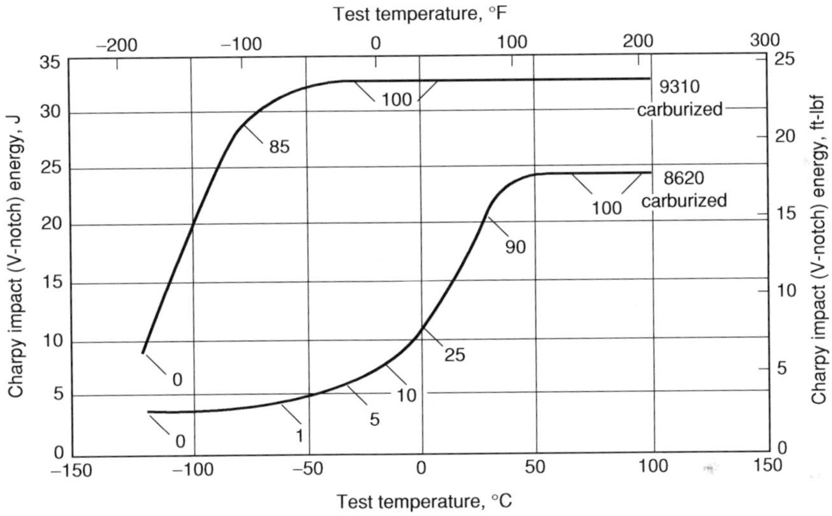

Fig. 85 Charpy V-notch toughness behavior of carburized SAE 8620 and 9310 steels with 0.38 mm (0.015 in.) case depth. Numbers adjacent to curves are percent fibrous fracture. Specimens were finished and notched before carburizing. Source: Ref 36

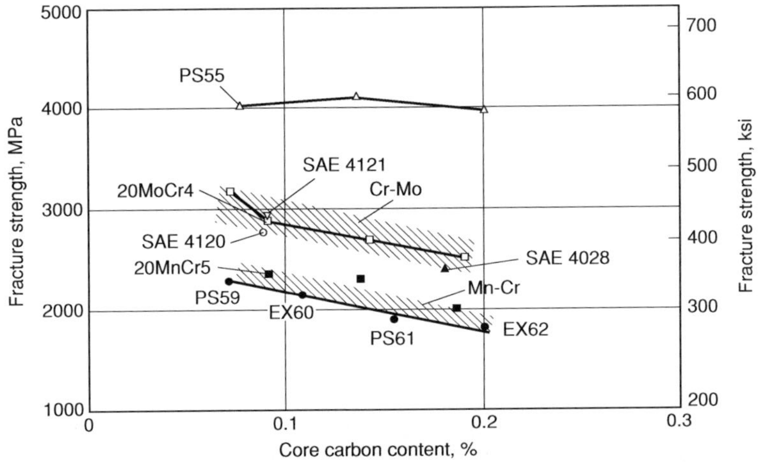

Fig. 86 Effect of core carbon content and alloy content on impact fracture strength of a series of steels carburized at 925 °C (1700 °F), cooled to 840 °C (1550 °F), and oil quenched and tempered at 150 °C (340 °F). Source: Ref 36

increasing carbon content as it did with the other test specimens.

Fracture Toughness

The fracture toughness properties of carburized steels have largely been inferred from measurements on through-hardened steels of medium- or high-carbon content, although some measurements of fracture toughness have been made on carburized specimens (Ref 111, 112). For example, Diesburg (Ref 111) studied the fracture toughness of several steels carburized to produce case depths (at 0.5% C) between 0.75 and 1.0 mm (0.03 and 0.04 in.). Fracture toughness was determined as a function of distance from the carburized and hardened surface using Charpy impact specimens with fatigue precracks generated first by electrodischarge machining a notch, then by propagating the crack beyond the notch about 0.13 mm (0.005 in.) by fatiguing the specimen in conventional high-cycle fatigue equipment.

Once precracked, the specimens were broken in slow-bend tests. The fracture load and crack length were used to calculate fracture toughness as described in ASTM E 399. Each data point in Fig. 87 represents one specimen with the tip of the precrack at the indicated depth below the carburized surface.

In Fig. 87, the fracture toughness values for the higher nickel steels SAE PS55, PS32, 9310, and 4820 are shown to be quite similar and much higher than the values observed for the lower nickel SAE 8620 steel at all depths below the surface. Close to the carburized surface, very little difference could be observed among the steels tested.

ACKNOWLEDGMENTS

Portions of this chapter were adapted from:

- C.A. Stickels, Gas Carburizing, *Heat Treating,* Vol 4, *ASM Handbook,* ASM International, 1991, p 312–324
- G. Krauss, Microstructures and Properties of Carburized Steels, *Heat Treating,* Vol 4, *ASM Handbook,* ASM International, 1991, p 363–375
- Evaluation of Carbon Control in Processed Parts, *Heat Treating,* Vol 4, *ASM Handbook,* ASM International, 1991, p 587–600

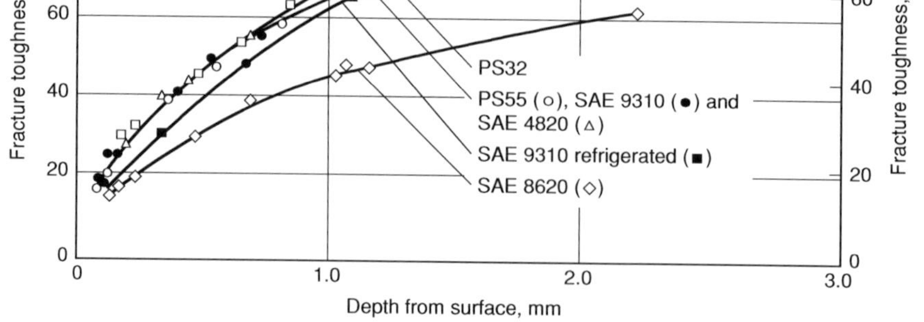

Fig. 87 Fracture toughness in carburized steels as a function of distance below the surface. The SAE PS55, 9310, and 8620 steels were commercial heats; the SAE PS32 and 4820 steels were laboratory heats. The PS32 and 4820 steels were quenched directly after carburizing at 925 °C (1700 °F) into 170 °C (340 °F) oil; other steels were cooled from 925 °C to 840 °C (1550 °F) before quenching into 65 °C (150 °F) oil. Data are also shown for 9310 steel that was refrigerated after quenching and before tempering. Source: Ref 111

- G. Krauss, Martensite, Austenite, Austenite and Fatigue, and Oxidation and Inclusions, a 4-part series of articles published in *Advanced Materials & Processes,* May, July, Sept, and Dec 1995
- G. Krauss, Bending Fatigue of Carburized Steels, *Fatigue and Fracture,* Vol 19, *ASM Handbook,* ASM International, 1996, p 680–690
- C.A. Stickels, Carburizing, *Friction, Lubrication, and Wear Technology,* Vol 18, *ASM Handbook,* 1992, p 873–877

REFERENCES

1. C.A. Stickels, C.M. Mack, and J.A. Pieprzak, *Metall. Trans. B.* Vol 13, 1982, p 613–623
2. Y. Shimosato, *Heat Treatment and Surface Engineering: New Technology and Practical Applications, Proceedings of the Sixth International Congress on Heat Treatment of Materials.* G. Krauss, Ed., ASM International, 1988, p 401–404
3. F.E. Harris, *Met. Prog.,* Vol 44, 1943, p 265–272
4. T.J. Schultz, T.J. Kuhn, and D.E. Goodman, Carburizing—Stepping into the Future, *Carburizing: Processing and Performance, Proceedings of an International Conference,* G. Krauss, Ed., ASM International, 1989
5. J. Crank, *The Mathematics of Diffusion,* 2nd ed., Oxford University Press, 1975
6. G.L. Tibbetts, *J. Appl. Phys.,* Vol 51, 1980, p 4813–4816
7. B. Uhrenius, *Scand. J. Metall.,* Vol 6, 1977, p 83–89
8. C.A. Stickels, *J. Heat Treat.,* Vol 1, 1979, p 31–41
9. D.R. Stull and H. Prophet, *JANAF Thermochemical Tables,* 2nd ed., NSRDS-NBS 37, U.S. Department of Commerce, June 1971
10. D.J. Bradley, J.M. Leitnaker, and F.H. Horne, *High Temp. Sci.,* Vol 12, 1980, p 1–10
11. A.R. Marder and G. Krauss, *Trans. ASM,* Vol 60, 1967, p 651–660
12. R.L. Brown, H.J. Rack, and M. Cohen, *Mat. Sci. Eng.,* Vol 21, 1975, p 25–34
13. E.S. Rowland and S.R. Lyle, *Trans. ASM,* Vol 37, 1946, p 27–47
14. G. Krauss, Microstructure and Performance of Carburized Steel, Part I: Martensite, *Adv. Mater. Process.,* May 1995, p 40Y–40BB
15. J.L. Pacheco and G. Krauss, Microstructure and High Bending Fatigue Strength in Carburized Steel, *J. Heat Treat.,* Vol 7 (No. 2), 1989, p 77–86
16. C.A. Stickels, Carburizing, *Friction, Lubrication, and Wear Technology,* Vol 18, *ASM Handbook,* ASM International, 1992, p 873–877
17. D.L. Williamson, K. Nakazawa, and G. Krauss, A Study of the Early Stages of Tempering in an Fe-1.2% C Alloy, *Metall. Trans. A,* Vol 10, 1979, p 1351–1363
18. R.A. Grange, C.R. Hribal, and L.F. Porter, Hardness of Tempered Martensite in Carbon and Low-Alloy Steels, *Metall. Trans. A,* Vol. 8, 1977, p 1775–1785
19. G. Krauss, Microstructure and Performance of Carburized Steel, Part II: Austenite, *Adv. Mater. Process.,* July 1995, p 48U–48Y
20. K.W. Andrews, Empirical Formulae for the Calculation of Some Transformation Temperatures, *J. Iron Steel I.,* Vol 203, 1965, p 721–727
21. D.P. Koistinen and R.E. Marburger, A General Equation Prescribing the Extent of the Austenite-Martensite Transformation in Pure Iron-Carbon Alloys and Plain Carbon Steels, *Acta Metall.,* Vol 7, 1959, p 59–60
22. C.A. Apple and G. Krauss, Microcracking and Fatigue in a Carburized Steel, *Metall. Trans.,* Vol 4, 1973, p 1195–1200
23. K.D. Jones and G. Krauss, Effects of High-Carbon Specimen Corners on Microstructure and Fatigue of Partial Pressure Carburized Steels, *Heat Treatment '79,* The Metals Society, London, 1979, p 188–193
24. G. Parrish and G.S. Harper, *Production Gas Carburizing,* Pergamon Press Inc., 1985
25. J.A. Sanders, "The Effects of Shot Peening on the Bending Fatigue Behavior of a Carburized SAE 4320 Steel," M.S. thesis, Colorado School of Mines, 1993
26. D.V. Doane and J.S. Kirkaldy, Ed., *Hardenability Concepts with Applications to Steel.* American Institute of Mining, Metallurgical, and Petroleum Engineers, 1978

27. C.A. Siebert, D.V. Doane, and D.H. Breen, *The Hardenability of Steels—Concepts, Metallurgical Influences, and Industrial Applications.* American Society for Metals, 1977
28. H. Treppschuh and R. Randak, Verfahren zur Herstellung Besonders Zaher, Borhaltiger Stahle, West German Patent 1608632, 1969
29. W.T. Cook, The Effect of Aluminum Treating on the Case-Hardening Response of Plain Carbon Steels, *Heat Treat. Met.,* Vol 11 (No. 1), 1984, p 21–23
30. D.V. Doane, Carburized Steel—Update on a Mature Composite, *Carburizing: Processing and Performance,* G. Krauss, Ed., ASM International, 1989, p 169–190
31. C.F. Jatczak, Hardenability of High Carbon Steels, *Metall. Trans.,* Vol 4, 1973, p 2267–2277
32. D.V. Doane and A.T. DeRetana, Predicting Hardenability of Carburizing Steels, *Met. Prog.,* Vol 100 (No. 3), 1971, p 65–69
33. J.M. Tartaglia and G.T. Eldis, Core Hardenability Calculations for Carburizing Steels, *Metall. Trans. A,* Vol 15, 1984, p 1173–1183
34. D.H. Breen, G.H. Walter, C.J. Keith, and J.T. Sponzilli, Computer Based System Selects Optimum Cost Steels, article series, *Met. Prog.,* Dec 1972, and Feb, April, June, Dec 1973
35. U. Wyss, Kohlenstoff und Harteverlauf in der Einsatzhartungsschicht Verschieden Legierter Einsatzstahle, *Härt.-Tech. Mitt.,* Vol 43, 1988, p 27–35; J.A. Halgren and E.A. Solecki, "Case Hardenability of SAE 4028, 8620, 4620, and 4815 Steels," Technical Paper 149A, Society of Automotive Engineers, 1960
36. *Modern Carburized Nickel Alloy Steels, Reference Book Series 11,005,* Nickel Development Institute, 1989
37. G. Krauss, *Steels: Heat Treatment and Processing Principles,* ASM International, 1990
38. G. Krauss, The Microstructure and Fracture of a Carburized Steel, *Metall. Trans. A,* Vol 9, 1978, p 1527–1535
39. G. Krauss, The Relationship of Microstructure to Fracture Morphology and Toughness of Hardened Hypereutectoid Steels, *Case Hardened Steels: Microstructure and Residual Stress Effects,* TMS/AIME, 1984, p 33–56
40. H.K. Obermeyer and G. Krauss, Toughness and Intergranular Fracture of a Simulated Carburized Case in EX-24 Type Steel, *J. Heat Treat.,* Vol 1 (No. 3), 1980, p. 31–39
41. T. Ando and G. Krauss, The Effect of Phosphorus Content on Grain Boundary Cementite Formation in AISI 52100 Steel, *Metall. Trans. A,* Vol 12, 1981, p 1283–1290
42. J.I. Goldstein and H. Yakowitz, *Practical Scanning Electron Microscopy,* Plenum Press, 1975, p 87–91
43. H. Ohtani and C.J. McMahon, Jr., Modes of Fracture in Temper Embrittled Steels, *Acta Metall.,* Vol 23, 1975, p 337–386
44. T. Ando, "Isothermal Growth of Grain Boundary Allotriomorphs of Cementite in Ternary Fe-C-Cr Austenite," Ph.D. thesis, Colorado School of Mines, 1982
45. T. Ando and G. Krauss, The Isothermal Thickening of Cementite Allotriomorphs in a 1.5Cr-1C Steel, *Acta Metall.,* Vol 29, 1981, p 351–363
46. D. Wicke and J. Grosch, Das Festigkeitsverhalten von Legierten Einsatzstahlen bei Sclagbeanspruchung, *Härt.-Tech. Mitt.,* Vol 32, 1977, p 223–233
47. B. Thoden and J. Grosch, Crack Resistance of Carburized Steel under Bend Stress, *Carburizing: Processing and Performance,* G. Krauss, Ed., ASM International, 1989, p 303–310
48. A.R. Marder, A.O. Benscoter, and G. Krauss, Microcracking Sensitivity in Fe-C Plate Martensite, *Metall. Trans.,* Vol 1, 1970, p 1545–1549
49. A.R. Marder and A.O. Benscoter, Microcracking in Fe-C Acicular Martensite, *Trans. ASM,* Vol 61, 1968, p 293–299
50. M.G. Mendiratta, J. Sasser, and G. Krauss, Effect of Dissolved Carbon on Microcracking in Martensite of an Fe-1.39 pct C Alloy, *Metall. Trans.,* Vol 3, 1972, p 351–353
51. R.P. Brobst and G. Krauss, The Effect of Austenite Grain Size on Microcracking in Martensite of an Fe-1.22C Alloy, *Metall. Trans.,* Vol 5, 1975, p 457–462
52. A.H. Rauch and W.R. Thurtle, Microcracks in Case Hardened Steel, *Met. Prog.,* Vol 69, 1956, p 73–76
53. L. Jena and P. Heich, Microcracks in Carburized and Hardened Steel, *Metall. Trans.,* Vol 3, 1972, p 588–590

54. K.D. Jones and G. Krauss, Microstructure and Fatigue of Partial Pressure Carburized SAE 8620 and EX24 Steels, *J. Heat Treat.,* Vol 1 (No. 1), 1979, p 64–71
55. R.F. Kern, Super Carburizing, *Heat Treat.,* Oct 1986, p 36–38
56. T. Ericsson, S. Sjostrom, M. Knuuttila, and B. Hildenwall, Predicting Residual Stresses in Cases, *Case-Hardened Steels: Microstructural and Residual Stress Effects,* D.E. Diesburg, Ed., TMS/AIME, 1984, p. 113–139
57. B. Scholtes and E. Macherauch, Residual Stress Determination, *Case-Hardened Steels: Microstructural and Residual Stress Effects,* D.E. Diesburg, Ed., TMS/AIME, 1984, p 141–151
58. J.A. Burnett, Prediction of Residual Stresses Generated during Heat Treating of Case Carburized Parts, *Residual Stresses for Designers and Metallurgists,* American Society for Metals, 1981, p 51–69
59. L.J. Ebert, The Role of Residual Stresses in the Mechanical Performance of Case Carburized Steel, *Metall. Trans. A,* Vol 9, 1978, p 1537–1551
60. D.P. Koistinen, The Distribution of Residual Stresses in Carburized Steels and Their Origin, *Trans. ASM,* Vol 50, 1938, p 227–241
61. B. Hildenwall and T. Ericsson, Residual Stresses in the Soft Pearlite Layer of Carburized Steel, *J. Heat Treat.,* Vol 1 (No. 3), 1980, p 3–13
62. C. Kim, D.E. Diesburg, and R.M. Buck, Influence of Sub-Zero and Shot-Peening Treatment on Impact and Fatigue Fracture Properties of Case-Hardened Steels, *J. Heat Treat.,* Vol 2 (No. 1), 1981, p 43–53
63. M.A. Panhans and R.A. Fournelle, High Cycle Fatigue Resistance of AISI E9310 Carburized Steel with Two Different Levels of Surface Retained Austenite and Surface Residual Stress, *J. Heat Treat.,* Vol 2 (No. 1), 1981, p 55–61
64. R. Chatterjee-Fischer, Internal Oxidation during Carburizing and Heat Treating, *Metall. Trans. A,* Vol 9, 1978, p 1553–1560
65. I.S. Kozlovskii, A.T. Kalinin, A.J. Novikova, E.A. Lebedeva, and A.I. Festanova, Internal Oxidation during Case-Hardening of Steels in Endothermic Atmospheres, *Met. Sci. Heat Treat.,* No. 3, 1967, p 157–161
66. C. Van Thyne and G. Krauss, A Comparison of Single Tooth Bending Fatigue in Boron and Alloy Carburizing Steels, *Carburizing: Processing and Performance,* G. Krauss, Ed., ASM International, 1989, p 333–340
67. R.E. Cohen, P.J. Haagensen, D.K. Matlock, and G. Krauss, "Assessment of Bending Fatigue Limits for Carburized Steel," Technical Paper 910140, SAE International, 1991
68. H.O. Fuchs and R.I. Stephens, *Metal Fatigue in Engineering,* John Wiley & Sons, 1980
69. R.W. Landgraf and R.H. Richman, Fatigue Behavior of Carburized Steel, *Fatigue of Composite Materials,* STP 569, American Society for Testing and Materials (ASTM), 1975, p 130–144
70. H. Brugger and G. Kraus S, Influence of Ductility on the Behavior of Carburizing Steel during Static and Dynamic Bend Testing, *Arch. Eisenhuttenwes.,* Vol 32, 1961, p 529–539
71. R.E. Cohen, D.K. Matlock, and G. Krauss, Specimen Edge Effects on Bending Fatigue of Carburized Steel, *J. Mater. Eng. Perform.,* Vol 1 (No. 5), 1992, p 695–703
72. D.H. Breen and E.M. Wene, Fatigue in Machines and Structures—Ground Vehicles, *Fatigue and Microstructure,* American Society for Metals, 1979, p 57–99
73. Z. Zurn and C. Razim, On the Fatigue Strength of Case Hardened Parts, *Carburizing: Processing and Performance,* G. Krauss, Ed., ASM International, 1989, p 239–248
74. M.B. Slane, R. Buenneke, C. Dunham, M. Semenek, M. Shea, and J. Tripp, "Gear Single Tooth Bending Fatigue," Technical Paper 821042, SAE International, 1982
75. D. Medlin, G. Krauss, D.K. Matlock, K. Burris, and M. Slane, "Comparison of Single Gear Tooth and Cantilever Beam Bend Fatigue Testing of Carburized Steel," Technical Paper 950212, SAE International, 1995
76. R.S. Hyde, R.E. Cohen, D.K. Matlock, and G. Krauss, "Bending Fatigue Crack Characterization and Fracture Toughness of Gas Carburized SAE 4320 Steel,"

Technical Paper 920534, SAE International, 1992
77. M.A. Zaccone, J.B. Kelley, and G. Krauss, Strain Hardening and Fatigue of Simulated Case Microstructures in Carburized Steel, *Carburizing: Processing and Performance,* G. Krauss, Ed., ASM International, 1989, p 249–265
78. M.M. Shea, "Impact Properties of Selected Gear Steels," Technical Paper 780772, SAE International, 1978
79. R.S. Hyde, "Quench Embrittlement and Intergranular Oxide Embrittlement: Effects on Bending Fatigue Initiation of Gas-Carburized Steel," Ph.D. dissertation, Colorado School of Mines, 1994
80. L. Magnusson and T. Ericsson, Initiation and Propagation of Fatigue Cracks in Carburized Steel, *Heat Treatment '79,* The Metals Society, London, 1979, p 202–206
81. K.A. Erven, D.K. Matlock, and G. Krauss, Effect of Sulfur on Bending Fatigue of Carburized Steel, *J. Heat Treat.,* Vol 9, 1991, p 27–35
82. R.S. Hyde, G. Krauss, and D.K. Matlock, Phosphorus and Carbon Segregation: Effects on Fatigue and Fracture of Gas-Carburized Modified 4320 Steel, *Metall. Trans. A,* Vol 25, 1994, p 1229–1240
83. G. Krauss, Heat Treated Martensitic Steels: Microstructural Systems for Advanced Manufacture, *ISIJ Int.,* Vol 35 (No. 4), 1995, p 349–359
84. K.A. Erven, D.K. Matlock, and G. Krauss, Bending Fatigue and Microstructure of Gas-Carburized Alloy Steels, *Mat. Sci. Forum,* Vol 102–104, 1992, p 183–198
85. R.E. Cohen, G. Krauss, and D.K. Matlock, "Bending Fatigue Performance of Carburized 4320 Steel," Technical Paper 930963, SAE International, 1993
86. R.S. Hyde, D.K. Matlock, and G. Krauss, "The Effect of Reheat Treatments on Fatigue and Fracture of Carburized Steel," Technical Paper 940788, SAE International, 1994
87. S. Gunnarson, Structure Anomalies in the Surface Zone of Gas-Carburized Case-Hardened Steel, *Met. Treat. Drop Forg.,* Vol 30, 1963, p 219–229
88. T. Naito, H. Ueda, and M. Kikuchi, Fatigue Behavior of Carburized Steel with Internal Oxides and Nonmartensitic Microstructures Near the Surface, *Metall. Trans. A,* Vol 15, 1984, p 1431–1436
89. W.E. Dowling, Jr., W.T. Donlon, W.B. Copple, and C.V. Darragh, Fatigue Behavior of Two Carburized Low Alloy Steels, *1995 Carburizing and Nitriding with Atmospheres,* J. Grosch, J. Morral, and M. Schneider, Ed., ASM International, 1995, p 55–60
90. C. Razim, Uber den Einfluss von Restaustenit auf des Festigkeitsverhalten Einsatzgeharteter Probenkorper bei Schwingender Bean-spruchung, *Harterei-Tech. Mitt.,* Vol 23, 1968, p 1–8
91. R.H. Richman and R.W. Landgraf, Some Effects of Retained Austenite on the Fatigue Resistance of Carburized Steel, *Metall. Trans. A,* Vol 6, 1975, p 955–964
92. M. Zaccone and G. Krauss, Elastic Limit and Microplastic Response of Hardened Steels, *Metall. Trans. A,* Vol 24, 1993, p 2263–2277
93. G.B. Olson, Transformation Plasticity and the Stability of Plastic Flow, *Deformation, Processing, and Structure,* G. Krauss, Ed., ASM International, 1984, p 391–424
94. Final Report, Advanced Rotorcraft Transmission Program, National Aeronautics and Space Administration, Cleveland, Ohio
95. R.J. Johnson, The Role of Nickel in Carburizing Steels, *Met. Eng. Quart.,* Vol 15
96. J. Grosch and O. Schwarz, Retained Austenite and Residual Stress Distribution in Deep Cooled Carburized Microstructures, *1995 Carburizing and Nitriding with Atmospheres,* J. Grosch, J. Morral, and M. Schneider, Ed., ASM International, 1995, p 71–76
97. B. Liscic, State of the Art in Quenching, *Quenching and Carburizing,* The Institute of Materials, London, 1993, p 1–32
98. A.K. Hellier, M.B. McGirr, S.H. Alger, and M. Stefulji, Computer Simulation of Residual Stresses during Quenching, *Quenching and Carburizing,* The Institute of Materials, London, 1993, p 127–138
99. G. Totten, Ed., *Quenching and Distortion Control,* ASM International, 1992
100. G. Krauss, Microstructure, Residual Stresses and Fatigue of Carburized Steels, *Quenching and Carburizing,* The Insti-

tute of Materials, London, 1993, p 205–225

101. K. Naito, T. Ochi, T. Takahashi, and N. Suzuki, Effect of Shot Peening on the Fatigue Strength of Carburized Steels, *Proc. Fourth International Conference on Shop Peening,* The Japan Society of Precision Engineering, Tokyo, 1990, p 519–526
102. A. Inada, H. Yaguchi, and T. Inoue, The Effects of Retained Austenite on the Fatigue Properties of Carburized Steels, *Heat and Surface '92,* Japan Technical Information Service, Tokyo, 1992, p 409–412
103. R.L. Widner, Failures of Rolling-Element Bearings, *Failure Analysis and Prevention,* Vol 11, *ASM Handbook,* ASM International, 1986, p 490–513
104. L.E. Alban, Failures of Gears, *Failure Analysis and Prevention,* Vol 11, *ASM Handbook,* ASM International, 1986, p 586–601
105. H. Swahn, P.C. Becker, and O. Vingsbo, Martensite Decay during Rolling Contact Fatigue in Ball Bearings, *Metall. Trans. A,* Vol 7, 1976, p 1099–1110
106. J.A. Martin, S.F. Borgese, and A.D. Eberhardt, Microstructural Alterations of Roller-Bearing Steel Undergoing Cyclic Stresses, *J. Basic Eng.,* 1966, p 555–567
107. V. Bhargava, G.T. Hahn, and C.A. Rubin, Rolling Contact Deformation, Etching Effects, and Failure of High Strength Bearing Steel, *Metall. Trans. A,* Vol 21, 1990, p 1921–1931
108. C.A. Stickels, Rolling Contact Fatigue Tests of 52100 Bearing Steel Using a Modified NASA Ball Test Rig, *Wear,* Vol 98, 1984, p 199–210
109. L. Kiessling, Rolling-Contact Fatigue of Carburized and Carbonitrided Steels, *Heat Treat. Met.,* Vol 7 (No. 4), 1980, p 97–101
110. C. Razim, "Some Facts and Considerations of Trends in Gear Steels for the Automotive Industry," *Alloys for the Eighties,* Climax Molybdenum Company, 1980, p 9
111. D.E. Diesburg, "High-Hardenability Carburizing Steels for Rock Bits," *Micon 78: Optimization of Processing, Properties, and Service Performance Through Microstructural Control,* ASTM STP 672, 1979, American Society for Testing and Materials (ASTM), p 207
112. V.K. Sharma, G.H. Walter, and D.H. Breen, Factors Influencing Fracture Toughness of High-Carbon Martensitic Steels, *Gear Technol.,* Jan/Feb 1989, p 7–18

CHAPTER 3

Vacuum and Plasma Carburizing

VACUUM AND PLASMA CARBURIZING represent the state of the art of carburizing processes, because both methods offer proven metallurgical and environmental benefits over atmosphere (gas), liquid, and pack carburizing methods. This chapter examines the capabilities of vacuum/plasma and atmosphere carburizing methods and compares their advantages and disadvantages. Although atmosphere carburizing remains the most widely used carburizing process (Fig. 1a), vacuum/plasma processes are expected to command a greater share of the carburizing market in the future (Fig. 1b).

Vacuum Carburizing

Vacuum carburizing, also referred to as low-pressure carburizing, is a non-equilibrium, boost-diffusion-type carburizing process in which the steel being processed is austenitized in a rough vacuum, carburized in a partial pressure of hydrocarbon gas, diffused in a rough vacuum, and then quenched in either oil or gas. Compared to conventional atmosphere carburizing, vacuum carburizing offers excellent uniformity and repeatability because of the high degree of process control possible with vacuum furnaces; improved mechanical properties due to the lack of intergranular oxidation; and potentially reduced cycle times, particularly when the higher process temperatures possible with vacuum furnaces are used.

Process Overview

Vacuum carburizing a steel is typically a four-step process:

1. Heat and soak step at carburizing temperature to ensure temperature uniformity throughout steel
2. Boost step to increase carbon content of austenite
3. Diffusion step to provide gradual case-core transition
4. Quenching step. This may be carried out by direct quenching in oil or in a high-pressure gas quenching system.

Heat and Soak Step. The first step is to heat the steel being carburized to the desired carburizing temperature, typically in the range of 845 to 1040 °C (1550 to 1900 °F), and to soak at the carburizing temperature only long enough to ensure that the steel is uniformly at temperature. Oversoaking, particularly above 925 °C (1700 °F), can result in a reduction in toughness due to grain growth.

During the first step, surface oxidation must be prevented, and any surface oxides present must be reduced. In a graphite-lined heating chamber consisting of graphite heating elements, a rough vacuum in the range of 13 to 40 Pa (0.1 to 0.3 torr) is usually satisfactory. In a ceramic-lined heating chamber with silicon carbide heating elements, a partial pressure of approximately 40 to 67 Pa (0.3 to 0.5 torr) of hydrogen is effective. Steels with a high chromium content (M-50 NiL, X-2 Modified), a high silicon content (Pyrowear Alloy 53), or other high-oxygen affinity alloying elements usually require a higher vacuum level prior to carburizing but do not normally require preoxidizing.

Boost Step. Second is the boost step of the process. This step results in carbon absorption by the austenite to the limit of carbon solubility in austenite at the process temperature for the steel being carburized. The boost step is achieved by backfilling the vacuum chamber to a partial pressure with either a pure hydrocarbon gas (for example, propane or acetylene) or a mixture of hydrocarbon gases. Ammonia can be added if nitrogen alloying of the case is desired. An inert gas such as nitrogen can also be added to the gas or gas mixture.

Carbon transfer occurs by dissociation of the hydrocarbon gas on the surface of the steel, with direct absorption of the carbon by the austenite and hydrogen gas being liberated. The reaction with propane is:

$$C_3H_8 + 3Fe = 3Fe(C) + 4H_2 \quad \text{(Eq 1)}$$

At typical carburizing temperatures, this reaction proceeds rapidly from left to right of the equation. Because such reactions are difficult to measure in situ, they cannot be used to control carbon potential when vacuum carburizing. Because there is no oxygen present, the oxygen-base methods of carbon-potential control used in conventional atmosphere carburizing cannot be used either. However, at least one furnace manufacturer has designed a system for vacuum carburizing that measures and controls the carbon potential of the carburizing gas.

A minimum partial pressure of hydrocarbon gas is required to ensure rapid carburizing of the austenite. The minimum partial pressure required varies with the carburizing temperature, the carburizing gas composition, and the furnace construction. Above the minimum partial pressure, the partial pressure of carburizing gas used has no relationship to the carburizing potential of the atmosphere. Typical partial pressures vary between 1.3 and 6.6 kPa (10 and 50 torr) in furnaces of graphite construction and 13 and 25 kPa (100 and 200 torr) in furnaces of ceramic construction. Partial pressures in excess of 40 kPa (300 torr) are not normally recommended because of the excessive carbon deposition within the furnace that accompanies higher partial pressures.

Diffusion Step. Third in the process is the diffusion step. If a steel were hardened with the carbon gradient resulting from the boost step only, particularly if no means of carbon-potential control were employed during the boost step, an undesirable microstructure adjacent to the carburized surface and an extremely abrupt case-core interface would result. The diffusion step enables the diffusion of carbon inward from the carburized surface, resulting in a lower surface carbon content (relative to the limit of carbon solubility in austenite at the carburizing temperature) and a more gradual case-core transition. The diffusion step is usually performed in a rough vacuum of 67 to 135 kPa (0.5 to 1.0 torr) at the same temperature used for carburizing. If carbon-potential control was used during the boost step, the diffusion segment might be shortened or eliminated.

Oil Quenching Step. The fourth step of the process is quenching. If a reheat step is not going to be employed, and/or no further machining is required, the steel is directly quenched in oil, usually under a partial pressure of nitrogen.

When vacuum carburizing is performed at a higher temperature than is normally used with conventional atmosphere carburizing, cooling to a lower temperature and stabilizing at that temperature prior to quenching is usually

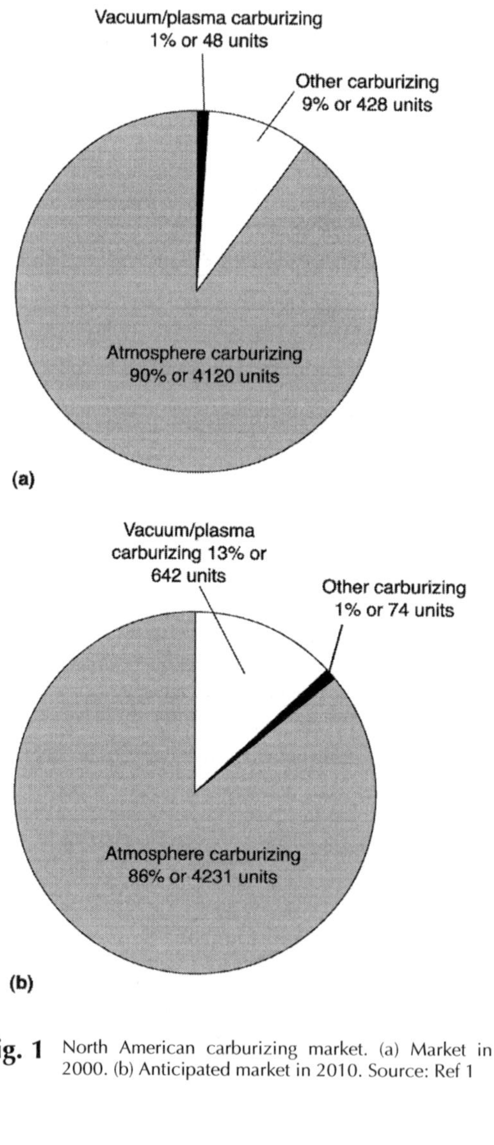

Fig. 1 North American carburizing market. (a) Market in 2000. (b) Anticipated market in 2010. Source: Ref 1

required. Alternatively, if a reheat step is going to be employed for grain refinement, and/or further machining is required, the steel is gas quenched from the diffusion temperature to room temperature, usually under a partial pressure of nitrogen.

Reheating usually consists of austenitizing in the 790 to 845 °C (1450 to 1550 °F) range followed by oil quenching. When aircraft-quality gearing or bearings are being processed, reheating is usually preceded by a subcritical anneal. A diagram of temperature and pressure versus time for a typical vacuum carburizing cycle with a reheat cycle is shown in Fig. 2.

High-Pressure Gas Quenching Step. Increasingly, vacuum carburizing is being carried out in conjunction with high-pressure gas quenching in 20 bar (2,000 kPa, or 300 psi) nitrogen or nitrogen-helium mixtures. High-pressure gas quenching is ideally suited for light loads, thin sections, and moderate-to-highly alloyed steels. An advantage of gas quenching compared to liquid quenchants (oil, water, and aqueous polymers) is that quenching with gas proceeds more uniformly, minimizing residual stresses and distortion. In addition to improved quench uniformity, gas quenching is a "clean" process, eliminating the need for a vapor degreasing step often used for oil quenching processes. Potential fire hazards and disposal problems are also eliminated.

Furnace Design

Vacuum carburizing is usually performed in a furnace specifically designed for this application. Vacuum carburizing units have been developed to operate in cell manufacturing operations found in commercial heat treating shops or just-in-time manufacturing plants. The standard line of furnaces consists of both one- and two-zone models. The double-zone model has one chamber for heating and the other for quenching.

The furnace can be of either graphite construction (graphite insulation and heating elements) or ceramic construction (refractory board insulation and silicon carbide heating elements). Graphite construction permits higher operating temperatures useful for a multipurpose furnace, whereas ceramic construction is well suited for vacuum carburizing, because it can be safely operated in air at process temperatures for die quenching or for facilitating soot removal. Figure 3 shows a typical continuous ceramic construction vacuum carburizing fur-

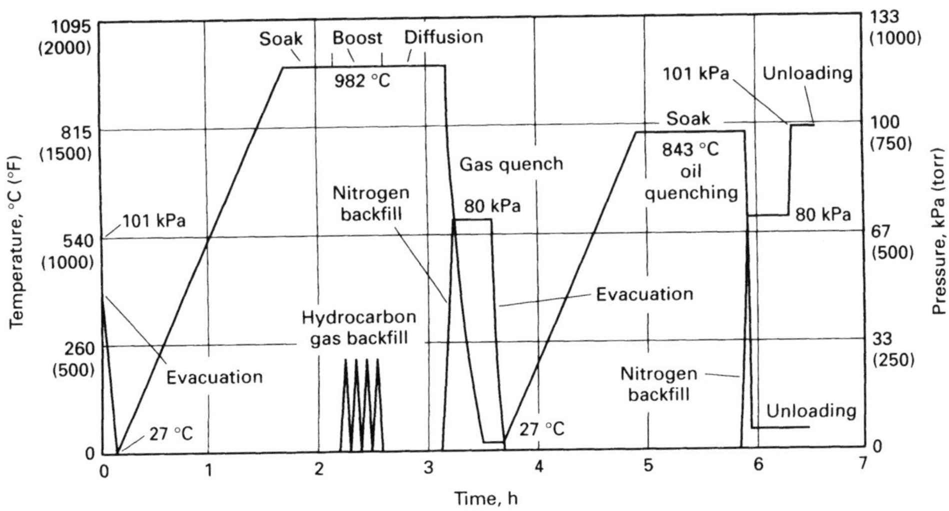

Fig. 2 Plot of temperature and pressure versus time for a typical vacuum carburizing process with a reheat cycle

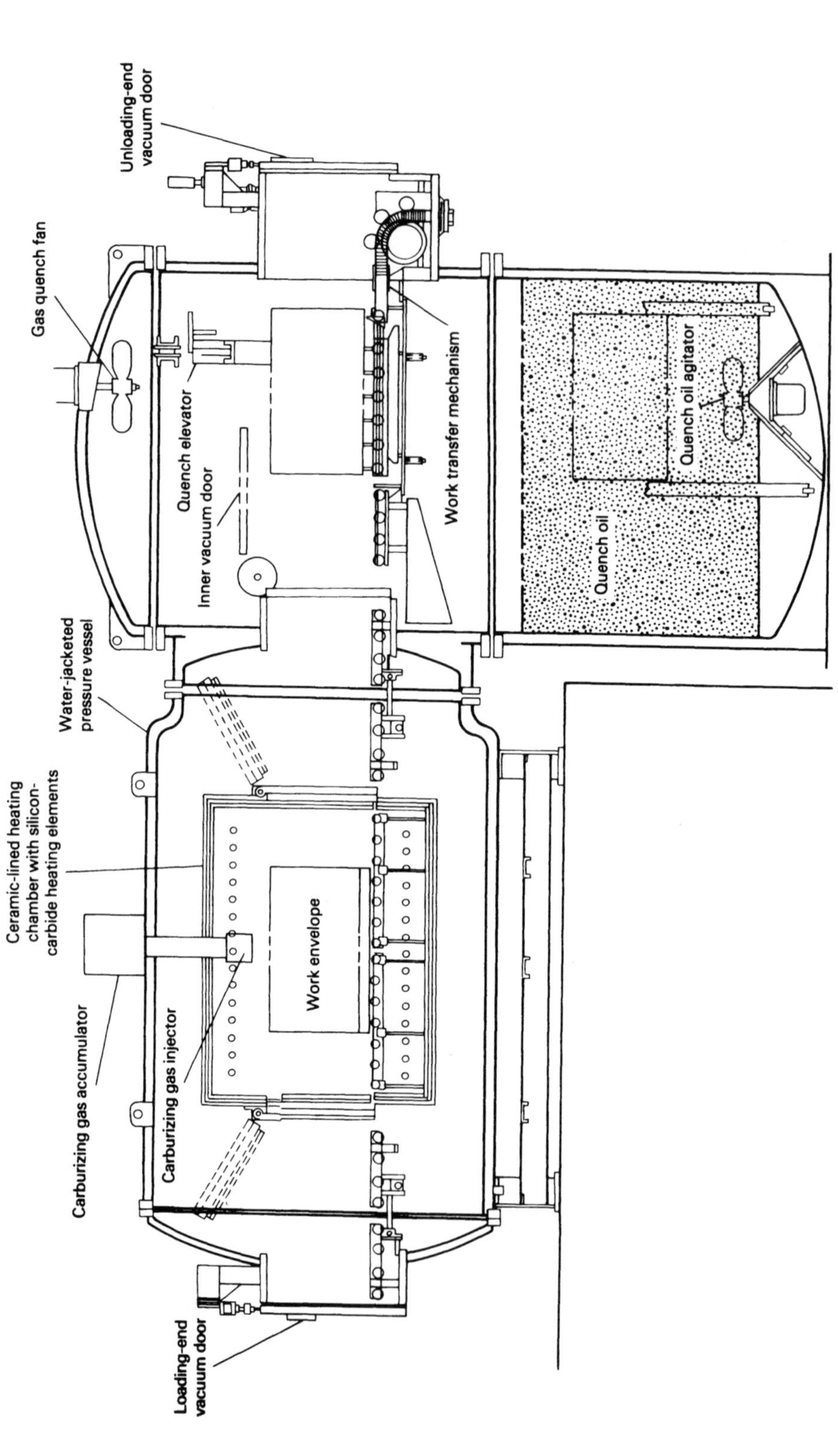

Fig. 3 A continuous ceramic vacuum carburizing furnace

nace. Figure 4 shows a typical batch-graphite construction vacuum furnace with carburizing capability.

High-Temperature Vacuum Carburizing

The reduction in carburizing time associated with a higher carburizing temperature has long been appreciated. However, typical atmosphere furnace construction generally restricts the maximum carburizing temperature to approximately 955 °C (1750 °F). The higher temperature capability of vacuum furnaces, as compared to typical atmosphere furnaces, permits the use of higher carburizing temperatures with correspondingly reduced cycle times.

High-temperature vacuum carburizing can significantly reduce the overall cycle time required to obtain effective case depths in excess of 0.9 to 1.0 mm (0.035 to 0.040 in.). For obtaining smaller case depths, high-temperature vacuum carburizing does not offer any advantages, because a grain-refining step is required, and the boost times tend to be too short for acceptable uniformity. Table 1 compares the time required to obtain 0.9 mm (0.035 in.) and 1.3 mm (0.050 in.) effective case depths via vacuum carburizing at both 900 °C (1650 °F) and 1040 °C (1900 °F) for an American Iron and Steel Institute (AISI) 8620 steel. As is apparent from Table 1, significant reductions in the total cycle time can be obtained by using high-temperature vacuum carburizing.

Metallurgists not familiar with high-temperature vacuum carburizing are often concerned that although reduced cycle times can be obtained by high-temperature carburizing, a degraded microstructure with reduced mechanical properties results. There is no evidence that any reduction in either monotonic or cyclic mechanical properties results from high-temperature vacuum carburizing, provided that the process is properly specified and controlled. One aircraft-quality gearing user has performed extensive work in the area of high-temperature vacuum carburizing and concluded that there is no loss of properties when either AISI 9310 or X-2 Modified are high-temperature vacuum carburized to aircraft-quality gearing process specifications.

There is also concern that the high process temperatures involved result in excessive distortion and size change. Although it is true that some geometries are sensitive to the ultimate process temperature used, distortion can be minimized by using proper preheating/heating techniques, minimizing times at temperature, proper fixturing, and quenching techniques that are only severe enough to result in the desired microstructure and do not develop excessive nonuniform stresses within the part. Gas pressure quenching helps alleviate residual stresses, particularly with the new moderate- to high-alloy grades of carburizing steels being developed. As far as any dimensional change greater than normal is concerned, the uniformity and repeatability of the vacuum carburizing process, even at elevated temperatures, allows for dimensional change during manufacturing planning.

Comparison of Atmosphere and Vacuum Carburizing

As indicated in Fig. 1, atmosphere or gas carburizing remains the most popular carburizing method, because it represents a good compromise between cost and performance. In recent years, improvements in the reliability of the vacuum carburizing process have allowed its benefits to be realized, and a number of papers have been published that compare the benefits of atmosphere and vacuum carburizing (Ref 1–4). This section reviews the various advantages and disadvantages associated with these carburizing methods and presents the relative technical merits of each process (Table 2). The section that follows describes a case study that compares the properties of a low-alloy gear steel processed by both atmosphere and vacuum carburizing.

Atmosphere Carburizing Characteristics (Ref 1). Atmosphere carburizing is an empirically based, time-proven process in which a carbon-rich atmosphere surrounding a workload is used to chemically react with the surface of the parts to allow an adequate quantity of carbon to be absorbed at the surface and diffuse into the material.

Advantages of atmosphere carburizing include:

- The lowest initial capital equipment investment cost
- Adequate process control; that is, all of the process variables are understood, and reliable

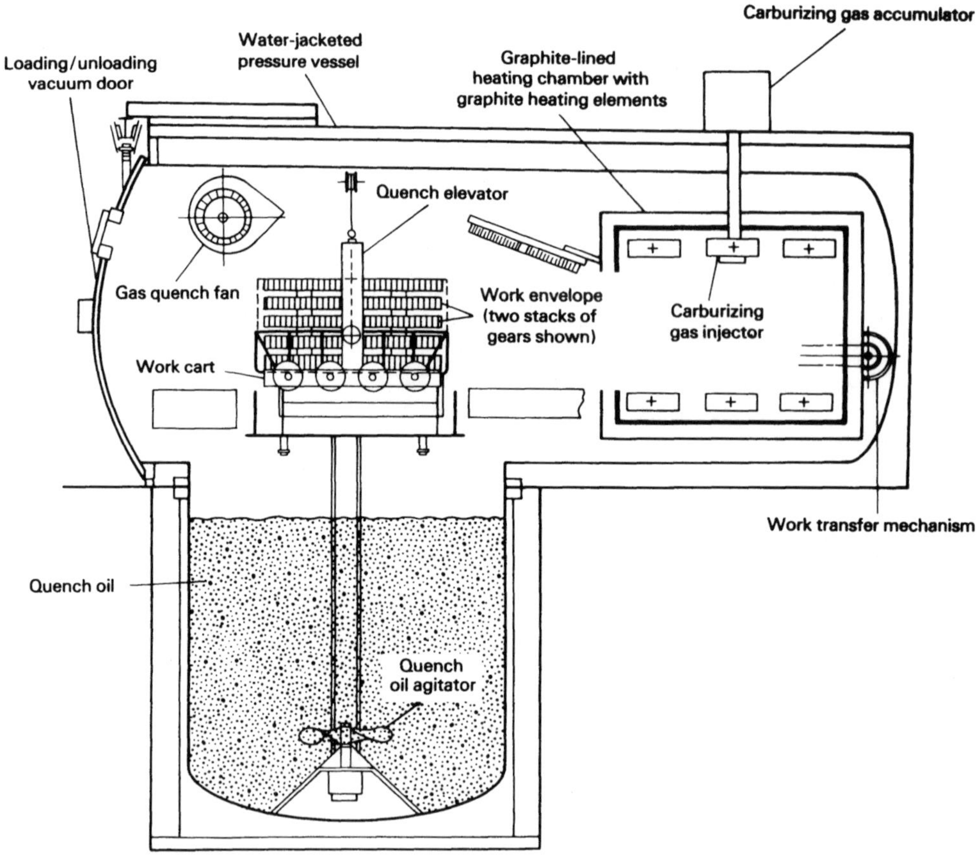

Fig. 4 A batch-graphite integral oil quench vacuum furnace with vacuum carburizing capability

Table 1 Comparison of time required to obtain a 0.9 mm (0.035 in.) and 1.3 mm (0.050 in.) effective case depth in an AISI 8620 steel at carburizing temperatures of 900 °C (1650 °F) and 1040 °C (1900 °F)

Effective depth		Carburizing temperature		Time, min								
mm	in.	°C	°F	Heating to carburizing temperature	Soaking prior to carburizing	Boost	Diffusion	Gas quench to 540 °C (1000 °F)	Reheat to 845 °C (1550 °F)	Soak at 845 °C (1550 °F)	Oil quench	Total
0.9	0.035	900	1650	78	45	101	83	(a)	(a)	(a)	15	>322
		1040	1900	90	30	15	23	20	22	60	15	275
1.3	0.050	900	1650	78	45	206	169	(a)	(a)	(a)	15	>513
		1040	1900	90	30	31	46	20	22	60	15	314

(a) Not available

Table 2 Comparison of atmosphere and vacuum carburizing technologies

Criteria	Atmosphere carburizing	Vacuum carburizing
Temperature range, °C (°F)	790–980 (1450–1800)	790–1100 (1450–2000)
Case uniformity, mm (in.) (a)	±0.25 (±0.010)	±0.05 (±0.002)
Carbon-transfer control	Yes	Limited to control of time and temperature
Load density, kg/m³ (lb/ft³) (b)	45–70 (100–150)	22.5–45 (50–150)
Carburizing time, min	*x* minutes	*x* minutes minus 10–20%
Carbonitriding(c)	NH_3 additions	NH_3 additions
Microstructure	Acceptable (in most cases)	Optimal (in most cases)
Internal oxidation, mm (in.)	0.0076–0.0127 (0.0003–0.0005) common	None
Carbides	Suppression difficult	Suppression possible
Dealloying	Yes(d)	None
Decarburization	Possible	None
Hydrogen pickup	Yes (at high temperature)	Slight (internal porosity diffusion)
Furnace conditioning	Required (4 h typical)	None
Shell temperature, °C, or °F	Warm (typically >65, or 150)	Cold (typically <<65, or 150)
Environmental impact	CO/NO_x emissions	Slight or none
Energy consumption	Low (~30%)	Lower (<30%)
Gas consumption	High (*x* cfh)	Low (1/3–1/6 *x* cfh)
Integration with cellular manufacturing	Difficult	Easy
Investment cost	Average	High

(a) Atmosphere and vacuum carburizing processes typically are different with respect to when gas additions are introduced, and this has the greatest impact on case uniformity. In atmosphere carburizing, enriching gas typically is added after the furnace set point is reestablished. Depending on the mass and configuration of the workload, a large temperature differential can exist between different locations within the workload. In the case of vacuum carburizing, a soak or stabilization period is built into the cycle to allow the workload to reach carburizing temperature prior to gas additions. (b) Loading density in vacuum carburizing equipment often is limited, due to the use of high-gas-pressure quenching chambers. (c) Techniques for vacuum carbonitriding using low pressure, <25 mbar (20 torr), are still being developed. (d) Due to oxidation at the part surface. Source: Ref 1

control devices are available to provide a measure of process repeatability.

- Capability of high-volume output using a wide variety of equipment styles, types, and workload sizes. Furnace types include box, pit, mechanized box (integral- and sealed-quench furnaces), pusher, conveyor (mesh belt and cast link belt), shaker hearth, rotary hearth, rotary drum (rotary retort), and car-bottom.
- Full automation capability, with recipe and part-number control of heat treat cycles
- Well-understood process problems allowing troubleshooting based on an established theoretical and empirical knowledge base

Disadvantages of atmosphere carburizing include:

- The need to "condition" equipment if idled or shut down prior to processing work
- A requirement of knowledge through empirically gained experience to achieve repeatable results. This is due to a wide variability in the type of equipment, its operation, maintenance, and constantly changing process conditions.
- The need for large material allowances for postprocessing operations due to accuracy and finish requirements. Case depths typically are specified in wide ranges (e.g., 0.75 to 1.25 mm, or 0.030 to 0.050 in.) to compensate for cycle-induced variability.
- Case depth quality issues; the best part of the case often is lost due to the amount of stock removal required.
- The need to constantly monitor environmental pollution issues, including air quality (for potentially hazardous gases such as CO and NO_x), water quality (for contamination concerns such as oil, minerals, etc.), waste disposal (quench oils), and safety issues (e.g., fire from combustible gases and quench oils, hot contact surfaces, and pinch points)

Vacuum Carburizing Characteristics (Ref 1). Vacuum carburizing is a proven method of *pure* carburizing and *pure* diffusion in which carbon penetrates into the surface of the steel being processed without interference from external influences, such as gas chemistry or surface contaminants.

Advantages of vacuum carburizing include:

- Easy integration into manufacturing. The process is clean, safe, simple to operate, and easy to maintain. Also, working conditions are excellent (that is, there are no open flames, heat, and pollution).

- Full automation capability using recipe or part-number control of heat treating cycles
- Capability of higher temperatures and flexible cycles, due to the type of equipment and the nature of the process
- Precise process control achieved using computer simulations, which allow adjustments to established cycles
- Consumption of energy by the equipment and process only when needed, due to the nature of the vacuum operation

Disadvantages of vacuum carburizing include:

- Higher initial capital equipment cost than atmosphere carburizing equipment
- Empirical process control, which requires processing loads to determine optimal settings or to fine tune simulator
- Formation of soot and tar, which occur due to the type, pressure, and quantity of hydrocarbon gas introduced

It is important to note that research during recent years has succeeded in finding combinations of pressure, gas type (e.g., acetylene), and flow parameters to minimize soot and tar formation as a concern in the vacuum carburizing process.

Case Study: Property Comparisons of a Gear Steel Processed by Atmosphere and Vacuum Carburizing (Ref 2)

Case Study Overview. The purpose of the study described in this section was to investigate whether vacuum carburizing could be used to improve the fatigue life of steels used for off-road vehicle transmission gearing. Fatigue is a major cause of gear failure, where the primary failure modes are gear tooth root bending and tooth pitting. Test samples were:

- Atmosphere carburized and oil quenched
- Vacuum carburized and oil quenched
- Vacuum carburized and high-pressure gas quenched

The effects of postheat grinding and shot peening were also examined.

The methods used to compare the vacuum and atmosphere carburizing processes in this study were x-ray diffraction (XRD) and microhardness testing.

Coupons of AISI 8620 low-alloy steel were heat treated using the different carburizing methods and subjected to identical post-heat treat grinding and shot peening operations.

X-ray diffraction was selected as an evaluation tool, because it can be used to measure residual stresses. Residual stresses are additive with applied stress, which makes their level an important factor in fatigue-critical components such as gears. Residual compressive stresses are desirable, because they oppose the applied, repetitive, and undesirable tensile stresses that cause fatigue failure. For gears, the areas of most concern are the flanks, which are subjected to contact loads that could cause pitting fatigue, and the roots, which experience tensile bending fatigue loads.

The greater the magnitude and depth of residual compressive stress, the greater the ability to improve fatigue properties. To enhance resistance to fatigue crack initiation, it is particularly important to have a higher compressive stress level at the outer surface. Also note that a deeper layer of compressive stress provides resistance to fatigue crack growth for a longer time than a shallower layer.

Carburizing Process Basics. Carburizing of a metal surface is a function of both the rate of carbon absorption into the steel and the diffusion of carbon away from the surface and into the interior of the part. Once a high concentration of carbon has developed at the surface during what is commonly called the boost stage, the process normally introduces a diffuse stage, where solid-state diffusion occurs over time. This step results in a change in the carbon concentration gradient between the carbon-rich surface and the interior of the steel. The result is a reduction of carbon concentration at the surface of the part, accompanied by an increase in the depth of carbon absorption.

The carburization process also induces desirable residual compressive stresses through the case-hardened layer. This stress state results from the delayed transformation and volume expansion of the carbon-enriched surface of the steel.

In atmosphere carburizing, parts are heated to austenitizing temperature in a neutral or carrier gas atmosphere that contains approximately 40% H_2, 40% N_2, and 20% CO. Small percentages of carbon dioxide (CO_2, up to 1.5%), water vapor (H_2O, up to 1%), and methane (CH_4, up to 0.5%), along with trace amounts of oxygen (O_2), also are present. The carburizing process also requires the addition of a hydrocarbon enriching gas, usually natural gas.

Of the 180 chemical equations that describe the reactions occurring during atmosphere carburizing, one of the most important is the water-gas reaction:

$$CO + H_2O = CO_2 + H_2 \quad \text{(Eq 2)}$$

Control of the atmosphere carburizing process is done by looking at the CO/CO_2 and H_2O/H_2 ratios of this equation using instruments such as dewpoint analyzers, infrared analyzers, and oxygen (carbon) probes.

In atmospheres containing CO and H_2, carbon transfer is dominated by the CO adsorption (ad) and the oxygen desorption reactions:

$$CO \rightarrow CO_{ad} \rightarrow [C] + O_{ad} \quad \text{(Eq 3)}$$

$$O_{ad} + H_2 \rightarrow H_2O \quad \text{(Eq 4)}$$

These two reactions yield an alternate form of the water-gas reaction:

$$CO + H_2 = [C] + H_2O \quad \text{(Eq 5)}$$

Thus, the transfer of carbon in atmospheres containing CO and H_2 is connected with a transfer of oxygen, giving rise to an oxidation effect in steel containing oxide-forming alloying elements such as silicon, chromium, and manganese. This phenomenon is known as internal or intergranular oxidation of steel (see Chapter 2, "Gas Carburizing," for details).

Atmosphere Carburized and Oil Quenched Hardness Profile. Figure 5 shows hardness profiles for an atmosphere carburized and oil quenched AISI 8620 steel gear.

Atmosphere carburizing to a depth of 0.36 mm (0.014 in.) produced a hardness of 58 HRC at both the gear tooth pitch line and root. From this depth, the hardness values quickly diverge. The effective case depth (at 50 HRC) is 0.76 mm (0.030 in.) in the root and 1.33 mm (0.0525 in.) at the pitch diameter. These values are typical of the vast majority of carburized gears currently in service.

For resistance to bending fatigue, it is desirable to achieve a deeper case in the root. This produces a deeper level of high-hardness, high-strength material with the benefit of residual compressive stress.

Vacuum carburizing, by comparison, does not use a carrier gas atmosphere but instead uses vacuum pumps to remove the atmosphere from the chamber before the process begins. For carburizing to take place in a vacuum furnace, all that is needed is a small, controlled addition of a hydrocarbon gas.

Unlike atmosphere carburizing, the breakdown of hydrocarbons in vacuum carburizing is via nonequilibrium reactions. This means that the carbon content at the surface of the steel is very rapidly raised to the saturation level of carbon in austenite. By repeating the boost and diffuse steps, any desired carbon profile and case depth can be achieved.

Today, vacuum carburizing is best performed using low-pressure techniques under 20 torr (25 mbar) and typically at temperatures between 790 and 1040 °C (1455 and 1900 °F). Hydrocarbon gases currently being used for vacuum carburizing are acetylene (C_2H_2), propane (C_3H_8), and, to a lesser degree, ethylene (C_2H_4). Methane (CH_4) is not used, because it is nearly nonreactive at these low pressures, unless the temperature is at or above 1040 °C (1900 °F).

Carbon is delivered to the steel surface in vacuum carburizing via reactions such as these:

$$C_2H_2 \rightarrow 2C + H_2 \quad \text{(Eq 6)}$$

$$C_3H_8 \rightarrow CH_4 + C_2H_4 \rightarrow C + 2CH_4 \quad \text{(Eq 7)}$$

$$C_2H_4 \rightarrow C + CH_4 \quad \text{(Eq 8)}$$

In the past, propane has been the primary hydrocarbon gas used for vacuum carburizing; however, propane dissociation occurs before the gas comes in contact with the surface of the steel, thus producing free carbon or soot. This uncontrolled soot formation results in poor carbon transfer to the part and loss of up-time productivity due to the need for additional heat treat equipment maintenance.

Development work done in the past few years has demonstrated that acetylene is a good performing gas for vacuum carburizing. This is because the chemistry of acetylene (Eq 6) is vastly different from that of propane or ethylene (Eq 7 and 8). Dissociation of acetylene delivers two carbon atoms to the one produced by dissociation of either propane or ethylene and avoids formation of nonreactive methane.

Control of the vacuum carburizing process is on a time basis. Carbon transfer rates are a function of temperature, gas pressure, and gas flow rate. Simulation programs have been written to determine the boost and diffuse times of the cycle.

Vacuum Carburized and Oil Quenched Hardness Profile. Figure 6 shows hardness profiles for a vacuum carburized and oil quenched AISI 8620 steel gear.

The overall case depth of maximum hardness for the vacuum carburized part is noticeably deeper than that of the atmosphere carburized part in Fig. 5. The vacuum carburized case depth of approximately 0.81 mm (0.032 in.) at 58 HRC is more than double that obtained with atmosphere carburizing, while the effective case depths (depth at 50 HRC) are similar. Also note the much greater consistency in root and pitch line hardnesses through a depth of 0.81 mm (0.032 in.) for vacuum carburizing versus atmosphere carburizing (Fig. 6 versus Fig. 5).

Vacuum Carburized and Gas Quenched Hardness Profile. The hardness profiles shown in Fig. 7 are for an AISI 8620 steel gear that has been vacuum carburized and then high-pressure gas quenched (HPGQ) in 20 bar (2,000 kPa, or 300 psi) nitrogen.

A comparison of Fig. 6 and 7 shows that use of HPGQ instead of oil quenching in vacuum carburizing results in a more uniform case depth between gear pitch line and root. The absence of a vapor layer in gas quenching results in a more uniform cooling rate along the gear tooth and root profile.

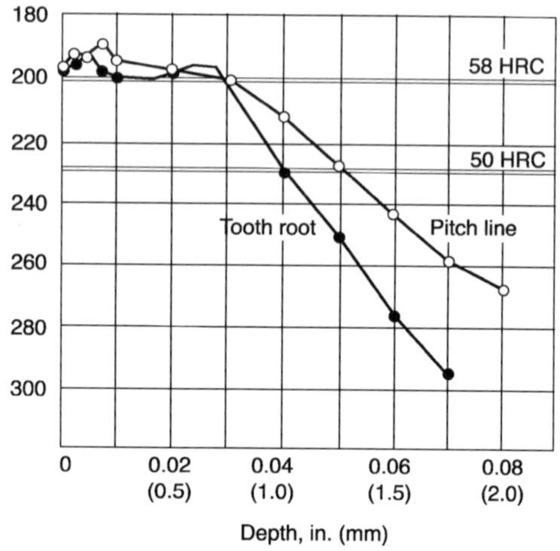

Fig. 6 Microhardness profiles at pitch line and tooth root for a vacuum carburized and oil quenched AISI 8620 gear. The overall case depth of maximum hardness is deeper than that of the atmosphere carburized part in Fig. 5.

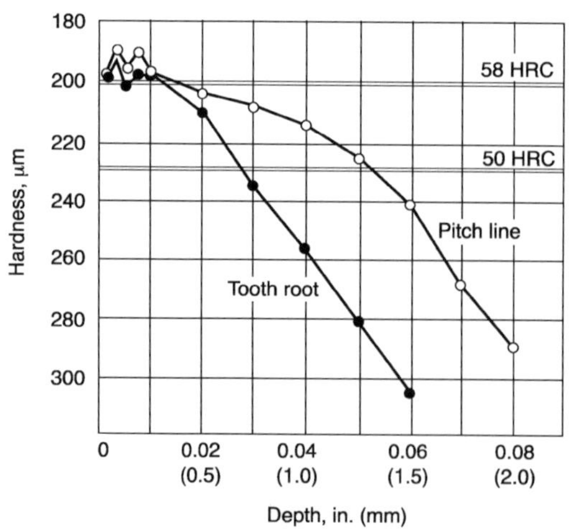

Fig. 5 Microhardness profiles at pitch line and tooth root for an atmosphere carburized and oil quenched AISI 8620 gear. For resistance to bending fatigue, it is desirable to achieve a deeper case in the root.

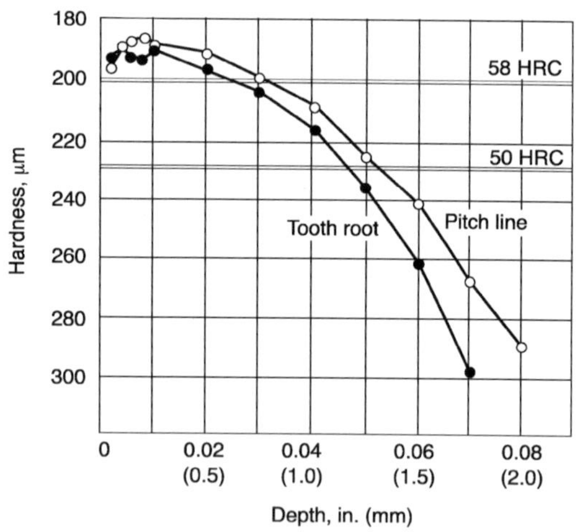

Fig. 7 Microhardness profiles at pitch line and tooth root for a vacuum carburized and high-pressure gas quenched AISI 8620 steel gear. Use of gas quenching instead of oil quenching (Fig. 6) results in a more uniform case depth between pitch line and root.

The Test Procedure. The following procedure was used to properly evaluate the effect of different heat treatments and post-heat treatment processes on residual stress in coupons of AISI 8620 low-alloy gear steel:

- Five coupons from the same heat lot of AISI 8620 were cut to size: 76 by 19 by 13 mm, ± 0.05 mm (3.00 by 0.75 by 0.505 in., ± 0.002 in.). The coupons were stamped, and a separate manufacturing process was defined for each (Table 3).
- Coupons were sent out for heat treatment—vacuum or atmosphere carburizing—according to the parameters in Table 4. Required surface hardness: 59 to 61 HRC. Vacuum carburized coupons were nitrogen gas quenched, while atmosphere carburized coupons were oil quenched.
- Heat treated coupons were ground to 12.7 ± 0.013 mm (0.5000 ± 0.0005 in.), removing no more than 0.15 mm (0.006 in.) from the nonstamped side where XRD was to take place.
- Three of the five coupons were sent out for shot peening.
- All five coupons were sent out for XRD on the nonstamped side.

Table 3 Test coupon manufacturing processes

Coupon	Identification	Process
1	EX2470	Vacuum carburize (VC)
2	EX2470-1	Vacuum carburize and shot peen (VC and SP)
3	EX2470-2	Atmosphere carburize (AC)
4	EX2470-3	Atmosphere carburize and shot peen (AC and SP)
5	EX2470-4	Vacuum carburize and dual shot peen (VC and DSP)

Table 4 Test parameters for atmosphere and vacuum carburized coupons

Parameter	Atmosphere	Vacuum
Temperature, °C (°F)	940 (1725)	940 (1725)
Boost time, min	300	32
Diffusion time, min	120	314
Hardening temperature, °C (°F)	845 (1550)	845 (1550)
Quenching method	Oil at 60 °C (140 °F)	Nitrogen gas at 20 bar
Tempering temperature, °C (°F)	175 (350)	175 (350)
Tempering time, h	2	2

The Benefits of Shot Peening. The primary purpose of shot peening gears is to enhance their fatigue life by inducing a high residual compressive stress at the surface of the tooth roots. Shot peening is most effective for parts subject to high-cycle fatigue loading.

A basic explanation is provided by the graph in Fig. 8, a typical stress-number of cycles (*S-N*) curve. It plots (tensile) stress, *S*, versus the num-

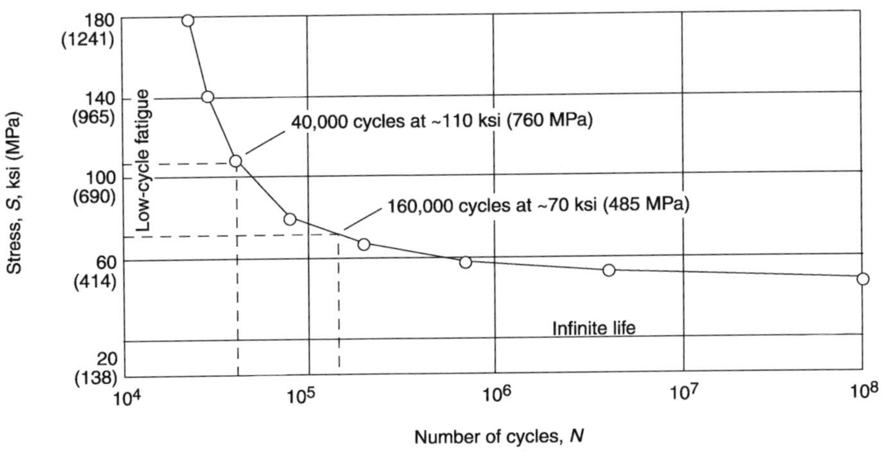

Fig. 8 Typical *S-N* curve, or plot of (tensile) stress, *S*, vs. number of load cycles, *N*. The primary purpose of shot peening gears is to enhance their fatigue life by inducing a high residual compressive stress at the surface of the tooth roots. Shot peening is most effective for parts subject to high-cycle fatigue loading (>10^4 to 10^5 cycles).

ber of load cycles, N. It is important to note that the vertical scale is linear, whereas the horizontal scale is logarithmic. This means that as tensile stress is reduced, fatigue life improves exponentially. An ~35% reduction of stress from 760 MPa (110 ksi) to 485 MPa (70 ksi) results in an improvement in fatigue life from 40,000 cycles to 160,000 cycles (400%). Additional reductions in tensile stress result in significantly more fatigue enhancement. At 415 MPa (60 ksi), for example, the anticipated fatigue life is ~400,000 cycles.

The residual compressive stresses produced by shot peening counteract applied tensile stresses. The compressive stresses are induced by impacts of small, spherical media (shot). The impact of each individual shot stretches the surface enough to yield it in tension. Because the surface cannot fully restore itself due to the mechanical yielding that has taken place, it is left in a permanent compressed state.

Shot peening results in a residual compressive stress at the surface—where most fatigue cracks initiate—that is ~55 to 60% that of the material ultimate tensile strength. For carburized gears, the surface compression is typically 1170 to 1725 MPa (170 to 250 ksi), which results in a significant improvement in fatigue properties.

The grinding process is applied to components so often and in so many forms (automatic, manual, with and without coolant) that it is often overlooked from a residual stress standpoint. However, its influence should not be discounted, especially when dealing with fatigue-critical parts.

During grinding, residual tensile stress may be created from generation of excessive, localized heat. The localized surface area being ground heats from friction and attempts to expand but cannot, because it is surrounded by cooler, stronger metal. If the temperature generated from grinding is high enough, however, the metal yields in compression due to the resistance to its expansion and reduced mechanical properties at elevated temperature. On cooling, the yielded material attempts to contract. The surrounding material resists this contraction, thus creating residual tensile stress. Because heat is the major cause of residual tensile stress from grinding, the importance of coolant for controlling these stresses is paramount.

X-Ray Diffraction Residual Stress Measurements. X-ray diffraction was used to measure the residual stresses at surface and subsurface locations. The technique measures strain by measuring changes in atomic distances. It is a direct, self-calibrating method that measures tensile, compressive, and neutral strains equally well. Strains are converted to stresses by multiplying by elastic constants appropriate for the alloy and atomic planes measured.

For this study, chromium Kα radiation was used to diffract the (211) planes at approximately 156° 2θ. The area measured was nominally 4 mm (0.16 in.) in diameter. Because only a few atomic layers are measured, the technique is considered a surface analysis technique. The subsurface measurements were made by electrochemically removing small amounts of material. These subsurface measurements were subsequently corrected for stress gradient and layer removal effects using standard analytical calculations.

Comparing the Processes. Hardness profiles for vacuum carburized (coupon 1) and atmosphere carburized (coupon 3) AISI 8620 steel coupons are compared in Fig. 9. A major advantage of vacuum carburizing over atmosphere carburizing is a deeper case of high hard-

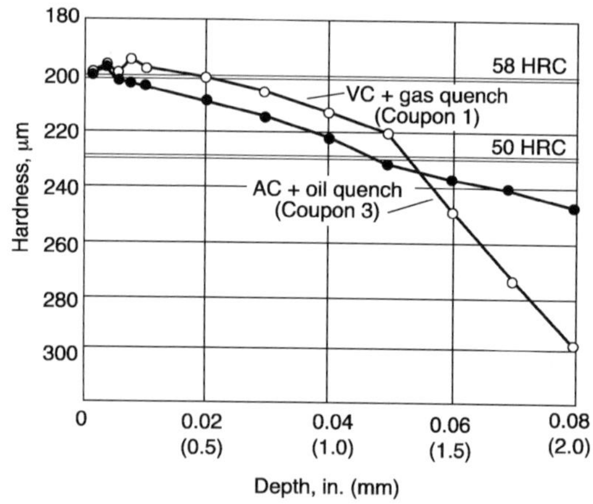

Fig. 9 Microhardness profiles for vacuum carburized and gas quenched (coupon 1) and atmosphere carburized and oil quenched (coupon 3) AISI 8620 steel coupons. A major advantage of vacuum carburizing is a deeper case of high hardness. VC, vacuum carburized; AC, atmosphere carburized

ness. Hardness values for the two carburizing processes are given in Table 5.

Effect of Peening. Residual stress distributions in the three carburized, ground, and shot peened coupons—coupons 2, 4, and 5—are plotted in Fig. 10.

From a fatigue standpoint, the solid layer of compression demonstrated for all three coupons implies excellent resistance to initiation and growth of fatigue cracks. The tensile stress required for a fatigue crack to develop must first overcome compressive stresses that are ~1035 MPa (150 ksi) at the surface and ~1515 MPa (220 ksi) at 0.05 mm (0.002 in.) below the surface. A tensile stress of 1035 MPa (150 ksi) produces a net stress of 0 MPa (0 ksi) at the surface when added to the residual compressive stress.

Coupons 2 (vacuum carburized and shot peened) and 4 (atmosphere carburized and shot peened) were shot peened at an Almen intensity of 14 to 16. (Almen intensity is a measure of the energy of the shot stream.) The steel shot had a hardness of 55 to 62 HRC and a nominal diameter of 0.58 mm (0.023 in.).

The residual stress curves in Fig. 10 have shapes typical of shot peened material. All three have a similar maximum compressive stress of ~1515 MPa (220 ksi). This value is ~55 to 60% that of the steel ultimate tensile strength at the surface. Because all three coupons were hardened to 59 to 62 HRC, they also had similar tensile strengths (at the surface).

The depth of the compressive stress layer is a function of the Almen intensity. It can be increased by increasing shot size and/or velocity. The depth is the location where the residual stress versus depth curves would cross the neutral axis (into tension) if the positively sloped lines were extended. A greater depth of compression is desired, because this layer is what resists fatigue crack growth. Coupons 2 and 4 were shot peened to the same intensity, so that the depth of their compressive stress layers is also essentially the same at ~0.18 to 0.20 mm (0.007 to 0.008 in.).

Table 5 Comparison of atmosphere and vacuum carburizing results

Property	Atmosphere	Vacuum
Depth to 58 HRC, mm (in.)	0.20 (0.008)	0.58 (0.023)
Surface hardness before grinding, HRC	59	60
Surface hardness after removal of 0.1 mm (0.004 in.) stock by grinding, HRC	58	62

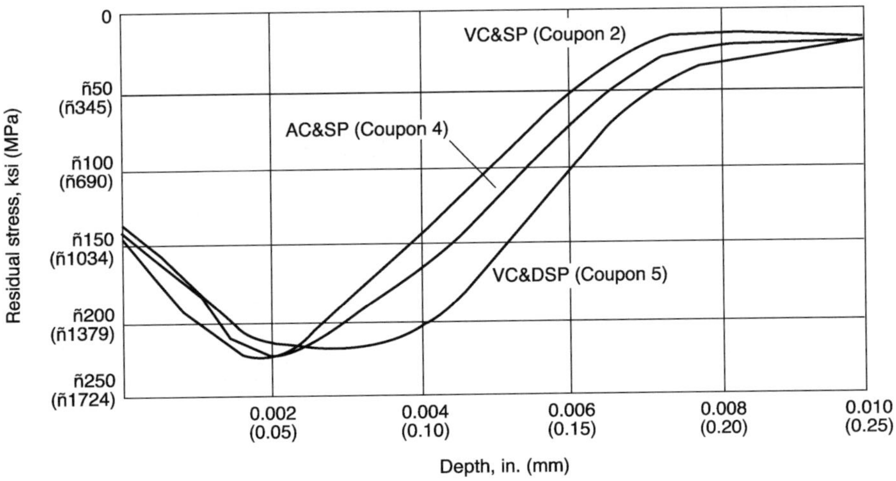

Fig. 10 Residual stress distributions in the three carburized, ground, and shot peened coupons (coupons 2, 4, and 5) of AISI 8620. Each contains a solid layer of compression that implies excellent resistance to initiation and growth of fatigue cracks. Note that the residual stress curve for dual-peened coupon 5 is –0.025 to 0.05 mm (0.001 to 0.002 in.) deeper than those for the single-shot-peened coupons. See Table 3 for definition of process abbreviations for coupons.

Dual Peening. The trade-off to increasing shot peening intensity is that there is additional cold work and material displacement at the point of shot impact. This generally results in less compression right at the surface (depth = 0) and a more aggressive surface finish.

Dual peening is performed to make up for the reduced compression resulting from high-intensity peening. The technique consists of shot peening the same surface twice—peening at a higher intensity is followed by peening at a lower intensity, usually with smaller media. The second peening reduces the degree of cold work at the surface, improving the surface finish, which, in turn, makes the surface more compressed.

Coupon 5 was dual peened. The process specified: MI-230H shot at 18 to 20 Almen followed by MI-110H shot at 8 to 10 Almen. The residual stress curve for this coupon (Fig. 10) is ~0.025 to 0.05 mm (0.001 to 0.002 in.) deeper than the curves for the coupons single shot peened at 14 to 16 Almen using MI-230H shot.

This would be expected for a carburized gear steel. The surface stress of coupon 5 (at depth = 0) is the same as that of the other two shot peened coupons. What most likely occurred is that it was less compressed after the first peening step. When the second was performed, the surface became even more compressed, to the ~930 MPa (135 ksi) level shown in Fig. 10. Therefore, coupon 5 would be expected to have the best fatigue performance of the three, because it has the most compressive stress throughout its depth. This is particularly evident between 0.08 and 0.20 mm (0.003 and 0.008 in.) below the surface. At 0.10 mm (0.004 in.) below the surface, for example, there is still 1380 MPa (200 ksi) of compression for coupon 5, compared with 1170 MPa (170 ksi) for coupon 4 and 1000 MPa (145 ksi) for coupon 2.

Effect of Grinding. Testing of coupons that had been ground gave unexpected results that required further investigation. All coupons were ground at the same time. Grinding was performed using a wheel that had coolant flow. The operator was instructed to remove no more than 0.025 mm (0.001 in.) of stock per pass, for a total of 0.10 mm (0.004 in.) of material removed from each coupon. This appeared to be acceptable grinding practice, and little thought was given to the technique prior to testing.

X-ray diffraction measurements indicated that tensile stresses existed on the surface of the vacuum carburized coupon (coupon 1) as high as 255 MPa (37 ksi) at 0.013 mm (0.0005 in.) below the surface. At a depth of ~0.10 mm (0.004 in.), the values crossed the neutral axis into compression. These results were immediately questioned. However, retesting at several locations using XRD verified that the original values were correct.

The explanation lies in the fact that additional heat was generated when grinding vacuum carburized coupon 1. The coupon was 1 HRC point harder at the surface and 4 HRC points higher after 0.10 mm (0.004 in.) of stock removal. These values are higher than those for the atmosphere carburized specimen (coupon 3). Additional heat from an increase in friction resulted in the generation of residual tensile stresses on the vacuum carburized coupon.

This is an excellent example of why it is important to carefully evaluate the amount of heat generated when grinding fatigue-critical parts. It also demonstrates that XRD is an effective tool for determining the residual stress state of components before they enter service.

Test Results Reviewed. This study compared atmosphere and vacuum carburizing of AISI 8620 gear steel and evaluated the influence of the subsequent manufacturing operations of shot peening and grinding. The primary goal of the study was to determine which carburizing process was more suitable for heavy-duty transmission gears. Gears are subject to both sliding and rolling-contact stresses on their flanks in addition to bending stresses in tooth roots. To meet these demanding performance criteria, the steel gears ideally would be hardened for strength and contact properties and have residual compressive surface stresses for bending fatigue resistance.

It was concluded that vacuum carburizing is superior to atmosphere carburizing for heat treating heavy-duty transmission gears and enjoys the following advantages:

- *Higher hardness (XRD coupons):* Surface (before grinding), 1 HRC point higher (60 versus 59 HRC); subsurface (after 0.10 mm, or 0.004 in., stock removal), 4 HRC points higher (62 versus 58 HRC)
- *Greater depth of high, ≥58 HRC, hardness (XRD coupons):* 58 HRC depths were vac-

uum, 0.58 mm (0.023 in.); atmosphere, 0.20 mm (0.008 in.)

- *Greater depth of high, ≥58 HRC, hardness (pitch line and root of actual gears):* 58 HRC depths were vacuum, 0.81 mm (0.032 in.); atmosphere, 0.38 mm (0.015 in.)
- *Deeper effective case in tooth root (actual gears):* Vacuum, 1.0 mm (0.040 in.); atmosphere, 0.699 mm (0.0275 in.)
- *Higher surface residual compression (XRD of coupons without shot peening or grinding):* Vacuum, 135 MPa (19.6 ksi); atmosphere, 98 MPa (14.2 ksi)
- *Improved consistency between the case layer at the pitch line of the gear flank and gear roots (actual gears):* Vacuum, 0.28 mm (0.011 in.) variation; atmosphere, 0.648 mm (0.0255 in.) variation

Shot Peening Results Reviewed. Both the vacuum carburized and atmosphere carburized surfaces responded equally to shot peening:

- Maximum compressive stress: ~1515 MPa (220 ksi)
- Compressive layer depth: ~0.18 to 0.20 mm (0.007 to 0.008 in.)

Dual shot peening at first a higher and then a lower intensity resulted in a greater depth of compression by ~0.025 to 0.05 mm (0.001 to 0.002 in.). The surface stress of the dual-peened coupon was very similar, at ~930 MPa (135 ksi), to that of the conventionally shot peened coupons. The higher-intensity first peen would have produced a less compressed surface, but the second, lower-intensity peen would have restored compressive stress to the ~930 MPa (135 ksi) level. The dual-peened coupon should have significantly better high-cycle fatigue properties than the single-peened coupons.

Fatigue. In terms of fatigue performance, the additional 34.5 MPa (5 ksi) of compression measured for the vacuum carburized coupon (not shot peened or ground) should yield significant increases in gear life under high-cycle fatigue loading, compared with that for the atmosphere carburized coupon.

Plasma Carburizing

Plasma or ion carburizing is basically a vacuum process that uses glow-discharge technology to introduce carbon-bearing ions to the surface of steel for subsequent diffusion below the surface. Plasma carburizing is effective in increasing carburizing rates, because the process effectively bypasses several steps in the dissociation process that produces active soluble carbon. With methane (CH_4) gas, for example, active carbon for adsorption can be formed directly from methane due to the ionizing effect of the plasma. Plasma carburizing thus allows higher process rates than conventional gas carburizing, which involves several reaction steps in the dissociation of methane into active carbon.

Another advantage compared to gas carburizing in some applications of plasma carburizing is that higher temperatures are permissible, because the process is performed in an oxygen-free vacuum. This advantage is similar to the process of vacuum carburizing described earlier in this chapter. However, vacuum carburizing exhibits some potential disadvantages when compared to plasma carburizing. Because vacuum carburizing is conducted at very low pressures, and the rate of flow of the carburizing gas into the furnace is very low, the carbon potential of the gas in deep recesses and blind-holes is quickly depleted. Unless this gas is replenished in these areas, a great nonuniformity in case depth over the surface of the part is likely to occur. If, in an effort to overcome this problem, the gas pressure is increased significantly, another problem arises, namely, free-carbon formation or sooting. Thus, in order to obtain cases of reasonably uniform depth over a part of complex shape, the gas pressure must be periodically increased to replenish the depleted atmosphere in recesses and then reduced again to the operating pressure. Clearly, a delicate balance exists in vacuum carburizing, where the process conditions must be adjusted to obtain the best compromise between case uniformity, risk of sooting, and carburizing rate. Plasma carburizing overcomes both of these major problems, yet retains the desirable features of a simple atmosphere and a higher permissible operating temperature.

Diffusion Characteristics

Similar to vacuum carburizing, plasma carburizing is performed in an oxygen-free environment, which permits higher temperatures and thus higher diffusion rates. Higher temperatures also bring about some additional benefits arising from the increased solubility of carbon in austenite as the temperature is increased. As shown in the low-carbon region of the iron-carbon composition diagram in Fig. 11, by raising the temperature from the vicinity of 900 °C (1650 °F) (the normal carburizing temperature for conventional atmosphere carburizing) to 1040 °C (1900 °F), the limit of carbon solubility for carbon in austenite is increased from approximately 1.2 to approximately 1.6 wt% C (indicated by the arrows on the abscissa of Fig. 11). Because the surface of the hot steel part becomes saturated to this higher value very quickly in plasma carburizing, the diffusivity (and hence the carburizing rate) increases because of the effect of higher dissolved carbon concentrations on the diffusion coefficient (*D*) for carbon in austenite.

The increase in the diffusion coefficient (*D*) with increasing carbon concentration is shown in Fig. 12. It may be seen that for carbon concentrations above approximately 1 wt% C, the diffusion coefficient increases very rapidly. For concentrations in the neighborhood of 1.5 to 1.6 wt% (the limit of solid solubility of carbon in austenite at the temperatures allowed in plasma carburizing), the diffusion coefficient (and hence the diffusivity) is more than twice that for

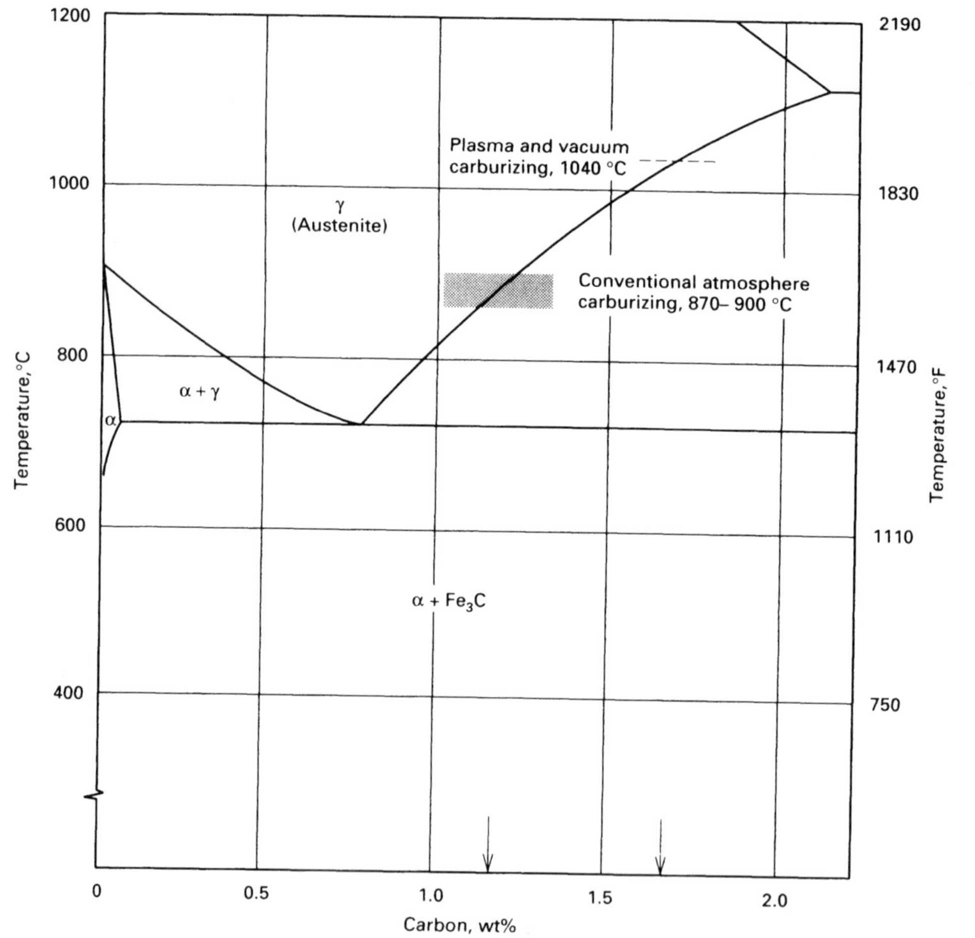

Fig. 11 Low-carbon region of the iron-carbon composition diagram. Arrows indicate the increase in the limit of carbon solubility for carbon in austenite.

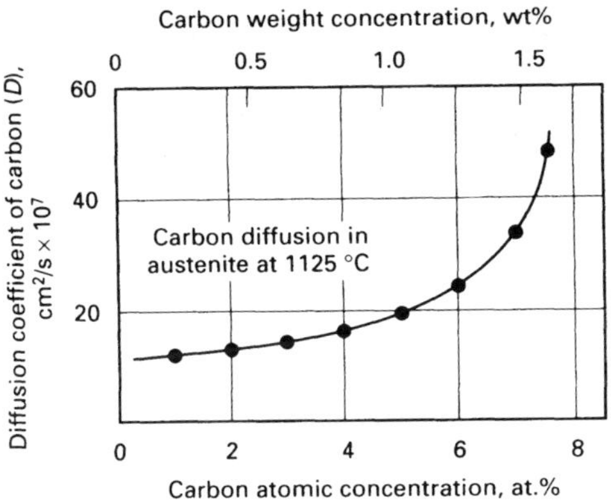

Fig. 12 Diffusion coefficient for carbon in austenite versus carbon concentration at 1125 °C (2060 °F). Source: Ref 5

a concentration below 1 wt%. Therefore, temperature has two effects acting simultaneously to increase the diffusion rate of carbon from the surface into the interior of the steel part: first, the effect of the increased temperature on the diffusivity, and second, the increased diffusivity brought about by the increased carbon solubility in the steel.

In addition, because the rate of adding carbon is increased during plasma carburizing, the combined effect is to bring about a profound increase in the rate of carburizing. For example, at a temperature of 1050 °C (1920 °F), enough carbon can be added in only 10 min (Fig. 13) to obtain a case depth of 1 mm (0.040 in.). While an additional diffusion step of 30 min is required to develop the effective case depth of 1

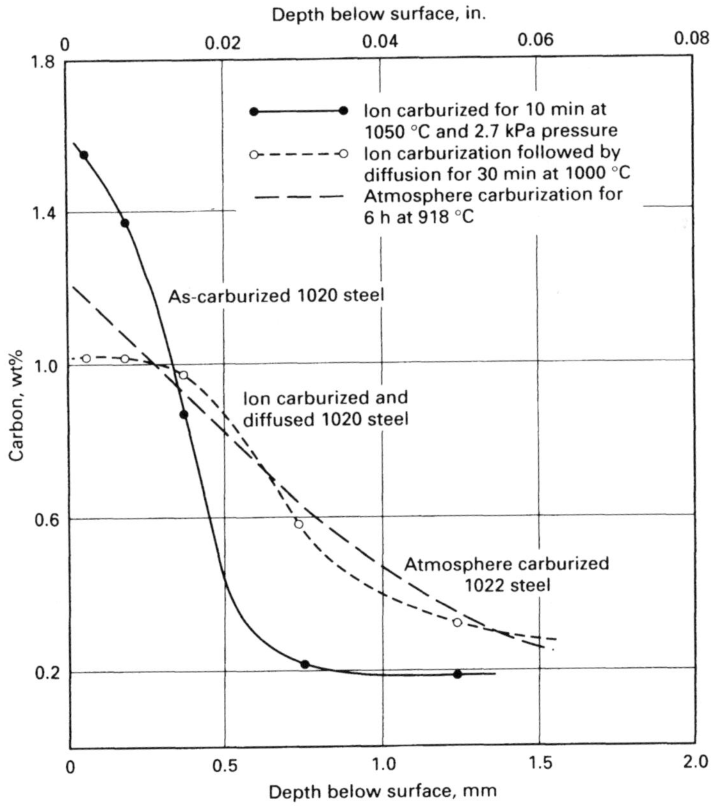

Fig. 13 Carbon concentration profile in AISI 1020 steel after ion carburizing for 10 min at 1050 °C (1920 °F) followed by vacuum diffusing for an additional 30 min at 1000 °C (1830 °F). A similar carbon concentration profile is obtained by atmosphere carburizing for 6 h at 910 °C (1685 °F). Source: Ref 6

mm (0.040 in.), all of the carbon needed was added in only 10 min from the methane plasma. As also shown in Fig. 13, 6 h is required to obtain a comparable carbon profile with conventional gas-atmosphere carburizing at 918 °C (1685 °F).

Plasma Carburizing Equipment

The physical arrangement of the apparatus and electrical circuitry required for plasma carburizing is shown schematically in Fig. 14. In this arrangement, the workpiece (cathode) is at ground potential and the positive potential needed to establish and maintain the glow discharge is fed into the vacuum enclosure through a suitable insulated lead-through to a counter-electrode (the anode). Auxiliary heating elements (either radiant or induction) surround the workpiece to heat it to the carburizing temperature, because the heat losses of the plasma are insufficient to heat the workload to the carburizing temperature (900 to 1000 °C, or 1650 to 1830 °F). Oil quench facilities are similar to those described for vacuum carburizing.

Advantages of Plasma Carburizing

Advantages of plasma (ion) carburizing include several technical factors (such as carburizing and case uniformity) as well as economic and environmental factors. These advantages are briefly discussed subsequently.

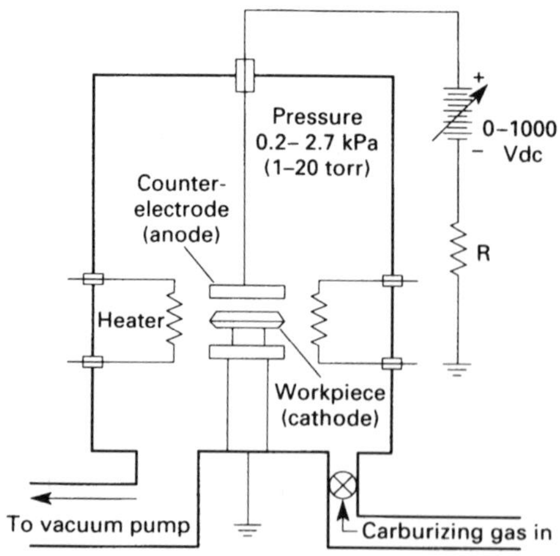

Fig. 14 Schematic of plasma (ion) carburizing apparatus

High Carburizing Rate. The marked increase in carburizing rate compared to gas or atmosphere carburizing is illustrated in Fig. 15. Shown are the carbon profiles obtained by plasma carburizing at 900 °C (1650 °F) for 10, 30, 60, and 120 min as well as that obtained by atmosphere carburizing for 240 min at the same temperature (Ref 6). It may be seen that the profile obtained by atmosphere carburizing for 240 min at 900 °C (1650 °F) may be obtained in one-half the time by plasma carburizing at the same temperature.

A similar 2-to-1 advantage is illustrated in Fig. 16 for AISI 8620 steel carburized at 980 °C (1800 °F). Plasma carburizing for 15 min at this temperature matches the carbon profile that required 30 min with atmosphere carburizing. Compared to vacuum carburizing for 30 min, the case depth at the 0.8 wt% C level is 50% deeper with plasma carburizing.

Typical carbon and hardness profiles through an ion carburized case in the as-carburized and in the carburized-and-diffused condition are shown in Fig. 17. At a carburizing temperature of 1000 °C (1830 °F), a 1 mm (0.040 in.) case is typically obtained with a 10 min carburize, 30 min diffuse cycle.

Improved Case Uniformity. Figure 18 illustrates the improvement in uniformity of case depth in gear-tooth profiles obtained with ion carburizing. Here, the case depth profile after ion carburizing at 980 °C (1800 °F) is compared to that obtained with atmosphere carburizing at the same temperature. The improved case depth uniformity obtained by ion carburizing on a part having deep recesses is illustrated in Fig. 19. In this figure, the case uniformity obtained with ion carburizing is compared to that obtained with vacuum carburizing.

Blind-Hole Penetration. Ion carburizing improves blind-hole penetration, or "down-hole" carburizing. In Fig. 20, for example, plasma carburizing achieves a uniform plateau of case depth of up to a length-to-diameter (*L/D*) ratio of approximately 12. Atmosphere carburizing achieved a uniform case depth for *L/D* ratios up to approximately 9 (Fig. 20).

Insensitivity to Steel Composition. The rate at which steel may be ion carburized is quite insensitive to alloy composition, as illustrated in Fig. 21. Essentially the same profile and depth case are obtained in three steels of different alloy content.

Ion carburizing is similarly insensitive to the hydrocarbon gas used as a carbon source. Figure

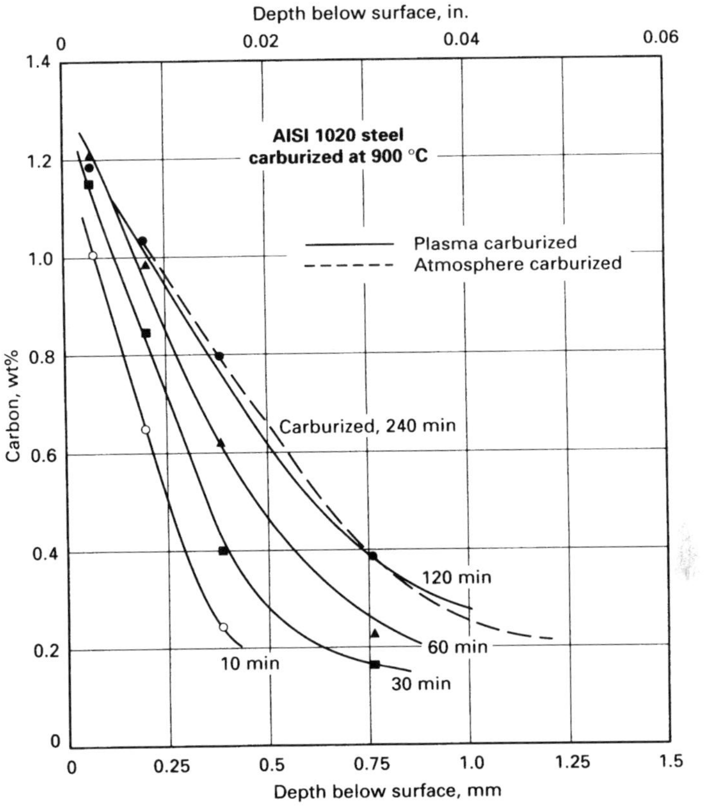

Fig. 15 Carbon concentration profiles in AISI 1020 steel after ion carburizing for 10, 20, 30, 60, and 120 min at 900 °C (1650 °F). Carbon profile after atmosphere carburizing for 240 min at 900 °C (1650 °F) shown for comparison. Source: Ref 6

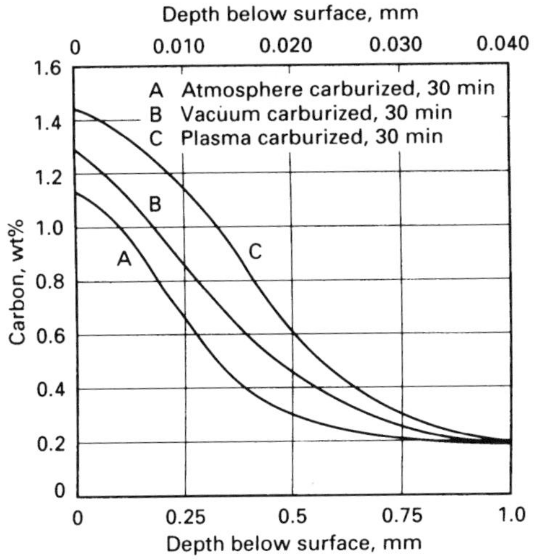

Fig. 16 Carbon gradient profile of atmosphere, vacuum, and plasma carburizing of AISI 8620 steel at 980 °C (1800 °F) saturation conditions for 30 min and followed by direct oil quenching

22 shows similar and comparable carbon and hardness profiles obtained by ion carburizing in methane, natural gas, and nitrogen-propane atmospheres.

Environmental Improvements. Plasma carburizing provides a much cleaner and safer environment than gas carburizing systems, and there is no fire hazard or toxic gases such as carbon monoxide from the flame screens on atmosphere furnaces. Because the carburizing zone is completely isolated from the surrounding work space, the furnace may also be placed in-line on the assembly floor. Furthermore, because the ambient temperature of the casing is near tap-water temperature, the furnace may be used in an air-conditioned environment.

Equipment and Operating Costs. Plasma carburizing is done in a special furnace using methane or propane gas at subatmospheric pressure for the carbon source, which eliminates the need for generated gas. Therefore, atmosphere

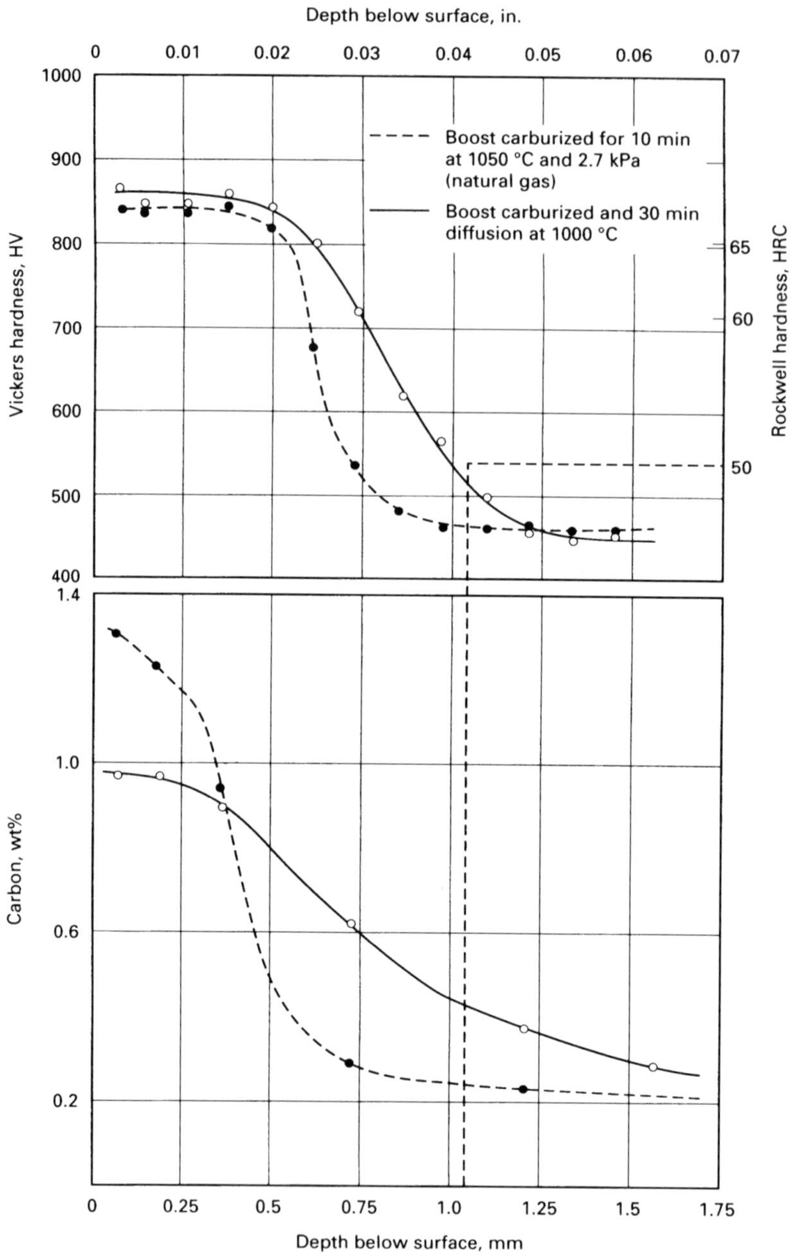

Fig. 17 Carbon concentration and hardness profiles in AISI 1020 steel after ion carburizing for 10 min at 1050 °C (1920 °F) followed by additional vacuum diffusing for 30 min at 1000 °C (1830 °F). Effective case depth is indicated by dotted line.

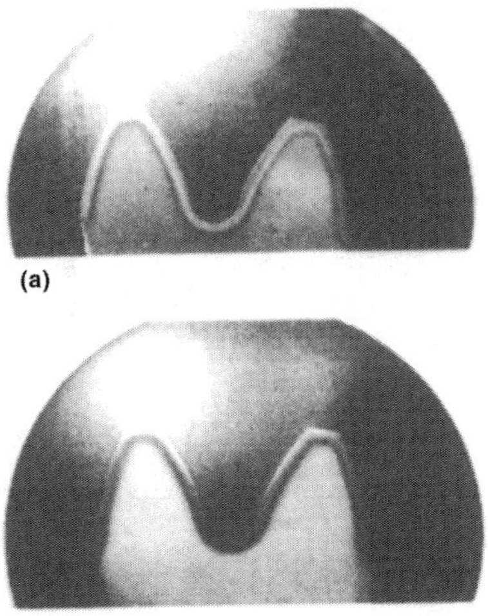
(a)

(b)

Fig. 18 Comparing uniformity of case depth over gear-tooth profiles. (a) Ion carburized at 980 °C (1800 °F). (b) Atmosphere carburized in a 980 °C (1800 °F) boost-diffuse cycle. Case depth in (a) exhibits more consistency, particularly in the root of the gear profile

generating equipment is not required, and simple gas-valve manifolds may be used with a resultant reduction in furnace capital equipment costs. However, when all ancillaries are taken into account, the costs of plasma versus atmosphere equipment are comparable.

Properties of Plasma Carburized Parts

Aside from the faster carburizing rates associated with the fast rate of reaching carbon saturation, plasma (ion) carburizing offers some additional metallurgical advantages. Because the glow-discharge parameters can be adjusted to give more control of the carburizing mechanism (carbon saturation and, in turn, diffusion), greater uniformity of the carburized case can be achieved.

Because carburizing and oil quench hardening occur under a vacuum, grain-boundary oxides do not form in plasma carburizing. This elimination of intergranular oxides improves fatigue performance in the un-ground condition (see, for example, Ref 7). In atmosphere carburizing, grain-boundary oxides to a depth of 0.05 to 0.12 mm (0.002 to 0.005 in.) are commonly

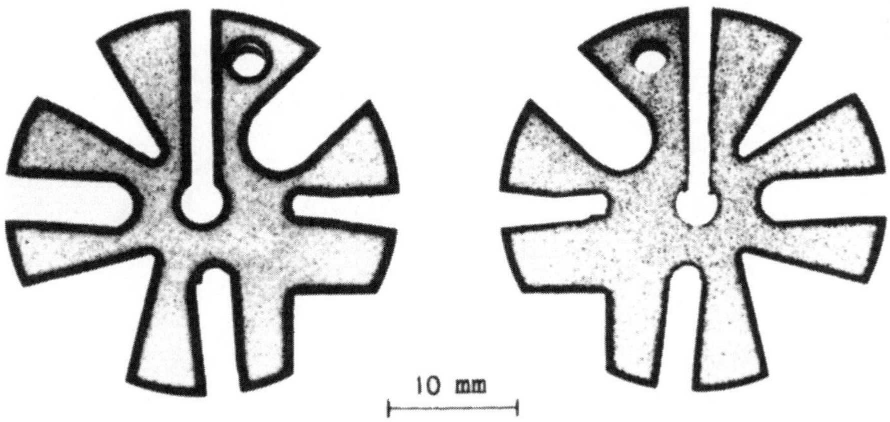

(a) (b)

Fig. 19 Comparing uniformity of case depth over a part with deep recesses. (a) Ion carburized. (b) Vacuum carburized

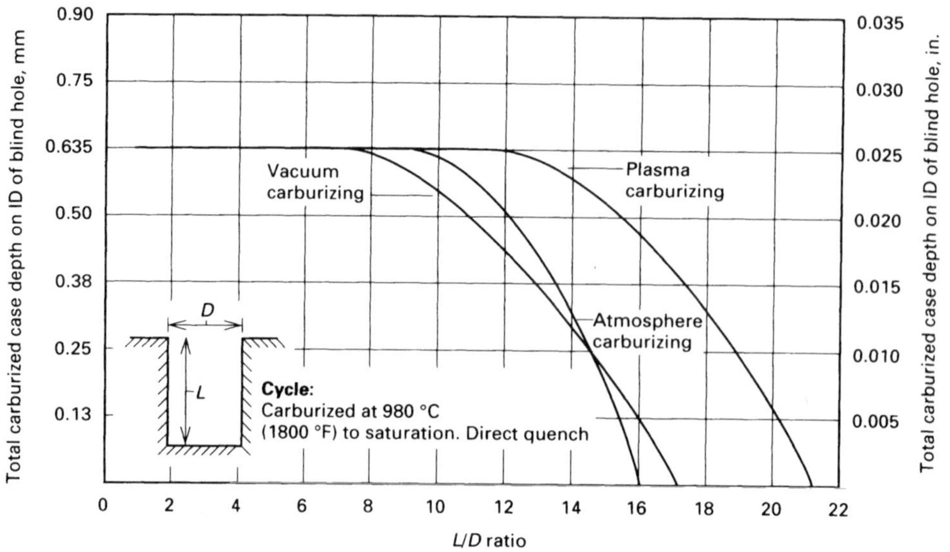

Fig. 20 Uniform case depth of 0.635 mm (0.025 in.) in a blind hole as a function of L/D ratios for plasma, vacuum, and gas carburizing. The curves were obtained by measuring case depths at different hole depths (L) in a 6.35 mm (0.25 in.) diam blind hole.

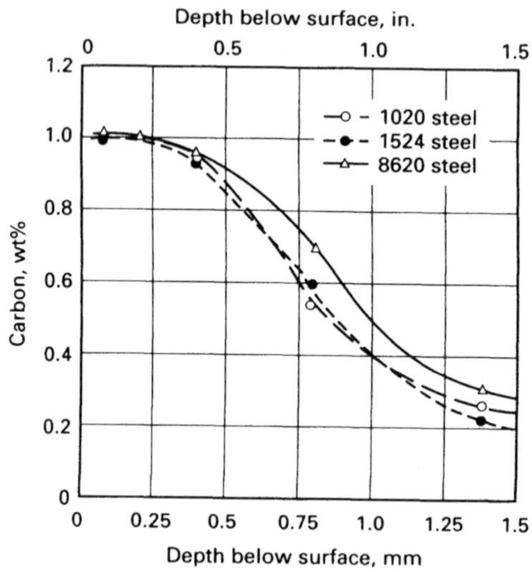

Fig. 21 Carbon concentration profiles in three carburizing steels after ion carburizing illustrating insensitivity to steel composition. Data are based on a boost-diffuse cycle of ion carburizing at 1040 °C (1900 °F) for 10 min followed by diffusion for 30 min at 1000 °C (1830 °F).

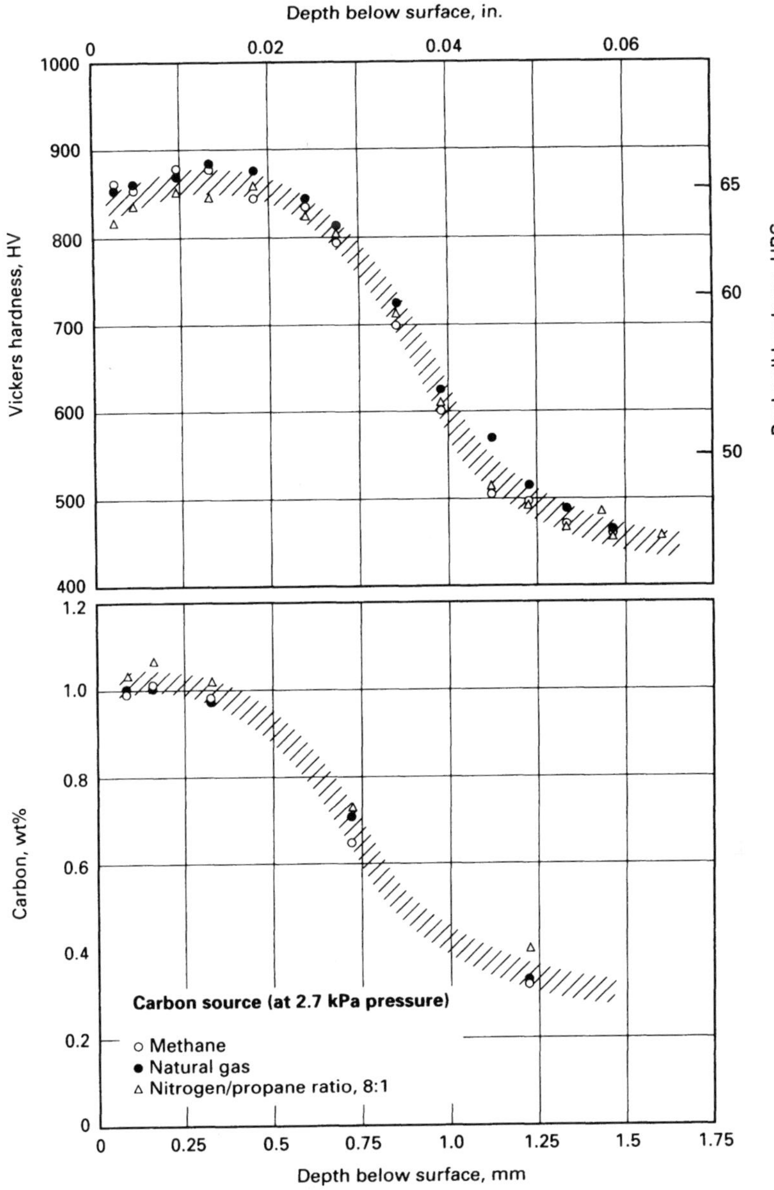

Fig. 22 Carbon concentration and hardness profiles through cases on AISI 1020 steel after ion carburizing in methane, natural gas, and in 8:1 nitrogen-propane combination. Data are based on a boost-diffuse cycle of ion carburizing for 10 min at 1050 °C (1920 °F) followed by 30 min of diffusion at 1000 °C (1830 °F).

found. The carbide level in the microstructure is no different than that found in atmosphere carburizing.

Parts in a load can be mechanically masked effectively by simply preventing the glow discharge from coming into contact with the areas not to be carburized, for example, by stacking or proper fixturing. Likewise, copper plating selected areas is effective for masking.

REFERENCES

1. D.H. Herring, Pros and Cons of Atmosphere and Vacuum Carburizing, *Ind. Heat.,* Jan 2002, p 45–48
2. G.D. Lindell, D.H. Herring, D.J. Breuer, and B.S. Matlock, Atmosphere versus Vacuum Carburizing, *Heat Treat. Prog.,* Nov 2001, p 33–41
3. W.J. Titus, Considerations when Choosing a Carburizing Process: Traditional Batch Atmosphere Carburizing versus Vacuum and Ion Carburizing, *Ind. Heat.,* Sept 1987
4. B. Edenhofer, Overview of Advances in Atmosphere and Vacuum Heat Treatments, *Heat Treat. Met.,* Vol 4, 1998, and Vol 1, 1999
5. C. Wells, W. Batz, and R.F. Mehl, *Trans. AIME,* Vol 188, 1950, p 553
6. W.L. Grube, *J. Heat. Treat.,* Vol 3 (No. 3), 1980, p 40–49
7. W.L. Wentland and J.Y. Yung, Plasma and Gas Carburizing of Fine Pitch AISI 9310 Gears, *Ion Nitriding and Ion Carburizing,* T. Spalvins and W.L. Kovacs, Ed., ASM International, 1990, p 245–248

SELECTED REFERENCES

- W.L. Grube and S. Verhoff, Plasma (Ion) Carburizing, *Heat Treating,* Vol 4, *ASM Handbook,* ASM International, 1991, p 352–362
- J. St. Pierre, Vacuum Carburizing, *Heat Treating,* Vol 4, *ASM Handbook,* ASM International, 1991, p 348–351

CHAPTER 4

Pack and Liquid Carburizing

PACK AND LIQUID CARBURIZING are carried out respectively in a solid or liquid at the carburizing temperature. In pack carburizing, the workpiece is heated in contact with a solid carburizing compound in a closed container. In liquid carburizing, the workpiece is immersed in a fused carburizing salt. The pack method is the oldest of the carburizing processes and was for many years the one most extensively employed. Its use in recent years, however, has decreased significantly due to improvements in alternative carburizing procedures (most notably gas carburizing) combined with the inherent limitations of the process. These include the time consumed in heating the charge, the high labor cost involved in packing and unpacking the containers, and the preclusion of direct quenching. Similarly, liquid carburizing does not enjoy the commercial importance it once had. This is primarily due to environmental concerns—spent salt baths are expensive and difficult to dispose of. This is particularly true of cyanide-containing baths. In addition, salt removal is very difficult for some parts. Despite their limitations, pack and liquid carburizing offer some advantages over other case hardening methods and continue to be offered by some heat treat shops. As such, they will be briefly reviewed here. More detailed information on these processes can be found in *Heat Treating,* Volume 4 of *ASM Handbook* and in earlier editions of *Metals Handbook.*

Pack Carburizing

Pack carburizing is a process in which carbon monoxide derived from a solid compound decomposes at the metal surface into nascent carbon and carbon dioxide. The nascent carbon is absorbed into the metal, and the carbon dioxide immediately reacts with carbonaceous material present in the solid carburizing compound to produce fresh carbon monoxide. The formation of carbon monoxide is enhanced by energizers or catalysts, such as barium carbonate ($BaCO_3$), calcium carbonate ($CaCO_3$), potassium carbonate (K_2CO_3), and sodium carbonate (Na_2CO_3), that are present in the carburizing compound. These energizers facilitate the reduction of carbon dioxide with carbon to form carbon monoxide. Thus, in a closed system, the amount of energizer does not change. Carburizing continues as long as enough carbon is present to react with the excess carbon dioxide. A schematic of the pack process is shown in Fig. 1.

Advantages and Disadvantages

Advantages. Among the principal advantages of pack carburizing are:

- It can make use of a wide variety of furnaces because it produces its own contained environment.

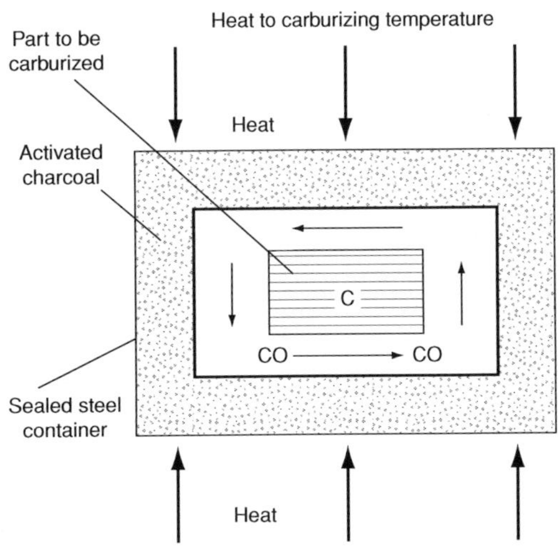

Fig. 1 Schematic of the pack carburizing process. C is carbon on the surface of the part; CO is carbon monoxide gas that is circulated around the part.

- It is ideally suited for slow cooling of work from the carburizing temperature, a procedure that may be advantageous for parts that are to be finish machined after carburizing and before hardening.
- Compared to gas carburizing, it offers a wider selection of stop-off techniques for selective carburizing.

Disadvantages. By its nature, pack carburizing is less clean and less convenient than other carburizing processes. Other disadvantages generally associated with pack carburizing include:

- It is not well suited to production of shallow case depths where strict case-depth tolerances are required.
- It cannot provide the degree of flexibility and accuracy of control over surface carbon content and carbon gradient that can be obtained in gas carburizing.
- It is not well suited for direct quenching or quenching in dies; thus, extra handling and processing are required for the hardening operation.
- More processing time is required for pack carburizing than for gas or liquid carburizing because of the necessity of heating and cooling the extra thermal mass associated with the compound and the container.
- It is labor intensive.
- It poses environmental problems associated with disposal of barium-containing carburizing compounds.

Because of the disadvantages and improvements in more controllable, less labor-intensive competing carburizing processes, pack carburizing is no longer a major commercial process.

Carburizing Compounds

The common commercial carburizing compounds are reusable and contain 10 to 20% alkali or alkaline earth metal carbonates bound to hardwood charcoal or to coke by oil, tar, or molasses. Barium carbonate is the principal energizer, usually comprising approximately 50 to 70% of the total carbonate content. The remainder of the energizer usually is made up of calcium carbonate, although sodium carbonate and potassium carbonate also may be used. It should be noted that barium carbonate, now designated by government regulations as a health hazard due to its toxicity and the disposal problems it presents, is gradually being phased out by U.S. manufacturers as a catalyst in pack-carburizing operations. Barium-free pack carburizing compounds produce deep cases without the disposal problems associated with barium-containing wastes.

Addition Rate. Because of losses associated with the use of pack-carburizing compounds, new compound usually is added to the used compound before it is returned to service. The loss in energizer normally is somewhat higher than loss of the rest of the compound. Therefore, somewhat larger percentages of new compound are used to ensure that the energizer level does not drop below approximately 5 to 8%. When direct quenching or severe mechanical handling methods are used, the addition rate may be as high as one part new compound to two parts used compound. When furnace cooling and careful handling methods are used, the addition rate may be one part new compound to three to five parts used compound.

Used compound often is screened to remove fines. The compound is then thoroughly mixed with the makeup material. Because many compounds, particularly those of the coated-charcoal type, are extremely friable, they require careful handling to minimize losses due to formation of dust or fines.

Process Control

In pack carburizing, as in other carburization processes, the carbon-concentration gradient obtained is a function of carbon potential, carburizing temperature and time, and the chemical composition of the steel. Two process-control attributes peculiar to pack carburizing are:

- There may be a variation in case depth within a given furnace load due to dissimilar thermal histories within the carburizing containers.
- Distortion of parts during carburizing may be reduced because the compound can be used to support the parts.

Carbon Potential. The carbon potential of the atmosphere generated by the carburizing compound, as well as the carbon content obtained at the surface of the work, increases directly with an increase in the ratio of carbon monoxide to carbon dioxide. Thus, more carbon is made available at the work surface by the use of energizers and carburizing materials that promote formation of carbon monoxide.

Temperature. Pack carburizing normally is performed at temperatures from 815 to 955 °C (1500 to 1750 °F). In recent years, the upper limits have been steadily raised, and carburizing temperatures as high as 1095 °C (2000 °F) have been used. Steelmaking processes have improved to the extent that fine grain size is maintained at temperatures approaching or exceeding 1040 °C (1900 °F). Above this temperature, the coarsening effect occurs only after prolonged periods of time, allowing high-temperature treatment without excessive grain coarsening.

The rate at which the carburized case is formed increases rapidly with temperature. If a factor of 1.0 is representative of 815 °C (1500 °F), the factor increases to 1.5 at 870 °C (1600 °F) and to more than 2.0 at 925 °C (1700 °F). Improved containers, fine-grain steels, and other improvements now permit the use of a wide variety of temperatures.

Time. The rate of change in case depth at a particular carburizing temperature is proportional to the square root of time. The rate of carburization is thus highest at the beginning of the cycle and gradually diminishes as the cycle is extended (see Fig. 2).

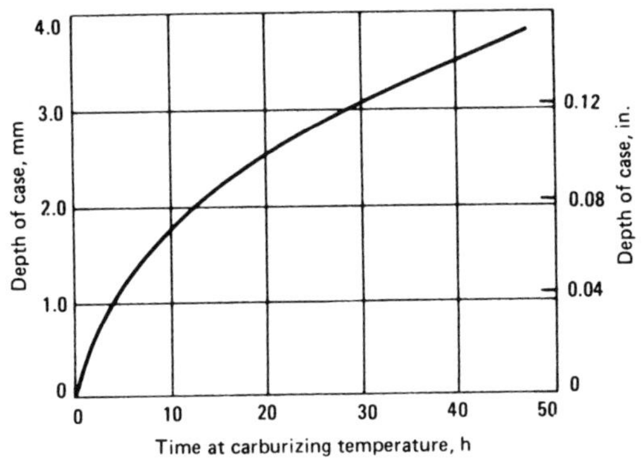

Fig. 2 Effect of time on case depth at 925 °C (1700 °F)

Pack-Hardenable Steels. Any carburizing grade of carbon or alloy steel is suitable for pack carburizing. It is generally agreed that the diffusion rate of carbon in steel is not markedly influenced by the chemical composition of the steel. Chemical composition does have an effect on the activity of carbon and thus can affect the carbon level at saturation for a particular temperature. Table 1 lists some of the carbon and alloy steels that are commonly pack carburized.

Depth of Case. Even with good process control, it is difficult to obtain parts with total case-depth variation of less than 0.25 mm (0.010 in.) from maximum to minimum in a given furnace load, assuming a carburizing temperature of 925 °C (1700 °F). Commercial tolerances for case depths obtained in pack carburizing begin at ±0.25 mm (±0.010 in.), and, for deeper case depths, increase to ±0.8 mm (±0.03 in.). Lower carburizing temperatures provide some reduction in case-depth variation because variation in the time required for all parts of the load to reach carburizing temperature becomes a smaller percentage of total furnace time. Because of the inherent variation in case depth and the cost of packing materials, pack carburizing normally is not used on work requiring a case depth of less than 0.8 mm (0.03 in.). Typical pack-carburizing temperatures selected to produce different case depths on a variety of production parts are given in Table 1.

Table 1 Typical applications of pack carburizing

	Dimensions(a)							Carburizing			
	OD		OA		Weight			Case depth to 50 HRC		Temperature	
Part	mm	in.	mm	in.	kg	lb	Steel	mm	in.	°C	°F
Mine-loader bevel gear	102	4.0	76	3.0	1.4	3.1	2317	0.6	0.024	925	1700
Flying-shear timing gear	216	8.5	92	3.6	23.6	52.0	2317	0.9	0.036	900	1650
Crane-cable drum	603	23.7	2,565	101.0	1,792	3,950	1020	1.2	0.048	955	1750
High-misalignment coupling gear	305	12.0	152	6.0	38.5	84.9	4617	1.2	0.048	925	1700
Continuous-miner drive pinion	127	5.0	127	5.0	5.4	11.9	2317	1.8	0.072	925	1700
Heavy-duty industrial gear	618	24.3	102	4.0	150	331	1022	1.8	0.072	940	1725
Motor-brake wheel	457	18.0	225	8.9	104	229	1020	3.0	0.120	925	1700
High-performance crane wheel	660	26.0	152	6.0	335	739	1035	3.8	0.150	940	1725
Calender bull gear	2,159	85.0	610	24.0	5,885	12,975	1025	4.0	0.160	955	1750
Kiln-trunnion roller	762	30.0	406	16.0	1,035	2,280	1030	4.0	0.160	940	1725
Leveler roll	95	3.7	794	31.3	36.7	80.9	3115	4.0	0.160	925	1700
Blooming-mill screw	381	15.0	3,327	131.0	2,950	6,505	3115	5.0	0.200	925	1700
Heavy-duty rolling-mill gear	914	36.0	4038	159.0	11,800	26,015	2325	5.6	0.220	955	1750
Processor pinch roll	229	9.0	5385	212.0	1,700	3,750	8620	6.9	0.270	1050	1925

(a) OD, outside diameter; OA, overall (axial) dimension

Furnaces for Pack Carburizing

The suitability of a furnace for pack carburizing depends on its ability, at reasonable cost, to: provide adequate thermal capacity and temperature uniformity (furnaces must be controllable to within ±5 °C, or ±9 °F, and must be capable of uniform through heating to within ±8 to ±14 °C, or ±14 to ±25 °F) and provide adequate support for containers and workpieces at the required temperatures.

The three types of furnaces most commonly used for pack carburizing are the box, car-bottom, and pit types. Box furnaces are loaded by mechanical devices or by in-plant transportation equipment. Car-bottom furnaces provide for convenient loading of heavy units. A car-bottom furnace with a car at each end allows a second car to be loaded while the furnace is in use, which minimizes the heat loss and downtime between batches. Pit furnaces are general-purpose furnaces that may be used for carburizing and other heat-treating operations that require minimum floor space.

Carburizing Containers

Materials. Carburizing containers are made of carbon steel, of aluminum-coated carbon steel, or of iron-nickel-chromium heat-resisting alloys. Although uncoated carbon steel boxes scale severely during carburizing and have short lives, they often are the most economical for processing odd lots and unusual shapes. Aluminum coating can significantly extend the life of a carbon steel container, making this material potentially the lowest in cost per hour per unit weight carburized.

In the long run, heat-resisting alloys are the most economical container materials for carburizing large numbers of moderate-size parts. However, because heat-resisting alloys are considerably higher in initial cost than plain or aluminum-coated carbon steel, they must be used continuously if they are to approach the lowest possible prorated cost.

Packing

Intimate contact between compound and workpiece is not necessary; however, when properly packed, the compound will provide good support for the workpiece. The layer of compound surrounding the work must be heavy enough to allow for shrinkage and to maintain a high carbon potential during the entire cycle but not so heavy as to unduly retard heating of the workpiece to carburizing temperature.

Procedure. Packing of the workpieces in a compound is a dusty and disagreeable operation (one of the reasons this process is losing favor in industry). For this reason, grouping of boxes, workpieces, and compound should be carefully planned so as to minimize handling of the compound. If possible, workpieces should come to the packer already stacked and sorted, preferably on open trays or in pans.

First, a layer of compound from 13 to 50 mm (1/2 to 2 in.) deep is placed in the empty box. The part or parts are then stacked in the container, and, if necessary, metal or ceramic supports or spacers are applied and internal container supports are inserted.

Whenever possible, workpieces should be packed with the longest dimension vertical to the base of the container. This is extremely important in processing long parts such as shafts and rolls because it minimizes the tendency of these parts to sag.

Selective Carburizing

Stop-off techniques can be used to carburize only portions of a workpiece during a pack carburizing operation (see Chapter 2, "Gas Carburizing," for a discussion of stop-off techniques). In addition, it may be possible to permit any portion of a part that is not to be carburized to protrude from the carburizing container. Alternatively, an inert or slightly oxidizing material may be packed around those areas of a part that are not to be carburized.

Liquid Carburizing and Cyaniding

Liquid carburizing is a process used for case hardening steel or iron parts. The parts are held at a temperature above Ac_1 in a molten salt that will introduce carbon and nitrogen, or carbon alone, into the metal. Diffusion of the carbon from the surface toward the interior produces a case that can be hardened, usually by fast quenching from the bath. Carbon diffuses from the bath into the metal and produces a case comparable with one resulting from gas carburizing in an atmosphere containing some ammonia. However, because liquid carburizing involves faster heat-up (due to the superior heat-transfer characteristics of salt bath solutions), cycle times for liquid carburizing are shorter than those for gas carburizing.

Most liquid carburizing baths contain cyanide, which introduces both carbon and nitrogen into the case. One type of liquid carburizing bath, however, uses a special grade of carbon, rather than cyanide, as the source of carbon. This bath produces a case that contains only carbon as the hardening agent.

Liquid carburizing may be distinguished from cyaniding (which is performed in a bath containing a higher percentage of cyanide) by the character and composition of the case produced. Cases produced by liquid carburizing are lower in nitrogen and higher in carbon than cases produced by cyaniding. Cyanide cases are seldom applied to depths greater than 0.25 mm (0.010 in.); liquid carburizing can produce cases as deep as 6.35 mm (0.250 in.). For very thin cases, liquid carburizing in low-temperature baths may be employed in place of cyaniding.

Cyanide-Containing Liquid Carburizing Baths

Light case and *deep case* are arbitrary terms that have been associated with liquid carburizing in baths containing cyanide. There is necessarily some overlapping of bath compositions for the two types of case. In general, the two types are distinguished more by operating temperature than by bath composition. Hence, the terms "low temperature" and "high temperature" are preferred.

Low-temperature cyanide-type baths (light-case baths) are those usually operated in the temperature range from 845 to 900 °C (1550 to 1650 °F), although for certain specific effects this range is sometimes extended to 790 to 925 °C (1450 to 1700 °F). Low-temperature baths are best suited to formation of shallower cases. Low-temperature baths are generally of the accelerated cyanogen type containing various combinations and amounts of the constituents listed in Table 2 and differ from cyaniding baths

Table 2 Operating compositions of liquid carburizing baths

	Composition of bath, %	
Constituent	Light case, low temperature 845–900 °C (1550–1650 °F)	Deep case, high temperature 900–955 °C (1650–1750 °F)
Sodium cyanide	10–23	6–16
Barium chloride	. . .	30–55(a)
Salts of other alkaline earth metals(b)	0–10	0–10
Potassium chloride	0–25	0–20
Sodium chloride	20–40	0–20
Sodium carbonate	30 max	30 max
Accelerators other than those involving compounds of alkaline earth metals(c)	0–5	0–2
Sodium cyanate	1.0 max	0.5 max
Density of molten salt	1.76 g/cm^3 at 900 °C (0.0636 lb/in.^{3} at 1650 °F)	2.00 g/cm^3 at 925 °C (0.0723 lb/in.^{3} at 1700 °F)

(a) Proprietary barium chloride-free deep-case baths are available. (b) Calcium and strontium chlorides have been employed. Calcium chloride is more effective, but its hygroscopic nature has limited its use. (c) Among these accelerators are manganese dioxide, boron oxide, sodium fluoride, and sodium pyrophosphate.

Table 3 Compositions and properties of sodium cyanide mixtures

	Composition, wt%			Melting point		Specific gravity	
Mixture grade designation	NaCN	$NaCO_3$	NaCl	°C	°F	25 °C (75 °F)	860 °C (1580 °F)
96–98(a)	97	2.3	Trace	560	1040	1.50	1.10
75(b)	75	3.5	21.5	590	1095	1.60	1.25
45(b)	45.3	37.0	17.7	570	1060	1.80	1.40
30(b)	30.0	40.0	30.0	625	1155	2.09	1.54

(a) Appearance: white crystalline solid. This grade also contains 0.5% sodium cyanate (NaNCO) and 0.2% sodium hydroxide (NaOH); sodium sulfide (Na_2S) content, nil. (b) Appearance: white granular mixture in that the case produced by a low-temperature bath consists predominantly of carbon.

High-temperature cyanide-type baths (deep-case baths) are usually operated in the temperature range from 900 to 955 °C (1650 to 1750 °F).

High-temperature baths are used for producing cases 0.5 to 3.0 mm (0.020 to 0.120 in.) deep. In some instances, even deeper cases are produced (up to about 6 mm, or 0.250 in.), but the most important use of these baths is for the rapid development of cases 1 to 2 mm (0.040 to 0.080 in.) deep. These baths consist of cyanide and a major proportion of barium chloride (Table 2).

Cyaniding (Liquid Carbonitriding)

Cyaniding, or salt bath carbonitriding, is a heat-treating process that produces a file-hard, wear-resistant surface on ferrous parts. When steel is heated above Ac_1 in a suitable bath containing alkali cyanides and cyanates, the surface of the steel absorbs both carbon and nitrogen from the molten bath. When quenched in mineral oil, paraffin-base oil, water, or brine, the steel develops a hard surface layer, or case, that contains less carbon and more nitrogen than the case developed in activated liquid carburizing baths.

Bath Composition. A sodium cyanide mixture such as grade 30 in Table 3 containing 30% NaCN, 40% Na_2CO_3, and 30% NaCl is generally used for cyaniding on a production basis. This mixture is preferable to any of the other compositions given in Table 3. The inert salts sodium chloride and sodium carbonate are added to cyanide to provide fluidity and to control the melting points of all mixtures. The 30% NaCN mixture, as well as those containing 45, 75, and 97% NaCN, may be added to the operating bath to maintain a desired cyanide concentration for a specific application.

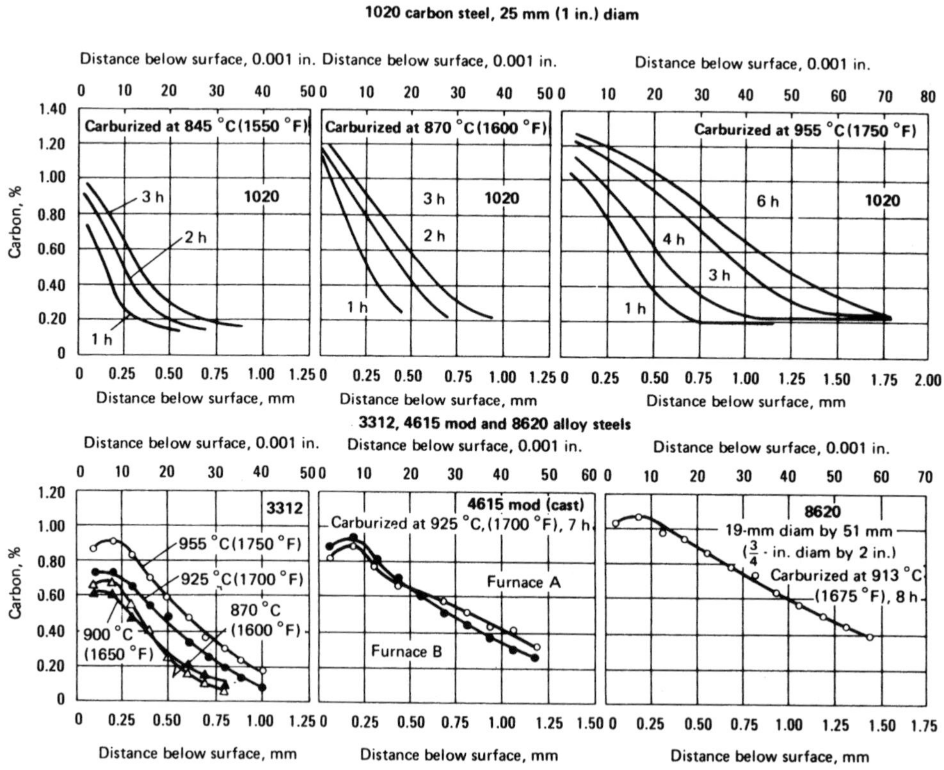

Fig. 3 Carbon gradients produced by liquid carburizing of carbon and alloy steels. Carbon gradients produced by liquid carburizing carbon and alloy steels in low-temperature and high-temperature baths. The 1020 carbon steel bars were carburized at 845, 870, and 955 °C (1550, 1600, and 1750 °F) for the periods shown. The data on 3312 alloy steel show the effect of four different carburizing temperatures on carbon gradient (time constant at 2 h). The data on modified 4615 steel castings indicate the slight differences in gradients obtained in two furnaces employing the same carburizing conditions (7 h at 925 °C, or 1700 °F). These data and the data on 8620 steel parts show a decrease in carbon content near the surface caused by diffusion of carbon during reheating to austenitizing temperature.

Cyaniding Temperature. Bath operating temperatures for cyanide hardening vary between 760 and 870 °C (1400 and 1600 °F). Temperatures near the lower end of this range are favored for minimizing distortion during quenching from the bath temperature.

When low-carbon and alloy steels are to be cyanided to produce a surface capable of resisting high-contact loads, the surface usually must be backed up with a fine-grain, tough core. This requires an operating bath temperature of about 870 °C (1600 °F).

Noncyanide Liquid Carburizing

Liquid carburizing can be accomplished in a bath containing a special grade of carbon instead of cyanide as the source of carbon. In this bath, carbon particles are dispersed in the molten salt by mechanical agitation, which is achieved by means of one or more simple propeller stirrers that occupy a small fraction of the bath. The basic composition and major operational features of this bath are as follows:

Composition	Alkali carbonates/chlorides; proprietary carbon cover
Decomposition	None
Contaminants	Metallics (Fe), SiO_2/Al_2O_3
Maintenance	Cover replenishment
Control	Physical: C1012 test-piece hardness. Chemical: H_2O insolubles (% C)
Correction	Add carbon; adjust agitation; sludge

The chemical reaction involved is not fully understood but is thought to involve adsorption of carbon monoxide on carbon particles. The carbon monoxide is generated by reaction between the carbon and carbonates, which are major ingredients of the molten salt. The adsorbed carbon monoxide is presumed to react with steel surfaces much as in gas or pack carburizing.

Operating temperatures for this type of bath are generally higher than those for cyanide-type baths. A range of approximately 900 to 955 °C (1650 to 1750 °F) is most commonly employed. Temperatures below approximately 870 °C (1600 °F) are not recommended, and may even

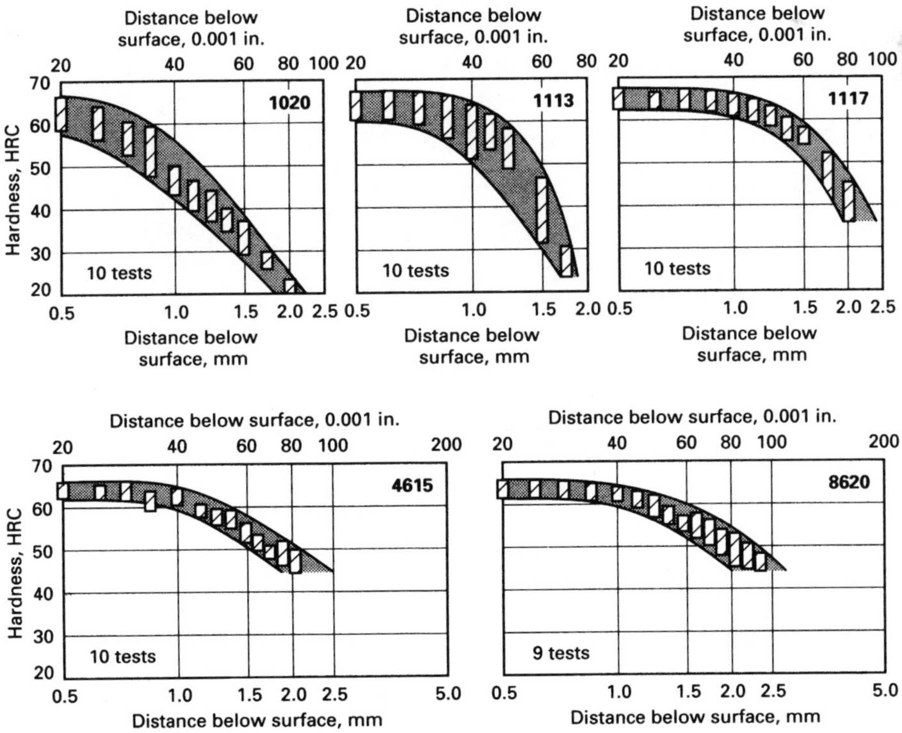

Fig. 4 Case-hardness gradients for selected steels showing scatter resulting from normal variations

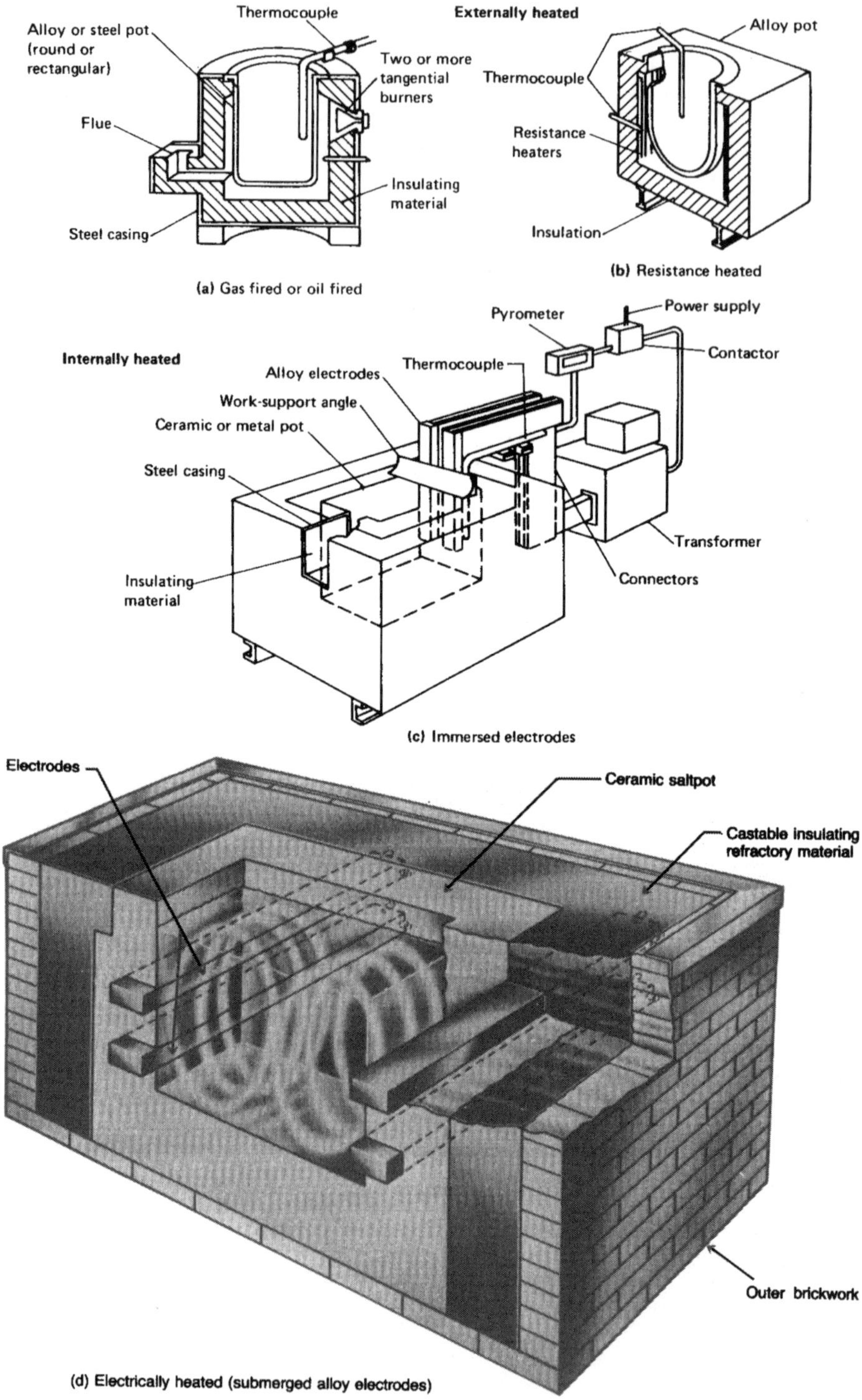

Fig. 5 Principal types of externally and internally heated salt bath furnaces used for liquid carburizing

lead to decarburization of the steel. The case depths and carbon gradients produced are in the same range as those for high-temperature cyanide-type baths (see Fig. 3), but there is no nitrogen in the case.

Temperatures above 955 °C (1750 °F) produce more rapid carbon penetration and do not adversely affect noncyanide baths because no cyanide is present to break down and cause carbon scum or frothing. Equipment deterioration is the chief factor that limits operating temperature.

Carbon Gradients

Figure 3 shows carbon gradients produced by liquid carburizing 1020 steel bars at 845, 870, and 955 °C (1550, 1600, and 1750 °F) for various lengths of time at carburizing temperature. Carbon-gradient data for two wrought alloy steels (3312 and 8620) and one cast alloy steel (4615 modified) are also shown. After carburizing, the 8620 steel parts were austenitized at 840 °C (1540 °F) and quenched in oil at 55 °C (130 °F). The 4615 modified steel parts were austenitized at 790 °C (1450 °F), quenched in salt at 190 °C (375 °F) for 3 min, and cooled in air.

Hardness Gradients

The indentation hardness data presented in Fig. 4 for five different steels indicate the effects of normal variations in practice on the hardness gradient. The shaded bands represent the scatter in results obtained from multiple tests of each steel. Although similar surface hardnesses are obtained with all five steels, depth of hardness varies with the alloy content of the steels. A comparison among the hardnesses of these five steels at a depth of 1 mm (0.040 in.) illustrates this variation.

Although a minimum case hardness of 60 HRC cannot be maintained to a depth of 1 mm (0.040 in.) with 1020 (0.30 to 0.60% Mn) steel, it can sometimes be achieved with 1113 (0.70 to 1.00% Mn) steel and can almost always be achieved with 1117 (1.00 to 1.30% Mn), 4615, and 8620 steels.

Furnaces and Equipment

Liquid carburizing is carried out in a salt bath furnace that may be heated either externally or internally (Fig. 5). In an externally heated furnace, heat is introduced into an annular space between the salt pot and the surrounding insulation, which usually is made of firebrick. In an internally heated furnace, heat is introduced directly into the salt. Both internally and externally heated furnaces generally have insulated lids that slide to open the bath and allow workpieces and fixtures to be positioned, usually with an overhead crane or with similar mechanized lifting equipment. For more detailed information, see the article "Salt Bath Equipment" in *Heat Treating,* Volume 4 of *ASM Handbook.*

Quenching

Most conventional quenching mediums, including water, brine, caustic solution, oil, and molten salts, are suitable for quenching parts that have been liquid carburized. However, the suitability of each medium must be related to specific parts and depends primarily on the hardenability of the steel, surface and core hardness requirements, and the amount of allowable distortion. Details on quenching practice following salt-bath processing can be found in the article "Liquid Carburizing and Cyaniding" in *Heat Treating,* Volume 4 of *ASM Handbook.*

Salt Removal (Washing)

The ease or difficulty with which salt can be removed from liquid carburized parts depends primarily on how simply or intricately shaped the parts are and whether they were quenched in water or in oil. To some extent, removal of salt may be complicated by the presence of insoluble residues. Water-quenched parts of simple design and with no blind holes or deep recesses usually are easy to clean. They may be rinsed thoroughly in water at approximately 80 °C (180 °F) and then coated with a rust-preventive fluid or soluble oil. Parts that are rinsed free of cyanide by immersion in a chloride salt and then isothermally quenched in a nitrate-nitrite salt are easily cleaned by agitated hot-water washing and rinsing. It is also possible to reclaim the nitrate-nitrite salt from the wash water.

Oil-quenched parts are more difficult to clean because the oil must be removed before the salts can be dissolved. Some salts may be insoluble. Use of power washers with hot water or emulsion cleaners is effective. An economical cleaning procedure begins with soaking of parts in hot water to float off the oil and remove the soluble salts. The parts may then be transferred to a hot agitated dilute alkaline cleaner having high sequestering properties. (Silicated cleaners and those containing carbonates or phosphates are not recommended, because of the formation

Table 4 Typical applications of liquid carburizing in cyanide baths

Part	Weight		Steel	Depth of case		Temperature		Time, h	Quench	Subsequent treatment	Hardness, HRC
	kg	lb		mm	in.	°C	°F				
Carbon steel											
Adapter	0.9	2	CR	1.0	0.040	940	1720	4	AC	(a)	62–63
Arbor, tapered	0.5	1.1	1020	1.5	0.060	940	1720	6.5	AC	(a)	62–63
Bushing	0.7	1.5	CR	1.5	0.060	940	1720	6.5	AC	(a)	62–63
Die block	3.5	7.7	1020	1.3	0.050	940	1720	5	AC	(a)	62–63
	1.1	2.5	CR	1.3	0.050	940	1720	5	AC	(a)	59–61
Disk	1.4	3	1020	1.3	0.050	940	1720	5	(b)	(b)	56–57
Flange	0.03	0.06	1020	0.4–0.5	0.015–0.020	845	1550	4	Oil	(c)	55 min(d)
Gage rings, knurled	0.09	0.2	1020	1.5	0.060	940	1720	6.5	AC	(a)	62–63
Hold-down block	0.9	2	CR	1.0	0.040	940	1720	4	AC	(a)	62–63
Insert, tapered	4.75	10.5	1020	1.3	0.050	940	1720	5	AC	(a)	62–63
Lever	0.05	0.12	1020	0.13–0.25	0.005–0.010	845	1550	1	Oil	(c)	(e)
Link	0.007	0.015	1018	0.13–0.25	0.005–0.010	845	1550	1	AC	...	...
Plate	0.007	0.015	1010	0.25–0.4	0.010–0.015	845	1550	2	Oil	(c)	(e)
Plug	0.7	1.6	CR	1.5	0.060	940	1720	6.5	AC	(a)	62–63
Plug gage	0.45	1	1020	1.5	0.060	940	1720	6.5	AC	(a)	62–63
Radius-cutout roll	7.7	17	CR	1.5	0.060	940	1720	6.5	AC	(a)	62–63
Torsion-bar cap	0.05	0.1	1022	0.02–0.05	0.001–0.002	900	1650	0.12	Caustic	(f)	45–47
Resulfurized steel											
Bushing	0.04	0.09	1118	0.25–0.4	0.010–0.015	845	1550	2	Oil	(c)	(e)
Dash sleeve	3.6	8	1117	1.1	0.045	915	1675	7	AC	(g)	58–63
Disk	0.0009	0.002	1118	0.13–0.25	0.005–0.010	845	1550	1	Brine	(c)	(e)
Drive shaft	3.6	8	1117	1.1	0.045	915	1675	7	AC	(h)	58–63
Guide bushing	0.2	0.5	1117	0.75	0.030	915	1675	5	(j)	...	58–63
Nut	0.04	0.09	1113	0.13–0.25	0.005–0.010	845	1550	1	Oil	(c)	(e)
Pin	0.003	0.007	1119	0.13–0.25	0.005–0.010	845	1550	1	Oil	(c)	(e)
Plug	0.007	0.015	1113	0.075–0.13	0.003–0.005	845	1550	0.5	Oil	(c)	(e)
Rack	0.34	0.75	1113	0.13–0.25	0.005–0.010	845	1550	1	Oil	(c)	(e)
Roller	0.01	0.03	1118	0.25–0.4	0.010–0.015	845	1550	2	Oil	(c)	(e)
Screw	0.003	0.007	1113	0.075–0.13	0.003–0.005	845	1550	0.5	Oil	(c)	(e)
Shaft	0.08	0.18	1118	0.25–0.4	0.010–0.015	845	1550	2	Oil	(c)	(e)
Spring seat	0.009	0.02	1118	0.25–0.4	0.010–0.015	845	1550	2	Oil	(c)	(e)
Stop collar	0.9	2	1117	1.1	0.045	925	1700	6.5	AC	(g)	60–63
Stud	0.007	0.015	1118	0.13–0.25	0.005–0.010	845	1550	1	Oil	(c)	(e)
Valve bushing	0.02	0.05	1117	1.3	0.050	915	1675	8	AC	(g)	58–63
Valve retainer	0.45	1	1117	1.1	0.045	915	1675	7	(j)	...	58–63
Washer	0.007	0.015	1118	0.25–0.4	0.010–0.015	845	1550	2	Oil	(c)	(e)
Alloy steel											
Bearing races	0.9–36	2–80	8620	2.3	0.090	925	1700	14	AC	(g)	61–64
Bearing rollers	0.20	0.5	8620	2.3	0.090	925	1700	14	AC	(g)	61–64
Coupling	0.03	0.06	8620	0.25–0.4	0.010–0.015	845	1550	2	Oil	(c)	(e)
Crankshaft	0.9	2	8620	1.0	0.040	915	1675	6.5	AC	(h)	60–63
Gear	0.34	0.75	8620	1.0	0.040	915	1675	6	AC	(g)	60–63
	0.03	0.06	8620	0.075–0.13	0.003–0.005	845	1550	0.5	Oil	(c)	(e)
Idler shaft	0.45	1	8620	0.75	0.030	915	1675	5	(i)	...	58–63
Pintle	4.5–86	10–190	8620	1.5	0.060	925	1700	12	(i)	...	58–63
Piston	0.20	0.5	8620	1.3	0.050	915	1675	8	AC	(g)	60–63
Plunger	0.45–82	1–180	8620	1.3	0.050	915	1675	8	(i)	...	58–63
Ram	2.3–23	5–50	8620	1.1	0.045	915	1675	7	(i)	...	58–63
Retainer	0.0009	0.002	9317	0.1–0.2	0.004–0.008	845	1550	0.33	Oil	(j)	(e)
Spool	0.45–54	1–120	8620	1.3	0.050	925	1700	7	(i)	...	58–63
Thrust cup	0.20	0.5	8620	1.1	0.045	915	1675	7	(i)	...	58–63
Thrust plate	5.4	12	8620	2.3	0.090	925	1700	14	AC	(g)	60–64
Universal socket	1.8	4	8620	1.5	0.060	915	1675	10	AC	(g)	58–63
Valve	0.01	0.03	8620	0.4–0.5	0.015–0.020	845	1550	4	Oil	(j)	60 min(d)
Valve seat	0.20	0.5	8620	1.1	0.045	915	1675	7	AC	(g)	60–63
Wear plate	0.45–3.6	1–8	8620	1.3	0.050	915	1675	7	AC	(g)	60–63

(a) Reheated at 790 °C (1450 °F), quenched in caustic, tempered at 150 °C (300 °F). (b) Transferred to neutral salt at 790 °C (1450 °F), quenched in caustic, tempered at 175 °C (350 °F). (c) Tempered at 165 °C (325 °F). (d) Or equivalent. (e) File-hard. (f) Tempered at 205 °C (400 °F). (g) Reheated at 845 °C (1550 °F), quenched in salt at 175 °C (350 °F). (h) Reheated at 775 °C (1425 °F), quenched in salt at 195 °C (380 °F). (i) Quenched directly in salt at 175 °C (350 °F). (j) Tempered at 165 °C (325 °F) and treated at –85 °C (–120 °F) of insoluble barium compounds in the presence of barium-containing salts.) If a white, powdery overlay of barium carbonate remains on the parts, it may be removed—following removal of all cyanide—by immersion in a dilute solution of acetic or inhibited hydrochloric acid.

Complex parts with blind holes, recesses, and threads are difficult to clean, particularly if they have been oil quenched. Liquid carburizing of parts that contain blind holes for which the depth exceeds twice the diameter is not recommended unless such holes can be plugged. Agitated hot water or a steam jet is probably the best solvent for salt held in recesses, crevices, and blind holes. Normally, these will remove all soluble salts and will soften insoluble residues. When part shape and tolerances permit, tumbling for 10 to 30 min in a mild alkali and a small quantity of sand is most effective in removing insoluble surface residues.

Washing of Cyanided Parts. Cyanide-hardened parts are easy to wash, even after oil quenching, because cyanide and sodium carbonate are good detergents and because all the components of the salt bath are water soluble. The work may be soaked in a tank of agitated boiling water, rinsed in clean hot water, and then rustproofed (if required). Power spray washers, using hot water in a two-stage system, also give satisfactory results.

Typical Applications

The applicability of liquid carburizing is evidenced by Table 4, in which all of the parts listed were treated on a production basis. For ease of reference, these parts have been separated according to type of steel (carbon, resulfurized, or alloy) and the parts in each group have been arranged in alphabetical order. This table also provides details, wherever they were available, regarding case depth, carburizing temperature and cycle time, method of quenching, subsequent treatment, and surface hardness. Although all of the parts listed in Table 4 were carburized in cyanide-type baths, noncyanide baths could have been used with slight adjustments in operating conditions.

Environmental and Safety Considerations

Precautions in the Use of Cyanide Salts. Cyanides cause violently poisonous reactions if allowed to come into contact with scratches or wounds (on the hands, for example); they are fatal if taken internally. Also, fatally poisonous fumes are evolved when cyanides are brought into contact with acids. The white deposits that form on hoods and cooler furnace parts consist mainly of sublimed sodium carbonate, with small amounts of sodium, potassium, and barium salts, but may contain some cyanide as the result of splashing.

Disposal of Cyanide Wastes. Cyanide wastes, whether dissolved in quench water or in the form of solid salt from pots, pose a serious disposal problem. The cyanide contents of these wastes must be altered chemically to render the material nonpoisonous before it is discharged into sewers or streams. Because of the toxicity of cyanide wastes, local ordinances and pollution authorities should be consulted regarding the proper disposal of wastes.

SELECTED REFERENCES

- R.W. Foreman, Pack Carburizing, *Heat Treating,* Vol 4, *ASM Handbook,* ASM International, 1991, p 325–328
- A.D. Godding, Liquid Carburizing and Cyaniding, *Heat Treating,* Vol 4, *ASM Handbook,* ASM International, 1991, p 329–347
- W.J. Laird, Jr., Salt Bath Equipment, *Heat Treating,* Vol 4, *ASM Handbook,* ASM International, 1991, p 475–483

CHAPTER 5

Carbonitriding

CARBONITRIDING is a modified carburizing process and not a form of nitriding. The process, which involves the diffusion of both carbon and nitrogen into the base steel, can be carried out in a salt bath (i.e., cyaniding, as described in Chapter 4, "Pack and Liquid Carburizing") or in a furnace gas atmosphere. More recently, plasma processing has also been used to carbonitride steel components. Emphasis in this Chapter is on the gas carbonitriding process, which consists of introducing ammonia (NH_3) into the carburizing atmosphere in order to add nitrogen to the carburized case as it is being produced (Fig. 1). Typically, carbonitriding is done at a lower temperature than carburizing—between 775 and 900 °C (1425 and 1650 °F) versus 870 and 1065 °C (1600 and 1950 °F) for carburizing—and for a shorter time. Combine this with the fact that nitrogen inhibits the diffusion of carbon and what generally results is a shallower case than is typical for carburized parts. A carbonitrided case is usually between 0.075 and 0.75 mm (0.003 and 0.030 in.) deep.

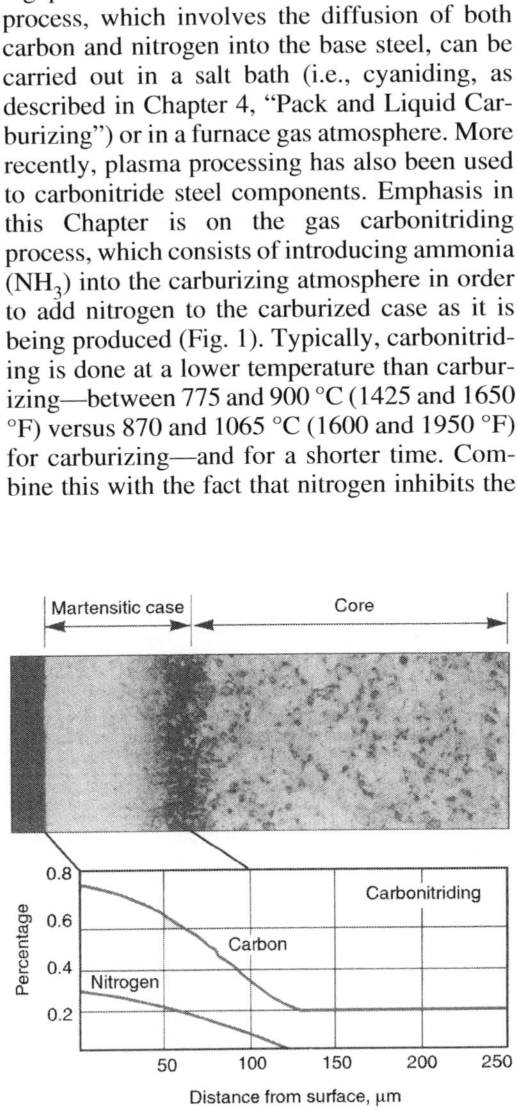

Fig. 1 Surface layers produced by carbonitriding of steel at 850 °C (1560 °F), where carbon predominates in the formation of a martensitic layer. Source: Ref 1

The nitrogen in carbonitrided steel enhances hardenability and makes it possible to form martensite in plain carbon and low-alloy steels that initially have low hardenability. Examples include American Iron and Steel Institute (AISI) 1018, 12L14, and 1117. Carbon is the primary element for improving hardness.

Nitrogen, similar to carbon, manganese, and nickel, is an austenite stabilizer, so retained austenite is a concern after quenching. Lowering the ammonia percentage reduces the amount of retained austenite and should be done if decreases in hardness or wear resistance cannot be tolerated. Another consequence of too-high nitrogen is the formation of voids or porosity. In general, the nitrogen content at the surface should be no greater than 0.40% (Fig. 1).

The nitrogen in the carbonitrided case also increases the resistance of steel to softening at slightly elevated temperatures. The greater the nitrogen content, the greater the resistance to softening. This is why higher tempering temperatures—up to 225 °C (440 °F)—are often used on carbonitrided parts. The resistance to softening manifests itself in improved wear resistance. Carbonitrided gears, for example, often exhibit better wear properties than carburized gears.

Applicable Steels and Applications

Steels commonly carbonitrided include those in the 1000, 1100, 1200, 1300, 1500, 4000, 4100, 4600, 5100, 6100, 8600, and 8700 series, with carbon contents up to approximately 0.25%. Also, many steels in these same series with a carbon range of 0.30 to 0.50% are carbonitrided to case depths up to approximately 0.3 mm (0.01 in.) when a combination of a reasonably tough, through-hardened core and a hard, long-wearing surface is required (shafts and transmission gears are typical examples). Steels such as 4140, 5130, 5140, 8640, and 4340 for applications such as heavy-duty gearing are treated by this method at 845 °C (1550 °F). High-density powder metallurgy steels are also often carbonitrided for improved wear resistance and toughness.

Applications. Although carbonitriding is a modified carburizing process, its applications are more restricted than those of carburizing. As has been stated previously, carbonitriding is largely limited to case depths of approximately 0.75 mm (0.03 in.) or less, while no such limitation applies to carburizing. There are two reasons for this: carbonitriding is generally done at temperatures of 870 °C (1600 °F) and below, whereas, because of the time factor involved, deeper cases are produced by processing at higher temperatures; and the nitrogen addition is less readily controlled than is the carbon addition, a condition that can lead to an excess of nitrogen and, consequently, to high levels of retained austenite and case porosity when processing times are too long.

The resistance of a carbonitrided surface to softening during tempering is markedly superior to that of a carburized surface. Other notable differences exist in terms of residual-stress pattern, metallurgical structure, fatigue and impact strength at specific hardness levels, and effects of alloy composition on case and core characteristics.

Various production parts that have been successfully carbonitrided are listed in Table 1. A review of this list suggests the range of applications for which carbonitriding has been found to be an advantage.

Composition of Case

The composition of a carbonitrided case depends on temperature, time, atmosphere composition, and type of steel. The higher the carbonitriding temperature, the less effective is the ammonia addition to the atmosphere as a nitrogen source, because the rate of spontaneous decomposition of ammonia to molecular nitrogen and hydrogen increases as the temperature is raised. Figure 2 shows that lower temperatures favor increased surface nitrogen concentrations. The addition of ammonia to a carburizing atmosphere has the effect of dilution by the following reaction:

$$2NH_3 \rightleftarrows N_2 + 3H_2 \quad \text{(Eq 1)}$$

Thus, as shown in Fig. 2, the carbon potential possible with a given carbon dioxide level is higher in a carburizing atmosphere than in a carbonitriding atmosphere. Dilution with nitrogen and hydrogen affects measurements of oxygen potential in a similar manner; the carbon potential possible with a given oxygen potential is higher in a carburizing atmosphere than in a carbonitriding atmosphere. Water vapor content, however, is affected much less by this dilution. Thus, the amount of dilution and its resulting effect on the atmosphere composition depends on the processing temperature, the amount of ammonia introduced, and the ratio of the total atmosphere gas flow rate to the volume of the furnace.

Carbonitriding can be carried out at such low temperatures as to produce a compound layer, so called because iron-carbon-nitrogen compounds are formed at the surface. In certain wear applications, this type of case structure is suitable. To produce this layer of compound, large percentages of ammonia are required. It is usually unnecessary to liquid quench parts carbonitrided in this manner. However, because the diffusion rate of nitrogen and the rate of formation of the compound are so slow at temperatures below 705 °C (1300 °F), such practice is economically applicable only to shallow cases in applications in which dimensional tolerances would be difficult to maintain if the parts were treated at higher temperatures.

Figure 3 shows carbon and nitrogen gradients and case hardness data for 1018 carbon steel and 8620 low-alloy steel that were carbonitrided for 4 h at 845 °C (1550 °F) in a batch-type radiant-tube furnace. These test data were obtained in a manufacturing plant under normal production conditions, employing a standard carbonitriding cycle. All test specimens were carbonitrided along with production loads of 22.7 kg (50 lb) of gears and shafts.

Depth of Case

Preferred case depth is governed by service application and by core hardness. Case depths of 0.025 to 0.075 mm (0.001 to 0.003 in.) are commonly applied to thin parts that require wear resistance under light loads. Case depths up to 0.75 mm (0.030 in.) may be applied to

Table 1 Typical applications and production cycles for carbonitriding

Part	Steel	Case depth, mm	Case depth, 0.001 in.	Furnace temperature, °C	Furnace temperature, °F	Total time in furnace	Quench
Carbon steels							
Adjusting yoke, 25 by 9.5 mm (1 by 0.37 in.)	1020	0.05–0.15	2–6	775 and 745	1425 and 1375	64 min	Oil
Bearing block, 64 by 32 by 3.2 mm (2.5 by 1.3 by 0.13 in.)	1010	0.05–0.15	2–6	775 and 745	1425 and 1375	64 min	Oil
Cam, 2.3 by 57 by 64 mm (0.1 by 2.25 by 2.5 in.)	1010	0.38–0.45	15–18	855	1575	2½ h	Oil
Cup, 13 g (0.46 oz)	1015	0.08–0.13	3–5	790	1450	½ h	Oil
Distributor drive shaft, 125 mm OD by 127 mm (5 by 5 in.)	1015	0.15–0.25	6–10	815 and 745	1500 and 1375	108 min	Gas(a)
Gear, 44.5 mm diam by 3.2 mm (1.75 by 0.125 in.)	1213(b)	0.30–0.38	12–15	855	1575	1¾ h	Oil(c)
Hex nut, 60.3 by 9.5 mm (2.4 by 0.37 in.)	1030	0.15–0.25	6–10	815 and 745	1500 and 1375	64 min	Oil
Hood-latch bracket, 6.4 mm diam (0.25 in.)	1015	0.05–0.15	2–6	775 and 745	1425 and 1375	64 min	Oil
Link, 2 by 38 by 38 mm (0.079 by 1.5 by 1.5 in.)	1022	0.30–0.38	12–15	855	1575	1½ h	Oil
Mandrel, 40 g (1.41 oz)	1117	0.20–0.30	8–12	845	1550	1½ h	Oil
Paper-cutting tool, 410 mm long	1117	~0.75	~30	...	...	...	...
Piston for torque converter, 1.8 kg (4 lb), 260 mm (10.2 in.) OD	1010	0.1–0.3	4–12	900	1650	45 min	Polymer
Retainer ring for torque converter, 0.32 kg (0.7 lb), 260 mm (10.2 in.) OD	1006	0.075–0.2	3–8	830	1525	45 min	Polymer
Segment, 2.3 by 44.5 by 44.5 mm (0.09 by 1.75 by 1.75 in.)	1010	0.38–0.45	15–18	855	1575	2½ h	Oil
Shaft, 4.7 mm diam by 159 mm (0.19 by 6.25 in.)	1213(b)	0.30–0.38	12–15	815	1500	2½ h	Gas(a)(d)
Shift collar, 59 g (2.1 oz)	1118	0.30–0.36	12–14	775	1430	5½ h	Oil(e)
Sliding spur gear, 66.7 mm OD (2.625 in.)	1018	0.38–0.50	15–20	870	1600	2 h(f)	Oil (g)
Spring pin, 14.3 mm OD by 114 mm (0.56 by 4.5 in.)	1030	0.25–0.50	10–20	815 and 745	1500 and 1375	144 min	Oil
Spur pinion shaft, 41.3 mm OD (1.625 in.)	1018	0.38–0.50	15–20	870	1600	2 h(f)	Oil(h)
Transmission shift fork, 127 by 76 mm (5 by 3 in.)	1040	0.25–0.50	10–20	815 and 745	1500 and 1375	162 min	Gas(a)
Alloy steels							
Helical gear, 82 mm OD (3.23 in.)	8617H	0.50–0.75	20–30	845	1550	6 h(f)	Oil(g)
Input shaft, 1.2 kg (2.6 lb)	5140	0.30–0.35	12–14	775	1430	5½ h	Oil(e)
Pinion gear, 0.2 kg (0.44 lb)	4047	0.30–0.35	12–14	775	1430	5½ h	Oil(e)
Ring gear, 0.9 kg (2 lb)	4047	0.20–0.30	8–10	760	1400	9 h	Oil(i)
Segment, 1.4 by 83 mm (0.055 by 3.27 in.)	8617	0.18–0.25	7–10	815	1500	1½ h	Gas(a)
Spur pinion shaft, 63.5 mm OD by 203 mm (2.5 by 8 in.)	5140H	0.05–0.20	2–8	845	1550	1 h(f)	Oil(j)
Stationary gear plate, 0.32 kg (0.7 lb)	5140	0.30–0.35	12–14	775	1430	5½ h	Oil(e)
Transmission main shaft sleeve, 38 mm OD by 25 mm (1.5 by 2 in.)	8622	0.15–0.25	6–10	815 and 745	1500 and 1375	108 min	Gas(a)
Transmission main shaft washer, 57 mm OD by 6.4 mm (2.25 by 0.25 in.)	8620	0.25–0.50	10–20	815 and 745	1500 and 1375	162 min	Gas(a)

(a) Modified carbonitriding atmosphere. (b) Leaded. (c) Tempered at 190 °C (375 °F). (d) Tempered at 150 °C (300 °F). (e) Tempered at 165 °C (325 °F). (f) Time at temperature. (g) Oil at 150 °C (300 °F); tempered at 150 °C (300 °F); for 1 h. (h) Oil at 150 °C (300 °F); tempered at 260 °C (500 °F) for 1 h. (i) Tempered at 175 °C (350 °F). (j) Oil at 150 °C (300 °F); tempered at 230 °C (450 °F) for 2 h. OD, outside diameter parts (such as cams) for resisting high compressive loads. Case depths of 0.63 to 0.75 mm (0.025 to 0.030 in.) may be applied to shafts and gears that are subjected to high tensile or compressive stresses caused by torsional, bending, or contact loads.

Medium-carbon steels with core hardnesses of 40 to 45 HRC normally require less case depth than steels with core hardnesses of 20 HRC or below. Low-alloy steels with medium-carbon content, such as those used in automotive transmission gears, are often assigned minimum case depths of 0.2 mm (0.008 in.).

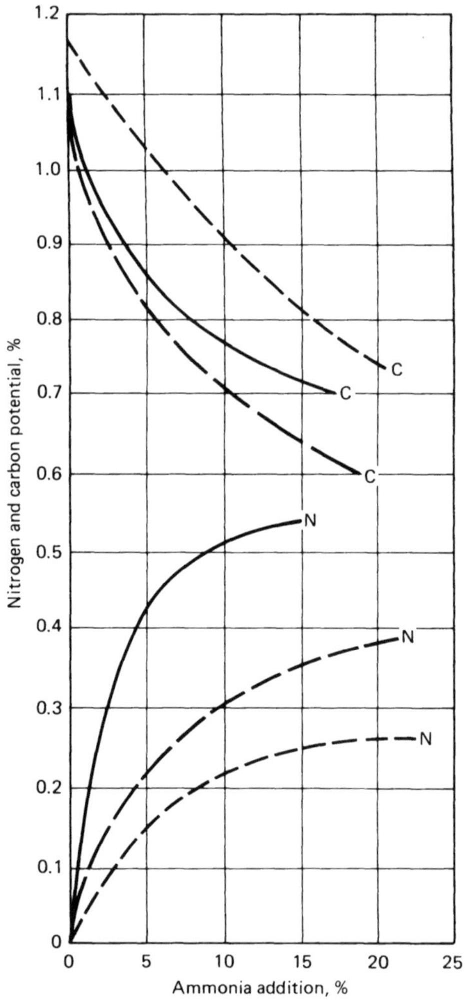

Fig. 2 Effect of ammonia additions on nitrogen and carbon potentials determined using low-carbon steel foil. For three sets of conditions: solid lines, 3 h at 850 °C (1560 °F) and 0.29% CO_2; broken lines, 1 h at 925 °C (1695 °F) and 0.13% CO_2; dashed lines, 1 h at 950 °C (1740 °F) and 0.10% CO_2. Source: Ref 2

Measurements of the case depths of carbonitrided parts may refer to effective case depth or total case depth, as with reporting case depths for carburized parts. For very thin cases, usually only the total case depth is specified. In general, it is easy to distinguish case and core microstructures in a carbonitrided piece, particularly when the case is thin and is produced at a low carbonitriding temperature; more difficulty is encountered in distinguishing case and core when high temperatures, deep cases, and medium-carbon or high-carbon steels are involved. Whether or not the core has a martensitic structure is also a contributing factor in case-depth measurements.

Effect of Time and Temperature. Based on a survey of industrial practice, Fig. 4 shows case depths for different combinations of total

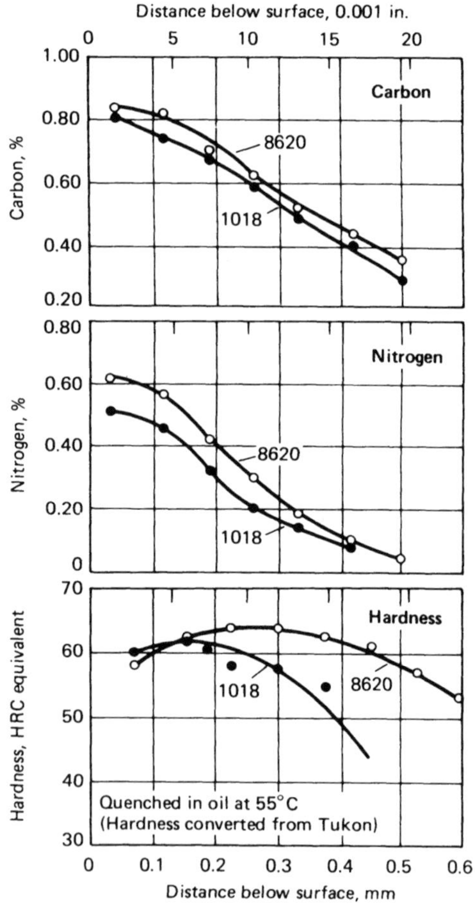

Fig. 3 Carbon, nitrogen, and hardness gradients for carburized 1018 and 8620 steels. For processing details, see text.

furnace treating time and temperature. (Note that all values given for case depth are for effective case depth unless otherwise stated.)

Figure 5(a) shows the effects of total furnace time on case depth for 1020 steel. Specimens were heated to 705, 760, 815, and 870 °C (1300, 1400, 1500, and 1600 °F) for periods of 15, 30, and 45 min. Figure 5(b) indicates the total case depths that can be obtained on a 1112 steel held

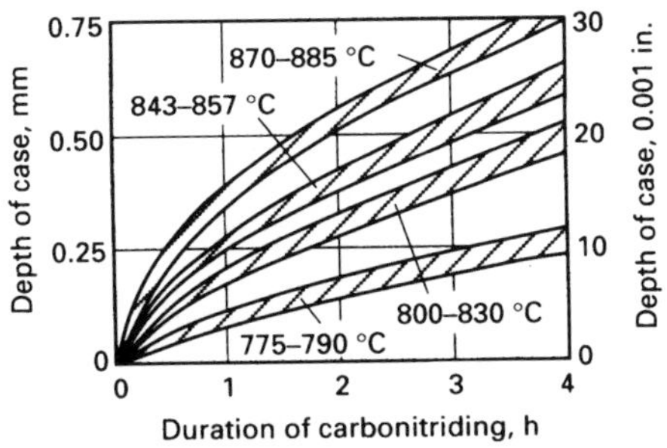

Fig. 4 Results of a survey of industrial practice regarding effects of time and temperature on effective case depth of carbonitrided cases

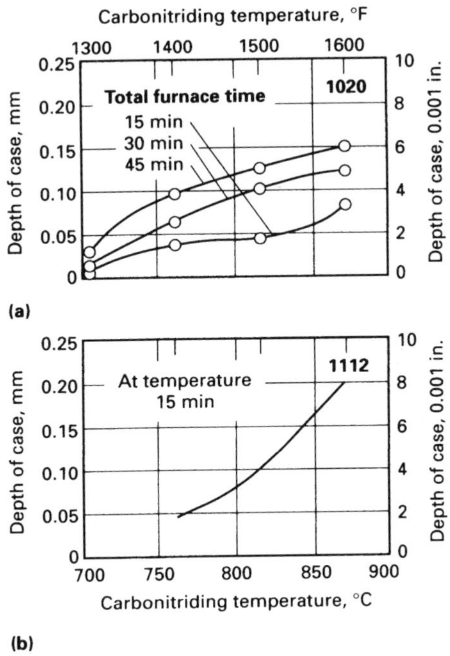

Fig. 5 Effects of temperature and of duration of carbonitriding on effective case depth. Both sets of data were obtained in the same plant. (a) 1020 steel; data given in terms of total furnace time. (b) 1112 steel; data given for 15 min at temperature

for 15 min at various temperatures between approximately 750 and 900 °C (1380 and 1650 °F). All data in Fig. 5 were obtained in a single plant.

Case-depth uniformity in carbonitriding depends on temperature uniformity within the furnace chamber, adequate circulation and replenishment of atmosphere, and distribution of the furnace charge so that it is uniformly exposed to the atmosphere.

Accurate control of treatment time is also a factor in controlling case-depth uniformity. All parts in a load should be at a uniform temperature prior to exposure to the carbonitriding atmosphere in order to achieve uniform results when the processing times are short.

Hardenability of Case

One major advantage of carbonitriding is that the nitrogen absorbed during processing lowers the critical cooling rate of the steel. That is, the hardenability of the case is significantly greater when nitrogen is added by carbonitriding than when the same steel is only carburized (Fig. 6). This permits the use of steels on which uniform case hardness ordinarily could not be obtained if they were only carburized and quenched. Where core properties are not important, carbonitriding permits the use of low-carbon steels, which cost less and may have better machinability or formability.

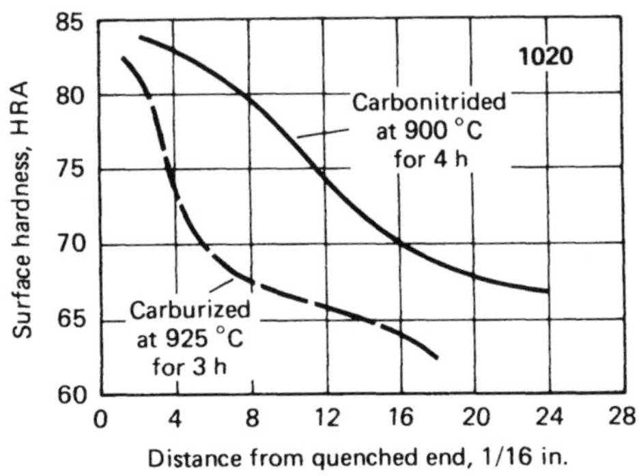

Fig. 6 End-quench hardenability curve for 1020 steel carbonitrided at 900 °C (1650 °F) compared with curve for the same steel carburized at 925 °C (1700 °F). Hardness was measured along the surface of the as-quenched hardenability specimen. Ammonia and methane contents of the inlet carbonitriding atmosphere were 5%; balance, carrier gas. Source: Ref 3

Because of the hardenability effect of nitrogen, carbonitriding makes it possible to oil quench steels such as 1010, 1020, and 1113 to obtain martensitic case structures. Because of lower processing temperatures and/or the use of less severe quenches, carbonitriding may produce less part distortion and better control of dimensions than carburizing and thus may eliminate the need for straightening or final grinding operations.

Hardness Gradients

Hardness at various levels in the case depends on the microstructure. Hardness gradients associated with the microstructures of 1117 steel are presented in Fig. 7. When the carbonitriding atmosphere was relatively high in ammonia (11% NH_3), the nitrogen content of the case was high, and enough austenite was retained after quenching to lower the hardness

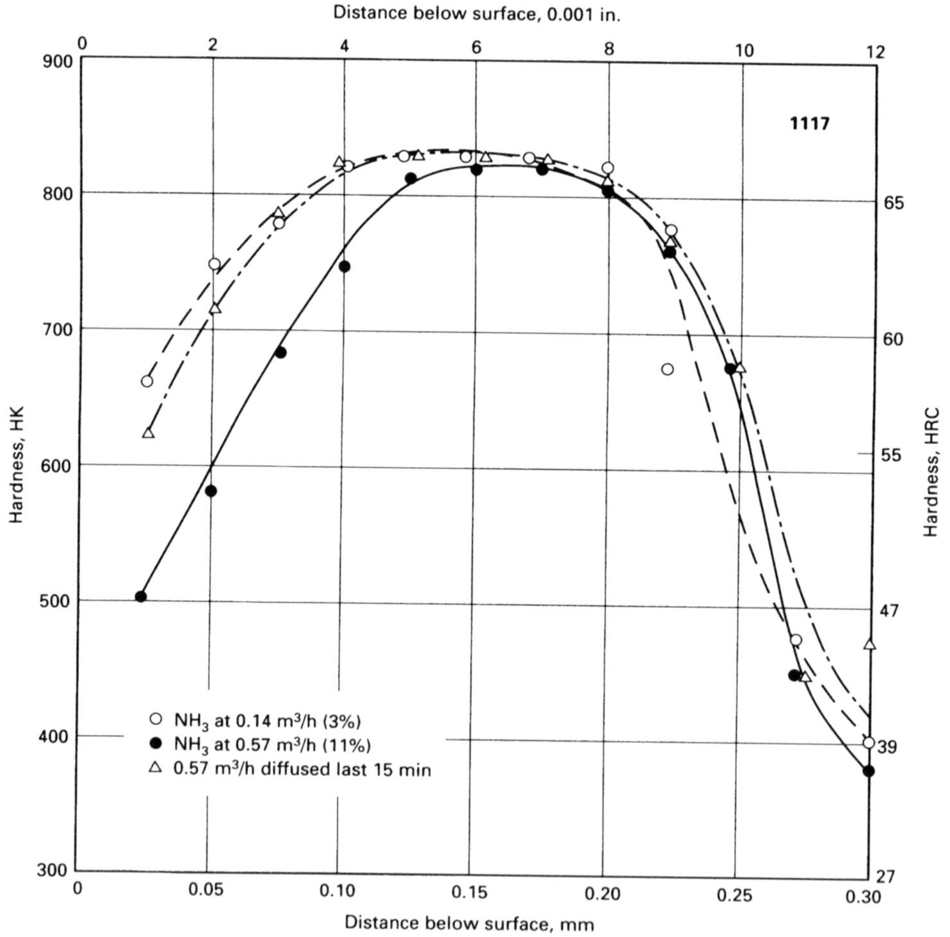

Fig. 7 Hardness gradients in 1117 steel carbonitrided at 815 °C (1500 °F) for $1\frac{1}{2}$ h and quenched in oil. Required minimum hardness of 630 HK (55 HRC) at 0.025 mm below the surface was met by reducing the percentage and flow rate of ammonia or by adding a diffusion period after carbonitriding, as indicated. Atmosphere consisted of endothermic carrier gas (dewpoint, –1 °C) at 4.25 m^3/h (150 ft^3/h), natural gas at 0.17 m^3/h (6 ft^3/h), and ammonia in the amounts indicated.

to 48 HRC, 500 g (1.1 lbf) load, at a depth of 0.025 mm (0.001 in.) below the surface. The amount of retained austenite was decreased, and hardness consequently was increased, either by lowering the ammonia flow rate from 0.57 to 0.14 m^3/h (20 to 5 ft^3/h), which reduced the ammonia content of the furnace atmosphere from 11 to 3%, or by introducing a 15 min diffusion period at the end of the carbonitriding operation. Either treatment increased the hardness to meet or exceed a specified minimum value of 55 HRC, 500 g (1.1 lbf) load at 0.025 mm (0.001 in.) below the surface.

Similar data relating ammonia content to hardness for 1018 steel carbonitrided at 790 °C (1455 °F) for $2^1/_2$ h and at 845 °C (1550 °F) for $2^1/_2$ h are shown in Fig. 8.

Void Formation

Subsurface voids or porosity in the case structure (Fig. 9) may occur in carbonitrided parts if the processing conditions are not adjusted properly. Although details of the mechanism of void formation are not completely understood, this problem has been related to excessive ammonia additions. Table 2 summarizes the factors that have been shown singly or in combination to contribute to void formation. No attempt has been made to quantify the interaction of the material and process variables presented in Table 2. Rather, this information should be used as a guide for avoiding or eliminating porosity problems. It should also be noted that the reprocessing of parts previously carbonitrided can, in many instances, also lead to void formation.

None of the data included in this article were obtained from parts or testpieces containing case porosity.

Furnaces and Furnace Atmospheres

Furnaces. Almost any furnace suitable for gas carburizing can be adapted to carbonitriding (see Chapter 2, "Gas Carburizing," for details). Whether dense or shallow (openly spaced) workloads are to be processed, the furnace must be equipped with a fan to circulate the atmosphere. For work that is to be clean and bright after quenching, the furnace must be equipped with protective-atmosphere vestibules to the quench area.

Furnace Atmospheres. The atmospheres used in carbonitriding generally comprise a mixture of carrier gas, enriching gas, and ammonia. Basically, the atmospheres used in carbonitriding are produced by adding from approximately 2 to 12% ammonia to a standard gas-carburizing atmosphere.

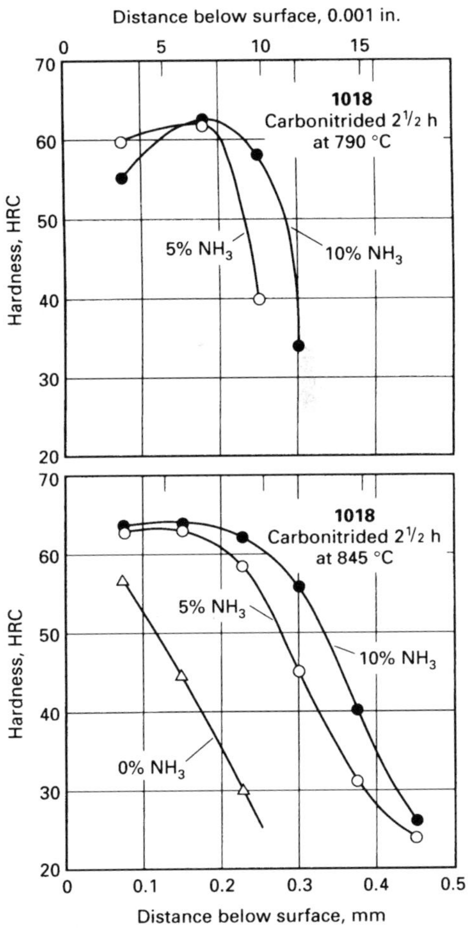

Fig. 8 Effect of ammonia content of carbonitriding gas on hardness gradient for 1018 steel carbonitrided at 790 or 845 °C (1455 or 1550 °F)

Control of Retained Austenite

Nitrogen lowers the transformation temperature of austenite. Therefore, a carbonitrided case usually contains more retained austenite than a carburized case of the same carbon content. The low indentation hardness resulting from retained austenite is undesirable in many applications. It can be extremely detrimental in components of close-fitting assemblies—for example, shaft and sleeve assemblies wherein the shaft is intended to rotate or reciprocate in the sleeve. The delayed transformation of austenite to martensite at ambient temperature results in a volume increase that may cause moving parts to bind or "freeze" in service.

The amount of retained austenite can be significantly decreased by cooling the quenched parts to –40 to –100 °C (–40 to –150 °F) or lower (cryogenic treatments as low as –196 °C, or –320 °F), as shown in Fig. 10. When close-tolerance ground parts are involved, this treatment should precede finish grinding. Subzero treatment of parts that are to be tempered should precede final tempering. Subzero treatment may cause microcracks in the case, particularly in coarse-grain steels. Because the amount of retained austenite is normally at a maximum near the steel surface, it can be removed from symmetrical contours by grinding. However, care must be exercised in grinding high retained-austenite surfaces because of the increased possibility of grinding burn or checking. If grinding is not required for any reason other than to remove retained austenite, it also may be considered an expensive operation. The most economical way to minimize retained austenite is by selection of preferred steels and control of the carbonitriding process.

Minimizing retained austenite in the carbonitrided case is assisted by modification of several processing factors:

- *Furnace temperature.* An increase in furnace temperature reduces the nitrogen content of the outer portions of the case, thus minimizing the amount of retained austenite. It is far better to reduce ammonia flow, rather than depend on increased temperature to lower the nitrogen content.
- *Carbon potential.* Lowering the carbon potential also helps reduce retained austenite.
- *Ammonia content* of the carbonitriding atmosphere should be restricted to the minimum

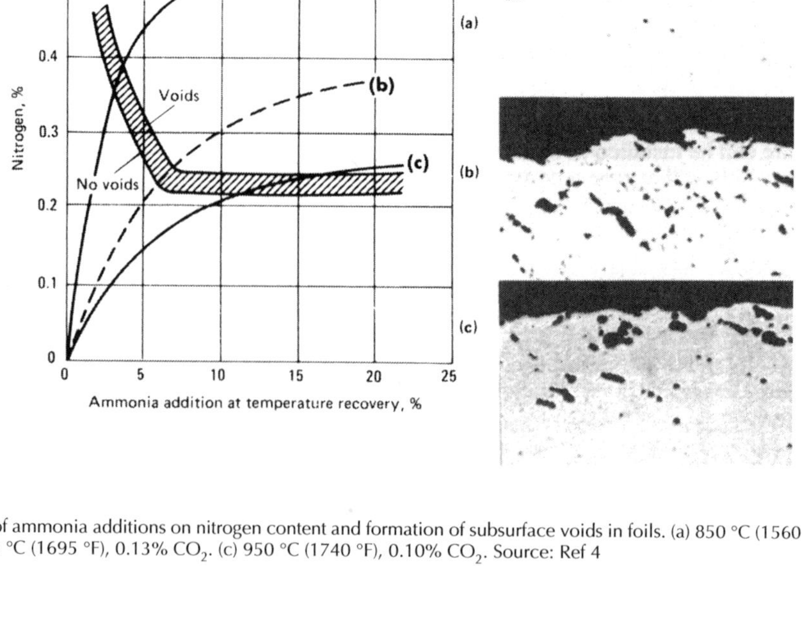

Fig. 9 Effect of ammonia additions on nitrogen content and formation of subsurface voids in foils. (a) 850 °C (1560 °F), 0.29% CO_2. (b) 925 °C (1695 °F), 0.13% CO_2. (c) 950 °C (1740 °F), 0.10% CO_2. Source: Ref 4

required to obtain the desired hardenability. A 1 to 5% NH_3 content in the inlet gas is usually a satisfactory starting point; a lower content decreases the rate of penetration but may be desirable to minimize retained austenite and to avoid case porosity.

Table 2 Effect of material/variables on the possibility of void formation in carbonitrided cases

Material/processing variables(a)	Possibility of void formation
Temperature increase	Increased
Longer cycles	Increased
Higher case nitrogen levels	Increased
Higher case carbon levels	Increased
Aluminum-killed steel	Increased
Increased alloy content of steel	Decreased
Severe prior cold working of material	Increased
Ammonia addition during heat-up cycle	Increased

(a) All other variables held constant. Sources: Based on data in Fig. 9 and Ref 3 and 4

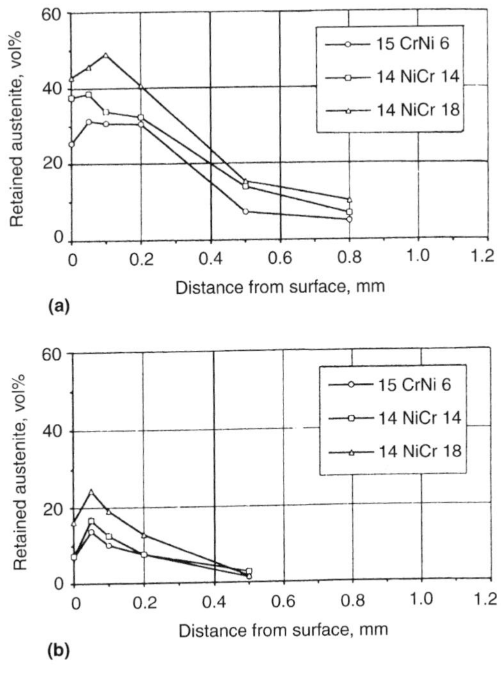

Fig. 10 Effect of cryogenic treatment on the amount of retained austenite in carbonitrided low-alloy steels. (a) Amount of retained austenite in the as-quenched (60 °C, or 140 °F) condition. (b) Amount of retained austenite after cryogenic treatment at −196 °C (−320 °F). Source: Ref 5

Quenching Media and Practices

Whether carbonitrided parts are quenched in water, oil, or gas depends on the allowable distortion, metallurgical requirements (such as case and core hardness), and type of furnace equipment employed.

Water Quenching. Depending on allowable distortion, parts made of low-carbon steel may be quenched in water. For example, shift-lever pins made of B1212 steel are commonly water quenched. Water quenching usually is restricted to those furnaces in which the work is transferred from the furnace into the air prior to quenching, thus avoiding possible contamination of the furnace atmosphere by water vapor. However, water quenching from a rotary-retort furnace is feasible, provided the quench chute is equipped with gas eductors and a water distribution system for condensing water vapor.

It should be noted that ammonia is extremely soluble in water and forms a product (NH_4OH) that is extremely corrosive to copper-base materials. In continuous operations where water is exposed to an ammonia-bearing atmosphere, brass agitators, copper tube bundles in heat exchangers, and similar copper alloy components should be avoided.

Oil Quenching. Quenching oil temperatures may vary from approximately 40 to 105 °C (100 to 220 °F). Special high-flash-point oils may be used at the higher temperatures to minimize distortion; sometimes, molten salt is used for the same reason. In the normal range of oil temperature (approximately 50 to 70 °C, or 120 to 160 °F), a mineral oil with a minimum flash point of 170 °C (335 °F) and a viscosity of 21×10^{-6} m^2/s at 38 °C (21 centistokes, or 100 Saybolt universal seconds, or SUS, at 100 °F) is commonly used. Special oils containing additives for increasing the quenching rate also may be used. To maintain maximum quenching effectiveness, quenching oils should have a low capacity for dissolving water. Quenching oils that dissolve even small amounts of water may lose effectiveness in 3 to 6 months; those that shed water completely may be used significantly longer.

Gas Quenching. Parts that have small mass (such as thin stampings) and that are subjected to sliding loads with low impact may be quenched in a stream of cooled atmosphere gas. Gas or atmosphere quenching principally serves to reduce distortion and to eliminate the high costs of straightening. (Usually, however, a gas-quenched case retains enough ductility to permit roller straightening, if required.)

In gas quenching, parts must be loaded into furnace trays carefully so that the surfaces of the parts can be cooled rapidly enough to produce desired hardness. Trays should be loaded and stacked so that the total mass of the load does not exceed that which can be satisfactorily quenched.

Polymer Quenching. Low distortion and dimensional accuracy of thin parts can be accomplished by means of an automated heat treating system that incorporates a quenching press. In one automotive plant, an integrated quench press-hardening machine is used to heat treat thin automotive torque-converter parts (Ref 6). The polymer quench medium is maintained at a 15% concentration, and the quench bath temperature is controlled at 50 °C (120 °F) using an external cooling device. Quench press forces up to 50 kN (11,240 lbf) are used.

Tempering

Many shallow-case carbonitrided parts are used without tempering. The presence of nitrogen in the carbonitrided case increases its resistance to softening, and the increase varies with the amount of nitrogen in the case. Resistance to tempering may be desirable where service operating temperatures are abnormally high or where hot straightening is employed.

Tempering data obtained on carbonitrided cases of 1018 steel are given in Fig. 11. The data relate temper resistance to both carbonitriding temperature and the ammonia content of the atmosphere. Figure 12 presents a summary of the effects of carbonitriding temperature and ammonia content on temper resistance, derived from the same specimens referred to in Fig. 11.

Because tempering a carbonitrided case at 425 °C (795 °F) and above results in a marked increase in notch toughness (see Table 3), parts that are to be subjected to repeated shock load-

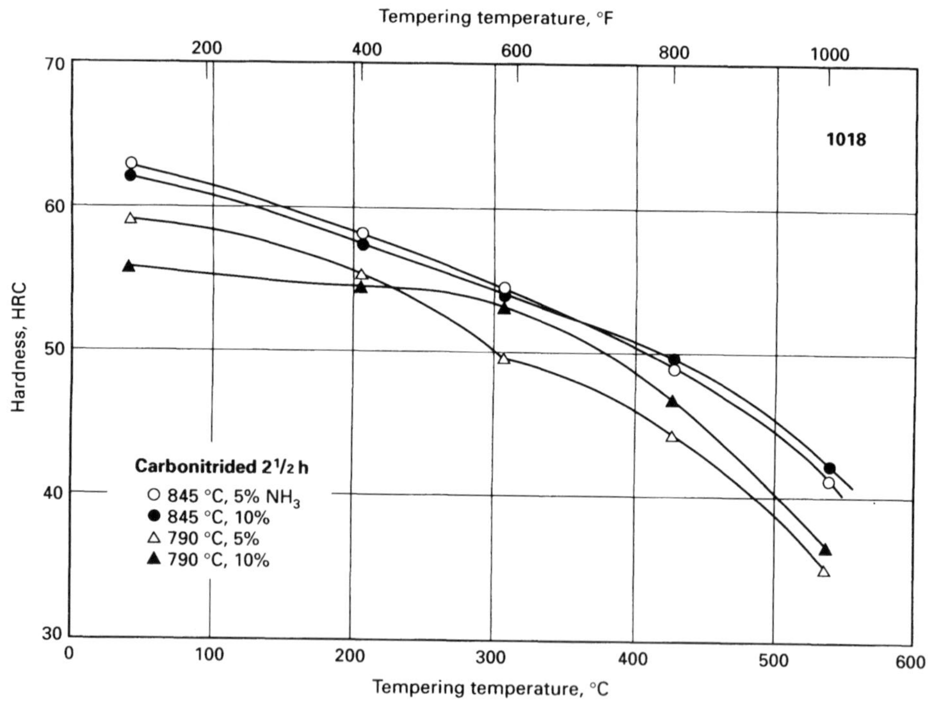

Fig. 11 Decrease of surface hardness with increasing temperature for specimens of 1018 steel carbonitrided under the conditions indicated. Rockwell C hardness converted from Rockwell 30-N. See also Fig. 12.

ing are invariably tempered to avoid impact and impact fatigue failures. Most carbonitrided gears are tempered at 190 to 205 °C (375 to 400 °F) to reduce surface brittleness and yet maintain a minimum case hardness of 58 HRC. Alloy steel parts that are to be surface ground are tempered to minimize grinding cracks. Low-carbon steel parts are frequently tempered at 135 to 175 °C (275 to 350 °F) to stabilize austenite and minimize dimensional variations. Tapping screws made of 1020 steel are tempered at 260 to 425 °C (500 to 795 °F) to reduce breakage in tapping holes in sheet metal. In contrast, parts that are case hardened primarily for wear resistance, such as dowel pins, brackets, and washers, need not be tempered.

Carbonitriding of Powder Metallurgy Parts

Carbonitriding is widely used as a process for case hardening parts made by powder metallurgy techniques from ferrous powders. Densities of the sintered compacts vary from approximately 6.5 g/cm³ (0.23 lb/in.³) up to those approaching that of wrought steel. Parts may or may not be copper infiltrated prior to carbonitriding.

Carbonitriding is effective in case hardening iron compacts made from electrolytic iron powders. Four characteristics of these compacts make case hardening by carburizing difficult: high martensite transformation temperature (M_s), very low hardenability, less surface oxidation, and inherent porosity, resulting in high rates of carbon penetration.

Carbonitriding at 790 to 815 °C (1450 to 1500 °F) solves these problems; lower rates of diffusion at these temperatures permit control of case depth and allow buildup of adequate carbon in the case. The effects of nitrogen in retarding the pearlite transformation result in sufficient hardenability to allow oil quenching.

File-hard cases (with microhardnesses equivalent to 60 HRC) and normal, predominantly martensitic structures can be consistently obtained. Shallow cases are obtainable, although the allowable range of case depth must be increased over that used for wrought steels. Typical ranges of case depth are 0.08 to 0.20

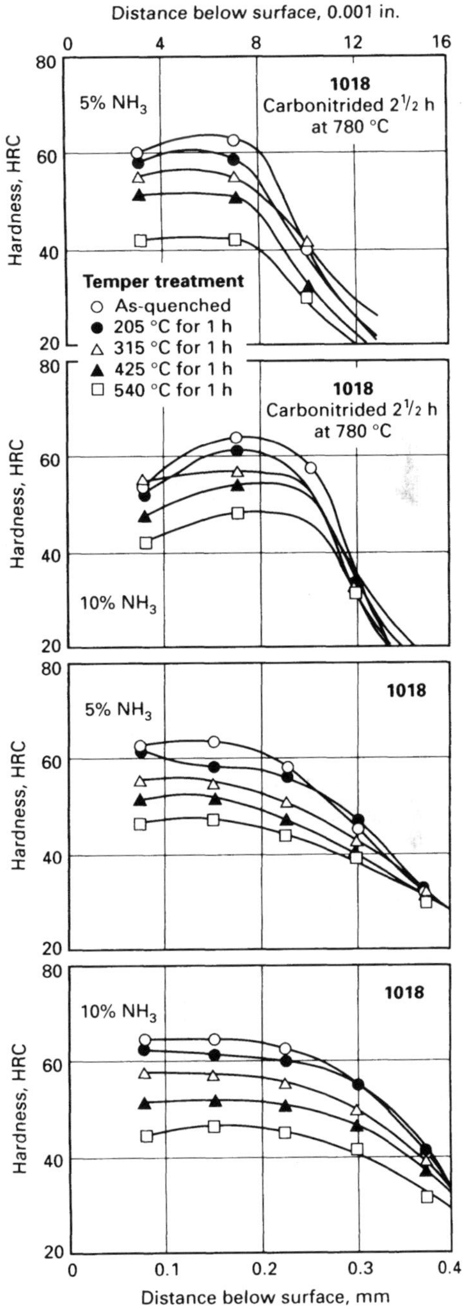

Fig. 12 Effect of tempering temperature on hardness gradients in carbonitrided cases. Rockwell C hardness converted from Vickers. Specimens were the same as Fig. 11 and were tempered as indicated.

mm (0.003 to 0.008 in.) and 0.15 to 0.30 mm (0.006 to 0.012 in.).

The high rate of carbon and nitrogen penetration that occurs as the result of porosity is demonstrated in Fig. 13 for parts made of iron powder conforming to ASTM B 310, class A. Although the rate of penetration decreases with increasing density, case depths for the higher densities (7.20 to 7.30 g/cm^3, or 0.260 to 0.264 lb/in.^{3}) are much deeper than those obtained with a wrought steel (7.87 g/cm^3 or 0.284 lb/in.^{3}). Most commercial iron powder compositions exhibit this type of response to carbonitriding; however, copper-infiltrated compacts are considerably more resistant to the penetration of carbon and nitrogen.

Tempering. Carbonitrided iron powder metallurgy parts are usually tempered, despite the fact that there is little danger of cracking untempered pieces. Tempering accomplishes the incidental result of facilitating tumbling and deburring operations. Although tempering is potentially capable of removing oil picked up and held in the pores in the part, air tempering of oil-quenched powder metallurgy parts is normally limited to temperatures not exceeding 205 °C (400 °F) because of the fire hazard at higher temperatures. Carbonitrided iron powder metallurgy parts are usually tempered at temperatures slightly higher than the temperatures used for carbonitrided wrought steel parts. Special cleaning procedures to remove oil, thus eliminating fire hazards, are incorporated in the processing steps when the tempering temperature exceeds 205 °C (400 °F).

Combined Carburizing/Carbonitriding Treatments

Often, carburizing and carbonitriding are used together to achieve much deeper case depths and better engineering performance for parts than could be obtained using only the carbonitriding process. This process is applicable particularly with steels with low case hardenability, that is, the 1000, 1100, and 1200 series steels. The process generally consists of carburizing at 900 to 955 °C (1650 to 1750 °F) to give the desired total case depth (up to 2.5 mm, or 0.100 in.), followed by carbonitriding for 2 to 6 h in the temperature range of 815 to 900 °C (1500 to 1650 °F) to add the desired carbonitrided case depth. The subject parts can then be oil quenched to obtain a deeper effective and thus harder case than would have resulted from the carburizing process alone. The addition of the carbonitrided surface increases the case residual compressive stress level and thus improves contact fatigue resistance as well as increasing the case strength gradient.

When the carburizing/carbonitriding processes are used together, the effective case depth (50 HRC) to total case depth ratio may vary from about 0.35 to 0.75 depending on the case hardenability, core hardenability, section size, and quenchant used. A more shallow effective or total case depth can be achieved with a given carbonitriding process by using fine grain steels containing higher amounts of aluminum or titanium. The nitrogen from the process

Table 3 Effect of tempering on Charpy V-notch impact strength of carbonitrided 1041 steel
Specimens were carbonitrided at 845 °C (1550 °F) for 3 h in an atmosphere containing 7% NH_3 and were oil quenched from the carbonitriding temperature. Specimens were copper plated before machining of V-notch to permit exposure of the notch to the carbonitriding atmosphere.

Test	Tempering temperature		Impact strength		Hardness, HRC(a)								
							Distance below surface, mm (in.)						
	°C	°F	J	ft · lb	Surface(b)	Core	0.075 (0.003)	0.15 (0.006)	0.25 (0.01)	0.38 (0.015)	0.64 (0.025)	1.0 (0.04)	1.4 (0.055)
1	As-quenched	As-quenched	1.4	1	60	53	63	64	64	63	61	61	58
2	370	700	2, 2	1.5, 1.5	47	46	57	57	55	54	49	50	50
3	425	800	29, 29	21.5, 21.5	42.5	43	57	57	56	55	49	47	47
4	480	900	69, 60	51, 44	38	38	54	54	52	50	42	38	38
5(c)	480	900	47, 52	35, 38	...	...	...	...	...	...	...	...	...
6	540	1000	78, 81	57.5, 60	35	32	49	50	50	47	36	33	32

(a) Converted from Vickers hardness. (b) Surface hardness is less than hardness at 0.075 mm below the surface because of retained austenite. (c) Tested at −18 °C (0 °F); all other tests at room temperature forms nitrides with the aluminum or titanium. The combined nitrogen does not improve case hardenability.

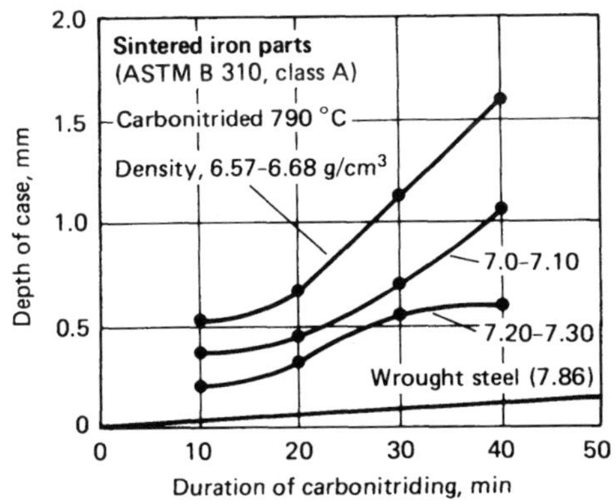

Fig. 13 Increase of case depth with decrease in density of iron powder metallurgy parts carbonitrided for various periods of time at 790 °C (1455 °F). The curve for the wrought steel represents the average case depth of a 5140 steel that was carbonitrided at 775 °C (1425 °F) for 8 h and quenched in oil at 75 °C (165 °F).

REFERENCES

1. D.H. Herring, Comparing Carbonitriding and Nitrocarburizing, *Heat Treat. Prog.,* April/May 2002, p 17–19
2. F.A. Clarkin and M.B. Bever, The Role of Water Vapor and Ammonia in Case Hardening Atmospheres, *Trans. ASM,* Vol 47, 1955, p 794–806
3. G.W. Powell, M.B. Bever, and C.F. Floe, Carbonitriding of Plain Carbon and Boron Steels, *Trans. ASM,* Vol 46, 1954, p 1359–1371
4. R. Davies and C.G. Smith, A Practical Study of the Carbonitriding Process, *Met. Prog.,* Vol 114 (No. 4), 1978, p 40–53
5. J. Grosch, Heat Treatment with Gaseous Atmospheres, *Steel Heat Treatment Handbook,* G.E. Totten and M.A.H. Howes, Ed., Marcel Dekker, Inc., 1997, p 663–719
6. P. Schobesberger, T. Streng, and S. Abbas, Low-Distortion Heat Treatment of Thin Parts, *Ind. Heat.,* Jan 2002, p 30–34

CHAPTER 6

Nitriding

NITRIDING is a surface hardening heat treatment that introduces nitrogen into the surface of steel at a temperature generally in the range of 500 to 575 °C (930 to 1065 °F) while it is in the ferritic condition. Nitriding, therefore, is similar to carburizing in that surface composition is altered, but different in that the nitrogen is added into ferrite instead of austenite. The fact that nitriding does not involve heating into the austenite phase field and a subsequent quench to form martensite means that nitriding can be accomplished with a minimum of distortion and excellent dimensional control.

Although nitriding can be carried out using a gaseous, liquid, or solid medium, gas nitriding using ammonia (NH_3) as the nitrogen-carrying species is the most commonly employed process. Steels that are nitrided are generally medium-carbon steels that contain strong nitride-forming elements such as aluminum, chromium, molybdenum, tungsten, and vanadium. Prior to nitriding, the steels are austenitized, quenched, and tempered. Tempering is performed at temperatures between 540 and 750 °C (1000 and 1380 °F), a range above which the nitriding is performed. Tempering above the nitriding temperature provides a core structure that is stable during nitriding.

Gas Nitriding

Gas nitriding is a case-hardening process whereby nitrogen is introduced into the surface of a solid ferrous alloy by holding the metal at a suitable temperature (below Ac_1, the austenite formation temperature, for ferritic steels) in contact with a nitrogenous gas, usually ammonia. Quenching is not required for the production of a hard case. The nitriding temperature for all steels is between 495 and 565 °C (925 and 1050 °F).

Application Factors

Principal reasons for nitriding are:

- To obtain high surface hardness
- To increase wear resistance and antigalling properties
- To improve fatigue life
- To improve corrosion resistance (except for stainless steels)
- To obtain a surface that is resistant to the softening effect of heat at temperatures up to the nitriding temperature
- Low distortion and deformation due to low processing temperature

Examples of typical gas nitriding applications and procedures are presented in Table 1. Table 2 lists examples of parts for which nitriding eliminated production or service problems that arose when the parts were case hardened by other methods.

Process Fundamentals and Case Composition/Microstructure (Ref 1)

Process Fundamentals. Gas nitriding is accomplished with ammonia gas that dissociates on the surface of the steel during heating, according to the following reactions:

$$NH_3 \rightarrow 3H + N \quad \text{(Eq 1)}$$

$$2N \rightarrow N_2 \quad \text{(Eq 2)}$$

$$2H \rightarrow H_2 \quad \text{(Eq 3)}$$

The atomic nitrogen and hydrogen components shown in Eq 1 are unstable and unite with other like atoms to form molecules, as shown in Eq 2 and 3. It is while they are in the atomic state that diffusion takes place. Nitriding atmospheres are in non-equilibrium while at the process temperature, which means that at the steel surface a high degree of nitrogen activity takes place from the ammonia. It is usual to have a 100% flow of ammonia into the container within the furnace, meaning that some of the ammonia dissociates into nitrogen and hydrogen gases when subjected to the furnace temperature, as shown earlier.

The ammonia that does not dissociate is known as undissociated ammonia. The undissociated ammonia is used to measure the nitriding activity taking place within the process container. The degree of decomposition can be expressed as:

$$\frac{\text{Amount of decomposed } NH_3}{\text{Added amount of } NH_3} \times 100\%$$

The decomposition rate of the ammonia gas is usually kept between 10 and 35%, depending on the steel being treated and the gas exchange rate.

Case Composition and Microstructure. The formed nitrided case has the following two distinct zones:

- *The compound layer at the surface of the steel.* This is also referred to as the white layer, because the layer is seen as a white zone when etched with nital. It is made up of two phases known as $Fe_{2-3}N$ epsilon (ε) and Fe_4N gamma prime (γ′) phases. The concentration of each phase is dependent on the chemistry of the steel, primarily the carbon concentration and the nitriding activity. In a typical nitriding steel with a carbon content of approximately 0.4 wt%, the formation of ε to γ′ is roughly equal using the gas nitriding process. The higher the carbon content of the steel, the greater the ε phase in the compound layer. The compound layer is usually up to 20 to 25 μm thick, the thickness varying with time, temperature, and gas composition. An excessively thick compound layer is very hard

Table 1 Nitriding applications and procedures

Part	Dimensions or weight of part	Steel	Nitriding time, h
Single-stage nitriding			
Hydraulic barrel	50 mm (2 in.) OD, 19 mm (3/4 in.) ID, 150 mm (6 in.) long	AMS 6470	48
Trigger for pneumatic hammer	...	AMS 6470	40
Governor push button	6 mm (1/4 in.) diam	AMS 6470	30
Tachometer shaft	380 mm (15 in.) long	AMS 6475	25
Helical timing gear	205 mm (8 in.) OD (4.5 kg or 10 lb)	4140	24
Gear	50 mm (2 in.) OD, 6 mm (1/4 in.) thick	4140	24
Generator shaft	25 mm (1 in.) OD, 355 mm (14 in.) long	4140	24
Rotor and pinion for pneumatic drill	22 mm (7/8 in.) diam	4140	9
Sleeve for pneumatic tool clutch	38 mm (1 1/2 in.) diam	4140	9
Marine helical transmission gear	635 mm (25 in.) OD (227 kg, or 500 lb)	4142	32
Oil-pump gear	50 mm (2 in.) OD, 180 mm (7 in.) long	4340	25
Loom shuttle	150 × 25 × 25 mm (6 × 1 × 1 in.)	410 stainless	8
Double-stage nitriding			
Ring gear for helicopter main transmission	380 mm (15 in.) OD, 350 mm (13.8 in.) ID, 64 mm (2.5 in.) long	AMS 6470(a)	60(b)
Aircraft cylinder barrel	180 mm (7 in.) OD, 305 mm (12 in.) long	AMS 6470	35(c)
Bushing	10 kg (23 lb)	AMS 6470	90
Cutter spindle	3 kg (7 lb)	AMS 6470	45
Plunger	75 mm (3 in.) OD, 1525 mm (60 in.) long	AMS 6475	72
Crankshaft	205 mm (8 in.) OD (journals), 4 m (13 ft) long	4130	65
Piston ring	150 mm (6 in.) OD, 4.25 m (14 ft) long	4130	65
Clutch	1 kg (2 lb)	4140	45
Double helical gear	50 kg (108 lb)	4140	97
Feed screw	4 kg (9 lb)	4140	45
Pumper plunger	0.5 kg (1 lb)	4140	127
Seal ring	9.5 kg (21 lb)	4140	90
Stop pin	3 kg (7 lb)	4140	90
Thrust collar	3.6 kg (8 lb)	4140	90
Wear ring	40 kg (87 lb)	4140	90
Clamp	7 kg (15 lb)	4150	90
Die	21 kg (47 lb)	4340	90
Gib	10 kg (23 lb)	4340	49
Spindle	122 kg (270 lb)	4340	90
Torque gear	62.5 kg (138 lb)	4340	90
Wedge	1.8 kg (4 lb)	4340	42
Pumper plunger	1.4 kg (3 lb)	420 stainless	127

Note: OD, outer diameter; ID, inner diameter; AMS, Aerospace Material Specification. (a) Vacuum melted. (b) 9 h at 525 °C (975 °F), 51 h at 545 to 550 °C (1015 to 1025 °F). (c) 6 h at 525 °C (975 °F), 29 h at 565 °C (1050 °F)

Table 2 Examples of parts for which nitriding proved superior to other case-hardening processes for meeting requirements

Part	Requirement	Material and process originally used	Resultant problem	Solution
Gear	Good wear surface and fatigue properties	Carburized 3310 steel 0.4 to 0.6 mm (0.017 to 0.025 in.) case	Difficulty in obtaining satisfactory case to meet a reliability requirement	AMS 6470 substituted for 3310 and double-stage nitrided for 25 h
High-speed pinion (on gear motor)	Provide teeth with minimum (equivalent) hardness of 50 HRC	8620 steel gas carburized at 900 °C (1650 °F) to 0.5 mm (0.02 in.) case, direct quenched from 845 °C (1550 °F), and tempered at 205 °C (400 °F)	Distortion in teeth and bore caused high rejection rate	4140 steel, substituted for 8620, was heat treated to 255 HB; parts were rough machined, finish machined, nitrided(a)
Bushings (for conveyor rollers handling abrasive alkaline material)	High surface hardness for abrasion resistance; resistance to alkaline corrosion	Carburized bushings	Service life of bushings was short because of scoring	Substitution of Nitralloy 135 type G (resulfurized) heat treated to 269 HB and nitrided(b)
Spur gears (in train of power gears; 10-pitch, tip modified)	Sustain continuous Hertz stress of 1035 MPa (150 ksi) (overload of 1550 MPa, or 225 ksi), continuous Lewis stress of 275 MPa (40 ksi) (overload of 725 MPa, or 105 ksi)(c)	Carburized AMS 6260	Gears failed because of inadequate scuff resistance, also suffered property losses at high operating temperatures	Substitution of material of Hll type, hardened and multiple tempered (3 h + 3 h) to 48 to 52 HRC, then double-stage nitrided(d)

Note: AMS, Aerospace Material Specification. (a) Single-stage nitrided at 510 °C (950 °F) for 38 h. Cost increased 5%, but rejection rate dropped to zero. (b) Single-stage nitrided at 510 °C (950 °F) for 38 h. Case depth was 0.46 mm (0.018 in.), and hardness was 94 HR15-N; parts had three times the service life of carburized parts. (c) Must withstand operating temperatures to 290 °C (550 °F). (d) 15 h at 515 °C (960 °F) (15 to 25% dissociation); then 525 °C (980 °F) (80 to 83% dissociation). Effective case depth (to 60 HRC), 0.25 to 0.4 mm (0.010 to 0.015 in.); case hardness, 67 to 72 HRC (converted from Rockwell 15-N scale) and brittle and may spall in use. Light grinding following the nitriding operation can remove the compound layer. In addition, there are special nitriding processes aimed at reducing this layer or at making it less brittle. A typical compound layer can be seen in Fig. 1.

- *The diffusion zone, the area immediately below the compound layer* (Fig. 1). This region is made up of stable nitrides formed by the reaction of nitrogen with the nitride-forming elements, such as aluminum, chromium, vanadium, molybdenum, and tungsten. The diffusion zone thickness is, again, time and temperature dependent, ranging up to 1 mm (0.040 in.). It is from this region that the fatigue and load-bearing strengths are determined.

The area below the diffusion zone is the core of the steel, which consists of tempered martensite (Fig. 1). The core hardness is considered to be the hardness achieved by the prehardening and tempering operations.

Nitridable Steels

Of the alloying elements commonly used in commercial steels, aluminum, chromium, vanadium, tungsten, and molybdenum are beneficial in nitriding, because they form nitrides that are stable at nitriding temperatures. Other alloying elements, such as nickel, copper, and manganese, have little, if any, effect on nitriding characteristics. Unalloyed carbon steels are not well suited to gas nitriding, because they form an extremely brittle case that spalls readily, and the hardness increase in the diffusion zone is small. The effects of specific alloying elements are discussed subsequently.

Aluminum forms very hard nitrides in the surface of the steel when nitrided. Generally, the maximum amount of aluminum permitted in the steel is in the region of 1.5%. Above 1% Al leads to surface cracking under extreme surface load conditions. This is because the core hardness of the material is usually very ductile. Thus, if severe loading is placed on the workpiece that has high ductility, then there is a very strong possibility that the surface of the case leads to crack propagation.

Molybdenum forms stable nitrides at the nitriding temperature and reduces the risk of surface embrittlement at the nitriding temperature.

Chromium also forms stable nitrides at the nitriding temperature; however, the high chromium contents found in some stainless steels make them more difficult to nitride. The

Fig. 1 Typical microstructure of a gas nitrided steel containing 0.4% C, 1.6% Cr, 0.35% Mo, and 1.13% Al. The structure consists of the white layer (WL), the diffusion zone (DZ), and the core (C), which has a base microstructure of tempered martensite of hardness 30 HRC.

reason for this is that chromium reacts with oxygen and forms a chrome oxide barrier on the surface of the material to be nitrided. It is necessary to break down the barrier of chrome oxide on the surface. Once the barrier has been broken down, then the nitriding is effective. It should be remembered that the higher the percentage of available chromium at the surface of the steel, the more difficult the steel is to nitride. The positive side of this is usually high surface hardness values.

Vanadium. The use of vanadium in the nitriding steel is once again conducive to forming stable nitrides in the formed case at the surface of the steel, in addition to which fine grain toughness is exhibited within the formed case.

Tungsten enables the steel to retain its hardness at high operating temperatures without any loss of surface hardness. Depending on the tungsten content and the general composition of the steel, the nitrided steel is able to operate at temperatures up to 590 °C (1100 °F) without any appreciable loss of surface hardness and has enhanced wear characteristics.

Silicon is considered to be a good nitride former. However, silicon is not usually present at high levels in steel. It is usually present as either an oxidizer or a stabilizer. It is not usually seen in sufficient volumes to be considered a strong nitride former.

Various Classes of Nitridable Steels. The following steels can be gas nitrided for specific applications:

- Aluminum-containing low-alloy steels, including the Nitralloy types containing 1% Al (Table 3)
- Medium-carbon, chromium-containing low-alloy steels of the 4100, 4300, 5100, 6100, 8600, 8700, and 9800 series
- Hot work die steels containing 5% Cr, such as H11, H12, and H13
- Low-carbon, chromium-containing low-alloy steels of the 3300, 8600, and 9300 series
- Air-hardening tool steels, such as A-2, A-6, D-2, D-3, and S-7
- High-speed tool steels, such as M-2 and M-4
- Nitronic stainless steels, such as 30, 40, 50, and 60
- Martensitic stainless steels of the 400 series, such as 422 and 440
- Austenitic stainless steels of the 200 and 300 series
- Precipitation-hardening stainless steels, such as 13-8 PH, 15-5 PH, 17-4 PH, 17-7 PH, A-286, AM350, and AM355

Aluminum-containing steels produce a nitrided case of very high hardness and excellent wear resistance. However, the nitrided case also has low ductility, and this limitation should be carefully considered in the selection of aluminum-containing steels. In contrast, low-alloy chromium-containing steels provide a nitrided case with considerably more ductility but with lower hardness. Nevertheless, these steels offer substantial wear resistance and good antigalling properties. Tool steels, such as H11 and D2, yield consistently high case hardness with exceptionally high core strength.

In addition to the steels discussed previously that are covered by North American standards and specifications, there are other international designations for nitriding steels. Examples of some of these are listed in Table 4.

Table 3 Nominal composition and preliminary heat treating cycles for aluminum-containing low-alloy steels commonly gas nitrided

Steel			Composition, %								Austenitizing temperature(a)		Tempering temperature(a)	
SAE	AMS	Nitralloy	C	Mn	Si	Cr	Ni	Mo	Al	Se	°C	°F	°C	°F
...	...	G	0.35	0.55	0.30	1.2	...	0.20	1.0	...	955	1750	565–705	1050–1300
7140	6470	135M	0.42	0.55	0.30	1.6	...	0.38	1.0	...	955	1750	565–705	1050–1300
...	6475	N	0.24	0.55	0.30	1.15	3.5	0.25	1.0	...	900	1650	650–675	1200–1250
...	...	EZ	0.35	0.80	0.30	1.25	...	0.20	1.0	0.20	955	1750	565–705	1050–1300

Note: SAE, Society of Automotive Engineers; AMS, Aerospace Material Specification. (a) Sections up to 50 mm (2 in.) in diameter, quenched in oil; larger sections may be water quenched.

Prior Heat Treatment and Surface Preparation

Prior Heat Treatment. All hardenable steels *must* be hardened and tempered before being nitrided. The tempering temperature must be high enough to guarantee structural stability at the nitriding temperature; the minimum tempering temperature is usually at least 30 °C (50 °F) higher than the maximum temperature to be used in nitriding.

In certain alloys, such as series 4100 and 4300 steels, hardness of the nitrided case is modified appreciably by core hardness; that is, a decrease in core hardness results in a decrease in case hardness. Consequently, in order to obtain maximum case hardness, these steels are usually provided with maximum core hardness by being tempered at the minimum allowable tempering temperature.

Surface Preparation of Parts to be Nitrided. After hardening and tempering, and before nitriding, parts should be thoroughly cleaned. Most parts can be successfully nitrided immediately after vapor degreasing. However, some machine-finishing processes, such as buffing, finish grinding, lapping, and burnishing, may produce surfaces that retard nitriding and result in uneven case depth and distortion. There are several methods by which the surfaces of parts finished by such methods may be successfully conditioned before nitriding.

One method consists of vapor degreasing parts and then abrasive cleaning them with aluminum oxide grit or other abrasives, such as garnet or silicon carbide, immediately prior to nitriding. Any residual grit must be brushed off before parts are loaded into the furnace. Parts should be handled with clean gloves.

A second method consists of preoxidizing the parts in an air atmosphere at approximately 330 °C (625 °F). This may be done as a separate operation, or it may be incorporated as part of the heating portion of the nitriding cycle if suitable precautions are taken.

Single-Stage and Double-Stage Nitriding

Either a single- or a double-stage process may be employed when nitriding with anhydrous ammonia. In the single-stage process, a temperature in the range of approximately 495 to 525 °C (925 to 975 °F) is used, and the dissociation rate ranges from 15 to 30%. This process produces a brittle, nitrogen-rich layer known as the white nitride layer at the surface of the nitrided case.

The double-stage process, known also as the Floe process (U.S. Patent 2,437,249), has the advantage of reducing the thickness of the white nitrided layer.

The first stage of the double-stage process is, except for time, a duplication of the single-stage process. The second stage may proceed at the nitriding temperature employed for the first stage, or the temperature may be increased from 550 to 565 °C (1025 to 1050 °F); however, at either temperature, the rate of dissociation in the second stage is increased to 65 to 80% (preferably, 75 to 80%). Generally, an external ammonia dissociator is necessary for obtaining the required higher second-stage dissociation.

The principal purpose of double-stage nitriding is to reduce the depth of the white layer produced on the surface of the case. Except for a reduction in the amount of ammonia consumed per hour, there is no advantage in using the double-stage process, unless the amount of white layer produced in single-stage nitriding cannot be tolerated on the finished part or unless the amount of finishing required after nitriding is substantially reduced. In Fig. 2, the amount of white layer formed during a 25 h double-stage nitriding treatment is compared to that formed

Table 4 British standard nitriding steels

Designation	Composition, %								
	C	Si	Mn	P	Cr	Mo	Ni	V	Al
EN 40 A	0.20–0.35	0.10–0.30	0.40–0.55	0.05 max	2.90–4.00	0.60–0.80	0.40 max	...	...
EN 40 B	0.20–0.30	0.10–0.35	0.40–0.65	0.05 max	2.90–3.50	0.40–0.70	0.40 max	0.10–0.30	...
EN 40 C	0.30–0.50	0.10–0.35	0.40–0.80	0.05 max	2.90–3.50	0.70–1.20	0.40 max	0.10–0.30	...
EN 41 A	0.25–0.35	0.10–0.35	0.65 max	0.05 max	1.40–1.80	0.10–0.25	0.40 max	...	0.90–1.30
EN 41 B	0.25–0.45	0.10–0.35	0.65 max	0.05 max	1.40–1.80	0.10–0.25	0.40 max	...	0.90–1.30

during a 24 h single-stage nitriding of the same material.

Figure 3 shows the effect of nitriding time on the depth of case developed on 4140 steel during double-stage nitriding at 525 °C (975 °F) for both stages. Use of a higher temperature during the second stage would have produced a deeper case of slightly lower hardness.

Hardness gradients obtained in double-stage nitriding of Society of Automotive Engineers (SAE) 7140 (Aerospace Material Specification, or AMS, 6470) are shown in Fig. 4. The hardness results shown in Fig. 4(a), (b), and (c) were obtained by grinding off progressively increasing amounts of case to form steps on which HR-15N readings were made. Data for Fig. 4(d), (e), and (f) were obtained from microhardness measurements, converted to HRC equivalents. Similar data, for double-stage nitriding of AMS 6475, are shown in Fig. 5.

To summarize, the use of a higher temperature during the second stage:

- Lowers the case hardness
- Increases the case depth
- May lower the core hardness depending on the prior tempering temperature and the total nitriding cycle time
- May lower the apparent effective case depth because of the loss of core hardness, depending on how effective case depth is defined

Operating Procedures

Furnace Purging. After loading and sealing the furnace at the start of the nitriding cycle, it is necessary to purge the air from the retort before the furnace is heated to a temperature above 150 °C (300 °F). This prevents oxidation of parts and furnace components, and, when ammonia is used as the purging atmosphere, avoids production of a potentially explosive mixture. Nitrogen is preferred in place of ammonia for purging, but the same precautions should be taken to avoid oxidation of parts, except when preoxidation is intentionally included as part of the cycle.

A typical purging cycle using anhydrous ammonia follows:

- Close furnace and start flow of anhydrous ammonia gas at as fast a flow rate as is practical with first step.
- Set furnace temperature control at 150 °C (300 °F) simultaneously. Heat furnace to this temperature but do not exceed.

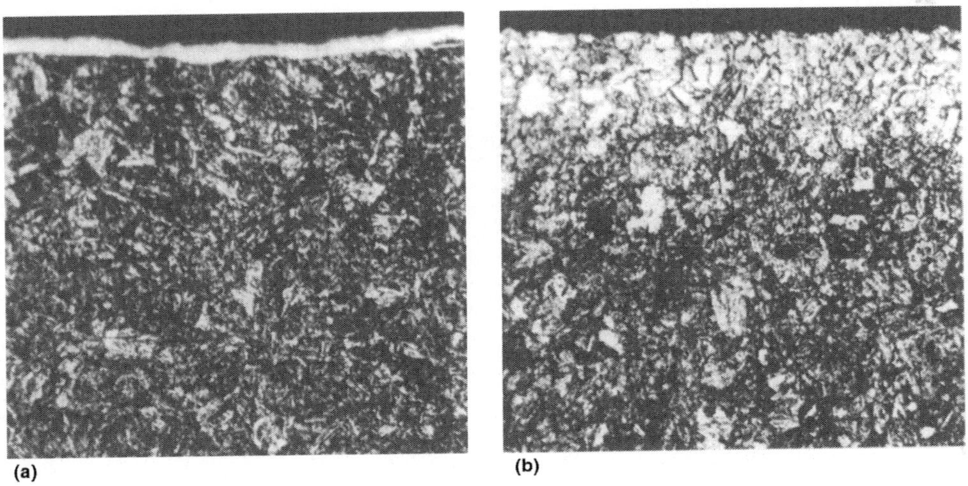

Fig. 2 Microstructure of quenched and tempered 4140 steel after (a) gas nitriding for 24 h at 525 °C (975 °F) with 20 to 30% dissociation and (b) gas nitriding for 5 h at 525 °C (975 °F) with 20 to 30% dissociation followed by a second stage of 20 h at 565 °C (1050 °F) with 75 to 80% dissociation. Both specimens were oil quenched from 845 °C (1550 °F), tempered for 2 h at 620 °C (1150 °F), and surface activated with manganese phosphate before nitriding. (a) Structure after single-stage nitriding 0.005 to 0.0075 mm (0.0002 to 0.0003 in.) white surface layer (Fe_2N), iron nitride, and tempered martensite. (b) The high second-stage dissociation caused absence of white layer, and the final structure had a diffused nitride layer on a matrix of tempered martensite. Both 2% nital. 400×

- When the furnace has been purged to the degree that 10% or less air and 90% or more ammonia are present in the retort, the furnace may be heated to the nitriding temperature.

It is not feasible to incorporate preoxidation as part of the cycle, unless nitrogen is available as a purging medium at the end of the 330 °C (625 °F) oxidizing stage. *Under no circumstances should ammonia be introduced into a furnace containing air at 330 °C (625 °F) because of the explosion hazard.*

Purging is employed also at the conclusion of the nitriding cycle when the furnace is cooled from the nitriding temperature. It is common practice to remove the ammonia remaining in the retort with nitrogen to reduce the amount of ammonia that would otherwise be released into the immediate area when the load is removed. Dilution of the ammonia lessens the discomfort to employees working near the furnace. The introduction of nitrogen into the retort can be delayed until the nitrided parts have cooled to below 150 °C (300 °F).

Dissociation Rates. The nitriding process is based on the affinity of nascent nitrogen for iron and certain other metallic elements. Nascent nitrogen is produced by the dissociation of gaseous ammonia when it contacts hot steel parts. Although various rates of dissociation can be used successfully in nitriding, it is important that the nitriding cycle begin with a dissociation rate of approximately 15 to 35% and that this rate be maintained for 4 to 10 h, depending on the duration of the total cycle; temperature should be maintained at approximately 525 °C (975 °F). Typically, ammonia is supplied at a flow rate to achieve a minimum of four (4) atmosphere changes in the retort per hour. This initial cycle develops a shallow white layer from which diffusion of nitrogen into the main case structure proceeds.

In most nitriding cycles, dissociation rates vary somewhat, even though the controlling factors—ammonia flow rate, surface area, and nitriding temperature—remain constant. Characteristically, the dissociation rate gradually increases as the cycle proceeds at a constant ammonia flow rate. This increase, however, usually is not enough to affect nitrided case characteristics significantly.

When nitriding with a dissociation rate of 15 to 35%, it is normal to control this rate entirely by the flow rate of ammonia. At a dissociation rate of 75 to 80%, however, it is necessary to introduce completely dissociated ammonia from an external dissociator or nitrogen to ensure adequate positive flow within the furnace.

Furnace Heating and Cooling. For reasons of economy, it is generally desirable to keep the total cycle time as short as possible by heating and cooling as rapidly as the equipment will permit and allowing for the following considerations. It may be an advantage to:

- Limit the heating rate, for example to 55 °C/h (100 °F/h) or slower, to allow time for any residual contaminants on the work and from stopoff point to be completely expelled before reaching the nitriding temperature.
- Both heat and cool at limited rates to minimize temperature gradients throughout large loads or in large and complex parts, and thus to minimize distortion.

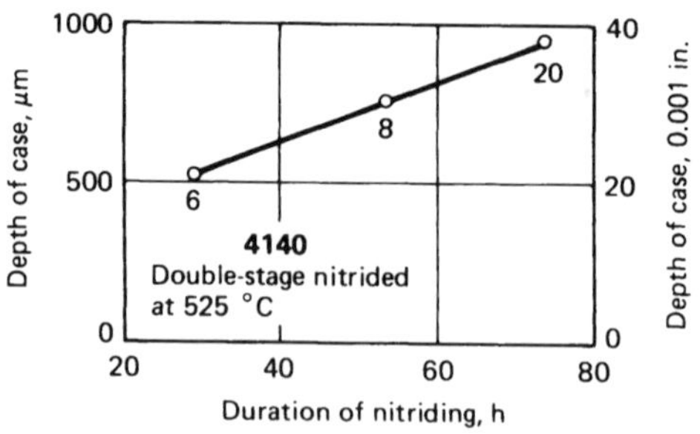

Fig. 3 Depth of case developed on 4140 steel during double-stage nitriding. Numbers indicate hours of nitriding at 15 to 25% dissociation. Remainder of cycle at 83 to 85% dissociation

Many pit-type nitriding furnaces are equipped with a heat exchanger that accelerates cooling of the furnace and work-load at the conclusion of the nitriding cycle. When an external water-cooled heat exchanger is used, the furnace heating elements are turned off when the nitriding cycle is completed, and the furnace temperature is allowed to drop approximately 55 °C (100 °F). At this point, the ammonia flow is approximately doubled, and the cooling water is turned on in the heat exchanger. The circulating blower of the heat exchanger also is turned on, and a gate valve is opened to permit circulation of the furnace atmosphere through the heat exchanger. Extreme care must be exercised to ensure a positive gas flow through the furnace, as evidenced by the exit gas bubbles. When gas flow through the furnace has been stabilized, the flow may be reduced to the minimum required for positive pressure. After cooling to 150 °C (300 °F) or below, the furnace may be opened.

Bell-type furnaces may be cooled with a cooling bell that is placed over the sealed nitriding retort after the heating bell has been removed. The following is a typical procedure for cooling a bell-type furnace with either raw ammonia or dissociated ammonia:

- Place cooling bell in position on base.

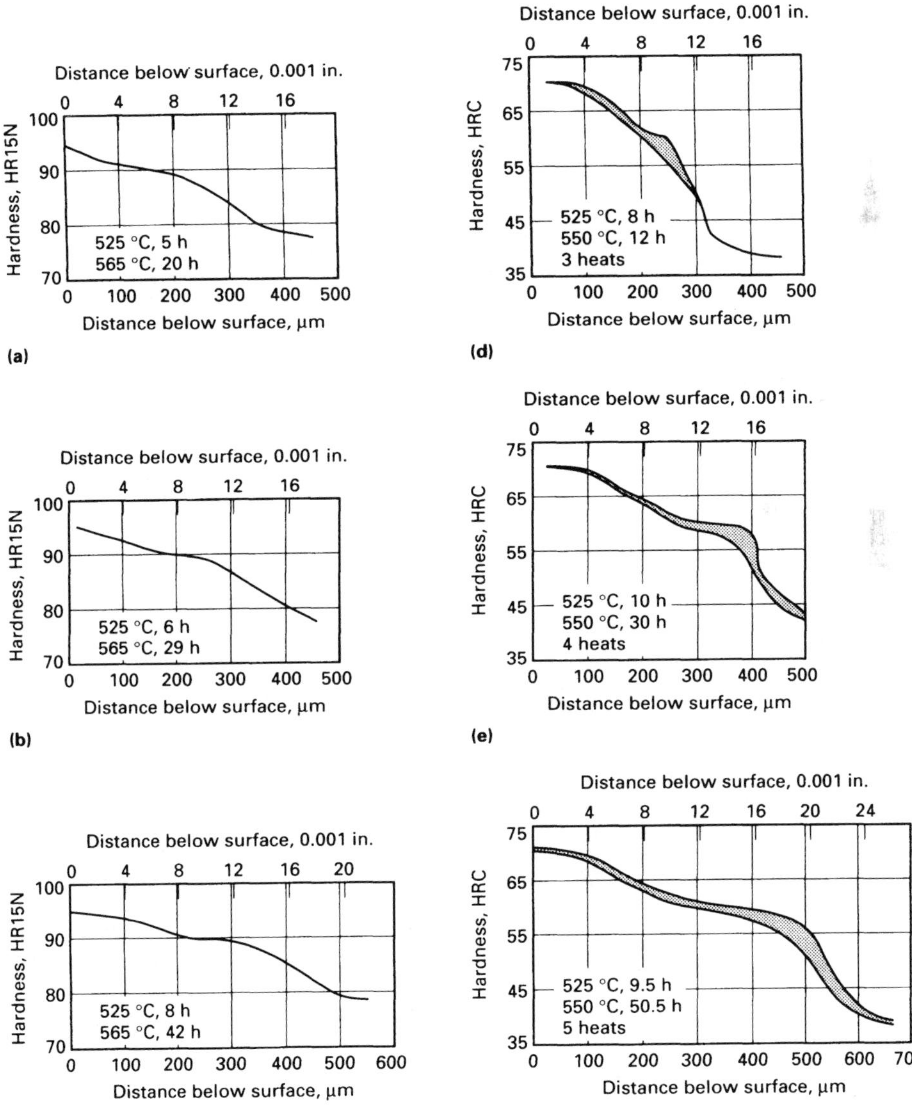

Fig. 4 Hardness gradients for double-stage nitrided SAE 7140 (AMS 6470) steel. (a)–(c) Material nitrided with dissociation rates of 15 to 20% during first stage and 60 to 70% during second stage. (d)–(f) Material nitrided with dissociation rates of 25 to 28% during first stage and 75 to 80% in second stage. Material produced by consumable-electrode vacuum arc remelting method

- Insert plug of cooling bell into receptacle.
- Turn on bell cooling fan.
- Cool furnace and load to not less than 315 °C (600 °F), as indicated on base recorder. Turn off flow of dissociated ammonia, and increase flow of raw ammonia to approximately 1.4 m^3/h (50 ft^3/h).
- When burette reading indicates a dissociation rate of 5% or less and temperature is 120 °C (250 °F) or below, shut off flow of ammonia to furnace and open air valve to furnace. If dissociation rate is 5% or less before temperature falls to 120 °C (250 °F), flow of raw ammonia can be reduced to 1.1 m^3/h (40 ft^3/h) for remainder of cooling time.
- Remove cooling bell.
- Open air valve to meter; allow 4.2 m^3/h (150 ft^3/h) of air into furnace, and open exhaust valve wide. (Level on manometer will drop nearly to zero.)
- Turn off base fan when burette reading increases to 65% dissociation. *(A mixture of 16 to 25% NH_3 in air is explosive. Therefore, fan must be shut off when ammonia level reaches 35%, or 65% burette reading.)* This eliminates hazard from sparks that may be generated by a moving fan.
- Continue to purge furnace until burette reading is 95% or higher. (This is not a safety precaution but is done to minimize discomfort to personnel nearby when furnace is opened.)
- Close air valve.
- Drain oil seal.

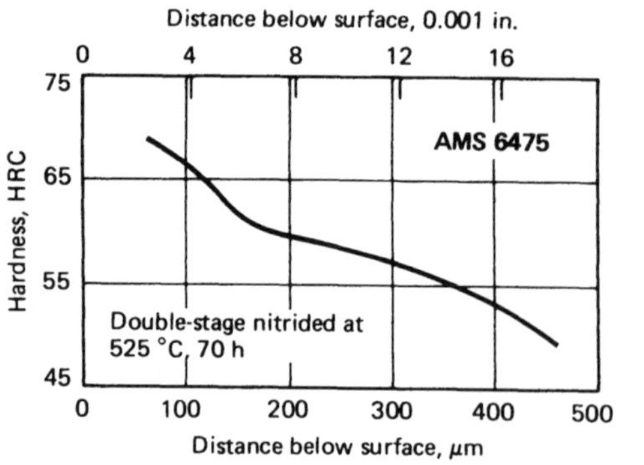

Fig. 5 Hardness gradient obtained for double-stage nitrided AMS 6475 steel. HRC hardness numbers were obtained by conversion from diamond pyramid hardness measurements. Core hardness after nitriding was 41.5 HRC. Data represent one air-melted heat of AMS 6475.

- Raise retort approximately 0.3 m (1 ft), and wipe off seal from retort lip before removing retort completely.
- Remove the charge from the furnace.

Control of Case Depth

Case depth and case hardness, the two criteria most commonly referred to in the control of case properties, vary not only with the duration and other conditions of nitriding but also with steel composition, prior structure, and core hardness.

Aluminum-Containing Steels. Of the aluminum-containing nitriding steels, the most widely used is SAE 7140 (AMS 6470 or Nitralloy 135M, as listed in Table 3). Figure 6 indicates the hardness gradients and case depths obtained with this steel, as a function of cycle time and nitriding conditions. Results were obtained in single-stage nitriding for various lengths of time up to 800 h and at temperatures ranging from 510 to 540 °C (950 to 1000 °F); several different dissociation rates are represented.

Chromium-Containing Low-Alloy Steels. Data relating case depth to nitriding time and conditions for chromium-containing low-alloy steels (principally 4140, 4337, 4340, and 8640) are given in Fig. 7 and 8. Of these steels, 4140 exhibits the best nitriding characteristics because of its higher chromium content and nickel-free composition. Although 4340 develops a heavier case than 8640 in the first 24 h of nitriding, this difference begins to decrease at the end of a 48 h cycle (Fig. 8).

Data for the AMS equivalents of 4337 and 4140 (AMS 6412 and 6382, respectively) in Fig. 7 are of particular interest, because they demonstrate the effect of core hardness on the hardness of the nitrided case. Core hardnesses as low as 21 to 23 HRC, and as high as 36 to 37 HRC, are considered.

Chromium-containing tool steels such as H11, H12, H13, and D2 provide high core strength with high case hardness, an excellent combination for applications involving severe impact or very high unit loading. Use of these steels is limited primarily by high cost and fabricating difficulties. Case depth results for these steels in single-stage nitriding at 525 °C (975 °F) and at 525 to 540 °C (975 to 1000 °F) are given in Fig. 8. The relatively shallow case depths obtained reflect the retarding effect of increased chromium content on the penetration of nitrogen. Hardness gradients for the same steels are shown in Fig. 9.

Distortion and Dimensional Changes

Distortion in nitriding may result from:

- Relief of residual stresses from prior operations, such as welding, hardening, machining, and so forth
- Stress introduced during nitriding due to inadequate support in the furnace or too rapid or nonuniform heating or cooling
- Stress introduced by the increase in volume that occurs in the case. This change causes a stretching of the core, which results in tensile stresses that are balanced by compressive stresses in the case after the parts have cooled to room temperature. The magnitude of the permanent set in the core and case is affected

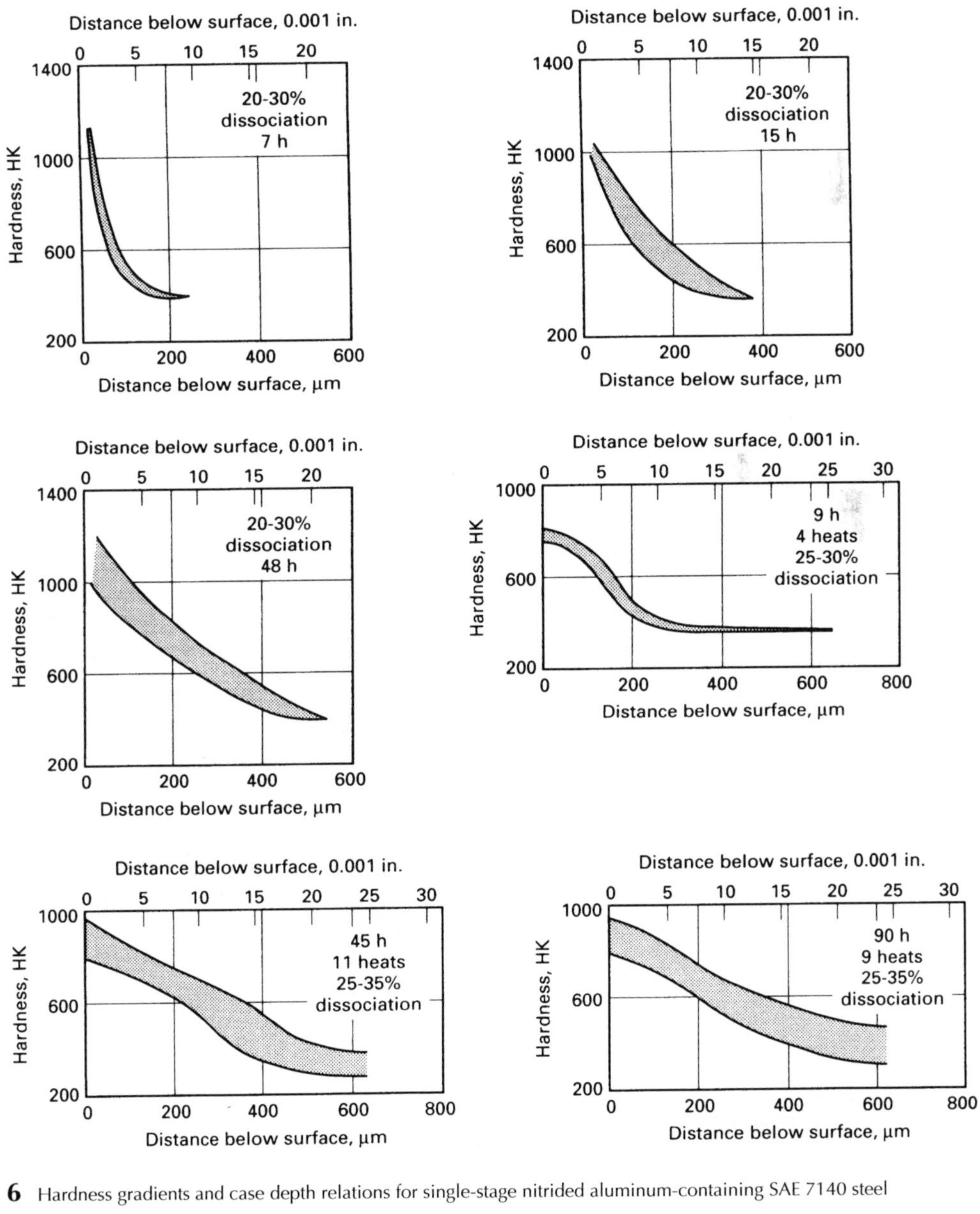

Fig. 6 Hardness gradients and case depth relations for single-stage nitrided aluminum-containing SAE 7140 steel

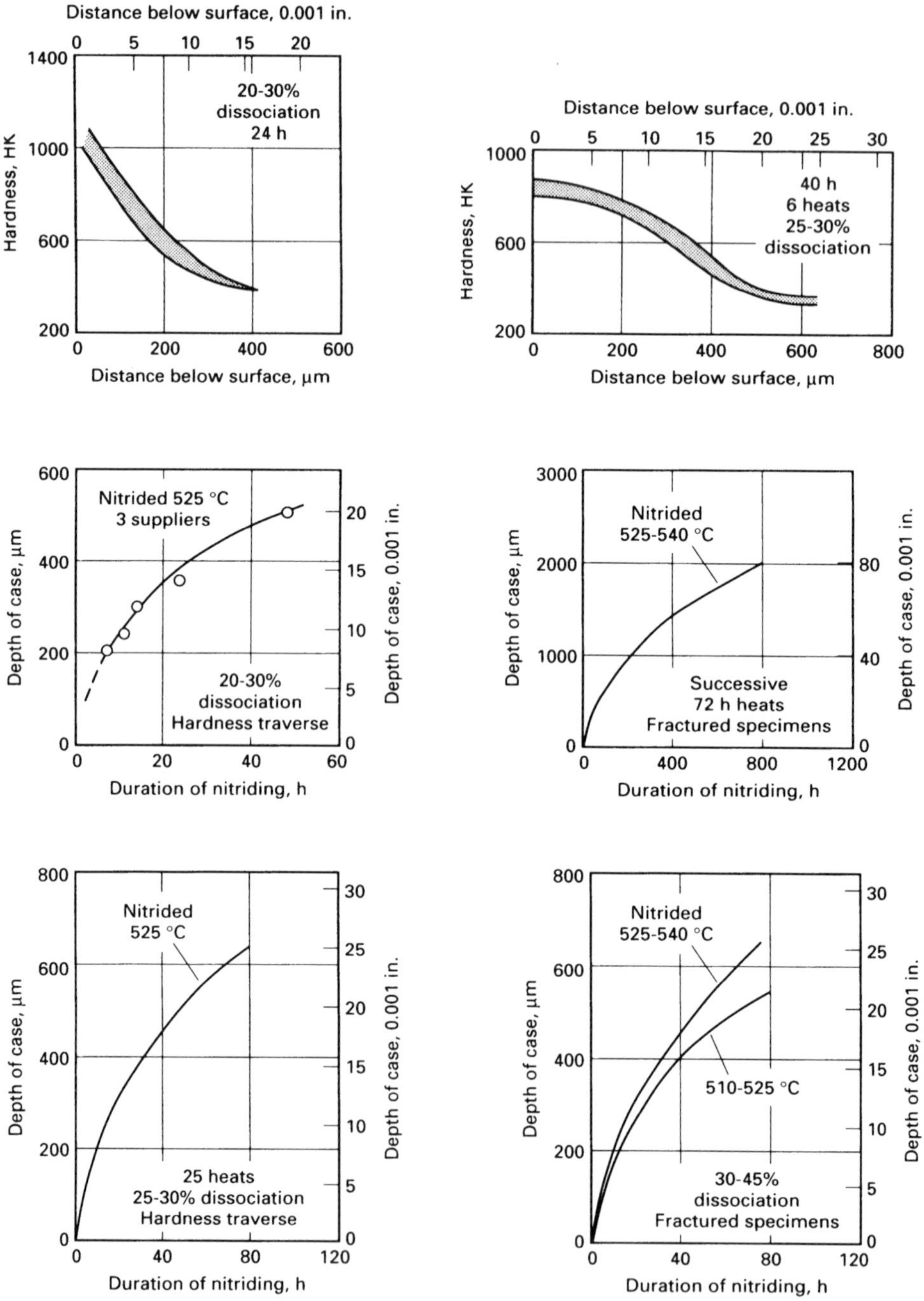

Fig. 6 **(continued)** Hardness gradients and case depth relations for single-stage nitrided aluminum-containing SAE 7140 steel

by yield strength of the material, thickness of the case, and by the amount and nature of the nitrides formed.

When the prior manufacturing practice and the mechanics of the nitriding cycle are properly controlled, growth becomes the primary cause of distortion. This is governed largely by composition, tempering temperature, time and temperature of nitriding, relative thickness of case and core, and shape of the part. Growth also is affected when some areas of the part are masked to prevent nitriding.

The amount of growth is usually constant for identical parts nitrided in different batches by a fixed processing cycle. Thus, after the amount of growth for a particular part has been determined experimentally, allowance for it can be made during final machining prior to nitriding. Before experiments are conducted to determine growth, parts must be thoroughly stress relieved.

An example of growth as a function of the wall thickness of hollow cylinders made of Nitralloy 135 is shown in Fig. 10. These data may be used as an approximation in estimating growth when nitriding by the double-stage process. They should be used as a guide for determining size changes only with respect to parts of this design, however; the growth that occurs in solid rounds of bars is of the order of a 0.4 mm (0.0015 in.) increase in diameter.

In some parts, the dimensional changes during nitriding involve both internal and external

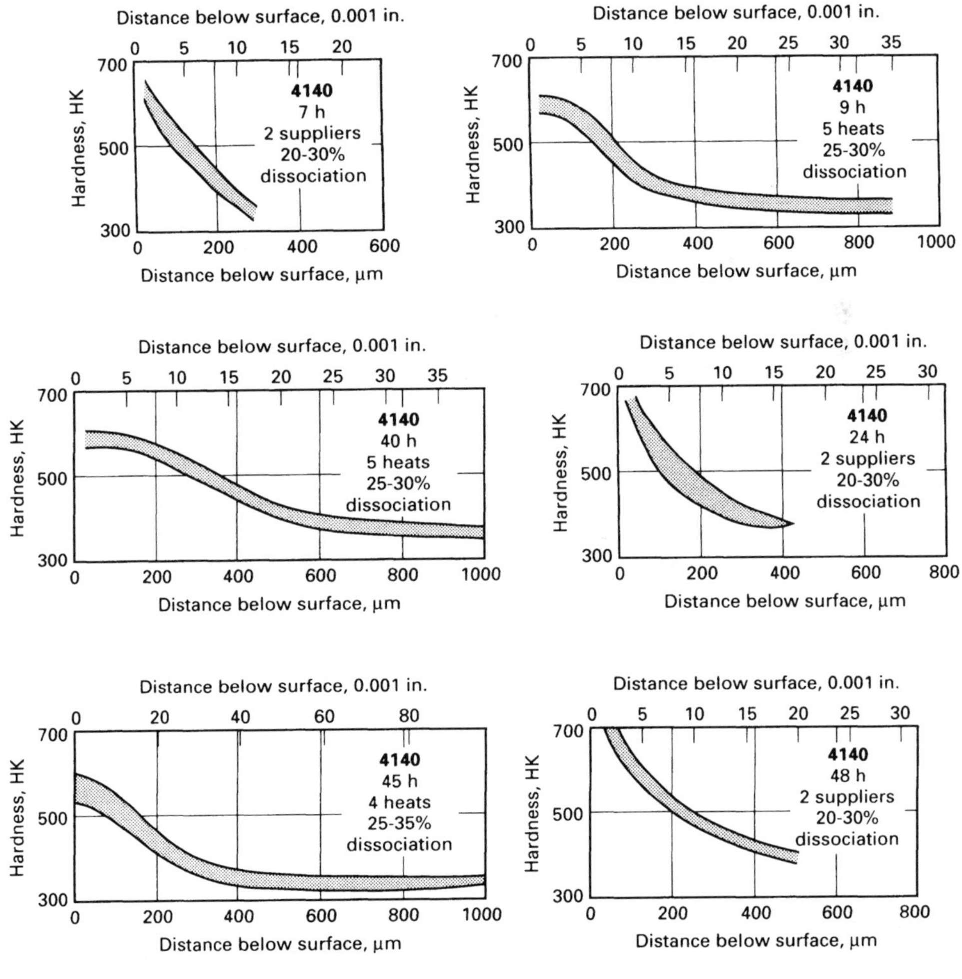

Fig. 7 Hardness gradients for nitrided chromium-containing low-alloy steels

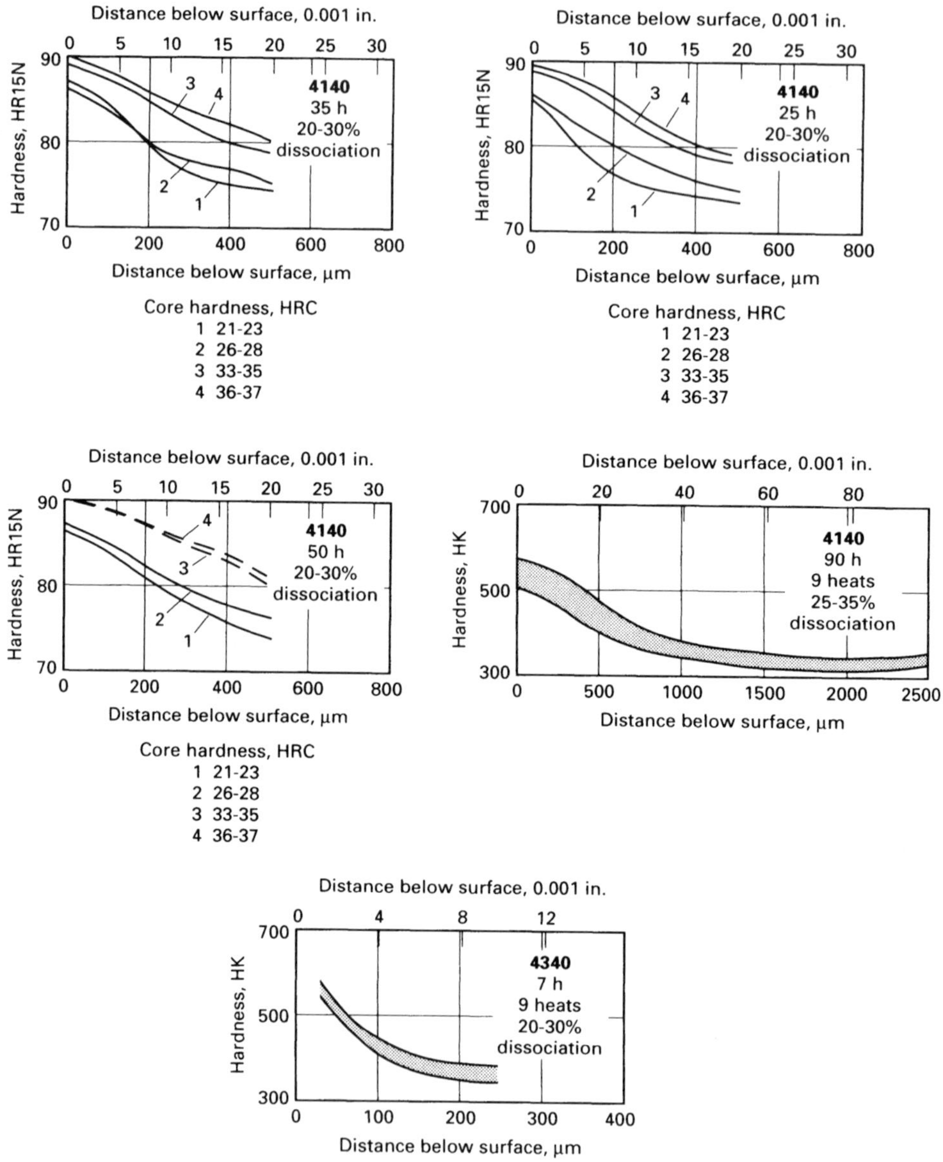

Fig. 7 **(continued)** Hardness gradients for nitrided chromium-containing low-alloy steels

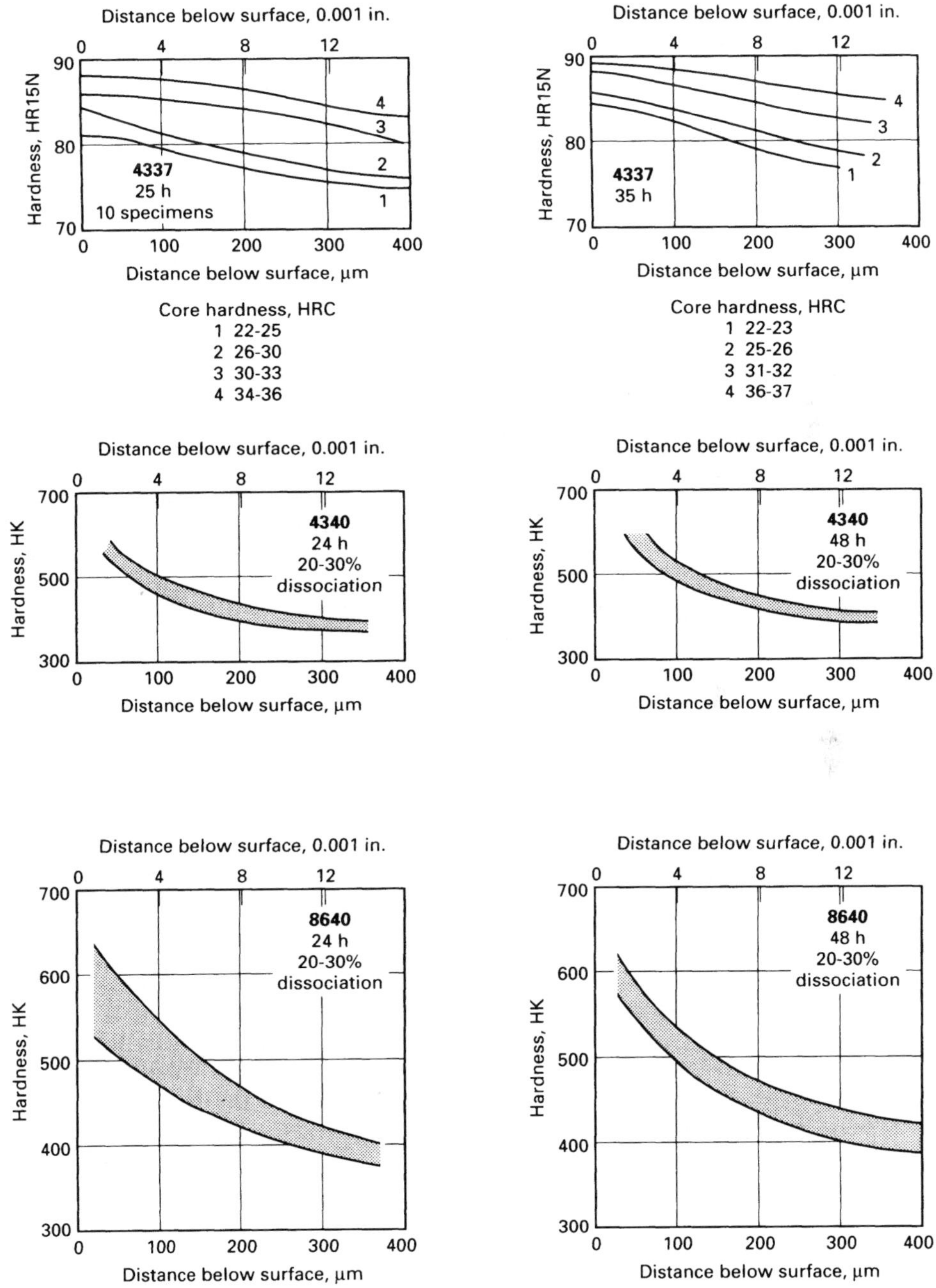

Fig. 7 (continued)

surfaces. For example, the bore diameter of a 300 mm (12 in.) diameter spur gear decreased as much as 0.025 mm (0.001 in.), whereas the overall gear dimension increased up to 0.1 mm (0.004 in.).

Sharp corners or edges should be avoided on parts to be nitrided, because the projections formed at sharp corners, as a result of the growth that takes place, are high in nitrogen content and susceptible to chipping. Similarly, sharp edges nitride throughout the section and have no supporting core. When sharp corners are unavoidable, brittleness may be reduced by nitriding one side only, if the other side is not a wearing surface. Frequently, the problems of growth are eliminated by nitriding only those surfaces that are subject to wear in service.

Stabilizing Treatment. In nitrided parts, there is a balance between compressive stresses in the case and tensile stresses in the core. If this balance is upset by grinding off a part of the case, slow dimensional changes may occur as the stresses approach equilibrium. (In some instances, slow dimensional changes resulting from stress redistribution during grinding have been erroneously attributed to wear.) To prevent these changes, nitrided parts are first ground almost to the final dimensions, then heated to 565 °C (1050 °F) for 1 h, and finally finish ground or lapped. Parts nitrided and not ground after nitriding have excellent dimensional stability.

Stress Relieving. Many standard procedures require that parts be rough machined, stress relieved, and finish machined before being nitrided. For many parts, this lengthy procedure may not be required. In general, it has been found that stress relieving after rough machining is required only for slender parts or parts with thin wall sections or large or complex sections, such as large welded gear assemblies.

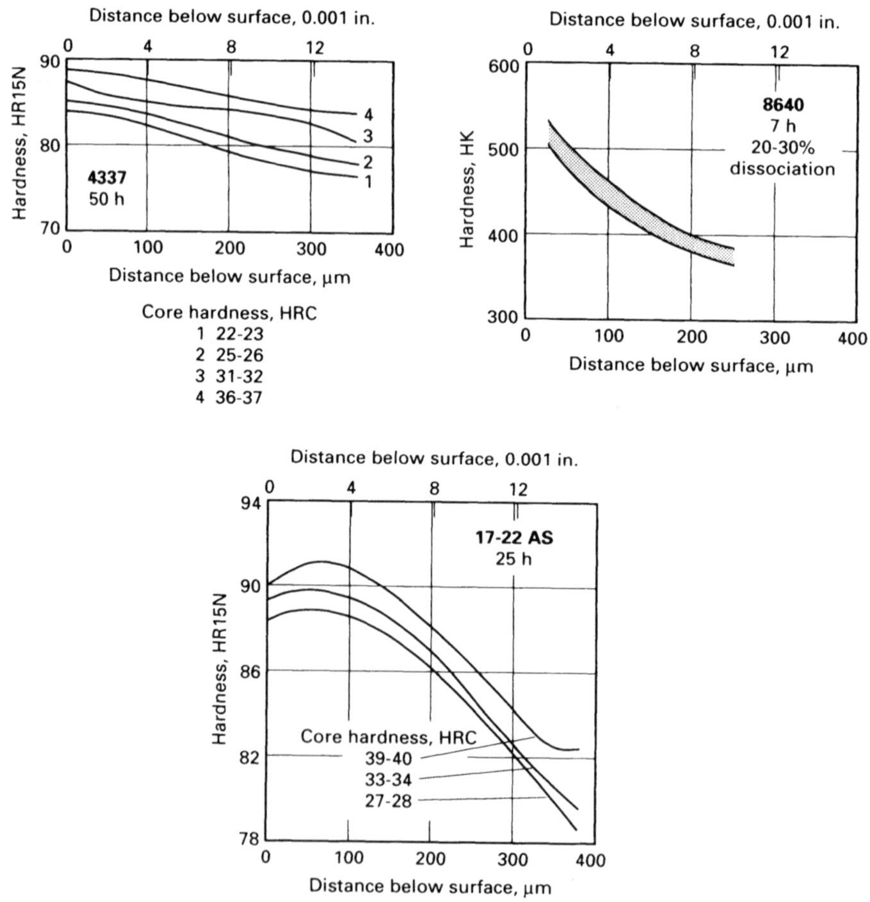

Fig. 7 **(continued)** Hardness gradients for nitrided chromium-containing low-alloy steels

When distortion is caused by the removal of induced machining stresses during nitriding, stress relieving at 620 °C (1150 °F) for 4 h prior to nitriding lessens or eliminates this problem.

Equipment

The gas nitriding furnace is a relatively simple furnace construction that, over the years, has been refined, mainly in the area of computerized process control rather than construction design (see, for example, the section "Controlled Nitriding" later in this chapter). Furnaces of several designs are in common use. Most of these are batch furnaces, which incorporate certain essential features:

- A means of sealing the charge to exclude air and other contaminants while containing the controlled atmosphere
- An inlet line for introducing atmosphere and an outlet line for exhausting used atmosphere
- A means of heating and appropriate temperature controls
- A means, such as a fan, for circulating atmosphere and equalizing temperature throughout the workload

Examples of commonly employed furnace types include the stationary vertical retort furnace (Fig. 11), the bell-type movable furnace (Fig. 12), and the box furnace. More detailed information on these furnace types can be found in Ref 2.

Common Gas Nitriding Problems

Problems that may disqualify nitrided parts from a particular service or seriously impair their performance can be classified into five categories, as shown in Table 5. The most commonly encountered problems originating during the nitriding process are:

- Low case hardness or shallow case
- Discoloration of workpieces
- Excessive dimensional changes
- Cracking and spalling of nitrided surfaces

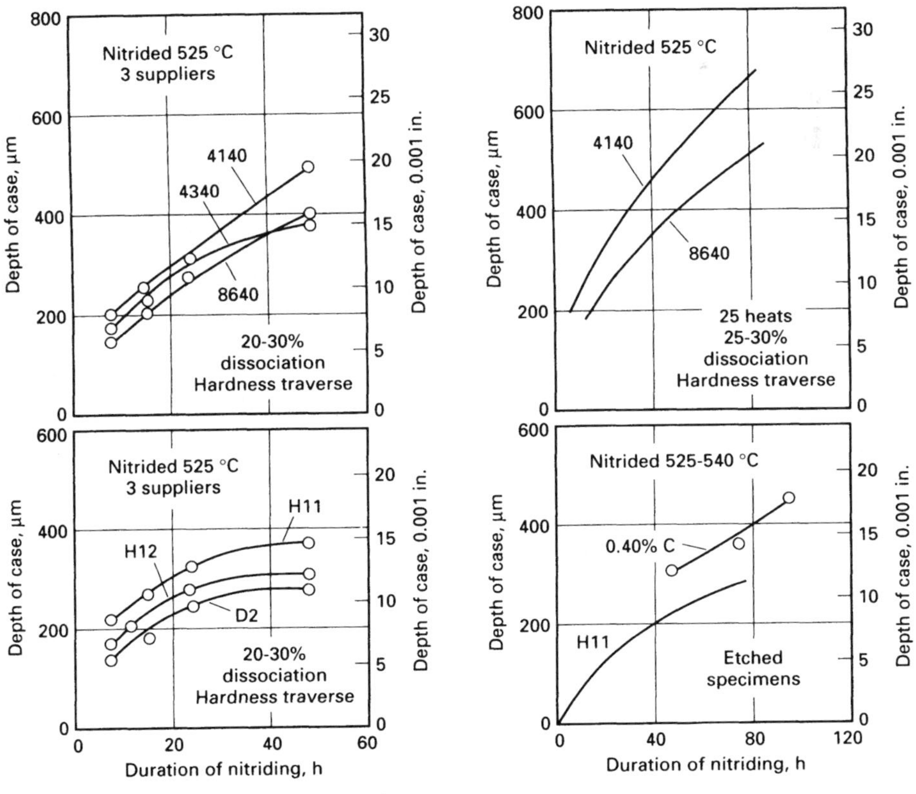

Fig. 8 Depth of case as a function of duration of nitriding for chromium-containing low-alloy and tool steels

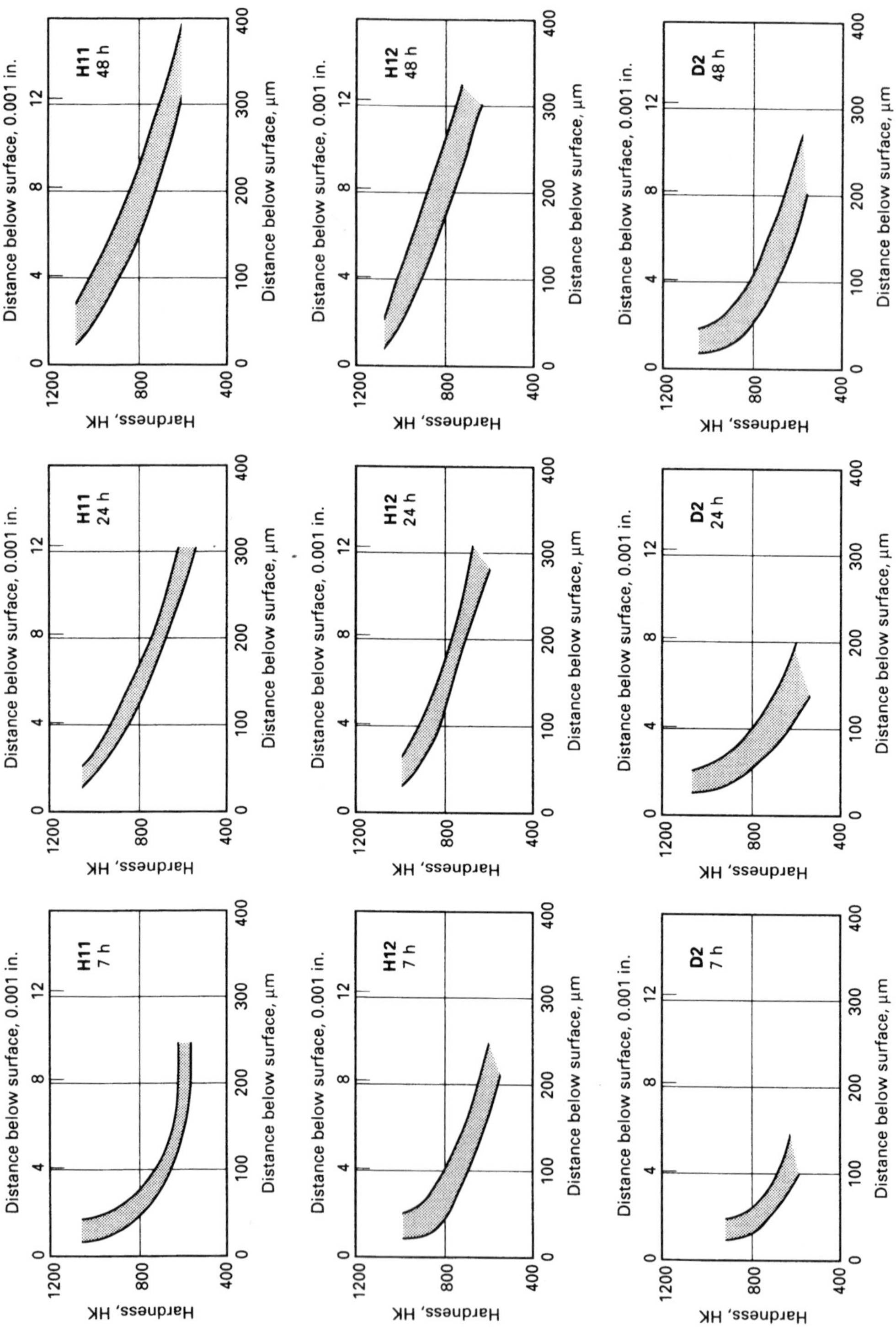

Fig. 9 Hardness gradients for chromium-containing tool steels

- Variations in percentage of ammonia dissociation
- White layer deeper than permitted
- Plugging of exhaust lines and pipette lines

A knowledge of the causes of these problems should be of assistance in avoiding, preventing, or correcting them. A number of possible causes are indicated subsequently and summarized in Table 6, which also suggests methods of prevention.

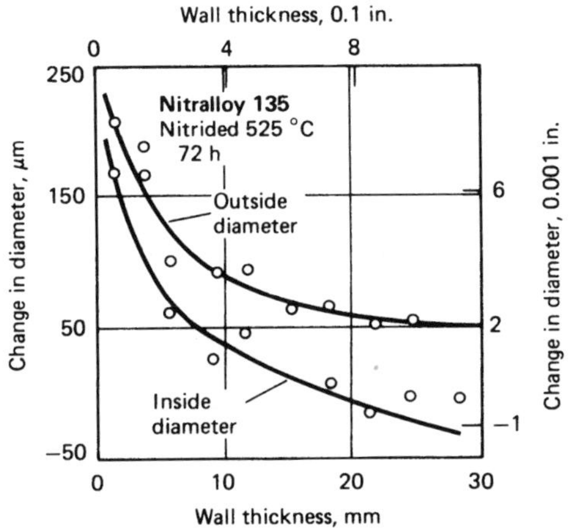

Fig. 10 Growth as a function of the wall thickness of 70 mm (2¾ in.) diam hollow cylinders double-stage nitrided for 72 h

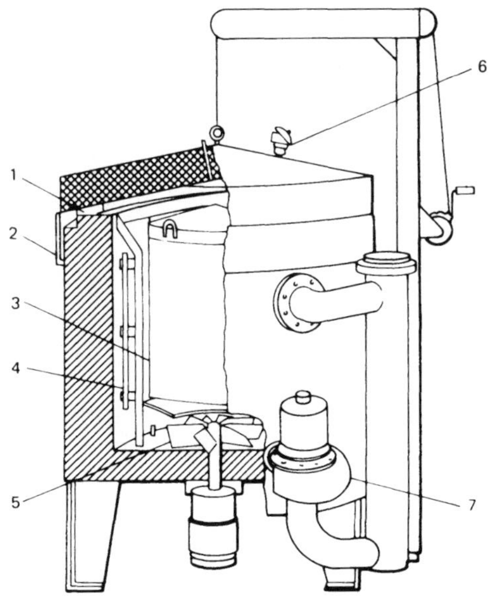

Fig. 11 Vertical retort nitriding furnace. 1, gasket; 2, oil seal; 3, work basket; 4, heating elements; 5, circulating fan; 6, thermocouple; and 7, cooling assembly. At end of cycle, a valve is opened and fan (not shown) incorporated in the external cooler circulates atmosphere through the water-jacketed cooling manifold.

Low case hardness or shallow case may be caused by the characteristics of the steel or faulty processing. The steel characteristics affecting case hardness and depth include:

- Composition unsuitable for nitriding
- Improper microstructure
- Failure to quench and temper prior to nitriding
- Low core hardness
- Surface passivation from machining, inadequate cleaning, or foreign matter

In terms of processing, a shallow case or low case hardness may be affected by:

- Excessively low or high nitriding temperature
- Insufficient ammonia flow
- Nonuniform circulation or temperature in furnace
- Prolonged exposure of furnace parts and work baskets to nitriding conditions such as ammonia (burnout required); see section on fixtures
- Insufficient time at temperature

Finally, low case hardness or shallow case may only be apparent—occurring as the result of inaccuracies in testing due to faulty adjustment of equipment, improper preparation or positioning of the test specimen, or the use of a test load excessive for the case depth.

Discoloration of workpieces may be caused by:

- Improper or inadequate prior surface treatment, including etching, washing, degreasing, and phosphate coating
- Oil, air, or moisture in the retort

Oil in the retort can occur because of:

- Inadequate cleaning of parts, especially those with deep holes and recesses
- Loss of pressure at seal or overheating of seal
- Leakage at the base, or other parts, of the furnace

Moisture in the retort can occur because of:

- Leakage from the cooling chamber

- Water being sucked in from water bottle during rapid cooling with inadequate gas flow

Air in the retort can occur because of:

- Inadequate seal
- Leakage due to inadequate sealing around pipes or thermocouple
- Introduction of air to purge ammonia while charge is at or above 175 °C (350 °F)

Excessive dimensional changes may be caused by:

- Inadequate stress relieving prior to nitriding
- Inadequate support of parts during nitriding

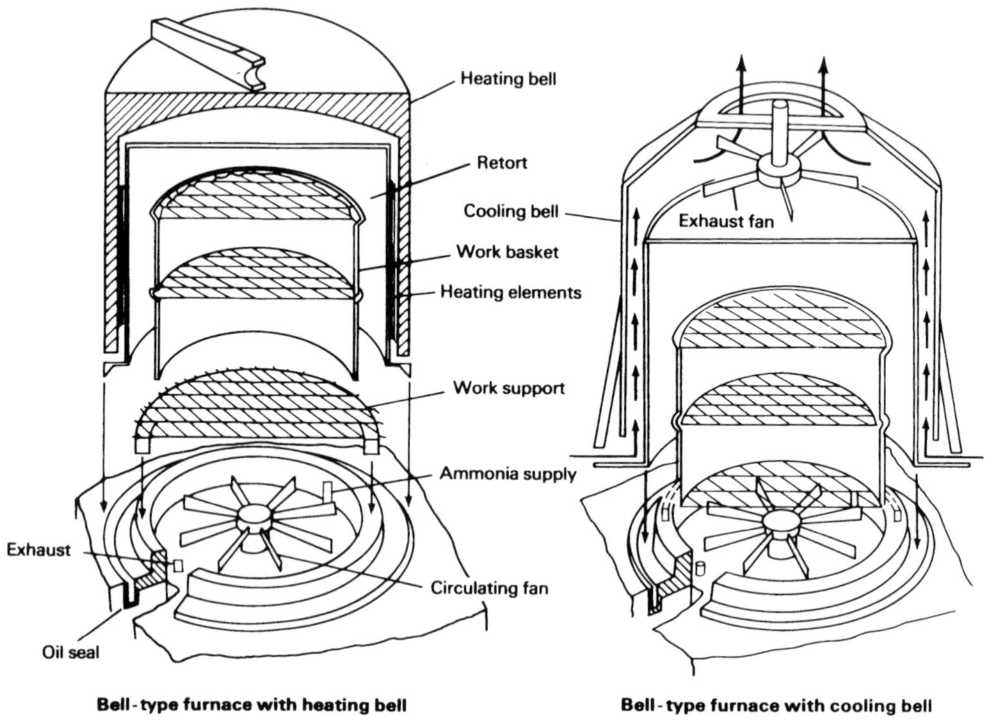

Fig. 12 Schematic of bell-type furnace showing stationary base surmounted by bell

Table 5 Classification of problems in nitrided steel components

Operation during which problem originates	Examples	Problems manifest after nitriding
Steelmaking process	Ferrite banding; carbide and/or sulfide segregation; laps; pipes; etc.	Nonuniform diffusion layer; microcracks at grain boundaries; surface cracks
Heat treatment preceding nitriding (hardening, tempering, stress relieving)	Cracks; decarburization; overheating; temper embrittlement	Cracks; low surface hardness; low core hardness; exfoliation of layer from surface
Machining prior to nitriding	Grinding cracks and burns; burrs; sharp corners; machining stresses	Surface cracks; portions of surface left unnitrided; oversaturation in fillets and corners; uneven, nonuniform, and no-continuous layer
Cleaning and activation	Solid precipitations on the surface; greasy film; oxidation	Soft spots on surface; partial or complete surface left unnitrided
Nitriding process	Nonuniformity of layer; excessive white layer; low surface hardness; thin diffusion case; oversaturation; excessive or insufficient porosity in white layer	Nonuniform and low surface hardness; brittleness; generally poor service properties; poor lubricity at surface

Source: Ref 3

- Inappropriate design of parts, including nonsymmetry of design and wide variations in section thickness
- Unequal cases on various surfaces of parts, resulting from nonuniform conditions (created by furnace design or manner in which parts are arranged in load) or variations in absorptive power of surfaces (resulting from stopoff practices or from variations in surface metal removed, surface finishing technique, or in degree of cleanliness)

Cracking and spalling of nitrided surfaces may be caused by dissociation in excess of 85% and also (especially for aluminum-containing steels) by:

- Design (particularly sharp corners)
- Excessively thick white layer
- Decarburization of surface in prior heat treatment
- Improper heat treatment

Variations in percentage of ammonia dissociation may be caused by:

- Charge being too small for furnace area
- Overactive surface of furnace parts and fixtures
- Leakage or loss of sample from burette
- Change in gas flow caused by buildup of pressure in furnace
- Variations in furnace temperature

White layer deeper than permitted may be caused by:

- Nitriding temperature being too low
- Percentage of dissociation below the recom-

Table 6 Typical problems originating during gas nitriding and methods of their prevention

Feature regulated by specification requirements	Type of fault	Root cause	Method of prevention
White layer thickness	White layer excessively thick	Atmosphere nitriding capability too aggressive	Use more diluted atmosphere, lower flow rate, and raise dissociation rate
Porous zone of white layer	Porous zone excessive	Atmosphere nitriding capability too aggressive, concentration too high	Use more diluted atmosphere, lower flow rate, and raise dissociation rate
Microstructure of diffusion case	Network of nitrides in diffusion layer	Atmosphere nitriding capability too aggressive, concentration too high	Use more diluted atmosphere, lower flow rate, and raise dissociation rate
Surface hardness	Surface hardness too low	Improper process temperature	Compliance with nitriding specifications; systematic control of temperature-measuring instruments; on alloy steels, for higher hardness use lower nitriding temperature
		Flow rate of nitriding atmosphere too low	Strict compliance of flow rates with specified values
		Perturbations in atmosphere circulation through load	Greater distances between nitrided components
		Nitriding time too short	Extend nitriding time
	Soft spots on nitrided surfaces	Insufficient purge of retort prior to nitriding process	Verify degree of purge of air from retort before switching heating on
Case microhardness profile	Failure to meet specified hardness at given depth	Atmosphere nitriding capability too low	Raise flow rate and lower dissociation rate; extend process time
Shape and size of part	Deformation	Improper fixturing or location of parts in retort	Slender parts should be suspended vertically
		Nonuniform layer thickness on different areas of surface	Prepare surface by thorough washing and/or activation prior to nitriding; raise atmosphere flow rate to achieve uniformity throughout retort
		Nitriding temperature too high	Nitride at temperature lower than that of preceding heat treat operation
Surface roughness	Excessive roughness	White layer too thick; lack of control at point of nucleation of white layer	Use less aggressive nitriding atmosphere; apply control at point of nucleation of white layer

Source: Ref 3

mended minimum (15%) during the first stage

- First stage held too long
- Percentage of dissociation too low during the second stage
- Fast purging with raw ammonia instead of cracked ammonia or nitrogen, above 480 °C (900 °F) during slow cooling

Plugging of exhaust lines and pipette lines is caused by precipitates that are formed by the reaction of ammonia with many of the various chemical compounds commonly present in ordinary domestic water. These precipitates may plug lines and prevent proper sampling or cause pressure to build up in the furnace by plugging exhaust lines or restricting valve openings.

Enlarging lines or treating them periodically with a dilute acid solution corrects this, especially if the solution is trapped in a low spot and drained. (The use of distilled water, or water of similarly low impurity, also eliminates this difficulty.)

In some installations, water from pipettes can leak down into exhaust lines, flushing scale and other foreign material into low spots or restrictions and thus plugging the lines. A drop leg to trap such products reduces trouble from this source, as does reduction of right-angle bends and elimination of pipes smaller than 19 mm ($^3/_4$ in.) in diameter, where possible.

Selective Nitriding

Many coatings are available as stopoffs to prevent gas nitriding of selected areas. The success of a coating depends on such variables as density and thickness of the coating, adhesion of coating to steel, surface finish of the part, and degree of leakage permitted.

Proprietary paints are effectively used in commercial heat treating operations. They are also used to touch up other coatings that have been inadvertently removed or damaged during processing. These paints usually consist of a tin base suspended in a vehicle of lacquer, aromatic hydrocarbon, or a water glass. It is important that the constituents be mixed in the proper proportions (thick coatings may run, and thin coatings are not completely effective) and that the paints be applied to uniform thickness. The surface to be painted must be very clean. Ground or polished surfaces may be difficult to wet uniformly with paint.

Plated deposits of bronze or copper are the most common stopoff coatings. Nickel (including electroless nickel), chrome, and silver are effective also, but their higher cost restricts their use to special applications.

Thickness and density of plated coatings are important in determining their effectiveness as stopoffs. Minimum thickness of bronze or copper plate should be 18 μm (0.7 mil) for ground surface finishes of 1.6 μm (64 μin.) or smoother, 25 μm (1.0 mil) for finishes between 1.6 and 3.2 μm (64 and 125 μin.), and 38 μm (1.5 mil) for finishes of 3.2 μm (125 μin.) and rougher. Compared to copper and bronze, nickel is a more effective stopoff; therefore, a thinner coating is permitted.

Electroplated silver is 100% effective when the plate thickness is a minimum of 38 μm (1.5 mil); it is 95% effective even during long nitriding cycles, when as little as 25 μm (1.0 mil) of plate is used.

Surface finish of the base metal also influences the thickness of the coating. A finish of 3 μm (120 μin.) requires a thicker coating than a finish of 1.5 μm (60 μin.). Usually, a finish of 1.5 μm (60 μin.) or smoother is recommended.

Processing Procedures. Several processing procedures are employed to accomplish selective nitriding. One of the most widely used consists of rough machining, plating, machining, or grinding areas to be nitrided, nitriding, then finish machining or grinding wherever required. In another procedure, the areas to be nitrided are masked to prevent plating. When masking is difficult, the plating material is applied to all surfaces and then selectively stripped from the areas to be nitrided.

Fine threads (external or internal) on precision parts can be protected by a tin-lead solder. The threads should be cleaned and coated with a flux containing a tinning compound, then heated slowly until both solder and flux are melted. The excess solder and flux are blown out with compressed air, leaving a coating thin enough so that it does not run during nitriding and does not require cleaning or stripping after nitriding.

When the application does not permit the retention of any protective plate on the finished part after nitriding, selection of the coating is important from the standpoint of subsequent stripping. Copper and silver are the easiest to strip; bronze is more difficult. Nickel is very difficult to remove without detrimentally affecting the part. Stopoff paint residues may be reduced by brushing or washing or may be removed by lightly blasting with fine abrasives.

Nitriding of Stainless Steels

Because of their chromium content, all stainless steels can be nitrided to some degree. Although nitriding adversely affects corrosion resistance, it increases surface hardness and provides a lower coefficient of friction, thus improving abrasion resistance.

Austenitic and Ferritic Alloys. Austenitic stainless steels of the 300 series are the most difficult to nitride; nevertheless, types 301, 302, 303, 304, 308, 309, 316, 321, and 347 have been successfully nitrided. These nonmagnetic alloys cannot be hardened by heat treating; consequently, core material remains relatively soft, and the nitrided surface is limited as to the loads it can support. This is equally true of the nonhardenable ferritic stainless steels. Alloys in this group that have been satisfactorily nitrided include types 430 and 446. With proper prior treatment, these alloys are somewhat easier to nitride than the 300-series alloys.

Hardenable Alloys. The hardenable martensitic alloys are capable of providing high core strength to support the nitrided case. Hardening, followed by tempering at a temperature that is at least 15 °C (25 °F) higher than the nitriding temperature, should precede the nitriding operation. Precipitation-hardening alloys, such as 17-4 PH, 17-7 PH, and A-286, also have been successfully nitrided.

Prior Condition. Before being gas nitrided, 300-series steels and nonhardenable ferritic steels should be annealed and relieved of machining stresses. The normal annealing treatments generally employed to obtain maximum corrosion resistance are usually adequate. Microstructure should be as nearly uniform as possible. Observance of these prior conditions prevents flaking or blistering of the nitrided case. Martensitic steels, as previously noted, should be in the quenched and tempered condition.

A special pretreatment for 410 stainless is hardening from a lower-than-normal temperature; this results in a very uniform nitrided case with reduced internal stresses. Cracking or spalling of the case is avoided; formation of brittle grain-boundary carbonitrides is suppressed. Austenitizing at 860 °C (1580 °F), followed by tempering at 595 °C (1100 °F) uniformly distributes carbides and provides low residual stress. Case growth is accommodated by a hardness of approximately 25 HRC.

Surface Preparation. The nitriding of stainless steels requires certain surface preparations that are not required for nitriding low-alloy steels. Primarily, the film of chromium oxide that protects stainless alloys from oxidation and corrosion must be removed. This may be accomplished by dry honing, wet blasting, pickling, chemical reduction in a reducing atmosphere, or submersion in molten salts, or by one of several proprietary processes. Surface treatment must precede placement of the parts in the nitriding furnace. If there is any doubt of the complete and uniform depassivation of the surface, further reduction of the oxide may be accomplished in the furnace by means of a reducing hydrogen atmosphere or halogen-based proprietary agents. Of course, hydrogen must be dry (free of water and oxygen).

Before being nitrided, all stainless parts must be perfectly clean and free of embedded foreign particles. After depassivation, care should be exercised to avoid contaminating stainless surfaces with fingerprints. Sharp corners should be replaced with radii of not less than 1.6 mm ($^1/_{16}$ in.).

Nitriding Cycles. In general, stainless steels are nitrided in single-stage cycles at temperatures from approximately 495 to 595 °C (925 to 1100 °F) for periods ranging from 20 to 48 h, depending on the depth of case required. Dissociation rates for the single-stage cycle range from 20 to 35%; a two-stage cycle using 15 to 30% in first phase and 35 to 45% in the second phase is also used. Thus, except for the prior depassivation of the metal surface, the nitriding of stainless steels is similar to the single-stage nitriding of low-alloy steels.

Nitriding Results. Hardness gradients are given in Fig. 13 for types 302, 321, 430, and 446. These data are based on a 48 h nitriding cycle at 525 °C (975 °F), preceded by suitable annealing treatments. A general comparison of the nitriding characteristics of series 300 and 400 steels is presented in Fig. 14; the comparison reflects the superior results that are obtained with series 400 steels, as well as the effects of nitriding temperature on depth of case. Data are plotted for single-stage nitriding at temperatures of 525 and 550 °C (975 and 1025 °F). For steels of both series, greater case depths were obtained at the higher nitriding temperature.

Applications. Although nitriding increases the surface hardness and wear resistance of stainless steels, it decreases general corrosion resistance by combining surface chromium with nitrogen to form chromium nitride. Consequently, nitriding is not recommended for applications in which the corrosion resistance of stainless steel is of major importance. For example, a hot-air valve made of cast type 347 and used in the cabin-heating system of a jet plane was nitrided to improve its resistance to wear by the abrading action of a sliding butterfly. When the valve remained in the closed position for an extended period, the corrosive effects of salt air froze the valve into position so that it could not be opened.

In contrast, a manufacturer of steam-turbine power-generating equipment has successfully used nitriding to increase the wear resistance of types 422 and 410 stainless steel valve stems and bushings that operate in a high-temperature steam atmosphere. Large quantities of these parts have operated for 20 years or more without difficulty. In a few instances, a light-blue oxide film has formed on the valve stem diameter, causing it to "grow" and thus reduce the clearance between stem and bushing; the growth condition, however, was not accompanied by corrosive attack.

Nitrided stainless is also being used in the food-processing industry. In one application, nitrided type 321 was used to replace type 302 for a motor shaft used in the aeration of orange juice. Because the unhardened 302 shaft wore at the rubber-sealed junction of the motor and the juice, leaks developed within three days. The nitrided 321 shaft ran for 27 days before wear at the seal resulted in leakage. In machinery used in the preparation of dog foods, nitrided type 420 gears have replaced gears made of an unhardened stainless and have exhibited a considerable increase in life.

Modern synthetic fibers, several of which are highly abrasive, have increased the wear of textile machinery. Mechanical parts in textile machines are subjected to high humidity, absence of lubrication, high-speed movements with repeated cycling, and the abrasive action of fibers traveling at high speeds. A shear blade made of hardened, 62 to 64 HRC, 1095 steel experienced a normal life of approximately one million cuts (four weeks of service) in cutting synthetic fibers at the rate of 90 cuts per minute. In contrast, a nitrided type 410 blade with 0.04

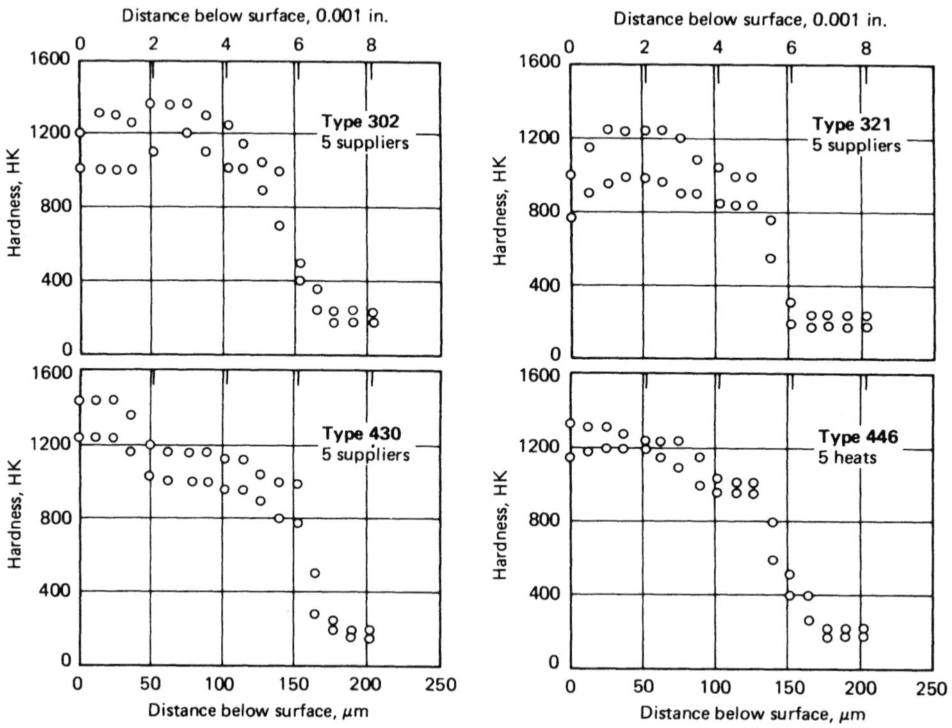

Fig. 13 Hardness range as a function of depth of case for four stainless steels that were annealed prior to nitriding. Annealing temperatures: type 302 and type 321, at 1065 °C (1950 °F); type 430, at 980 °C (1800 °F); and type 446, at 900 °C (1650 °F)

mm (0.0015 in.) case depth showed less wear after completion of five million cuts.

With nitrided stainless steels, the case almost always has lower corrosion resistance than the base material; nevertheless, the corrosion resistance of the case can be adequate for certain applications. For example, nitrided types 302 and 410 stainless steel resist attack from warp conditioner and size in the textile industry but do not resist attack from the acetic acid used in dyeing liquors.

Nitrided stainless is not resistant to mineral acids and is subject to rapid corrosion when exposed to halogen compounds. However, a nitrided type 302 piston lasted for more than five years in a liquid-ammonia pump; it replaced a piston made of an unnitrided 300-series alloy that lasted approximately six months. Nitrided 17-4 PH impellers have performed satisfactorily and without corrosion in various types of hydraulic pumps.

Modified Gas Nitriding Processes

The gas nitriding processes described thus far in this Chapter have used ammonia as the processing gas. The ammonia cracks on contact with the heated steel surface, and the degree of cracking, or dissociation rate, depends on temperature and gas flow rate and has been the only factor controlling the nitriding capability of the atmosphere. However, ammonia-only atmospheres do not offer sufficient flexibility, and, in some cases, nitrided steel surfaces have been oversaturated with nitrogen, resulting in the presence of brittle layers that may require post-process removal.

In recent years, many nitriding shops have switched to mixed-gas atmospheres, for example, ammonia and hydrogen (the dilution method of nitriding) or ammonia carefully mixed with various gas components (controlled nitriding), to reduce the problems associated with traditional gas nitriding with ammonia. Both of these are briefly reviewed.

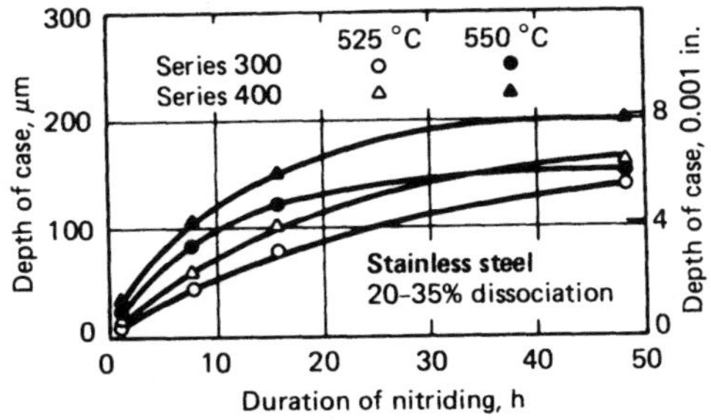

Fig. 14 Comparison of nitriding characteristics of series 300 and 400 stainless, single-stage nitrided at 525 and 550 °C (975 and 1025 °F)

The Dilution Method of Nitriding (Ref 1)

When hydrogen is added to the flow of ammonia, the nitrogen component of the ammonia is diluted. The compound zone and its quality at the surface of the steel are controlled by the carbon in the steel, which, if at a high enough level, begins the nucleation of the ε phase in the compound zone. It is also influenced by nitrogen, which, at high enough concentrations, encourages the nucleation of the γ′ phase of the compound zone. It can be seen that very careful control of the gas chemistry is of great importance.

Care should be taken when using hydrogen as a diluent gas because of its high inflammability. Simple precautions should be taken when using hydrogen:

- Use good effective retort seals.
- Do not allow oxygen to mix with hydrogen, because this constitutes a serious explosion or fire risk.
- Ensure that when the cycle is complete, the retort has been given sufficient time to cool down before opening it to the atmosphere. Cool to below 150 °C (300 °F), then purge with nitrogen to remove all traces of hydrogen.

These precautions also apply to conventional gas nitriding.

Diluted atmospheres have proven to be advantageous, giving the process more flexibility and producing compound layers with greater ductility by reducing the amount of active and available nitrogen.

Controlled Nitriding (Ref 3, 4)

Controlled nitriding is a further development of the traditional gas nitriding in which all of the process parameters are computer controlled. The key to the success of the process lies in its ability to effectively control the concentration of nitrogen in the surface layer of the treated parts. By controlling the activity of nitrogen in the gas atmosphere, it is possible to control the activity of nascent nitrogen, a factor that determines the nitriding capability of the atmosphere. The technology involves measuring and adjusting what is termed the nitriding potential, N_p, which is expressed by the formula:

$$N_p = \frac{P_{NH_3}}{(P_{H_2})^{1/2}}$$

where P_{NH_3} is the partial pressure of ammonia, and P_{H_2} is the partial pressure of hydrogen in the furnace atmosphere.

It should be noted that the term *potential* is in no way equivalent to what is meant by carbon potential in carburizing atmospheres. It is, rather, the reaction constant for the dissociation of ammonia. Its control allows the formation of predictable case structures and depths for a range of steels, including some tool and stainless grades. Higher N_p values produce higher surface concentrations of nitrogen and steeper concentration gradients. Lower potentials allow the development of nitrided cases without any brittle-compound (white) layer in high-alloy steels. This can be of particular advantage in certain applications, such as aerospace or other components subject to high compact stresses. In these applications, the brittleness and/or unpredictability of the white layer produced by conventional nitriding is detrimental.

The controlled nitriding process differs from conventional gas nitriding in that it does not use pure ammonia. Instead, other gas components—including nitrogen, previously dissociated ammonia, hydrogen, argon, and oxygen—are added in a controlled way to conventional ammonia. The effect of these additions, with the exception of argon, is to reduce the nitriding potential of the atmosphere; argon has no effect on N_p. Controlled gas nitriding expands on the advantages of nitriding, such as corrosion resistance and low distortion, and expands the range of metals suitable for this surface-hardening process.

How the Process Works. This system makes it possible to adjust the availability of atomic nitrogen at the gas-metal interface and to precisely control the properties by changing the composition of the gas mix. This is done not only by selecting the correct atmosphere composition for a given material and process temperature but also by maintaining the preset nitriding potential value for each stage of the process.

Temperature and gas-flow controls, linked by a computer connected to a nitriding potential-sensing device, allow the programming of the complete process cycle to produce predictable and reproducible surface-layer characteristsics for a given material. The control system can store a number of process recipes that may be selected for specified materials and applications. When the operator selects a process from the software's "library," the time, temperature, atmosphere composition, and nitriding potential for each stage of the cycle are automatically controlled.

The computer program also controls functions such as opening and closing of valves, switching electric motors on and off, and maintaining optimal pressure inside the retort. The operator may manually select less critical factors, such as printout frequency and the temperature at which the process is to stop after cooling.

Process Advantages. The control of the nitrogen concentration in the surface zone by maintaining a preset nitriding potential of the atmosphere offers several advantages over traditional gas nitriding and other competitive surface treatment methods:

- Improved ductility and toughness of the compound layer, reducing the risk of cracking or spalling
- Improved surface roughness values
- Improved microhardness profiles (Fig. 15)
- Cost-effectiveness through optimized cycle times
- Often, even lower temperatures than conventional nitriding, thus further minimizing distortion
- A broader range of treatable steel grades
- Controlled size and amount of porosity of the outer layer, an advantage for parts requiring self-lubricity
- Enhanced corrosion resistance through the introduction of additional compounds into the surface layer

In addition, by eliminating the white layer where it is not wanted, the process expands on the traditional advantage of nitriding not needing mechanical finishing operations. It also limits the surface roughness that accompanies nitriding by appropriately structuring the nitriding stages. The process is ecologically more "friendly" than conventional gas nitriding, because less ammonia is used overall.

Pack Nitriding

Pack nitriding (U.S. Patent 4,119,444), which is a process analogous to pack carburizing (see Chapter 4, "Pack and Liquid Carburizing"), employs certain nitrogen-bearing organic compounds as a source of nitrogen. On heating, the compounds used in the process form reaction products that are relatively stable at temperatures up to 570 °C (1060 °F). Slow decomposition of the reaction products at the nitriding temperature provides a source of nitrogen. Nitriding times of 2 to 16 h can be employed. Parts are packed in glass, ceramic, or aluminum containers with the nitriding compound, which is often dispersed in an inert packing media. Containers are covered with aluminum foil and heated by any convenient means to the nitriding temperature.

Liquid Nitriding

Liquid nitriding (nitriding in a molten salt bath) employs the same temperature range as gas nitriding, that is, 510 to 580 °C (950 to 1075 °F). The case-hardening medium is a molten, nitrogen-bearing, fused-salt bath containing either cyanides or cyanates. Unlike liquid carburizing and cyaniding, which employ baths of similar compositions, liquid nitriding is a subcritical (that is, below the critical transformation temperature) case-hardening process; thus, processing of finished parts is possible, because dimensional stability can be maintained. Also, liquid nitriding adds more nitrogen and less carbon to ferrous materials than that obtained through higher-temperature diffusion treatments.

The liquid nitriding process has several proprietary modifications and is applied to a wide variety of carbon, low-alloy steels, tool steels, stainless steels, and cast irons.

Liquid Nitriding Applications

Liquid nitriding processes are used primarily to improve wear resistance of surfaces and to increase the endurance limit in fatigue. For many steels, resistance to corrosion is improved. These processes are not suitable for many applications requiring deep cases and hardened cores, but they have successfully replaced other types of heat treatment on a performance or economic basis. In general, the uses of liquid nitriding and gas nitriding are similar and, at times, identical. Gas nitriding may be preferred in applications where heavier case depths and dependable stopoffs are required. Both processes, however, provide the same advantages: improved wear resistance and antigalling properties, increased fatigue resistance, and less distortion than other case-hardening processes employing through heating at higher temperatures. Four examples

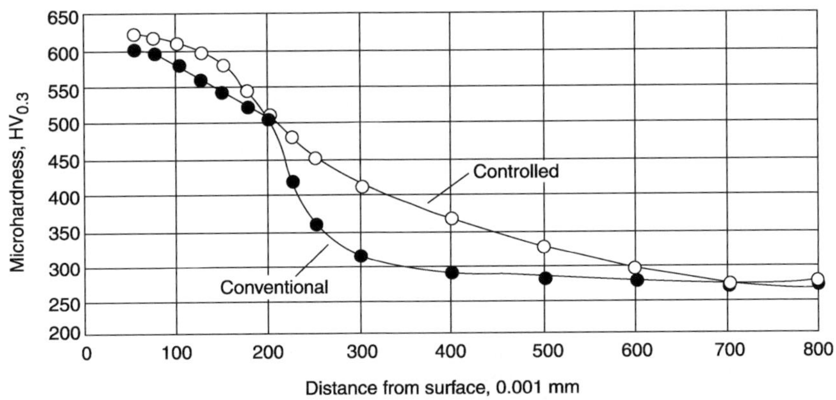

Fig. 15 Microhardness profiles of EN 19 chromium-molybdenum steel produced by controlled and conventional gas nitriding processes. Source: Ref 4

of parts for which liquid nitriding was selected over other case-hardening methods appear in Table 7.

The degree to which steel properties are affected by liquid nitriding may vary with the process used and the chemical control maintained. Thus, critical specifications should be based on prior test data or documented information.

Liquid Nitriding Systems

The term *liquid nitriding* has become a generic term for a number of different fused-salt processes, all of which are performed at subcritical temperature. Operating at these temperatures, the treatments are based on chemical diffusion and influence metallurgical structures primarily through absorption and reaction of nitrogen rather than through the minor amount of carbon that is assimilated. Although the different processes are represented by a number of commercial trade names, the basic subclassifications of liquid nitriding are those presented in Table 8.

A typical commercial bath for liquid nitriding is composed of a mixture of sodium and potassium salts. The sodium salts, which comprise 60 to 70% (by weight) of the total mixture, consist of 96.5% NaCN, 2.5% Na_2CO_3, and 0.5% NaCNO. The potassium salts, 30 to 40% (by weight) of the mixture, consist of 96% KCN, 0.6% K_2CO_3, 0.75% KCNO, and 0.5% KCl. The operating temperature of this salt bath is 565 °C (1050 °F). With aging (a process described subsequently in the section "Operating Procedures"), the cyanide content of the bath decreases, and the cyanate and carbonate contents increase (the cyanate content in all nitriding baths is responsible for the nitriding action, and the ratio of cyanide to cyanate is critical). This bath is widely used for nitriding tool steels, including high-speed steels, and a variety of low-alloy steels, including the aluminum-containing nitriding steels.

Another bath for nitriding tool steels has a composition as follows:

Component	Amount, %
NaCN	30.00 max
Na_2CO_3 or K_2CO_3	25.00 max
Other active ingredients	4.00 max
Moisture	2.00 max
KCl	bal

A proprietary nitriding salt bath has the following composition:

Component	Amount, wt%
NaCN	60
K_2CO_3	15
KCl	24
Moisture	1

Several special liquid nitriding processes employ proprietary additions, either gaseous or

Table 7 Automotive parts for which liquid nitriding proved superior to other case-hardening processes for meeting service requirements

Component	Requirement	Material and process originally used	Resultant problem	Solution
Thrust washer	Withstand thrust load without galling and deformation	Bronze, carbonitrided 1010 steel	Bronze galled, deformed; steel warped	1010 steel nitrided 90 min in cyanide-cyanate bath at 570 °C (1060 °F) and water quenched(a)
Shaft	Resist wear on splines and bearing area	Induction harden through areas	Required costly inspection	Nitride for 90 min in cyanide-cyanate salt bath at 570 °C (1060 °F)
Seat bracket	Resist wear on surface	1020 steel, cyanide treated	Distortion; high loss in straightening(b)	1020 nitrided 90 min in cyanide-cyanate salt bath and water quenched(c)
Rocker arm shaft	Resist water on surface; maintain geometry	SAE 1045 steel, rough ground, induction hardened, straightened, finish ground, phosphate coated	Costly operations and material	SAE 1010 steel liquid-nitrided 90 min in low-cyanide fused salt at 570 to 580 °C (1060 to 1075 °F)(d)

(a) Resulted in improved product performance and extended life, with no increase in cost. (b) Also, brittleness. (c) Resulted in less distortion and brittleness, and elimination of scrap loss. (d) Eliminated finish grinding, phosphatizing, and straightening solid, that are intended to serve several purposes, such as accelerating the chemical activity of the bath, increasing the number of steels that can be processed, and improving the properties obtained as a result of nitriding.

Cyanide-free liquid nitriding salt compositions have also been introduced. However, in the active bath, a small amount of cyanide, generally up to 5.0%, is produced as part of the reaction. This is a relatively low concentration, and these compositions have gained widespread acceptance within the heat treating industry, because they do contribute substantially to the alleviation of a potential source of pollution.

Liquid Pressure Nitriding

Liquid pressure nitriding is a proprietary process in which anhydrous ammonia is introduced into a cyanide-cyanate bath. The bath is sealed and maintained under a pressure of 7 to 205 kPa (1 to 30 psi). The ammonia is piped to the bottom of the retort and is caused to flow vertically. The percentage of nascent nitrogen in the bath is controlled by maintaining the ammonia flow rate at 0.6 to 1 m^3/h (20 to 40 ft^3/h). This results in ammonia dissociation of 15 to 30%.

The bath contains sodium cyanide and other salts, which permits an operating temperature of 525 to 565 °C (975 to 1050 °F). Because the molten salts are diffused with anhydrous ammonia, a new bath does not require aging and may be put into immediate operation employing the recommended cyanide-cyanate ratio, namely, 30 to 35% cyanide and 15 to 20% cyanate. Except for dragout losses, maintenance of the bath within the preferred ratio range is greatly simplified by the anhydrous ammonia addition, which serves continuously to counteract bath depletion.

The retort cover may be opened without causing complete interruption of the nitriding process. Loss of pressure within the retort results in a reduction in the nitriding rate. However, when the retort is sealed and pressure is reinstated through the resumption of ammonia gas flow, nitriding proceeds at the normal rate.

Depth of case depends on time at temperature. The average nitriding cycle is 24 h, although total cycle time may vary between 4 and 72 h. To stabilize core hardness, it is recommended that all parts be tempered at a temperature at least 28 °C (50 °F) higher than the nitriding temperature before they are immersed in the nitriding bath.

Hardness gradients and case depths resulting from pressure nitriding of 410 stainless steel, American Iron and Steel Institute (AISI) type D2, and SAE 4140 are shown in Fig. 16, 17, and 18.

Aerated Bath Nitriding

Aerated bath nitriding is a proprietary process (U.S. Patent 3,022,204) in which measured amounts of air are pumped through the molten bath. The introduction of air provides agitation and stimulates chemical activity. The cyanide content of this bath, calculated as sodium cyanide, is maintained at preferably approximately 50 to 60% of the total bath con-

Table 8 Liquid nitriding processes

Process identification	Operating range composition	Chemical nature	Suggested post-treatment	Operating temperature °C	Operating temperature °F	U.S. patent number
Aerated cyanide-cyanate	Sodium cyanide (NaCN), potassium cyanide (KCN) and potassium cyanate (KCNO), sodium cyanate (NaCNO)	Strongly reducing	Water or oil quench; nitrogen cool	570	1060	3,208,885
Casing salt	Potassium cyanide (KCN) or sodium cyanide (NaCN), sodium cyanate (NaCNO) or potassium cyanate (KCNO), or mixtures	Strongly reducing	Water or oil quench	510–650	950–1200	...
Pressure nitriding	Sodium cyanide (NaCN), sodium cyanate (NaCNO)	Strongly reducing	Air cool	525–565	975–1050	...
Regenerated cyanate-carbonate	Type A: potassium cyanate (KCNO), potassium carbonate (K_2CO_3)	Mildly oxidizing	Water, oil, or salt quench	580	1075	4,019,928
	Type B: potassium cyanate (KCNO), potassium carbonate (K_2CO_3), 1–10 ppm, sulfur (S)	Mildly oxidizing	Water, oil quench, or salt quench	540–575	1000–1070	4,006,643

tent, and the cyanate is maintained at 32 to 38%. The potassium content of the fused bath, calculated as elemental potassium, is between 10 and 30%, preferably approximately 18%. The potassium may be present as the cyanate or the cyanide, or both. The remainder of the bath is sodium carbonate.

This process produces a nitrogen-diffused case 0.3 mm (0.012 in.) deep on plain carbon or low-alloy steels in a $1\frac{1}{2}$ h cycle. The surface layer (0.005 to 0.01 mm, or 0.0002 to 0.0004 in. deep) of the case is composed of ε Fe_3N and a nitrogen-bearing Fe_3C; the nitrided case does not contain the brittle Fe_2N constituent.

Beneath the compound zone of Fe_3N, a diffusion zone exists that consists of a solid solution of nitrogen in the base iron. Depth of nitrogen diffusion in 1015 steel as a function of nitriding time at 565 °C (1050 °F) is shown in Fig. 19. The outer compound layer provides wear resistance, while the diffusion zone improves fatigue strength.

It should be noted that only chromium-, titanium-, and aluminum-alloyed steel respond well to conventional bath nitriding. Plain carbon (nonalloyed) steels respond well to aerated bath nitriding but not to conventional nitriding. Thus, the aerated process should be specified for nitriding all plain carbon steels, because test data show that plain carbon steel does not develop adequate hardness in a nonaerated nitriding bath. However, the full effect of nitriding is not realized unless an alloy steel is selected.

Aerated Cyanide-Cyanate Nitriding. Another aerated process for liquid nitriding is a high-cyanide, high-cyanate system that is proprietary (U.S. Patent 3,208,885). The cyanide content of the fused salt is maintained in the

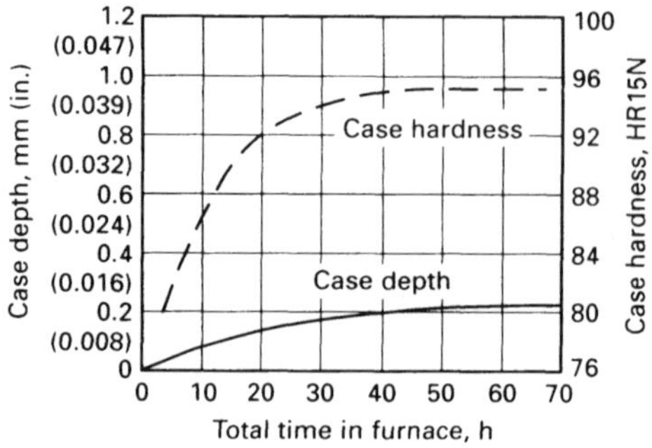

Fig. 16 Results of liquid pressure nitriding on type 410 stainless steel (composition, 0.12C-0.45Mn-0.41Ni-11.90Cr; core hardness, 24 HRC)

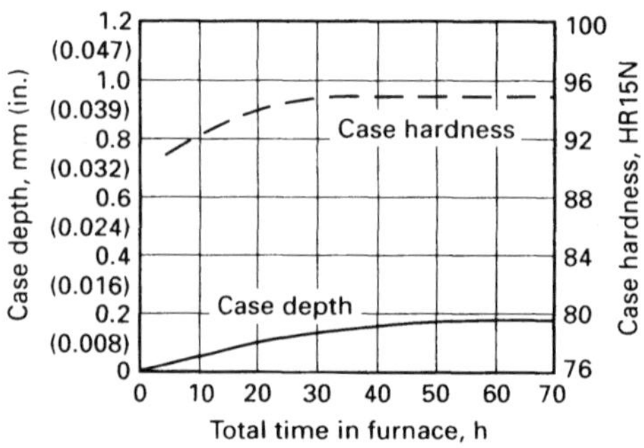

Fig. 17 Results of liquid pressure nitriding on AISI type D2 tool steel (composition, 1.55C-0.35Mn-11.50Cr-0.80Mo-0.90V; core hardness, 52 HRC)

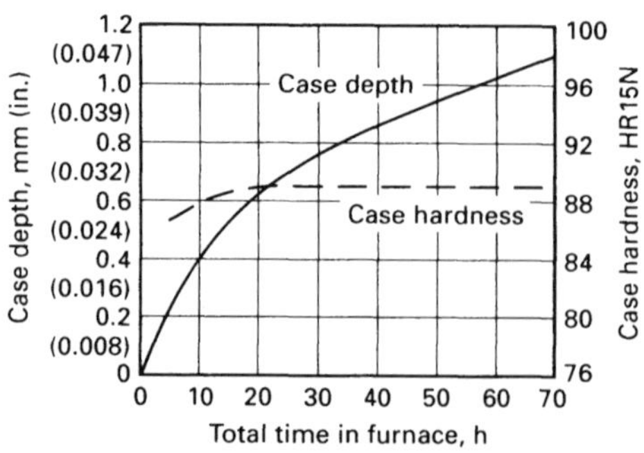

Fig. 18 Results of liquid pressure nitriding on SAE 4140 low-alloy steel (composition, 0.38C-0.89Mn-1.03Cr-0.18Mo; core hardness, 35 HRC)

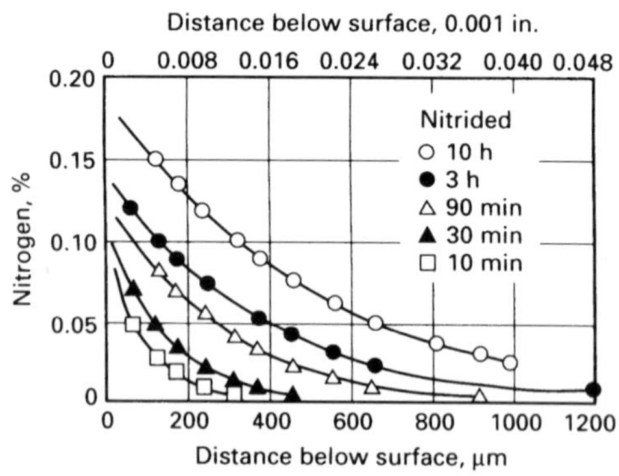

Fig. 19 Nitrogen gradients in 1015 steel as a function of time of nitriding at 565 °C (1050 °F), using the aerated bath process

range of 45 to 50% calculated as potassium cyanide, and the cyanate content is maintained in the range of 42 to 50% calculated as potassium cyanate. Makeup salt consists of a precise mixture of sodium and potassium cyanides that are oxidized by aeration to the mixed cyanate. The ratio of sodium ions to potassium ions is important in duplicating the integrity of the compound zone and the diffusion zone.

The process is performed in a titanium-lined container, and it produces a compound zone of ε iron nitride to a depth of 0.010 to 0.015 mm (0.0004 to 0.0006 in.) and a diffusion zone of 0.356 to 0.457 mm (0.014 to 0.018 in.) in plain carbon steels with a 90 min treating time, as shown in Fig. 20. The surface hardness of the compound zone may vary between 300 and 450 HK if carbon or low-alloy steels are being treated. Surface hardness of stainless steels treated by this process may reach 900 HK at 200 gf load.

Aerated Low-Cyanide Nitriding. Environmental concerns have led to the development of cyanide-free processes for liquid nitriding. In these proprietary processes, the base salt is supplied as a cyanide-free mixture of potassium cyanate and a combination of sodium carbonate and potassium carbonate, or sodium chloride and potassium chloride. Minor percentages of cyanide develop during use in these compositions. The problem is overcome in one process (U.S. Patent 4,019,928) by quenching in an oxidizing quench salt that destroys the cyanide and cyanate compounds (which have pollution capabilities) and produces less distortion than that resulting from water quenching. An alternate method used by U.S. Patent 4,006,643 is the incorporation of lithium carbonate plus minute amounts of sulfur (1 to 10 ppm) in the base salt to hold cyanide formation to below 1.0%.

Fig. 20 Nitrided case and diffusion zone produced by cyanide-cyanate liquid nitriding. The characteristic needle structure is seen only after a 300 °C (570 °F) aging treatment.

These low-cyanide processes have been shown in tests to produce the same results as those developed in the previously mentioned liquid nitriding processes. The diffusion curves and case depths are quite similar to those shown in Fig. 16, 17, and 18. Because a high cyanate (65 to 75% KCNO) level in the absence of cyanide would be expected to produce iron nitride compound zones slightly lower in carbon and slightly higher in nitrogen, it is good practice to develop new tests and operational data when converting to one process from another.

AMS 2753 for Low-Cyanide Liquid Nitriding

The aerated low-cyaniding treatment described previously (U.S. Patent 4,006,643) is specified under AMS 2753. Excerpts from this specification follow.

Hardening. Parts requiring core hardness shall be heat treated to the required core hardness before processing. Tempering to produce the specified core hardness shall be at a temperature not lower than 590 °C (1090 °F), except when tempering is conducted in conjunction with nitriding.

Stress Relief. Parts in which residual stresses may cause cracking or excessive distortion due to thermal shock or dimensional change because of metallurgical transformations during nitriding shall be stress relieved prior to final machining. Stress relieving shall be performed at a temperature not lower than 590 °C (1090 °F).

Cleaning. Parts, at the time of nitriding, shall be clean and free of scale or oxide, entrapped sand, core material, metal particles, oil, and grease, and shall be completely dry.

Preheating. Parts shall be preheated in air at 260 to 345 °C (500 to 650 °F) to maintain bath temperature and to avoid thermal shock on immersion in the nitriding salt.

Nitriding. Parts shall be immersed in an aerated cyanate bath, as indicated in Table 9.

Quenching. Following treatment, parts shall be quenched in fused salts, water, oil, soluble oil solution, or air. Parts, except those made of air-hardening tool steels, may be cooled to 290 to 400 °C (550 to 750 °F) prior to actual quenching, when permitted by the purchaser.

Depth of compound layer shall be determined in accordance with SAE J423, microscopic method, at magnification of 500×, as indicated in Table 10.

Quality of Compound Layer. Any continuous surface porosity present shall not extend deeper than one-half the observed depth of the compound layer, determined by examining specimens metallographically at 500× magnification.

Hardness of compound layer shall be determined by microhardness measurements in accordance with ASTM E 384 on the nitrided surface or on metallographically prepared cross sections of the nitrided case using Knoop or another appropriate hardness tester, as agreed on by purchaser and vendor (see Table 11).

Effects of Steel Composition

Although the properties of alloy steels are improved by the compound and diffusion layers, relatively greater improvement is achieved with plain carbon steels of low and medium carbon content. For example, the improvement in fatigue strength of unnotched test bars of 1015 steel nitrided by this process for 90 min at 565 °C (1050 °F) and water quenched (to further enhance fatigue properties) is roughly 100%. Improvement obtained with similarly treated test bars made of 1060 steel is approximately 45 to 50%.

The diffusion of nitrogen in carbon steels is directly affected by carbon content, as shown in Fig. 21. The nitride-forming alloying elements also inhibit nitrogen diffusion. For example, the inhibiting effect of chromium on diffusion is shown in Fig. 22, which compares nitrogen in a low-carbon steel (1015) and a chromium-containing low-alloy steel (5115).

Although the visible nitrogen diffusion zone shown by the Fe_4N needles in Fig. 20 can be measured under the microscope to a depth of approximately 0.41 mm (0.016 in.), actual nitrogen penetration can be measured up to 1.02 mm (0.040 in.), as shown in Fig. 23. This nitrogen is in solution, is under stress, and is precipitated as Fe_4N. It is responsible for the fatigue improvement resulting from liquid nitriding. The improvement is more apparent in plain carbon steels, resulting in the substitution of these steels for high-carbon and low-alloy steels in many applications (Table 12).

Case Depth and Case Hardness

Data indicating depth of case obtained in liquid nitriding various steels in a conventional bath at 525 °C (975 °F) for up to 70 h are shown in Fig. 24. The steels include three chromium-containing low-alloy steels (4140, 4340, and 6150), two aluminum-containing nitriding steels (SAE 7140 and AMS 6475), and four tool steels (H11, H12, M50, and D2). All were nitrided in a salt bath with an effective cyanide

Table 9 Recommended procedures for liquid salt bath nitriding in aerated low-cyanide baths

Material	Recommended time, h		Temperature	
	min	max	°C	°F
Carbon and low-alloy steels	1	2	580 ± 5	1075 ± 10
Tool and die steels (structural)	1/2	3	540–580	1000–1075
Tool steels (cutting)	1/12	1	540–580	1000–1075
Corrosion- and heat-resistant steels	1	2	580 ± 5	1075 ± 10
Ductile, malleable, and gray cast iron	1	4	580 ± 5	1075 ± 10
Powder metal products (ferrous)	1/2	2	580 ± 5	1075 ± 10

content of 30 to 35% and a cyanate content of 15 to 20%. Case depths were measured visually on metallographically prepared samples that were etched in 3% nital. Before being nitrided, samples were tempered to the core hardnesses indicated.

Figure 25 presents data on case hardness obtained in liquid pressure nitriding the follow-

Table 10 Depth of compound layer after liquid salt bath nitriding in an aerated low-cyanide bath

Material	Case depth(a) mm	Case depth(a) in.
Carbon and low-alloy steels	0.0038–0.02	0.00015–0.001
Tool and die steels (structural)	0.003–0.012	0.0001–0.0005
Tool and die steels (cutting)	0.003	0.0001
Corrosion- and heat-resistant steels	0.0038–0.02	0.00015–0.001
Ductile, malleable, and gray cast iron	0.0038–0.02	0.00015–0.001
Powder metal products (ferrous)	0.0038–0.02	0.00015–0.001

(a) Ranges show minimum and maximum case depth.

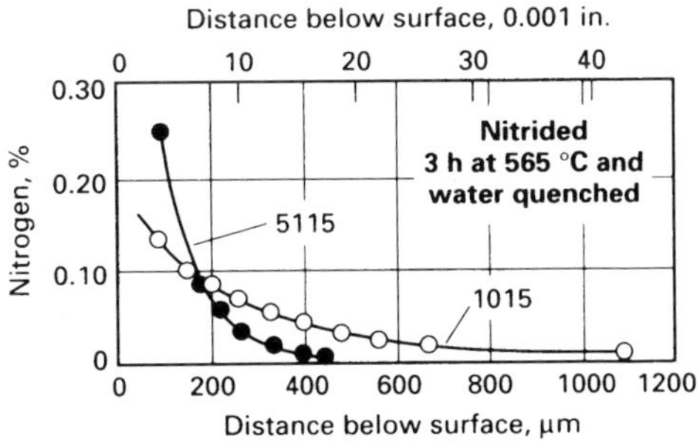

Fig. 22 Comparison of nitrogen gradients in a low-carbon steel and in a low-alloy steel containing chromium, both nitrided by the aerated bath process

Table 11 Hardness of the compound layer obtained after liquid salt bath nitriding in an aerated low-cyanide bath

Material	Hardness, min (HK at 100 gf load)
Carbon steels	300
Low-alloy steels	450
Tool and die steels	700
Corrosion- and heat-resistant steels	900
Ductile, malleable, and gray cast iron	600
Powder metal products (ferrous)	600

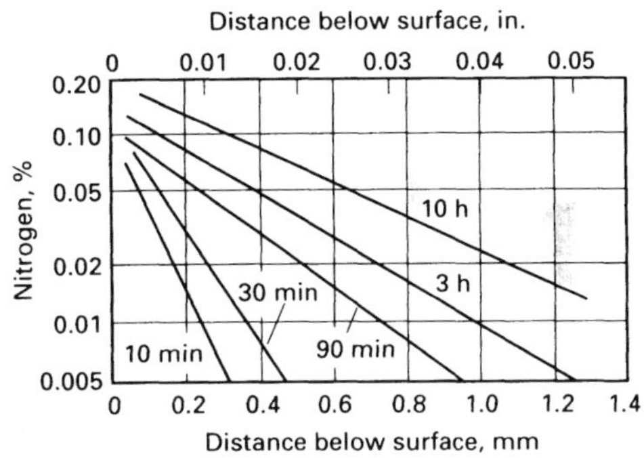

Fig. 23 Nitrogen diffusion in AISI 1015 steel

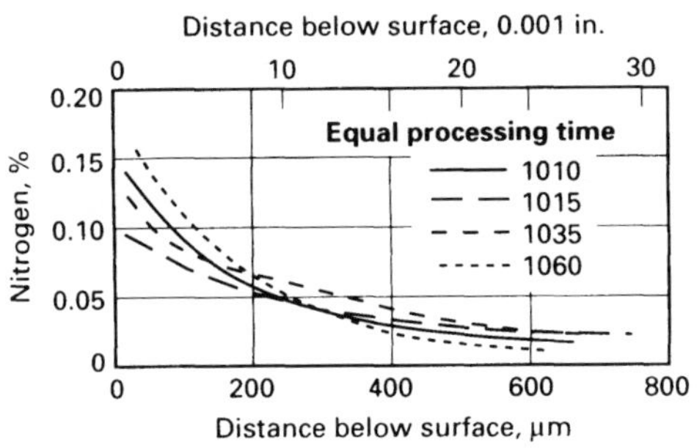

Fig. 21 Effect of carbon content in carbon steels on the nitrogen gradient obtained in aerated bath nitriding

Table 12 Improvement in fatigue properties of low-temperature liquid-nitrided ferrous materials

Steel type	Property improvement, %
Low-carbon steels	80–100
Medium-carbon steels	60–80
Stainless steels	25–35
Low-carbon, chrome manganese steels	25–35
Chrome alloy, medium-carbon steels	20–30
Cast irons	20–80

ing alloy steels and tool steels: SAE 7140, AMS 6475, 4140, 4340, medium-carbon H11, low-carbon H11, H15, and M50. The various core hardnesses, nitriding temperatures, and cycle times were as noted in the graphs in Fig. 25. Case depth and hardness results are comparable to those obtained in single-stage gas nitriding.

High-Speed Steels (Ref 5). Liquid nitriding is preferred to gas nitriding for high-speed steel cutting tools, because it is capable of producing a more ductile case with a lower nitrogen content.

Although any of the liquid nitriding baths or processes may be used to nitride high-speed steel, the commercial bath consisting of 60 to 70% sodium salts and 30 to 40% potassium salts is most commonly employed. The nitriding cycle for high-speed steel is of relatively short duration, seldom exceeding 1 h; in all other respects, however, the procedures and equipment are similar to those used for low-alloy steels.

The cyanide baths employed in liquid nitriding introduce both carbon and nitrogen into the surface layers of the nitrided case. Normally, the highest percentages of both elements are found in the first 0.025 mm (0.001 in.) surface layer. For carbon and nitrogen gradients, see the section "Liquid Nitriding."

The effect of time in a liquid nitriding bath at 565 °C (1050 °F) on the nitrogen content of the first 0.025 mm (0.001 in.) surface layer of a T1 high-speed steel is shown in Table 13. A nitrogen content of 0.06% was obtained in the first 3 min at temperature, and it gradually increased to 1.09% at the end of a 6 h cycle at this temperature.

As shown in Table 14, carbon also was absorbed by the steel, at nitriding temperatures as low as 455 °C (850 °F). In a 30 min nitriding cycle, the carbon content of the first 0.025 mm (0.001 in.) surface layer increased with an increase in the nitriding temperature. However, it was reported that only a portion of the carbon was absorbed by the steel, most of the carbon being mechanically attached to the surface, filling microscopic pits. (This pitting is not dangerous under normal conditions, because the pits are shallower than ordinary grinding or machining marks.)

High-speed steel tools that are nitrided in fresh baths or for short times show steep nitrogen and hardness gradients. To avoid these steep gradients, which are believed responsible for the brittleness of the case after such treatments, the use of longer immersion time, higher temperature, or a thoroughly aged bath is recommended. To avoid brittleness of case when relatively short immersion times are used, the cyanate content of the bath should exceed 6%. These conditions often will lower the surface hardness as well as the hardness gradient.

Figure 26 compares the hardness gradients obtained on specimens of T1 high-speed steel nitrided at 565 °C (1050 °F) for 90 min in a new bath and for various lengths of time in an aged bath.

Nitriding of decarburized high-speed steel tools should be avoided, because it results in a brittle surface condition. For those surfaces that have been softened from grinding, nitriding is frequently employed as an offsetting corrective measure.

Liquid nitriding provides high-speed steel tools with high hardness and wear resistance and a low coefficient of friction. These properties enhance tool life in two somewhat related ways. The high hardness and wear resistance

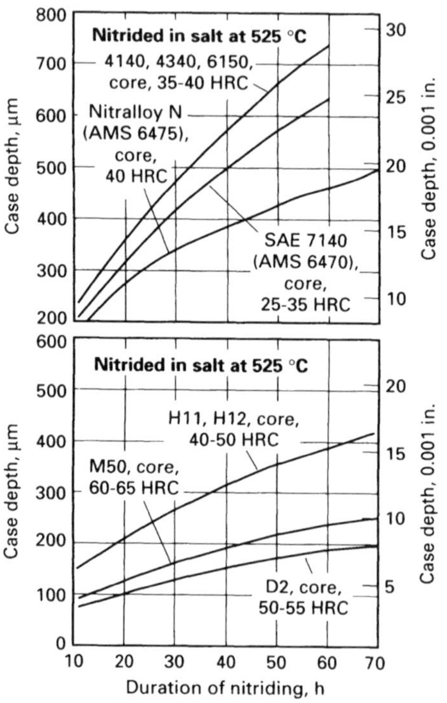

Fig. 24 Depth of case for several chromium-containing low-alloy steels, aluminum-containing steels, and tool steels after liquid nitriding in a conventional salt bath at 525 °C (975 °F) for up to 70 h

lower the abrading action of chips and work on the tool, and the low frictional characteristics serve to create less heat at and behind the tool point, in addition to assisting in the prevention of chip pickup.

Operating Procedures

Among the important operating procedures in liquid nitriding are the initial preparation and heating of the salt bath, aging of the molten salts (when required), and analysis and maintenance of salt bath composition. Virtually all steels must be quenched and tempered for core properties before being nitrided or stress relieved for distortion control. So, prior heat treatment may be considered an essential part of the operating procedure.

Prior Heat Treatment. Alloy steels usually are given a prior heat treatment similar to that preferred for gas nitriding. Maintenance of dimensional and geometric stability during liquid nitriding is enhanced by hardening of parts prior to nitride treatment. Tempering temperatures should be no lower than the nitriding temperature and preferably slightly above.

Depending on steel composition, the effect of core hardness is similar to that encountered in gas nitriding.

Starting the Bath. Case-producing salt compositions may vary with respect to manu-

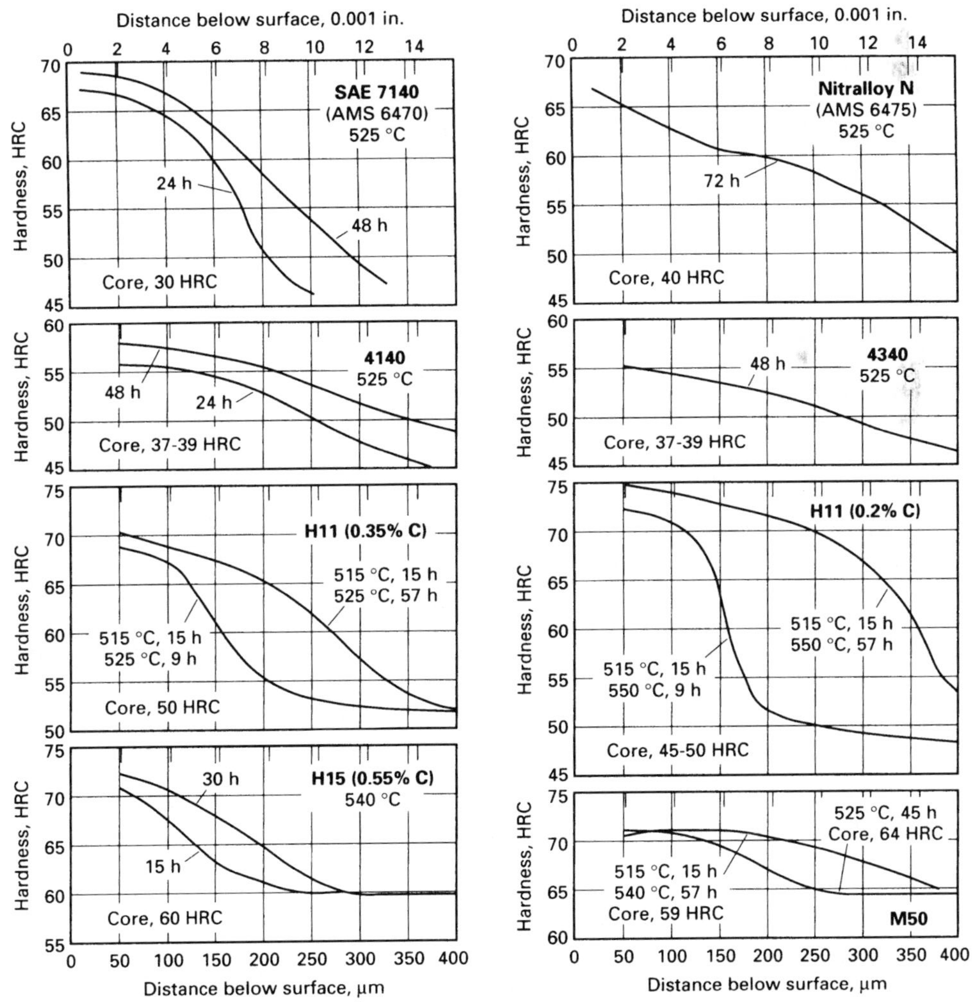

Fig. 25 Hardness gradients for several alloy and tool steels nitrided in salt by the liquid pressure process. Rockwell C hardness values are converted from Knoop hardness measurements made using a 500 g load. Temperatures are nitriding temperatures.

facturers, but they are basically sodium and potassium cyanides, or sodium and potassium cyanates. Cyanide, the active ingredient, is oxidized to cyanate by aging as described subsequently. The commercial salt mixture (60 to 70% sodium salts, 30 to 40% potassium salts) is melted at 540 to 595 °C (1000 to 1100 °F). *During the melting period, a cover should be placed over the retort to guard against spattering or explosion of the salt, unless the equipment is completely hooded and vented. It is mandatory that the salts be dry before they are placed in the retort; the presence of entrapped moisture may result in an eruption when the salt mixture is heated.*

Externally versus Internally Heated Salt Baths. Salt baths may be heated externally or internally. For externally heated salt baths, startup power should be limited to 37% of total capacity until signs of melting are apparent on all sides of the salt bath. For internally heated salt baths, natural gas flame torches having a moderate flame are effective in melting a pool of molten salt for a conductive path between electrodes.

Aging the Bath. Liquid nitriding compositions that do not contain a substantial amount of cyanate in the original melt must be aged before use in production. Aging is defined as the oxidation of the cyanide to cyanate. Aging is not merely a function of temperature alone but also depends on the surface-to-volume ratio of the molten bath. It is the surface air (oxygen)-to-salt contact that oxidizes cyanide to cyanate.

Molten salts in conventional baths should be aged by being held at 565 to 595 °C (1050 to 1100 °F) for at least 12 h, and no work should be placed in the bath during the aging treatment. Aging decreases the cyanide content of the bath and increases the cyanate and carbonate contents. Before nitriding is begun, a careful check of the cyanate content should be made. Nitriding should not be attempted until the cyanate content has reached at least the minimum operating level recommended for the bath.

Bath Maintenance. To protect the bath from contamination and to obtain satisfactory nitriding, all work placed in the bath should be thoroughly cleaned and free of surface oxide. An oxide-free condition is especially important when nitriding in low-cyanide salts. These compounds are not strong reducing agents and therefore are incapable of producing a good surface on any oxidized work. Either acid pickling or abrasive cleaning is recommended prior to

Table 13 Effect of nitriding time on surface nitrogen content of T1 high-speed tool steel

Nitrogen content of first 0.025 mm (0.001 in.) layer

Time at 565 °C (1050 °F)	Nitrogen, %
3 min	0.06
10 min	0.093
30 min	0.15
90 min	0.26
3 h	0.58
6 h	1.09

Table 14 Carbon content of nitrided T1 high-speed tool steel

Carbon content of the first 0.025 mm (0.001 in.) surface layer of steel originally containing 0.705% C. Some of the carbon was in pits on the surface, rather than diffused into the steel.

Nitriding			
Temperature			
°C	°F	Time, min	Surface carbon, %
455	850	30	0.85
510	950	30	0.99
565	1050	30	1.18
565	1050	360	1.63

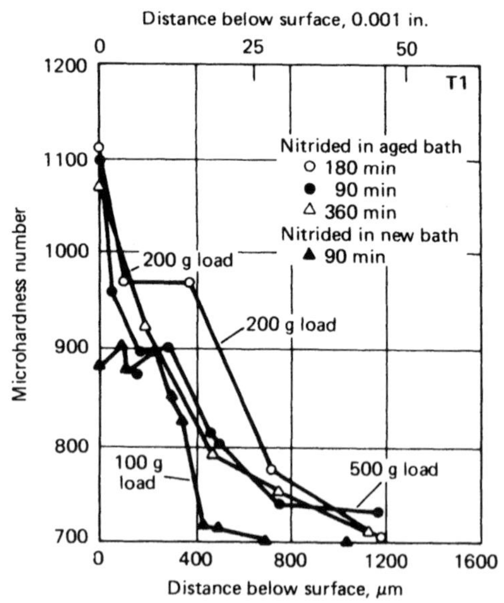

Fig. 26 Effect of bath condition and immersion time on hardness gradients in type T1 high-speed steel specimens nitrided at 565 °C (1050 °F)

nitriding. Finished clean parts should be preheated before being immersed in the bath to rid them of surface moisture.

A high cyanate content (up to approximately 25%) provides good results, but carbonate content should not exceed 25%. Carbonate content can be readily lowered by cooling the bath to 455 °C (850 °F) and allowing the precipitated salt to settle to the bottom of the salt pot. It can then be spooned from the bottom by means of a perforated ladle.

To minimize corrosion at the air-salt interface, salts should be completely changed every three or four months (replacement of salt is usually far more economical than replacement of the pot). When the bath is not in use, it should be covered; excessive exposure to air causes a breakdown of cyanide to carbonate and adversely affects pot life.

The ratio of cyanide content to cyanate content varies with the salt bath process and the composition of the bath. The commercial NaCN-KCN bath, after aging for one week, achieves a ratio of 21 to 26% cyanide to 14 to 18% cyanate. The bath used in liquid pressure nitriding operates with a cyanide content of 30 to 35% and a cyanate content of 15 to 20%. The aerated bath is controlled to a ratio of 50 to 60% cyanide to 32 to 38% cyanate. The aerated noncyanide nitriding process is controlled to a ratio of 36 to 38% cyanate to 17 to 19% carbonate.

Equipment

Salt bath furnaces used for nitriding may be heated by gas, oil, or electricity and are essentially similar in design to salt bath furnaces used for other processes. Although batch installations are most common, semi-continuous and continuous operations are feasible. Generally, the same furnace equipment can be used for other heat treating applications by merely changing the salt. For example, the liquid carburizing furnaces illustrated in Chapter 4, "Pack and Liquid Carburizing," can be used for liquid nitriding.

A variety of materials are used for the pots, electrodes, thermocouple protection tubes, and fixtures employed in salt bath nitriding, depending primarily on the salt mixture and process. For example, low-carbon steel is sometimes used for furnace liners, although titanium is recommended for one of the processes (U.S. Patent 3,208,885). Inconel 600 is presently being applied to the noncyanide process described in U.S. Patent 4,019,928. Type 430 stainless steel is recommended for a low-cyanide process described in U.S. Patent 4,006,643. Cast HT alloy is a satisfactory fixture material, and type 446 stainless steel has been used for fixtures and thermocouple protection tubes. One plant reports the successful use of Inconel pots in liquid pressure nitriding; the same plant reports also that electrodeposited nickel performs satisfactorily as a stopoff in the liquid pressure bath. In general, however, nickel-bearing materials are not recommended for nitriding salt baths.

Safety Precautions. The following safety precautions should be observed when operating salt bath furnaces for nitriding steels:

- Operating personnel must be carefully instructed in handling the poisonous cyanide-containing salts.
- All chemical containers must be clearly marked to indicate contents.
- Personnel should be provided with facilities for washing their hands thoroughly to prevent contamination by the cyanide salts.
- Shields, gloves, aprons, and eye protection should be worn by operating personnel.
- Parts and workpiece support fixtures should be preheated to drive off any moisture that may be present before they are immersed in the molten salt bath.
- Proper venting of furnace and rinse tanks to the outdoors is recommended in order to provide safety against fumes and spattering and to minimize corrosion in the work area.
- *Nitrate-nitrite salts must not come in contact with nitriding salts in the molten state. Contact will result in an explosion. Storage of these salts should be properly labeled and stored apart.*

Plasma (or Ion) Nitriding

Plasma, or ion, nitriding is a method of surface hardening using glow discharge technology to introduce nascent (elemental) nitrogen to the surface of a metal part for subsequent diffusion into the material. In a vacuum, high-voltage electrical energy is used to form a plasma, through which nitrogen ions are accelerated to impinge on the workpiece. This ion bombardment heats the workpiece, cleans the surface, and provides active nitrogen. Ion nitriding provides better control of case chemistry and uniformity and has other advantages, such as lower part distortion than conventional (gas) nitriding. A key difference between gas and ion nitriding is the mechanism used to generate nascent nitrogen at the surface of the work.

Case Structures and Formation

The case structure of a nitrided steel, which may include a diffusion zone with or without a compound zone (Fig. 27), depends on the type and concentration of alloying elements and the time-temperature exposure of a particular nitriding treatment. Moreover, because the formation of a compound zone and/or a diffusion zone depends on the concentration of nitrogen, the mechanism used to generate nascent nitrogen at the surface of the workpiece also affects the case structure. These factors are discussed subsequently, with emphasis on the differences between gas and plasma nitriding.

Diffusion Zone of a Nitrided Case. The diffusion zone of a nitrided case can best be described as the original core microstructure with some solid solution and precipitation strengthening. In iron-base materials, the nitrogen exists as single atoms in solid solution at lattice sites or interstitial positions until the limit of nitrogen solubility (≈0.4 wt% N) in iron is exceeded. This area of solid-solution strengthening is only slightly harder than the core. The depth of the diffusion zone depends on the nitrogen concentration gradient, time at a given temperature, and the chemistry of the workpiece.

As the nitrogen concentration increases toward the surface, very fine, coherent precipitates are formed when the solubility limit of nitrogen is exceeded. The precipitates can exist both in the grain boundaries and within the lattice structure of the grains themselves. These precipitates, nitrides of iron or other metals, distort the lattice and pin crystal dislocations and thereby substantially increase the hardness of the material.

In some ferrous alloys, the diffusion zone formed by nitriding cannot be seen in a metallograph, because the coherent precipitates are generally not large enough to resolve. In Fig. 28, for example, martensite in the diffusion zone cannot be visually distinguished from that in the core. In some materials, however, the nitride precipitate is so extensive that it can be seen in an etched cross section. Such is the case with stainless steel (Fig. 29), in which the chromium level is high enough for extensive nitride formation.

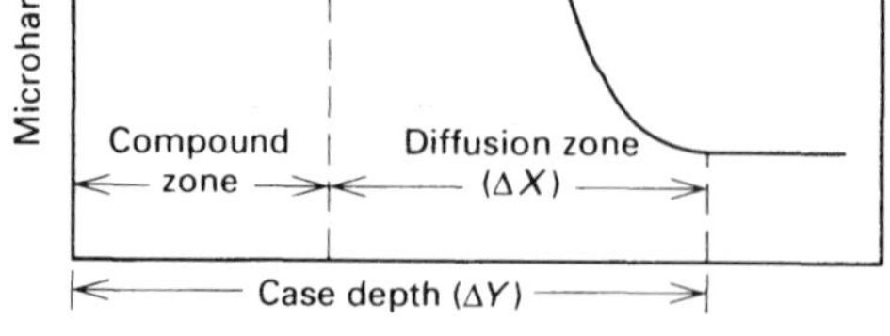

Fig. 27 Factors affecting the microhardness profile of a nitrided steel. The hardness of the compound zone is unaffected by alloy content, while the hardness of the diffusion zone is determined by nitride-forming elements (Al, Cr, Mo, Ti, V, Mn). ΔX is influenced by the type and concentration of alloying elements; ΔY increases with temperature and decreases with alloy concentration.

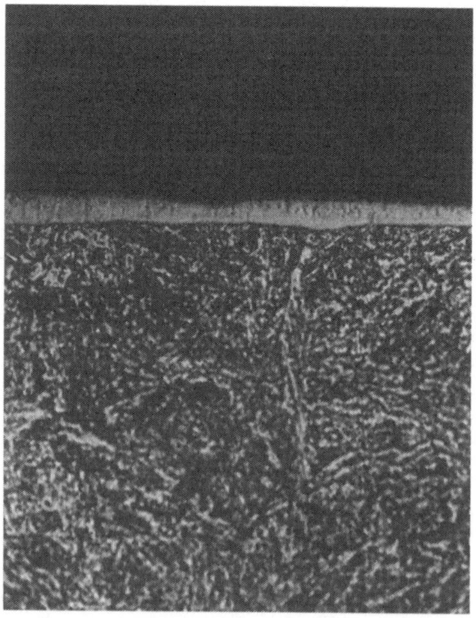

Fig. 28 Compound layer of γ′ (Fe_4N) on the ion-nitrided surface of quenched and tempered 4140 steel. The γ′ compound layer is supported by a diffused case, which is not observable in this micrograph. Nital etched. 500×

Compound Layers in Nitrided Steels. As discussed earlier in this Chapter, the compound zone is the region where the γ' (Fe_4N) and ε ($Fe_{2-3}N$) intermetallics are formed. Because carbon in the material aids ε formation, methane is added to the process gas when an ε layer is desired. Hydrogen also tends to catalyze Fe_2N formation. These compound layers are called white layers, because they appear white on a polished, etched cross section.

Structure of Gas-Nitrided Steel Case. Gas nitriding with ammonia produces a compound zone that is a mixture of the γ' and ε compounds; the mixture is due to the variability of ammonia dissociation, and therefore of nitriding potential, as the compound layer is formed. In conventional gas nitriding, the nascent nitrogen is produced by introducing NH_3 to a work surface that is heated to at least 480 °C (900 °F). Under these conditions, the ammonia, catalyzed by the metal surface, dissociates to release nascent nitrogen into the work and hydrogen gas into the atmosphere of the furnace. The nitriding potential, which determines the rate of introduction of nitrogen to the surface, is determined by the NH_3 concentration at the work surface and its rate of dissociation. This nitriding potential, which can vary significantly in the gas process, is responsible for the limited control of microstructure in the nitrided case.

Fig. 29 Observable diffusion zone on the unetched (white) portion of an ion-nitrided 416 stainless steel. Nital etched. 500×

X-ray diffraction has shown that from the outer surface to the beginning of the diffusion zone, the dominant compound changes from ε to γ'. However, both phases exist throughout the entire white layer, which is referred to as a dual-phase layer.

The dual-phase layer has two characteristics that make it susceptible to fracture:

- Weak bonding at the interface between phases
- Different thermal expansion coefficients in the two phases

Layers that are particularly thick or that are subjected to temperature fluctuation in service are particularly prone to failure.

Another mechanical weakness in the gas-nitrided white layer is porosity in the outer region of the layer. As the compound zone builds, ammonia dissociation becomes more sluggish without the catalytic action of the steel surface, and gas bubbles begin to form in the layer.

Structure of Ion-Nitrided Steel Case. In the ion nitriding process, nitrogen gas (N_2) can be used instead of ammonia, because the gas is dissociated to form nascent nitrogen under the influence of the glow discharge. Therefore the nitriding potential can be precisely controlled by the regulation of the N_2 content in the process gas. This control allows precise determination of the composition of the entire nitrided case, selection of a monophase layer of either ε or γ', or total prevention of white-layer formation (Fig. 30).

General Process Description

An ion nitriding system is shown in Fig. 31. The parts to be nitrided are cleaned, usually by vapor degreasing, loaded into the vacuum vessel, and secured. The subsequent process of plasma nitriding can be broken down into four steps: vessel evacuation, heating to nitriding temperature, glow-discharge processing at nitriding temperatures, and cooling.

Vessel evacuation is performed by a roughing pump or roughing pump-blower combination so that pressure is reduced to a level of 0.05 to 0.1 torr (mm of mercury). This is necessary to remove most of the initial air and any contaminants. Harder vacuum levels can be achieved but are not necessary for most materials.

The method of heating the load to nitriding temperature has evolved over the years. In the past, loads were heated only by the glow discharge itself. This method presented some difficulty, because moisture and other impurities on the work surface tended to cause arcing to the parts in the early stages of the heating cycle. The methods applied to extinguish or prevent arcs also tended to lengthen the heating cycle significantly.

Today, resistance heaters or cathode shields are normally used to bring the load to nitriding temperatures (375 to 650 °C, or 700 to 1200 °F) before glow discharge. Heating of the load can be with glow discharge only, using a cathode preheating shield arrangement up to an interme-

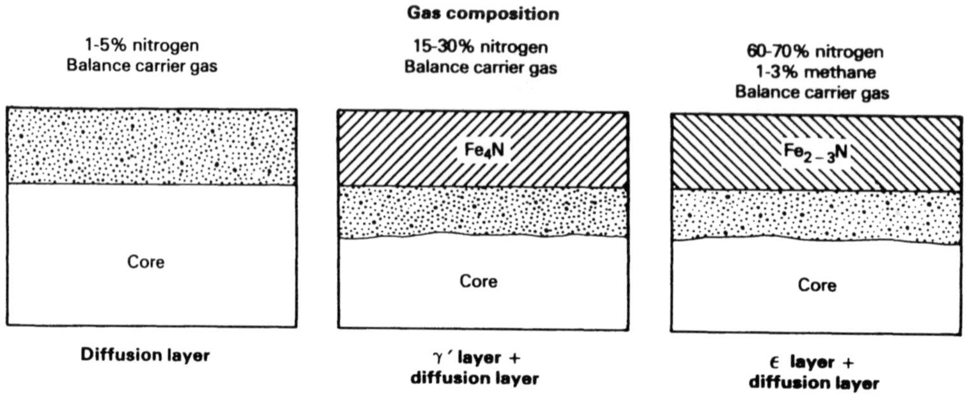

Fig. 30 Typical gas compositions and the resulting metallurgical configurations of ion-nitrided steel

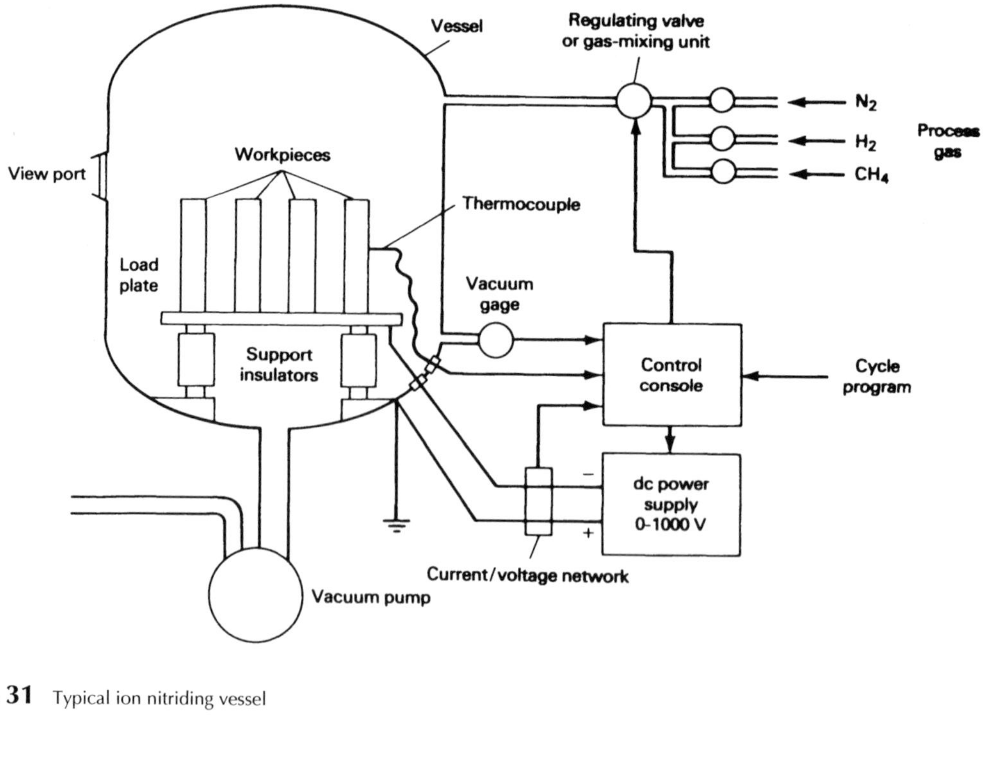

Fig. 31 Typical ion nitriding vessel

diate temperature, and then switching to glow discharge on the parts using resistance heating elements or convection. The most common approach is with resistance heating. While heating, the pressure is increased so that the glow seam does not get too thick and cause localized overheating.

Glow-Discharge Process. After the work load is heated to desired temperature, process gas is admitted at a flow rate determined by the load surface area. Pressure is regulated in the 1 to 10 torr range by a control valve just upstream from the vacuum pump. The process gas is normally a mixture of nitrogen, hydrogen, and, at times, small amounts of methane.

In the presence of this process gas, the load is maintained at a high negative direct current (dc) potential (500 to 1000 V) with respect to the vessel, which is grounded. Under the influence of this voltage, the nitrogen gas is dissociated, ionized, and accelerated toward the workpiece (the cathode). Within a short distance of the workpiece, the positively charged nitrogen ion then acquires an electron from the cathode (workpiece) and thus emits a photon. This photon emission during the return of nitrogen ions to their atomic state results in the visible glow discharge that is characteristic of plasma techniques. On impact with the workpiece, the kinetic energy of the nitrogen atoms is also converted into heat, which can totally (or in combination with an auxiliary heating source) bring the load to nitriding temperature.

The glow discharge surrounding negatively charged workpieces forms at voltages of 200 to 1000 V (Fig. 32), with gas pressures of 1 to 10 torr. The thickness of the glow envelope (or glow seam) can be altered by pressure, temperature, gas mix composition, dc voltage, and current. Typically a large or thick glow envelope is created with higher temperature, lower pressure, high hydrogen concentration in the gas mix, and higher dc voltage and current. A desirable glow-discharge thickness is approximately 6 mm (0.25 in.), unless parts with holes or slots require a thinner glow envelope.

During the glow-discharge process, different alloy or iron atoms combine with the nitrogen as it diffuses into the material, forming a hardened surface and case. Figure 33 shows these mechanisms for iron. The nitriding current (proportional to the nitrogen ion flux), temperature, and process time determine the depth of the nitride case achieved. A uniform glow-discharge enve-

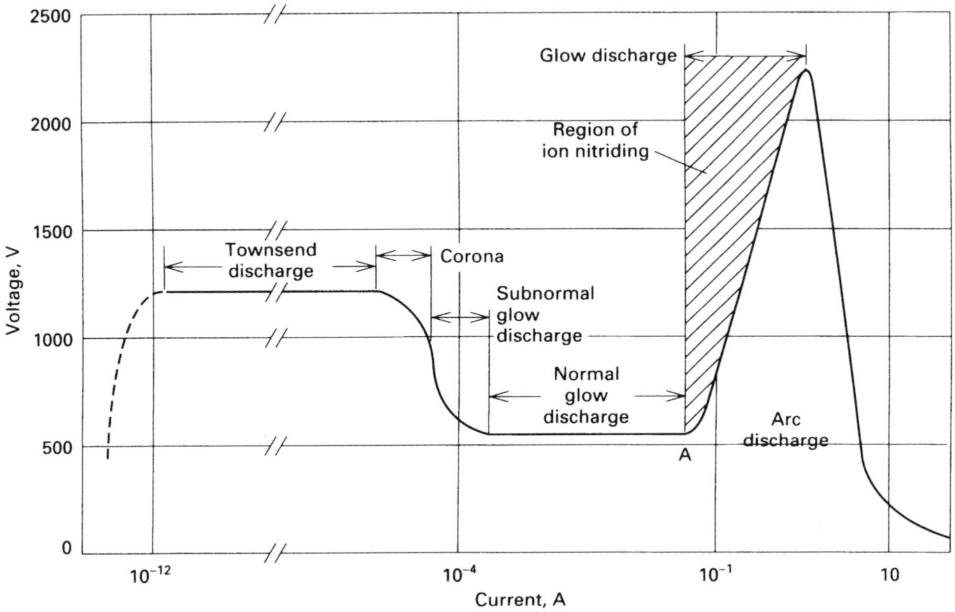

Fig. 32 Voltage versus current characteristics of different types of discharge in argon

lope is also necessary for proper case uniformity, especially when part geometry is complex (as with a gear or fuel injector).

Cooling. After the glow-discharge process, the voltage and process gas flow are terminated, and the load is cooled by inert-gas circulation. Cooling is accomplished by backfilling with nitrogen or other inert gases and recirculating the gas from the load to a cold surface such as the cold wall. From that point, the heat can be transferred and removed via the water in the cooling jacket.

Equipment

A basic ion nitriding system (Fig. 31) consists of a vacuum chamber, a power supply, and a process gas system with a gas-mixing panel or other mass flow controls. An isolated hearth or work support fixturing is also required to ensure electrical isolation between the workpiece and vacuum vessel. An auxiliary heating system and a rapid cooling system can also be included to improve cycle time.

Ion nitriding control systems may vary in complexity. Microprocessor systems are generally used to control or monitor several parameters. These include the work temperature, vessel wall temperature, vacuum (absolute pressure) level, glow-discharge voltage and current, auxiliary heating source voltage and current, and gas mix composition. The microprocessor also controls the various inputs and outputs necessary for activating/stopping or sequencing valves and motors.

Vessel Construction. The vessel is a vacuum chamber, which can be a hot-walled design or a dual-walled and water-cooled design. The vessel can be horizontally or vertically loaded in a drop-bottom, pit, or bell arrangement. Typically, no internal insulation is required because of the lower temperature (less than 650 °C, or 1200 °F) and the desire to create sufficient heat loss to support a steady dc power supply output to the workload.

The isolated-hearth arrangement is divided into three basic areas:

- High-voltage feed-through arrangement, which carries the voltage through the vessel wall while maintaining a good vacuum seal
- Load support insulators, which carry the actual load weight while providing good dielectric qualities
- Charge plate or fixture, which has the workpieces placed on it or provides mechanical masking if desired

Sight ports placed around the vessel provide a view of the ion nitriding process and are necessary for checking the load and ensuring that the selected parameters are accurate and that no detrimental hollow cathode disturbances (overlapping glow-discharge envelopes) have developed.

Power Supply and Control. The dc power supply is the most important component of an ion nitriding system. The power supply must provide an output voltage from 0 to 1000 V and an output current matched to the size of the vessel and workload. Typical current ratings range from 25 to 450 A (dc). The amount of power

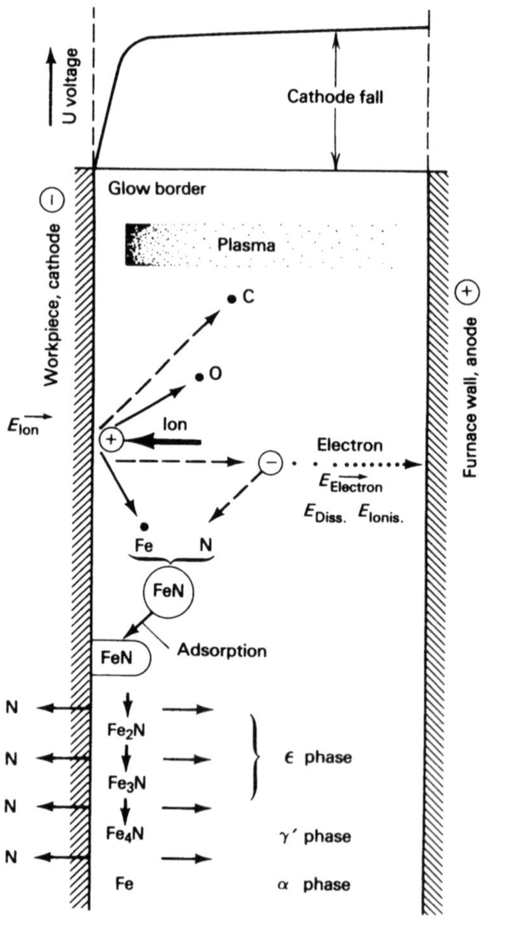

Fig. 33 Glow-discharge ion nitriding mechanisms (Koelbel's model)

applied to the load determines the temperature. Most power supplies provide proportional output control through silicon-controlled rectifiers (SCRs).

Another important design consideration in the power supply is arc detection/suppression controls. Arcing can occur because the glow-discharge process causes the removal of surface impurities, which are always present. The impurities are removed in the form of an arc, in which there is a sudden decrease in voltage and increase in current. Because of this, both minimum and maximum current levels and voltage rate of change (dV/dt) and voltage/current relationships (slope) must be constantly monitored. When an arc is detected, the power output is momentarily shut off and the existing power diverted from the workload to avoid any possible damage. This is accomplished by placing an inductive load (choke) in line with the output of the power supply and using a crowbar SCR to short the output and momentarily dissipate the power.

Atmosphere and Pressure Control. The gas-mixing panel is used for blending gases, usually nitrogen, hydrogen, and methane. A typical composition for a γ' compound layer would be 75% H and 25% N. For an ε compound layer, a typical gas mixture would be 70% N, 27% H, and 3% methane. The mixing can be accomplished by injecting the gases through orifices at an equal pressure and varying the time of flow to establish the correct percentage concentration, or mixing can be done with mass flow control systems.

Ion nitriding is generally performed at absolute pressure levels of 130 to 1300 Pa (1 to 10 torr), necessitating a means of controlling pressure levels. Control is accomplished in two stages. First, a motorized needle valve on the inlet line to the vessel, in series with the gas-mixing panel, proportionally controls the gas flow up to a maximum level. At this point, a reverse-acting valve on the evacuation line between the vessel and the vacuum pump controls the amount of gas being evacuated until the desired pressure set point is met.

Fixturing to hold or mask the workpieces mechanically can be designed to optimize load placement or performance. Fixturing must minimize gaps between areas of glow discharge to avoid overlapping (hollow cathode disturbances) of the glow envelope. Also, the cross section of masking should resemble that of the workpieces to allow better temperature uniformity.

Auxiliary Heating. If the workpiece is large, auxiliary heating is necessary when the glow-discharge process is insufficient for direct heating. Auxiliary heating can be accomplished in several ways, the most common of which are cathode preheating and resistance heating.

Cathode preheating, which occurs during the beginning stages of the ion nitriding process, requires an internal shield that is electrically isolated from the vessel wall. This shield is electrically charged and heats up and radiates the heat to the workload, allowing a faster heat-up time.

Resistance heating generally uses a low-voltage alternating current power supply such as a variable reactance transformer connected to graphite or alloy heating elements. As with cathode preheating, the elements heat up and radiate to the workpieces to speed the heat-up time.

Workpiece Factors

As mentioned in the section "Case Structures and Formation," the nitrogen concentration achieved during nitriding affects the depth and hardness of the case. In addition, the microstructure and resulting mechanical properties of a nitrided case also depend on the original composition and microstructure of the workpiece.

Suitability of Materials. In general, the response of a material to nitriding depends on the presence of strong nitride-forming elements. Plain carbon steels can be nitrided, but the diffused case is not significantly harder than the core. The strongest nitride formers are aluminum, chromium, molybdenum, vanadium, and tungsten. Because the white-layer constituents are only compounds of iron and nitrogen, the hardness of these layers is essentially independent of alloy content.

The premier nitriding steels are the Nitralloy series (see Table 3), which combine approximately 1 wt% Al with 1.0 to 1.5 wt% Cr. Other alloys that form excellent diffused cases are the chromium-bearing alloys, such as the 4100, 4300, 5100, 6100, 8600, 8700, 9300, and 9800 series. Other good nitriding materials include most of the tool and die steels, stainless steels, and precipitation-hardening alloys. Table 15 lists the various types of steels that are commonly processed by plasma nitriding.

Parts made by powder metallurgy (P/M) can also be ion nitrided, but precleaning is more

Table 15 Surface hardness ranges and case depths for ferrous alloys that are frequently plasma nitrided

Material group	Description	DIN material No.	Hardness, DPH	Nitride hardness depth, mm
Structural steel	St 37-2	1.0038	200–350	1.0 max
	St 60	1.0062	300–550	1.0 max
Free-cutting steel	9 S 20	1.0711	200–300	1.0 max
	9 S Mn 36	1.0715	. . .	max
	9 SMnPb 28	1.0718	200–350	1.0 max
	ETG 80	1.0727	350–450	1.0 max
	ETG 100	1.0727	450–650	1.0 max
	16 MnCrS 5	1.7139	600–750	1.0 max
Case-hardening steel	C 10 E	1.1121	. . .	max
	20 Mn 5	1.1133	. . .	max
	Ck 15	1.1141	300–400	1.0 max
	14 NiCr 10	1.5732	. . .	max
	15 CrNi 6	1.5919	650–750	1.0 max
	21 NiCrMo 2	1.6523	500–600	1.0 max
	17 CrNiMo 6	1.6587	650–750	0.8 max
	16 MnCr 5	1.7131	650–750	1.0 max
	20 MnCr 5	1.7147	650–750	1.0 max
Heat treatment steel (nonalloyed)	Ck 30	1.1178	300–450	1.0 max
	Ck 45	1.1191	30–500	1.0 max
	C 60 E	1.1221	350–500	1.0 max
Heat treatment steel (alloyed)	55 Cr 3	1.7176	. . .	max
	25 CrMo 4	1.7218	550–650	1.0 max
	42 CrMo 4	1.7225	550–650	1.0 max
	30 CrMoV 9	1.7707	850–950	0.8 max
	50 CrV 4	1.8159	500–650	0.8 max
Nitriding steel	34 CrAl 6	1.8504	950–1150	0.8 max
	34 CrAlMo 5	1.8507	950–1150	0.8 max
	31 CrMoV 9 V	1.8519	850–1000	0.8 max
	34 CrAlNi 7	1.8550	950–1150	0.8 max
Roller bearing steel	100 Cr 6	1.3505	500–650	1.0 max
	X 102 CrMo 17	1.3543	1000–1200	0.2 max
Spring steel	Ck 75	1.1248	500–600	1.0 max
	60 SiMn 5	1.5142	500–600	1.0 max
	58 CrV 4	1.8161	600–700	0.8 max
Nonalloyed tool steel	C 105 W 1	1.1545	550–650	1.0 max
	C 80 W 2	1.1625	550–650	1.0 max
High-speed tool steel	S 12-1-4	1.3302	1000–1200	0.2 max
	S 6-5-2	1.3343	1000–1200	0.2 max
	S 18-0-1	1.3355	1000–1200	0.2 max
Tool steel for cold work	X 165 CrV 12	1.2201	1000–1200	0.2 max
	29 CrMoV 9	1.2307	850–950	0.4 max
	40 CrMnMo 7	1.2311	600–700	0.8 max
	40 CrMnMoS 8-6	1.2312	600–700	0.8 max
	X 100 CrMoV 5-1	1.2363	800–900	0.4 max
	X 155 CrVMo 12-1	1.2379	1000–1200	0.2 max
	X 45 NiCrMo 4	1.2767	600–700	0.8 max
	90 MnCrV 8	1.2842	550–650	0.8 max
Tool steel for hot work	42 Cr 13	1.2083	1000–1200	0.2 max
	40 CrMoV 5-1	1.2344	900–1200	0.2 max
	60 WCrMoV 9-4	1.2622	800–900	0.2 max
	55 NiCrMoV 6	1.2713	500–600	0.6 max
Corrosion- and acid-resistant steel	X 30 Cr 13	1.4028	950–1200	0.2 max
	X 14 CrMoS 17	1.4104	950–1200	0.2 max
	X 90 CrMoV 18	1.4112	950–1200	0.2 max
	X 38 CrMoV 15	1.4117	950–1200	0.2 max
	X 5 CrNi 18 10	1.4301	950–1200	0.2 max
	X 10 CrNiS 18-9	1.4305	950–1200	0.2 max
	X 5 CrNiMo 17 12 2	1.4401	950–1200	0.2 max
	X 90 CrCoMoV 17	1.4535	950–1200	0.2 max
Cast iron	GG 25 CrMo	. . .	600–700	0.2 max
	GG 25	0.6025	300–400	0.3 max
	GG 30	0.6030	350–450	0.2 max
Nodular cast iron	GGG 40	0.7040	400–500	0.2 max
	GGG 60	0.7060	500–600	0.2 max
	GGG 70	0.7070	. . .	0.2 max
Sintered steel	Astaloy Mo	. . .	400–500	0.8 max
	Sint D 30	. . .	270–350	0.8 max

DIN, Deutsche Industrie-Normen; DPH, diamond pyramid hardness. Source: Nitrion GmbH critical than with wrought alloys because of the porosity characteristic. A baking operation should precede the ion nitriding of P/M parts to break down or release agents and/or to evaporate any cleaning solvents.

Significant hardening in the diffusion zone cannot be developed in carbon steels or cast iron. However, a compound zone can be formed and is often excellent for wear resistance in lightly loaded parts. Because the compound zone is supported by a relatively soft diffusion zone, applications involving high localized stresses should be avoided with these materials.

Effect of Prior Microstructure. As with other diffusion methods, the initial microstructure can also influence the response of a material to nitriding. In the case of alloy steels, a quenched and tempered structure is considered to produce the optimal nitriding results. The tempering temperature should be 15 to 25 °C (30 to 50 °F) above the anticipated nitriding temperature to minimize further tempering of the core during the nitriding process.

If the nitriding of a nonmartensitic matrix is desired, it is important that prior heat treatment be accompanied by as fast a cooling as possible to provide a relatively low-temperature austenite transformation and retain a high percentage of the nitride-forming element in solution for subsequent precipitation.

Hardness profiles for typical ion-nitrided alloys are shown in Fig. 34. The hardness increase of an ion-nitrided layer is virtually the same as for any nitriding process that provides the same nitrogen concentration profile. As previously mentioned, the hardness of the diffused case depends on precipitation hardening, while that of the white layer depends on the type and thickness of the compound formed. Because the white layers are compounds of only iron and nitrogen, the hardness of these layers is essentially independent of alloy content.

Case Hardness. The concentration and size of alloy nitride precipitates formed, together with parent material hardness, determine the hardness observed in a nitrided case. Figure 35 shows the results of ion nitriding a 0.32C-3Cr-1Mo-0.3V alloy steel at several temperatures, with time held constant. Case depth increases with temperature, and near-surface hardness is maximized near 450 °C (840 °F). Figure 36 shows a similar effect on M2 tool steel quenched and tempered to 62 HRC. Processing at temperatures just above the hardness maximum offers several advantages, such as:

- Higher core hardness can be retained by reducing tempering temperatures.

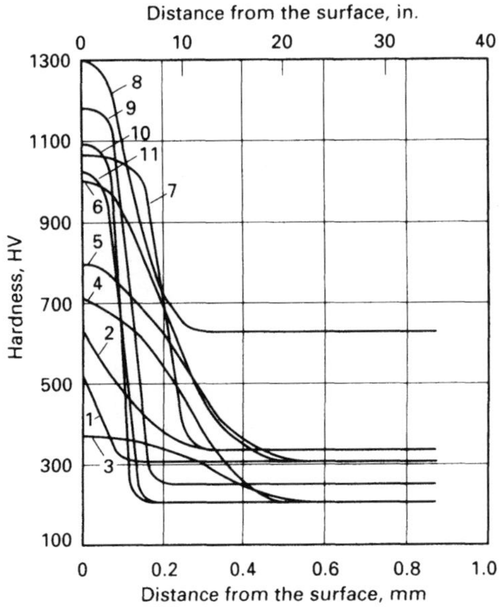

Fig. 34 Hardness profile for various ion-nitrided materials. 1, gray cast iron; 2, ductile cast iron; 3, AISI 1040; 4, carburizing steel; 5, low-alloy steel; 6, nitriding steel; 7, 5% Cr hot work steel; 8, cold-worked die steel; 9, ferritic stainless steel; 10, AISI 420 stainless steel; 11, 18-8 stainless steel

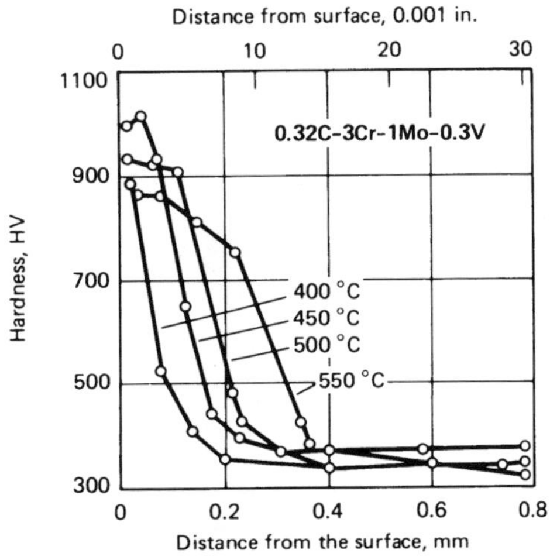

Fig. 35 Influence of treatment temperature on hardness profile

- The possibility of distortion is reduced.
- Parts with low surface roughness remain virtually unchanged.

White-Layer Properties. In general, case depth and white-layer composition should be selected for the anticipated operating conditions of the nitrided component. The ε layer is best for wear and fatigue applications that are relatively free of shock loading or high localized stresses.

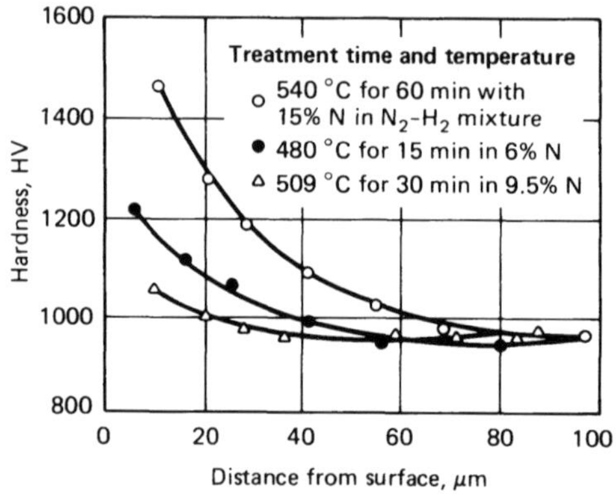

Fig. 36 Microhardness profile of nitrided layer in quenched and tempered M2 tool steel (tempered to 62 HRC) after various plasma nitriding conditions

The γ′ layer is somewhat softer and less wear resistant but is tougher and more forgiving in severe loading situations. The white layer also provides increased lubricity. In addition to mechanical properties, the white layer, which is relatively inert, provides increased corrosion resistance in a variety of environments.

Fatigue strength, in addition to hardness and wear resistance, is significantly improved by nitriding (Fig. 37). The formation of precipitates in the diffused case results in lattice expansion. The core material, in an attempt to maintain its original dimension, holds the nitrided case in compression. This compressive stress essentially lowers the magnitude of an applied tensile stress on the material and thus effectively increases the endurance limit of the part.

Applications and Advantages and Disadvantages

Applications. Various alloy steel and cast iron wear components, including gears, crankshafts, cylinder liners, and pistons, are excellent candidates for the ion process. In one case, P/M transmission gears are being ion nitrided to improve mechanical properties. In general, case depth and white-layer composition should be selected for the anticipated operating conditions of the nitrided component.

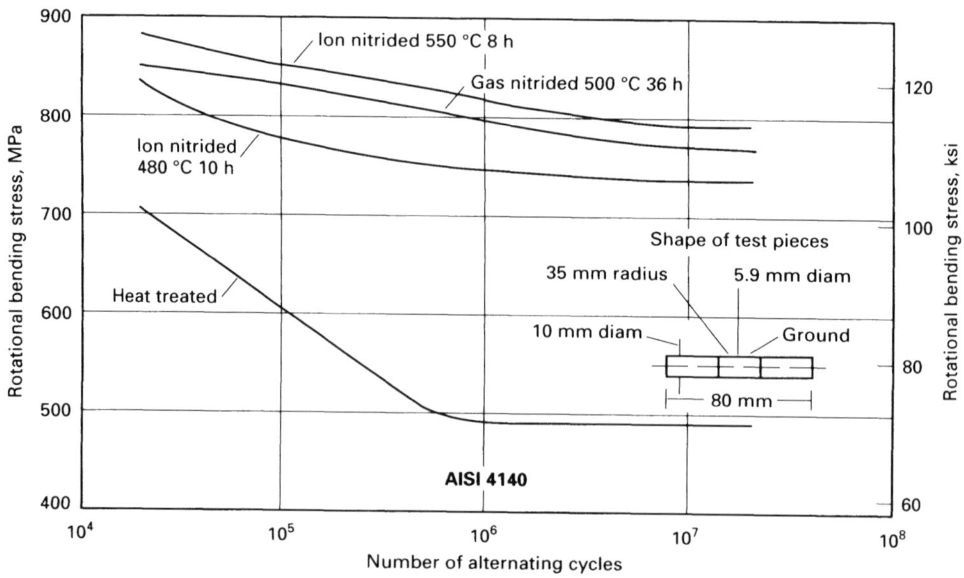

Fig. 37 Effect of nitriding on fatigue strength

One rapidly growing area of ion nitriding is the fuel systems industry. Components used in fuel injection systems experience erosive wear from the fuel and fatigue from rapid cycling of fuel pressure. Ion nitriding greatly enhances the resistance to both of these effects.

The increased lubricity of white layers combined with hardness and fatigue strength has generated significant growth of the ion process in the tool and die industry. Hot work dies, which usually fail by thermal fatigue and sticking, have particularly benefited from ion nitriding following quenching and tempering.

Advantages. Ion nitriding, when compared to conventional (gas) nitriding, offers more precise control of the nitrogen supply at the workpiece surface and the ability to select either an ε or a γ' monophase layer or to prevent white-layer formation entirely (Fig. 30). Other advantages of ion nitriding are:

- Improved control of case thickness
- Lower temperatures (as low as 375 °C, or 700 °F, due to plasma activation, which does not exist in gas nitriding)
- Lower distortion
- No environmental hazard (freedom from handling ammonia)
- Reduced energy consumption
- Ability to automate
- Ability to shield areas where nitriding is not desired by simple mechanical masking

Disadvantages associated with the plasma nitriding process include:

- High capital cost
- The need for precision fixturing with electrical connections
- Long processing times compared to other short-cycle nitrocarburizing processes
- Lack of feasibility of liquid quenching for carbon steels

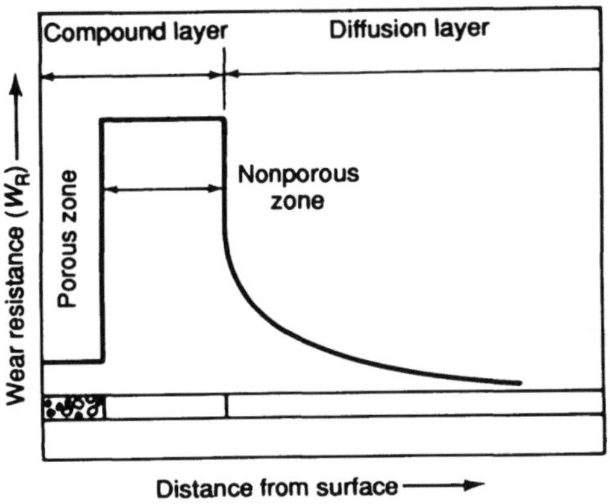

Fig. 38 Plot of adhesive and abrasive wear resistance versus distance from the surface for a nitride layer

Wear Resistance of Nitrided Steels

There are hundreds of articles in the literature that describe wear problems and their solution by different nitriding or nitrocarburizing processes. However, few of them describe the investigated nitride layers sufficiently. Most fail to specify the nitriding process, nitriding parameters, phase composition, compound layer thickness, and chemical composition and microstructure of the nitrided material. Therefore, the wear behavior of nitrided parts is somewhat difficult to ascertain from the literature. In this section, trends were deduced from recent, albeit contradictory, results in order to make generalized statements on wear behavior.

The wear resistance of a nitride layer changes with distance from the surface (Fig. 38). Wear

Table 16 Effect of nitride layer on the wear resistance and the interaction parameter of metals as a function of wear mechanism

	Interaction parameters		Relative wear resistance(b)	
Wear mechanism	Compound layer (a)	Diffusion layer (a)	Compound layer	Diffusion layer
Adhesion	Structure	Formation of a protection layer (pl)	XXX	X
Tribo-oxidation	$v_{pl,form} > v_{pl,sol}$	$v_{pl,form} > v_{pl,sol}$	X	X
	$v_{pl,form} < v_{pl,sol}$	$v_{pl,form}$ $v_{pl,sol}$	(c)	(c)
Abrasion	Structure	Solid solution and precipitation hardening	XX	X
Surface fatigue	Structure	Solid solution and precipitation hardening	X	XXX

(a) $v_{pl,form}$, formation rate of friction-induced protective layer; $v_{pl,sol}$, solution rate of friction-induced protective layer. (b) Increase relative to nonnitrided surface. (c) Decrease in wear resistance

resistance is reduced in the porous zone of the compound layer because of the lower fatigue strength, reduced density, and the notch effect of the pores. The wear resistance of the poreless zone of the compound layer is significantly higher than that of the diffusion zone and the core material. If the compound layer has a homogeneous structure, wear resistance is constant throughout. In the diffusion layer, wear resistance decreases to the value of the core material with increasing distance from the surface; this is because of the reduction of the density of the nitride precipitates and the supersaturation of the matrix with nitrogen.

The wear behavior of a nitride layer is often assumed to be an integral reaction of the nitriding layer to wear loading. This view does not explain wear mechanisms and their interaction during loading. Therefore, wear resistance and mechanisms must be discussed in conjunction with the structure of the nitriding layer. Table 16 gives the wear resistance of the compound layer and the diffusion layer in terms of the four major wear mechanisms discussed subsequently.

Adhesion. In adhesive wear, the wear resistance of the compound layer is very high compared with nonnitrided parts. This is because the inclination to microwelding decreases due to the change in the electron configuration in the outer region of the part. In the diffusion layer, wear resistance can also increase because of the formation of protective nitrogen-containing layers.

Tribo-oxidation. The wear resistance of the compound layer against tribo-oxidation depends on the formation of protective layers in the contact zone. If the formation rate, $v_{pl,form}$, of the friction-induced protective layer is greater than the concurrent solution rate, $v_{pl,sol}$, wear resistance increases. If $v_{pl,sol} > v_{pl,form}$, wear resistance decreases. The wear resistance of the diffusion layer seems to behave similarly.

Abrasion. The compound layer is very resistant to abrasive wear, because the structure of the compound layer allows only very low plastic deformation. Compared with nonnitrided steel, the wear resistance of the diffusion layer is higher—a result of the higher fatigue strength obtained by solid-solution strengthening and precipitation hardening.

Surface Fatigue. It is assumed that the wear resistance of nitrided parts to surface fatigue is higher than that of nonnitrided parts, as the changed lattice structure of the compound layer prevents plastic deformation. The diffusion zone has very high resistance to surface fatigue wear; here plastic deformations are very small because of solid-solution strengthening and precipitation hardening.

Influence of Variables on Wear Resistance of Nitrided Parts

Compound Layer. Variables that influence the wear resistance of the compound layer of nitrided parts are illustrated in Fig. 39 to 45. A

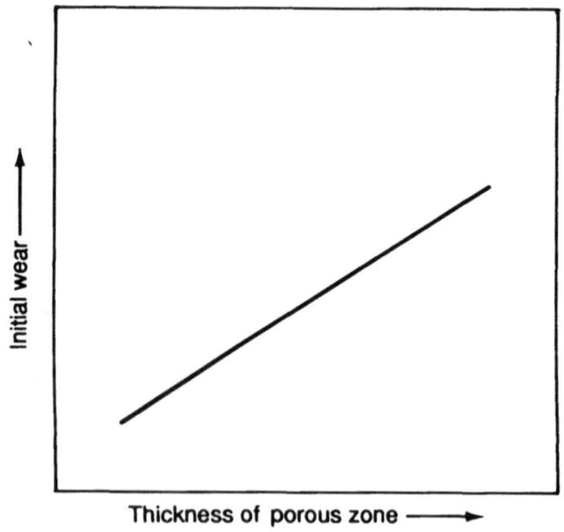

Fig. 39 Effect of porous zone thickness on adhesive wear resistance of the compound layer of a nitrided part

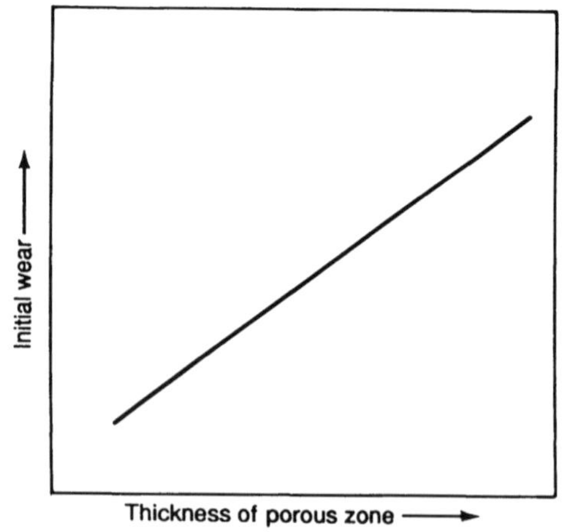

Fig. 40 Effect of porous zone thickness on abrasive wear resistance of the compound layer of a nitrided part

porous zone will raise the initial wear in the case of adhesive and abrasive wear (Fig. 39 and 40). If tribo-oxidation is the main wear mechanism, a large increase in wear resistance can occur (Fig. 41). This may be due to absorption of the lubrication medium and subsequent formation of a lubrication layer. In the case of surface fatigue (Fig. 42), compound layer thickness has no influence on wear resistance if the porous zone of the compound layer is small and the maximum of the true stress (σ_v) lies in the deeper regions of the compound layer or below it. If the thickness of the porous zone increases the distance from the surface of the maximum stress, wear resistance to surface fatigue will rapidly decrease.

The structure and composition of the compound layer also influence its wear resistance. Investigations have shown that the adhesive wear resistance of the compound layer is strongly affected by the volume of ε-nitrides. In most cases, resistance increases with increasing ε-nitride content (Fig. 43). Similar behavior exists when nitrided parts are abrasively stressed (Fig. 44a). However, there is a difference between layers consisting only of nitrides and those containing carbonitrides. At constant volume of ε-nitride, carbonitride layers show significantly higher wear resistance than nitride layers (Fig. 44b). For surface fatigue, a higher content of ε-nitride in the compound layer will result in increased wear resistance if the surface pressure is held constant (Fig. 45).

Diffusion Layer. Variables that affect the wear resistance of the diffusion layer of nitrided parts are illustrated in Fig. 46 to 53. Hardening of the alloy by precipitation methods such as solid-solution strengthening will raise the adhesive, abrasive, and surface fatigue wear resistance (Fig. 46–48). Greater nitriding depths also increase wear resistance (Fig. 49–51). Increases in the nitriding depth allow initial high wear resistance to abrasion to be upheld for a longer time, thus increasing component life (Fig. 50).

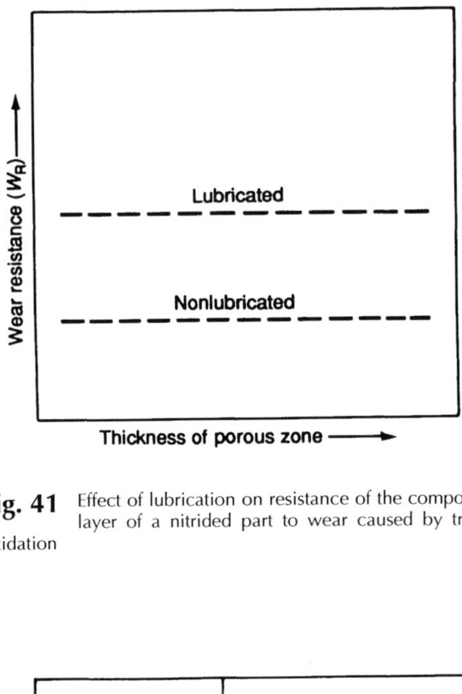

Fig. 41 Effect of lubrication on resistance of the compound layer of a nitrided part to wear caused by tribo-oxidation

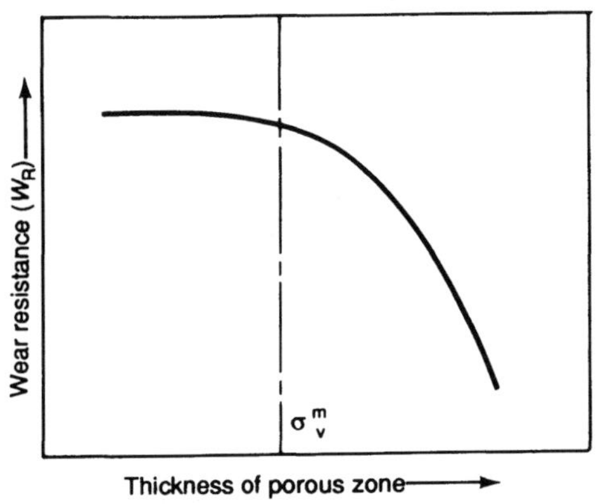

Fig. 42 Effect of porous zone thickness and maximum stress on resistance of the compound layer of a nitrided part to wear caused by surface fatigue

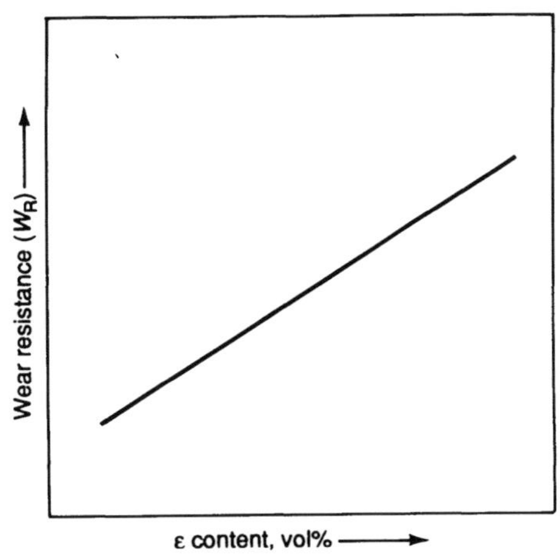

Fig. 43 Effect of ε-nitride content on adhesive wear resistance of the compound layer of a nitrided part

Behavior is similar for surface fatigue stressing: increases in nitriding depth allow higher surface pressures and constant wear resistance (Fig. 51).

Nitride type and distribution in the diffusion layer can be changed by a postnitriding aging, strongly influencing wear resistance. For example, when nitrided parts are age hardened at constant temperature for different times, a characteristic hardness profile with a distinct maximum will result. Adhesive or abrasive wear stressing of such parts also results in a maximum wear resistance; however, the maximum shifts to higher aging times. Thus, both types of wear are sensitive to the type and distribution of the nitride precipitation in the diffusion layer: a matrix with finely dispersed $Fe_{16}N_2$ nitrides has a lower wear resistance than one with a coarse distribution of Fe_4N nitrides (Fig. 52). The same is true for parts subjected to wear caused by surface fatigue (Fig. 53).

Influence of the Nitriding/ Nitrocarburizing Process on Wear Behavior

Investigations of the influence of the nitriding process on wear behavior can be divided into two groups: process oriented and materials science oriented. The objective of process-oriented investigations is often to show the clear advantages of a certain process. Materials-science-oriented investigations are usually more varied. Evaluation of published results of experimental work indicates that nitriding process variables do not exert much influence if nitriding layers are compared and are optimized to the actual wear stress.

Process variables do have an influence, however, if changes in geometry, which depend on the process, are considered. Typical process-dependent changes in geometry are shown in Table 17. Parts with elevations at their corners will have shorter lifetimes because of the higher surface pressures at the corners than parts with constant surface pressures over the entire contact zone.

Optimization of Wear by Process Technology

The influence of nitriding process parameters on wear resistance is not very high, but, with limitations, specific wear properties can be fitted to loading conditions by process optimizing. Table 18 summarizes possible enhancements of wear resistance for various wear mechanisms.

Adhesive or Abrasive Wear. In adhesive or abrasive wear, the properties of the compound layer must be optimized by:

- Changing γ′-nitride to ε-nitride by higher nitriding potentials (nitriding numbers)
- Changing nitrides to carbonitrides by adding carbon donors such as Endogas and CO_2 or

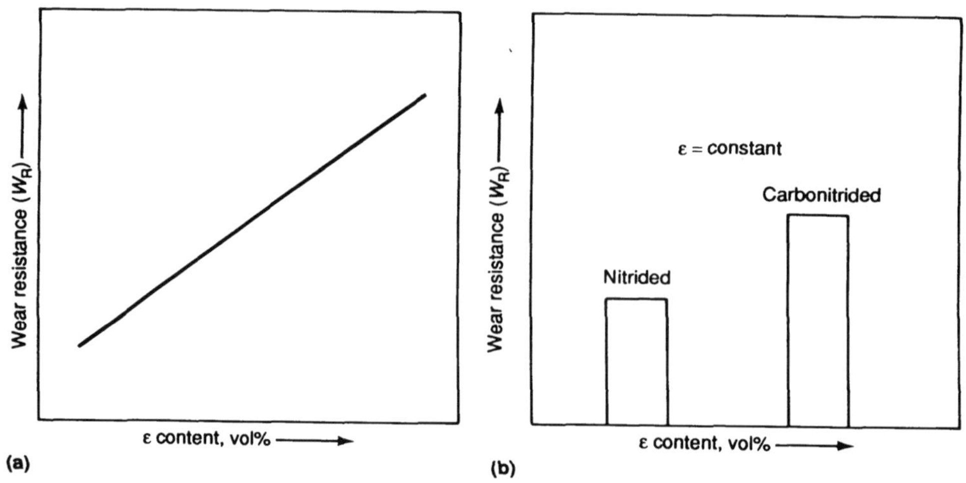

Fig. 44 Effect of ε-nitride content on abrasive wear resistance of the compound layer of a nitrided part. (a) Resistance increases with increasing ε content. (b) At constant volume of ε, carbonitride layers show significantly higher resistance than nitride layers.

by using a material with a higher carbon content
- Minimizing the porous zone by polishing

Porosity can also be influenced by the atmosphere (for example, lower nitriding potential or lower content of carbon donors), but this may worsen the properties of the compound layer. However, a small porous zone in the compound layer may be advantageous when the shape of the surface can be optimized by running-in (initial) wear.

In gas nitriding, the structure of the compound layer can be changed by a definite nitriding atmosphere—that is, a definite nitriding number [ratio $p(NH_3)/p(H_2)^{1.5}$] or by additional carbon-containing gases such as CO_2 or Endogas. In plasma nitriding, additional parameters such as pressure and glow discharge conditions are beneath the atmosphere. In bath nitriding, the concentration of the cyanate and cyanide content of the bath can be changed.

A reaction layer can be formed by an oxidation process after nitriding. Usually, these processes are performed in salt baths or in gaseous atmospheres.

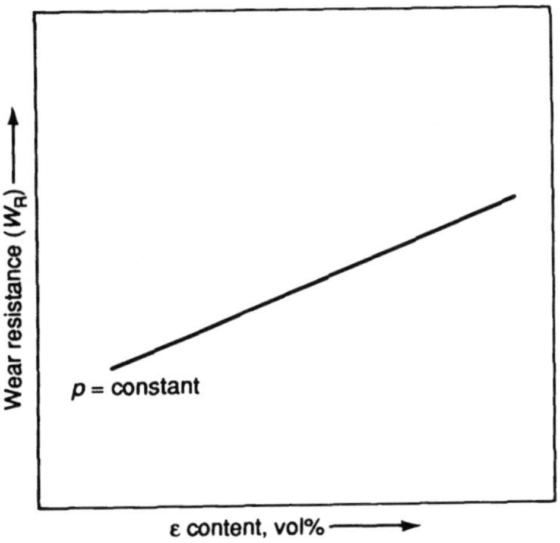

Fig. 45 Effect of ε-nitride content on resistance of the compound layer of a nitrided part to wear caused by surface fatigue at constant pressure

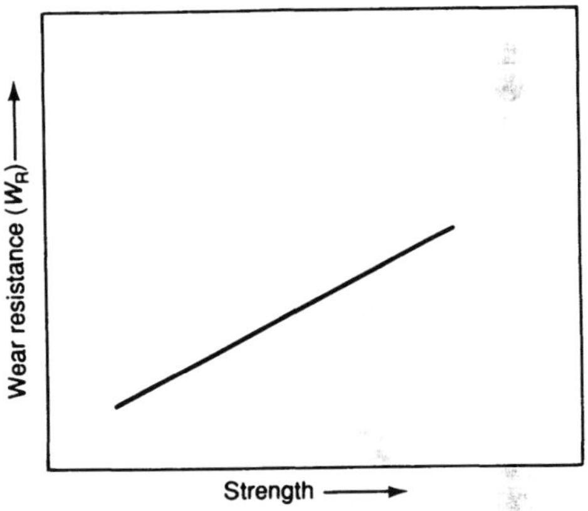

Fig. 47 Effect of material strength on abrasive wear resistance of the diffusion layer of a nitrided part

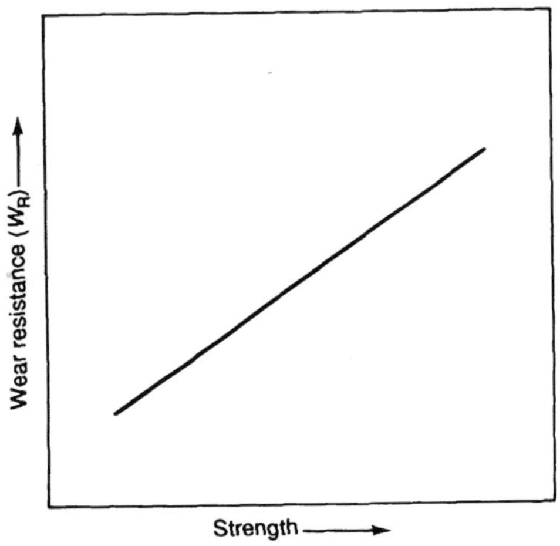

Fig. 46 Effect of material strength on adhesive wear resistance of the diffusion layer of a nitrided part

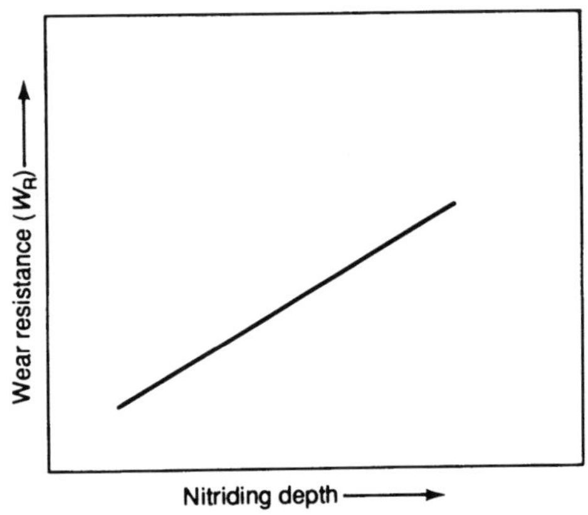

Fig. 48 Effect of material strength on resistance of the diffusion layer of a nitrided part to wear caused by surface fatigue

The adhesive and abrasive wear resistance of the diffusion layer can be enhanced by strengthening the diffusion layer. The most effective means of this are by proper material selection (for example, high content of nitride-forming elements) and by heat treatment before nitriding (for example, hardening and tempering instead of normalizing).

Tribo-oxidation. If tribo-oxidation is the primary wear mechanism, only limited improvements can be made by changing the nitriding conditions. In this case, it is necessary, if

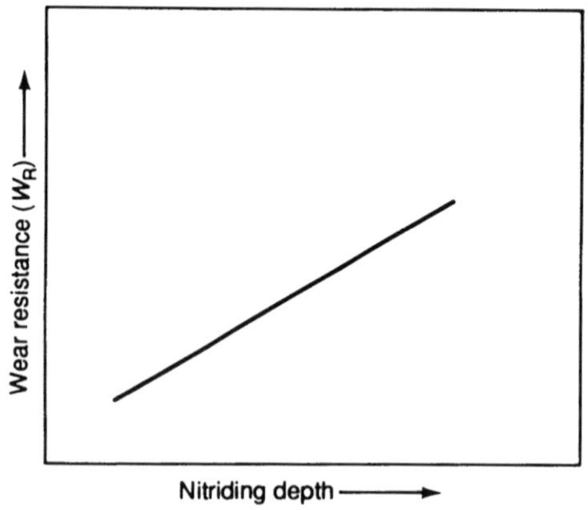

Fig. 49 Effect of nitriding depth on adhesive wear resistance of the diffusion layer of a nitrided part

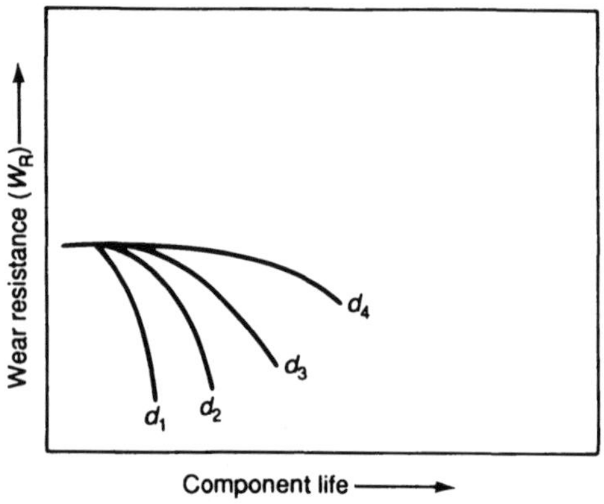

Fig. 50 Effect of nitriding depth on abrasive wear resistance of the diffusion layer of a nitrided part. Increases in nitriding depth increase component life. Nitriding depth, d: $d_1 < d_2 < d_3 < d_4$

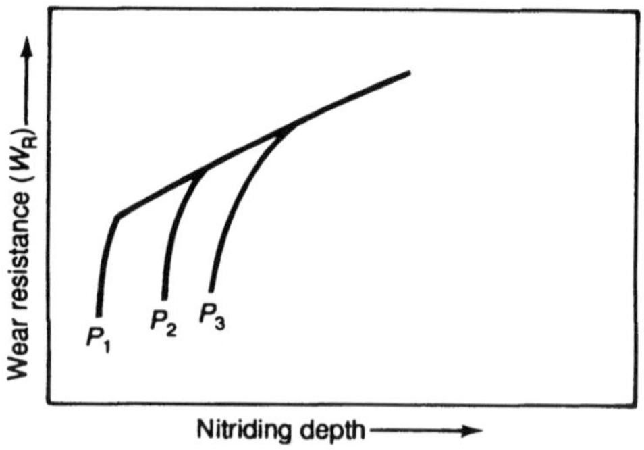

Fig. 51 Effect of nitriding depth on resistance of the diffusion layer of a nitrided part to wear caused by surface fatigue. Increases in nitriding depth allow the part to withstand higher surface pressures without sacrificing wear resistance. $p_1 < p_2 < p_3$

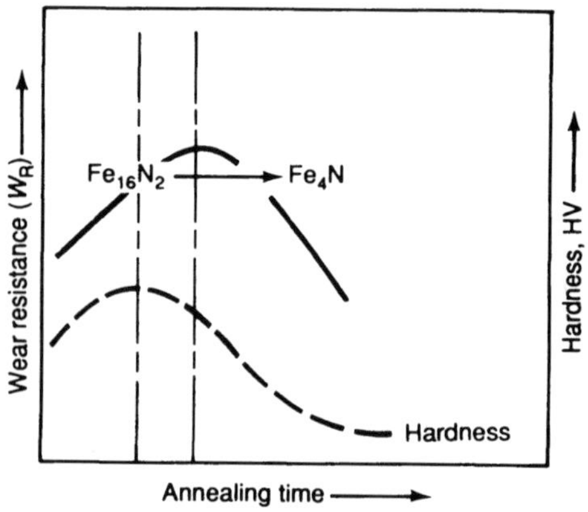

Fig. 52 Hardness profile showing effect of annealing time on abrasive and adhesive wear resistance of the diffusion layer of a nitrided part. See text for discussion of the effect of type and distribution of nitrides.

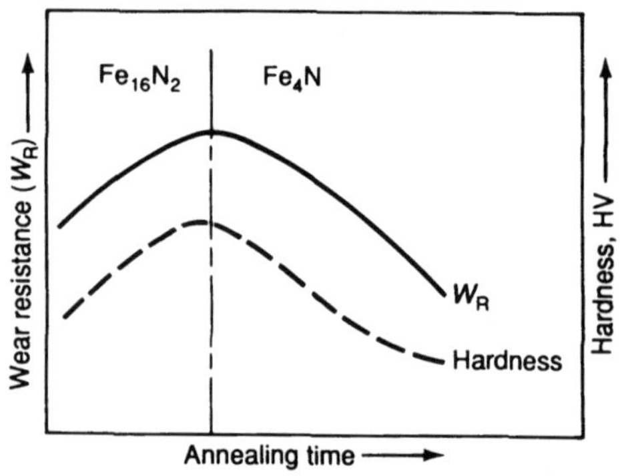

Fig. 53 Hardness profile showing effect of annealing time on resistance of the diffusion layer of a nitrided part to wear caused by surface fatigue. See text for discussion of the effect of type and distribution of nitrides.

possible, to change the conditions of the tribosystem.

Surface Fatigue. If controlling surface fatigue is the objective, measures to optimize the diffusion layer can be effective—in particular, fitting the nitriding hardness profile to the stressing profile and influencing the nitride precipitates (type, form, size, and distribution). The choices of material and heat treatment before nitriding depend on the required core strength. Nitriding of alloyed steels often results in thick deposits of carbides and/or nitrides in the grain boundaries, which may promote crack initiation. If the parts are to be under high load, such deposits should be avoided by changing the heat treatment process.

The nitriding depth can be increased by increasing process temperature and/or treating time. If merely prolonging the nitriding time does not result in sufficient nitriding depths, a two-step process can be performed. In the first step, nitriding is done at low temperatures to generate a finely dispersed nitride distribution. The second nitriding step is done at higher tempera-

Table 17 Effect of nitriding process on surface topography and surface finish

Nitriding process	Material cross section showing surface topography	Nitriding parameters: Temperature °C (°F)	Nitriding parameters: Duration, h	Total roughness peak-to-valley (Rt): µm	Total roughness peak-to-valley (Rt): µin.	Arithmetic average surface roughness (Ra): µm	Arithmetic average surface roughness (Ra): µin.
Before nitriding		...	...	1	40	0.21	8.4
Plasma nitrided (bearing surface nitrided)		530 (985)	24	5	200	0.58	23
Plasma nitrided (totally nitrided)		550 (1020)	2	6	240	0.62	25
		550 (1020)	24	6	240	0.67	27
Salt-bath nitrided		570 (1060)	4	11	440	1.22	49
Gas nitrided		500 (930)	36	11	440	1.06	42
		500 (930)	84	13	520	1.21	48

Table 18 General guidelines for improving the wear resistance of nitrided steels as a function of wear mechanism and surface layer type

Layer	Wear mechanism(s)	Action required to improve wear resistance	Recommended change in process parameter
Compound layer	Abrasion, adhesion	Optimization of structure: • Transform γ′ to ε microstructure by using higher nitriding potentials • Convert from nitriding to carbonitriding to add additional carbon donors • Minimize porous zone region by polishing surface	Modify nitriding atmosphere: • Salt bath: change content of base from CN^- to CNO^- • Gaseous: change nitriding number, flow rate, and carbon donors
		Formation of reactive layers	Add oxidation process after nitriding process is completed
	Tribo-oxidation	Modification of tribosystem	...
	Surface fatigue	Reduction in size of porous zone	Modify nitriding atmosphere
Diffusion layer	Abrasion, adhesion	Increase strength of material	Select alternate materials; add heat treatment prior to nitriding (use hardening and tempering operations instead of normalizing)
	Tribo-oxidation	Modification of tribosystem	...
	Surface fatigue	Optimization of hardness profile	Increase temperature and duration of nitriding process; select alternate materials; add heat treatment prior to nitriding; increase cooling rate after nitriding; add annealing treatment after nitriding
		Modification of composition (type, form, particle size, and distribution of nitrides) to control precipitation of nitrides	...

tures to raise the nitriding depth. This procedure avoids coarsening of the nitride distribution and thus a decrease in hardness. The cooling rate from the nitriding temperature and postnitriding treatments can be used to influence the type, size, and distribution of the nitrides.

Corrosive Wear. Nitriding improves resistance to corrosive wear. Resistance can be further improved by oxidation of the compound layer after nitriding (oxynitriding). To achieve a very smooth layer capable of sustaining high loads, it is recommended that the oxidized layer be polished and then oxidized again.

ACKNOWLEDGMENTS

Portions of this chapter were adapted from:

- Gas Nitriding, *Heat Treating*, Vol 4, *ASM Handbook*, ASM International, 1991, p 387–409
- Liquid Nitriding, *Heat Treating*, Vol 4, *ASM Handbook*, ASM International, 1991, p 410–419
- Nitriding and Nitrocarburizing, *Friction, Lubrication, and Wear Technology*, Vol 18, *ASM Handbook*, ASM International, 1992, p 878–883
- Plasma (Ion) Nitriding, *Heat Treating*, Vol 4, *ASM Handbook*, ASM International, 1991, p 420–424

REFERENCES

1. D. Pye, Nitriding Techniques and Methods, *Steel Heat Treatment Handbook*, G.E. Totten and M.A.H. Howes, Ed., Marcel Dekker, 1997, p 721–764
2. Gas Nitriding, *Heat Treating*, Vol 4, *ASM Handbook*, ASM International, 1991, p 387–409
3. G.J. Tymowski, W.K. Liliental, and C.D. Morawski, Typical Nitriding Faults and Their Prevention Through the Controlled Nitriding Process, *Ind. Heat.*, Jan 1995, p 39–44
4. G.J. Tymowski, W.K. Liliental, and C.D. Morawski, Take the Guesswork Out of Nitriding, *Adv. Mater. Process.*, Dec 1994, p 52–54
5. Heat Treating of Specific Classes of Tool Steels, *Heat Treating*, Vol 4, *ASM Handbook*, ASM International, 1991, p 734–760

CHAPTER 7

Nitrocarburizing

NITROCARBURIZING, also referred to as ferritic nitrocarburizing, is a modified form of nitriding, not a form of carburizing. In the process, nitrogen and carbon are simultaneously introduced into the steel while it is in the ferritic condition, that is, at a temperature below which austenite begins to form during heating. Nitrocarburizing treatments have been successfully applied to most ferrous materials, including wrought and powder metallurgy plain carbon steels, free-machining steels, microalloyed steels, alloy steels, tool steels, stainless steels, and cast irons. Engineering components such as rocker-arm spacers, textile machinery gears, pump cylinder blocks, and jet nozzles have been treated for wear resistance, while components such as crankshafts and drive shafts have been treated for improved fatigue properties. The treatment can be carried out in a liquid salt bath or a gaseous atmosphere. Plasma nitrocarburizing is also becoming increasingly popular in view of the ease of control of the process and its environmental friendliness.

The Compound Layer and Diffusion Zone

A complex sequence is involved in the formation of the nitrocarburized case. Of importance here is that normally a very thin layer of single-phase epsilon (ε) iron-carbonitride, $Fe_{2-3}(N,C)$, is formed between 450 and 590 °C (840 and 1095 °F). Although the thickness of this compound or white layer is a function of temperature, gas composition, and gas volume (flow), it is generally between 10 and 40 μm for most applications (Fig. 1). Associated with the compound layer is an underlying diffusion zone containing iron (and alloy) nitrides and absorbed nitrogen. The total case depth of the compound layer and diffusion zone can reach 1 mm (0.040 in.).

The hard (60 to 72 HRC) compound layer has excellent wear and antiscuffing properties and is produced with minimum distortion. The diffusion zone, provided it is substantial enough, improves fatigue properties such as endurance limit, especially in carbon and low-alloy steels. The diffusion zone is also responsible for some of the increased hardness of the nitrocarburized case, especially in the more highly alloyed steels that contain strong carbide formers.

Porosity. It is not uncommon to observe porosity in the compound layer due to a carburizing reaction at the steel surface. This reaction influences the nitriding kinetics and therefore the degree and type of porosity at the surface of the ε-carbonitride layer. Three types of layer can be produced: no porosity, sponge porosity, or columnar porosity. Some applications require deep, nonporous ε layers. Other applications where, for example, optimal corrosion resistance is needed benefit from the presence of sponge porosity. Still others benefit from columnar porosity, where oil retention can enhance wear resistance.

Liquid Nitrocarburizing

Liquid or salt bath nitrocarburizing was first established in the late 1940s when high-cyanide nitrocarburizing salt baths were introduced. Environmental considerations and the increased cost of detoxification of cyanide-containing effluents have led to the development of low-cyanide nontoxic salt bath nitrocarburizing treatments.

Cyanates are the active nitriding constituent of both high-cyanide and low-cyanide nitrocarburizing baths. Reduction of the cyanide content permits markedly higher cyanate concentrations in the low-cyanide baths; this results in greatly increased nitriding activity. Unlike the reducing high-cyanide baths, the nominal cyanate and carbonate composition of the low-cyanide baths is oxidizing. The baths are composed of primarily potassium salts with some sodium salts. During nitriding, cyanates yield nitrogen to the steel and form carbonates. Cyanate concentration is maintained by the use of organic regenerators, which supply nitrogen to reform cyanates from carbonates.

Process Benefits (Ref 2). Salt bath nitrocarburizing offers many benefits, including:

- *High-quality components:* Superior corrosion resistance by using oxidative cooling; consistent, repeatable results both with low and high throughputs; uniform, rapid heat transfer via the salt melt; uniform nitrocarburizing effect, even on components having narrow openings; and very good running-in wear behavior due to the formation of a pore zone
- *Easy-to-use process:* Simplified precleaning and monitoring (only a few parameters, such as temperature, treatment time, and bath composition need to be monitored); simplified plant technology; and good part quality regardless of the size or make up of the load
- *High flexibility:* Parts requiring different treatment times can be processed at the same time; different materials can be processed together in one load; treatment time and run-through time is very short; modular unit design allows easy matching to suit varying throughputs; use of media having different cooling rates (water, salt bath, air blast, nitrogen, and vacuum); and treatment temperature range of 480 to 630 °C (895 to 1165 °F)

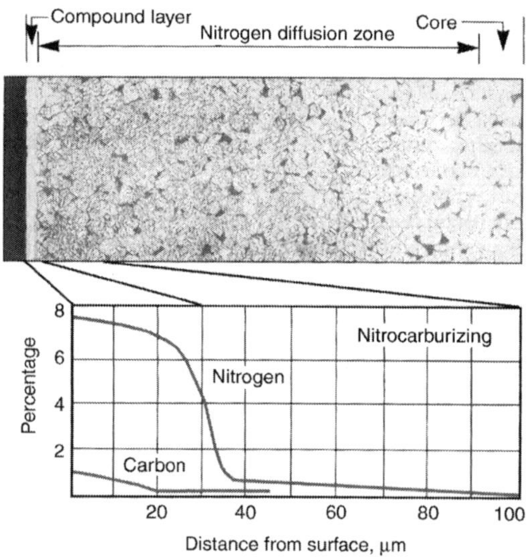

Fig. 1 Surface layer produced by (ferritic) nitrocarburizing at 570 °C (1060 °F), where nitrogen is the predominant element in the epsilon (ε) carbonitride layer. Source: Ref 1

Process Variations

Typically, the salt bath nitrocarburizing process consists of:

- Preheating parts in air to a temperature of 350 °C (660 °F)
- Nitrocarburizing in a salt bath at a temperature of 570 to 580 °C (1060 to 1075 °F) for approximately 1 to 2 h
- Intermediate cooling to a temperature of 400 °C (750 °F)
- Cooling to room temperature
- Cleaning in water

However, there are a number of proprietary processes that differ from the steps outlined previously. For example, there are low-temperature and high-temperature variations. In addition, parts can be further processed by lapping or polishing and, if specified, given a post-salt-bath oxidizing treatment. The post-treatment enhances corrosion resistance over the nitrocarburized-only condition, in some instances providing corrosion resistance superior to that of chromium- and nickel-plated parts. Polymeric coatings can be applied after oxidizing treatments to provide even greater corrosion resistance.

Examples of four process variations carried out at the intermediate nitrocarburizing temperature of 570 to 580 °C (1060 to 1075 °F) are described subsequently. High-cyanide nitrocarburizing baths have been in use since the late 1940s. Initially, the sulfur-containing variant was used to produce a wear-resistant surface of iron sulfide (see process 2). A sulfur-free high-cyanide bath was developed in the mid-1950s, now known as aerated bath nitriding (process 1). This process and a low-cyanide variant of it (process 4) are commonly used.

Both processes 1 and 2 are similar in that components are typically preheated to approximately 350 to 480 °C (660 to 900 °F) and then transferred to the nitrocarburizing salt bath at 570 °C (1060 °F). The major components of the baths for both processes are normally alkali metal cyanide and cyanate. Salts are predominately potassium, with sodium.

Process 1: High Cyanide without Sulfur. At the treatment temperature of 570 °C (1060 °F), the process is controlled largely by two reactions: an oxidation reaction and a catalytic reaction. The oxidation reaction involves transformation of cyanide to cyanate:

$$4NaCN + 2O_2 \rightarrow 4NaCNO \quad \text{(Eq 1)}$$

$$2KCN + O_2 \rightarrow 2KCNO \quad \text{(Eq 1a)}$$

Though this reaction can proceed by natural oxidation of the cyanide bath, eventually leading to the desired cyanate content, the mechanism of natural aging does not provide the higher cyanate level made possible with aeration. To provide agitation and stimulate chemical activity, therefore, dry air is introduced into the bath.

The catalytic reaction involves breaking down cyanate in the presence of the steel components being treated, thus supplying carbon and nitrogen to the surface:

$$8NaCNO \rightarrow 2Na_2CO_3 + 4NaCN + CO_2 + (C)_{Fe} + 4(N)_{Fe} \quad \text{(Eq 2)}$$

$$8KCNO \rightarrow 2K_2CO_3 + 4KCN + CO_2 + (C)_{Fe} + 4(N)_{Fe} \quad \text{(Eq 2a)}$$

As a result of this treatment, a wear-resistant compound zone, rich in nitrogen and carbon, is formed on component surfaces (Fig. 2).

Process 2: High Cyanide with Sulfur. The same basic oxidation and catalytic reactions of process 1 also occur in this process. In addition, further reactions take place because of sulfites in the melt. These sulfites are reduced to sulfides, in conjunction with the oxidation of the cyanide to cyanate, as follows:

$$Na_2SO_3 + 3NaCN \rightarrow Na_2S + 3NaCNO \quad \text{(Eq 3)}$$

$$K_2SO_3 + 3KCN \rightarrow K_2S + 3KCNO \quad \text{(Eq 3a)}$$

Thus, the sulfur present in the bath acts as an accelerator, with the result that the cyanate is produced more readily than if the sulfur compounds were absent. Consequently, external aeration is not normally used in the process. Potassium and sodium cyanates produced by the reactions in Eq 1 and 3 catalytically decompose at the surface of ferrous materials to liberate carbon monoxide and nascent nitrogen. The carbon monoxide dissociates to liberate active carbon. The carbon, in conjunction with the nascent nitrogen, diffuses into the material being treated to form the compound zone.

The exact mechanism by which sulfur is impregnated into the material is not clear. Various sulfides react with the component being treated to form iron sulfide; this is the black deposit observed on the surface of components after treatment.

The compound layer formed on mild steel after a 90 min treatment, followed by water quenching, is shown in Fig. 3. The compound layer formed by cyanide salt bath nitrocarburizing treatments, and, in particular, by the sulfur-containing high-cyanide process, contains an outer region of microporosity. These pores, which readily absorb oil, may assist the antiscuffing properties of treated components under lubrication conditions.

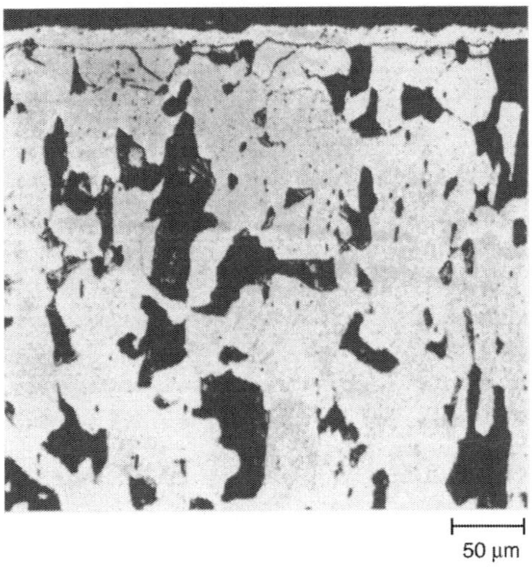

Fig. 2 Metallographic appearance of salt bath nitrocarburized mild steel after 1.5 h at 570 °C (1060 °F) followed by water quenching

Process 3: Low Cyanide with Sulfur. This patented process confers sulfur, nitrogen, and presumably, carbon and oxygen to surfaces of ferrous materials. The process is unique in that lithium salts are incorporated in the bath composition. Cyanide is held to very low levels: 0.1 to 0.5%. Sulfur species, present in the bath at concentrations of 2 to 10 ppm, cause sulfidation to occur simultaneously with nitriding. Sulfur levels near 10 ppm result in an apparently porous compound zone (Fig. 4); the dark areas are actually iron-sulfide nodules, not voids. This compound zone is similar to the high-cyanide, sulfur-containing nitrocarburizing process that has, however, columnar iron-sulfide inclusions.

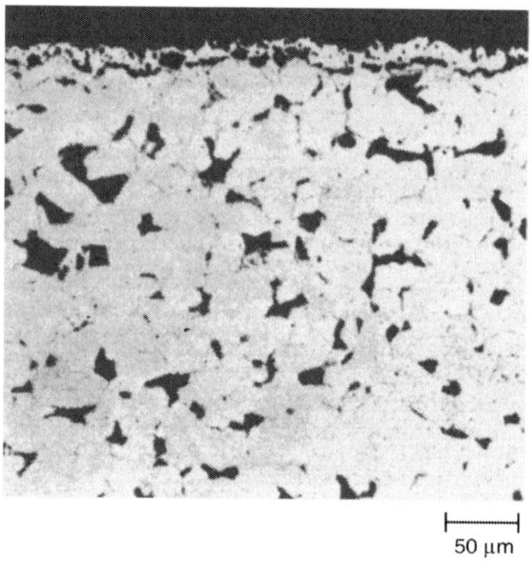

Fig. 3 Metallographic appearance of mild steel after similar treatment to Fig. 2. Iron-sulfide inclusions in the outer region of the compound zone are apparent after this treatment, in which sulfur acts as an accelerator.

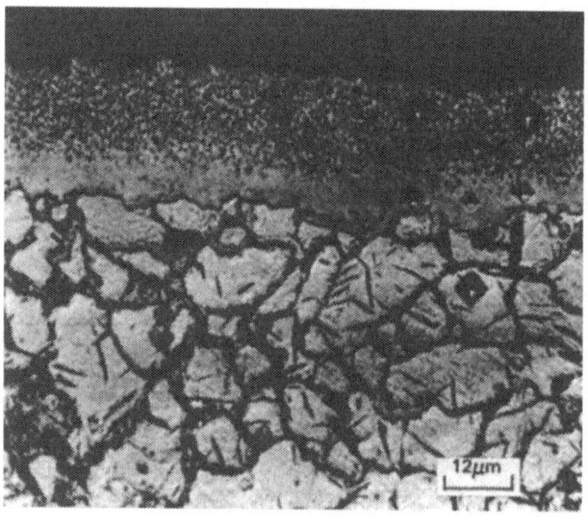

Fig. 4 Sample of plain carbon steel after low-cyanide salt bath nitrocarburizing treatment (process 3). The high level of apparent porosity is a characteristic of high sulfur content in the compound zone; dark areas are actually iron-sulfide nodules, not voids.

Bath composition can be adjusted to lower sulfur levels (2 ppm) to form a less porous layer with a lower iron-sulfide content.

A compound layer 20 to 25 μm (800 to 1000 μin.) thick forms in 90 min at 570 °C (1060 °F) on American Iron and Steel Institute (AISI) 1010 steel, compared with the 8 to 10 μm (320 to 400 μin.) layer formed by the high-cyanide sulfur-bearing nitrocarburizing process in the same time. Figure 5 shows the thickness of the compound layer as a function of the treatment time for the nontoxic and cyanide-based treatments.

Process 4: Low Cyanide without Sulfur. A low-cyanide alternative to the cyanide-based process 1 has been developed. This process, similar to process 3, is a cyanate bath with no lithium or sulfur compounds and very low cyanide levels (2 to 3%). Melon, an organic polymer, is used for bath regeneration.

When water quenching is employed, the low level of cyanide permits easier detoxification. Alternatively, quenching into a caustic-nitrate salt bath at 260 to 425 °C (500 to 795 °F) may be used for cyanide/cyanate destruction.

Processing temperature for process 4 is 570 to 580 °C (1060 to 1080 °F); the rate of compound zone formation is comparable to that of process 3. Metallurgical results are virtually identical with the cyanide-based process 1.

Wear and Antiscuffing Characteristics. The resistance to scuffing after salt bath nitrocarburizing treatments has been frequently tested with a Falex lubricant testing machine (Fig. 6). A 32 by 6.4 mm (1.25 by 0.25 in.) test-

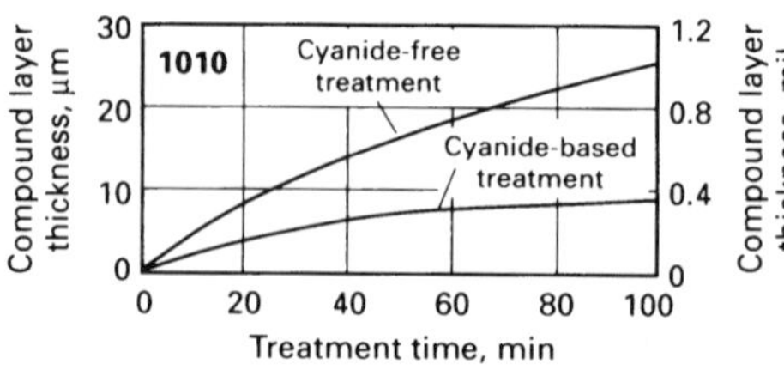

Fig. 5 Comparison of compound zone thickness produced by low-cyanide and cyanide-based treatments containing sulfur

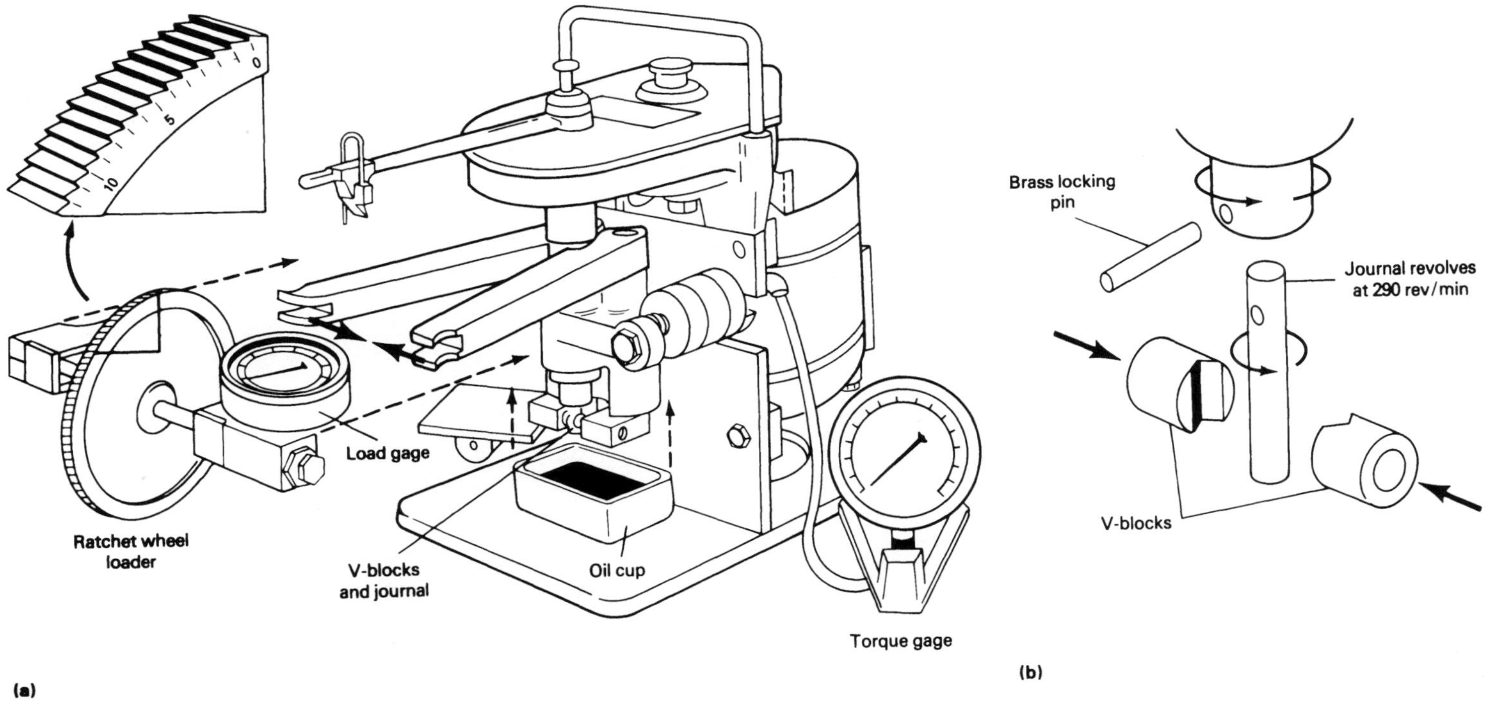

Fig. 6 Lubricant tester used to measure endurance (wear) life and load-carrying capacity of either dry solid-film lubricants or wet lubricants in sliding steel-on-steel applications. (a) Key components of instrument. (b) Exploded view showing arrangement of V-blocks and rotating journal

piece is attached to the main drive shaft by means of a shear pin, and two anvils or jaws having a 90° V-notch fit into holes in the lever arms. During testing, the jaws are clamped around the testpiece, which rotates at 290 rpm, and the load exerted by the jaws is gradually increased. Both test-pieces and jaws can be immersed totally in a small tank containing lubricant or other fluid, or tests can be carried out dry.

Table 1 lists results of a few representative Falex tests for plain low-carbon steels both before and after cyanide salt bath nitrocarburizing treatments. The untreated low-carbon steel specimens do not show any significant scuffing resistance even when tested under oil-lubricated conditions. After treatment, however, even when tested dry, there is a considerable improvement in antiscuffing properties. Specimens tested in the dry condition after salt bath nitrocarburizing generate so much heat that they eventually become red hot and are extruded under the applied load. Untreated testpieces seize at relatively low loads before becoming red hot, whereas treated samples, even after extrusion, show no signs of scuffing. During testing in oil, the specimens become highly polished. Similar Falex test results are reported for low-cyanide salt bath nitrocarburizing treatments.

Low-Temperature Salt Bath Nitrocarburizing (Ref 3)

As indicated earlier, most salt bath nitrocarburizing operations are carried out at a processing temperature of 570 to 580 °C (1060 to 1075 °F). However, some tool and other high-alloy steels are susceptible to reductions in core hardness after standard nitrocarburizing. With lower treatment temperatures, core hardness can be maintained or sometimes increased. The low-temperature nitrocarburizing process takes place at 480 °C (900 °F), although it can operate at 480 to 520 °C (900 to 970 °F). Specific advantages of this process are:

- Core hardness and tensile strength are maintained in the tempered condition.
- Very thin compound layers can be formed.
- Distortion is extremely low.
- Formation of a compound layer on high-speed steels can be suppressed.
- Hardness of surface and diffusion layers can be customized.

This low-temperature process is beneficial for high-alloy steels such as stainless, tool, die, and high-speed steels. For example, cutting tools of high-speed steel are often nitrocarburized at 580 °C (1075 °F) to generate a harder surface, but they must be treated for only a few minutes to avoid brittleness. Such treatment causes no reduction in core hardness, but it can be challenging to accurately control the processing time. By treating tools at a lower temperature, such as 520 °C (970 °F) for periods of 30 to 60 min, the necessary hard nitride layer is developed but without any brittleness.

Hardness Comparisons. The influence of the nitrocarburizing temperature on the core hardness of various steels in the hardened and tempered condition is shown in Fig. 7. Differences are especially significant in hardness measurements in the 420 and D2 steels. After the steel has been treated at 480 °C (900 °F), hardness actually increases slightly over the original condition, which can be attributed to an

Table 1 Comparison of plain carbon steels wear tested prior to and following cyanide salt bath nitrocarburizing

Condition of test pieces and jaws	Testing medium(a)	Applied load, kgf	Applied load, lbf	Condition of test pieces	Material
Untreated	SAE 30 oil	320	700	Scuffed	En32 (0–15% C)
Untreated	Water	270	600	Badly scuffed	En32 (0–15% C)
Untreated	Air	320	700	Scuffed	En32 (0–15% C)
Untreated	Air	205	450	Scuffed	AISI 1045
Treated(b)	SAE 30 oil	Limit of gage, 1150	Limit of gage, 2500	No scuffing	En32 (0–15% C)
Treated(b)	Water	450	1000	Scuffed	En32 (0–15% C)
Treated(b)	Air	760	1675	No scuffing, became hot and extruded	En32 (0–15% C)
Treated(c)	Air	660	1450	Extruded	AISI 1045

(a) Falex scuffing tests at 290 rpm in EN8 (0.4% C) jaws, 90 min running time. (b) Treatment 2, cyanide nitrocarburizing salt bath, with sulfur present as an accelerator. (c) Treatment 1, cyanide nitrocarburizing salt bath age-hardening effect. As expected, the temperature had no effect on the austenitic 302 steel or the HNV3 valve steel.

Surface and core hardness comparisons have been made for several steels based on the standard, 580 °C (1075 °F) nitrocarburizing treatment for 1.5 h and on low-temperature treatments for 1.5, 3, and 6 h at 480 °C (900 °F). The results follow:

- *Martensitic 420 steel (Fig. 8):* Standard temperature processing yields a surface hardness of 1200 HV, the same as after 6 h at the lower temperature. However, core hardness from the higher-temperature treatment drops dramatically down to 400 HV; with the lower temperature, it remains at 650 HV. Shorter treating periods also yield no loss of core hardness.
- *D2 tool steel (Fig. 9):* Results are similar to those of the 420 steel. Surface hardness ranges from 1000 to 1300 HV, with the lower

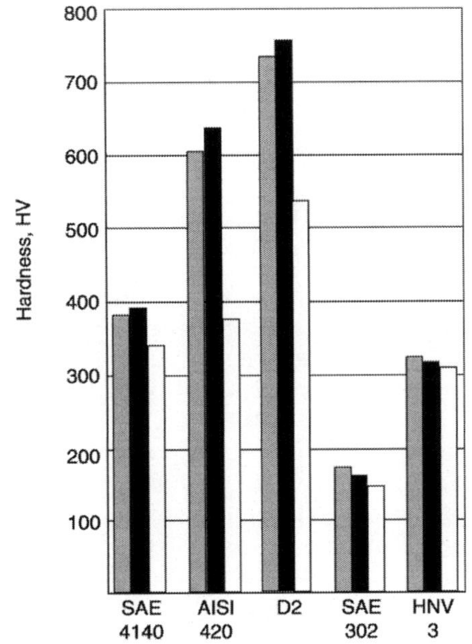

Fig. 7 Core hardness of various steels before and after salt bath nitrocarburizing. The gray columns represent the steel hardness before heating; the black columns represent hardness after the steel was held at 480 °C (900 °F) for 3 h; the white columns represent hardness after the steel was held at 580 °C (1075 °F) for 1.5 h. Source: Ref 3

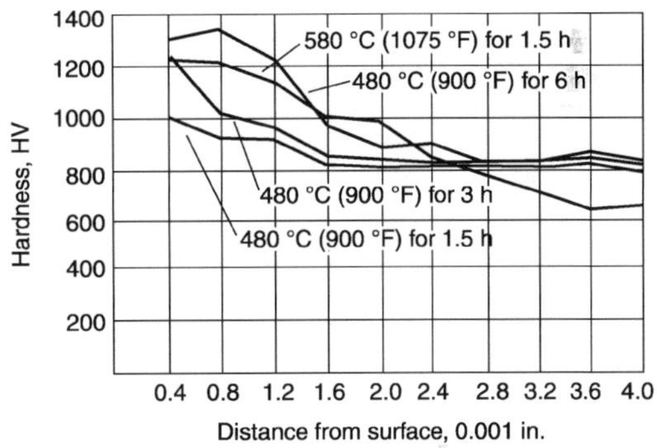

Fig. 9 Hardness profile for D2 steel at various temperatures. With the D2 steel, hardness ranges from 1000 to 1300 HV, with the lower temperature maintaining the core hardness at 800 HV, compared with a drop to 650 HV after treatment at 580 °C (1075 °F). Source: Ref 3

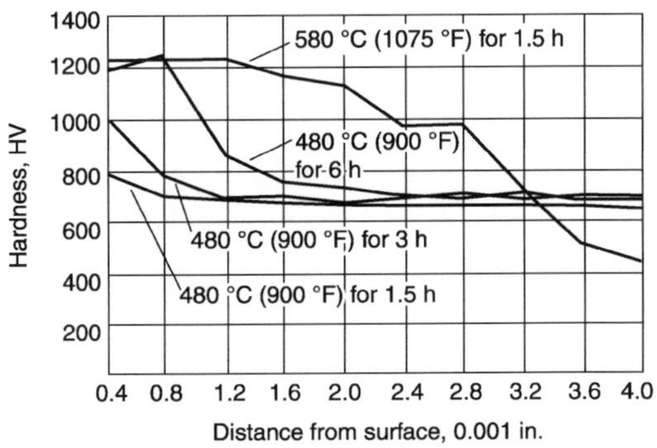

Fig. 8 Hardness profile for Society of Automotive Engineers (SAE) 420 stainless steel at various temperatures. Note that surface hardness of the 420 steel is 1200 HV after standard temperature treatment, while core hardness drops to 400 HV. At the lower temperature, core hardness remains at 650 HV. Source: Ref 3

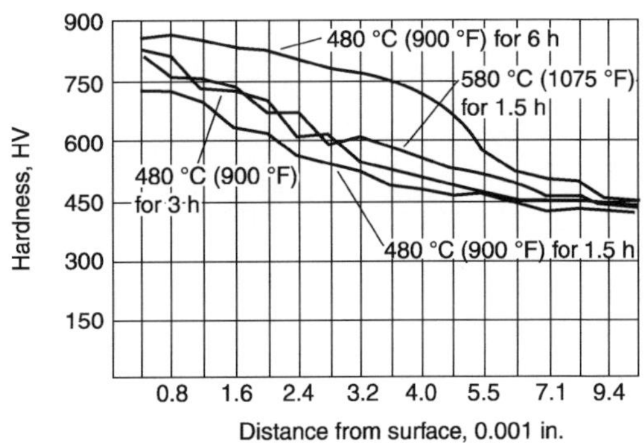

Fig. 10 Hardness profile for SAE 4140 steel. After 6 h at 480 °C (900 °F), hardness from the surface down to a depth of 175 μm is significantly higher than all other options. The lack of nitride-forming elements in the steel allows a deeper penetration of nitrogen at both temperatures than is possible with more highly alloyed steels. Source: Ref 3

temperature maintaining the core hardness at 800 HV, compared with a drop to 650 HV measured after treatment at 580 °C (1075 °F).

- *Medium-carbon alloy steel, Society of Automotive Engineers (SAE) 4140 (Fig. 10):* It is frequently processed at the standard temperature of 580 °C (1075 °F), with excellent results. Although normally no reason compels processing it at a lower temperature, the results are interesting. After 6 h at 480 °C (900 °F), the hardness from the surface down to a depth of 175 μm (0.007 in.) is significantly higher than all other options. The lack of nitride-forming elements in this steel allows a deeper penetration of nitrogen at both temperatures than is possible with higher-alloyed steels. The differences in the compound layers (lightly etched) and diffusion layers (darker etched) in 4140 and D2 steels depend on time and temperature, as shown in Fig. 11. Note the very thin but well-formed compound layers formed in both materials after 3h at 480 °C (900 °F).
- *HNV 3 valve steel (Fig. 12):* Of interest is the exceptional surface hardness of 1300 HV reached after 6h at low temperature, compared with only 1000 HV hardness reached after 580 °C (1075 °F) treatment. However, nitrogen penetration is less, infiltrating only to a depth of 40 μm (0.0016 in.) compared with 60 μm (0.0024 in.) depth at the higher temperature.
- *Austenitic stainless steels, such as 302:* Nitrocarburizing at 580 °C (1075 °F) works well, forming a closed compound layer with an increased hardness down to a depth of 70 μm (0.0028 in.). However, treatment at 480 °C (900 °F) is not appropriate in this case, because it results in a very thin, partially broken compound layer with no diffusion layer. Because resulting core hardness differs very little between the two process temperatures, the 580 °C (1075 °F) temperature is more suitable for these steels.

High-Temperature Salt Bath Nitrocarburizing (Ref 2)

As with low-temperature nitrocarburizing, treating at temperatures higher than conventional process temperatures has broadened the field of applications for salt bath nitrocarburizing. Of the three major considerations in the salt bath (treatment time, bath composition, and temperature), temperature plays the most important role. When the nitrocarburizing temperature is increased from 580 to 630 °C (1075 to 1165 °F), a compound layer with approximately twice the thickness can be obtained at equivalent treatment times (Fig. 13).

In addition to forming a thicker compound layer, a sublayer forms on unalloyed and low-alloy steels. Depending on the cooling method used after the treatment, the sublayer consists of carbon-nitrogen bainite or carbon-nitrogen mar-

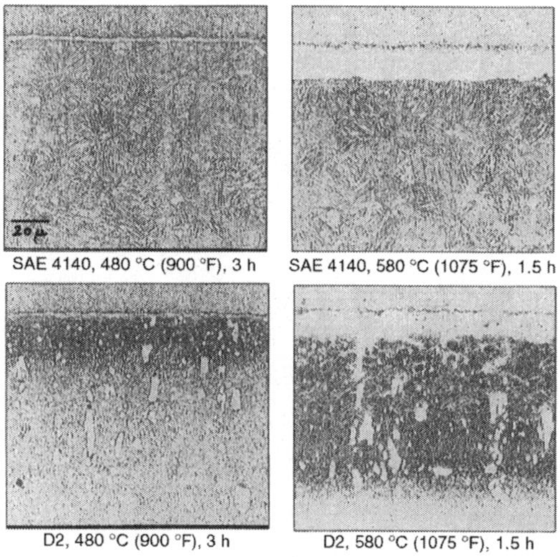

Fig. 11 Effect of time and temperature on compound layer (lightly etched) and diffusion zone (darker etched) formation. Source: Ref 3

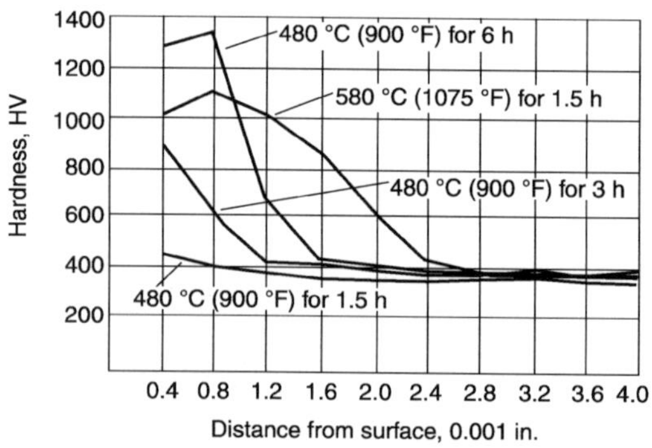

Fig. 12 Hardness profile of HNV3 steel at various temperatures. Hardness of 1300 HV is reached on this valve steel after 6 h at low temperature, compared with 1000 HV after treatment at 580 °C (1075 °F). However, nitrogen penetration is less, with depth of 40 μm versus 60 μm at the higher temperature. Source: Ref 3

tensite, with a high percentage of undercooled austenite. In nitrocarburizing, the transformation to austenite begins at the point of highest nitrogen concentration, which is the phase boundary between the compound layer and diffusion layer.

From this point, the austenitic intermediate layer grows into the diffusion layer during nitrocarburizing. With increasing alloy content, the transformation temperature of the austenite changes. In steels containing over 5% Cr, the transformation temperature is above 650 °C (1200 °F).

High-temperature treatment produces virtually no change in the corrosion resistance of C45 (AISI 1045) and 42CrMo4 (AISI 4140), but significantly improves the corrosion resistance of steels having a higher chromium content. This appears to be attributed to the thicker compound layer. However, it is not possible to improve the corrosion resistance of austenitic steels.

In many cases, fatigue strength and wear resistance can be enhanced by treatment at higher temperatures. This allows achieving the same layer thickness in a shorter treatment time.

A further variant of high-temperature nitrocarburizing is a two-stage process consisting of heating the workpiece to a higher temperature (for example, 630 °C or 1165 °F), then lowering the temperature to 580 °C (1075 °F), which produces good results. The compound-layer thicknesses achievable on C45 at 630 and 580 °C (1165 and 1075 °F) (holding at each temperature for 30 min) are similar to those produced using the standard treatment at 580 °C (1075 °F) for 90 min. However, corrosion resistance is increased to 1000 h using the two-stage process compared with approximately 500 h for conventionally treated material (salt spray test Deutsche Industrie-Normen (DIN) 50 021, or ASTM B 117).

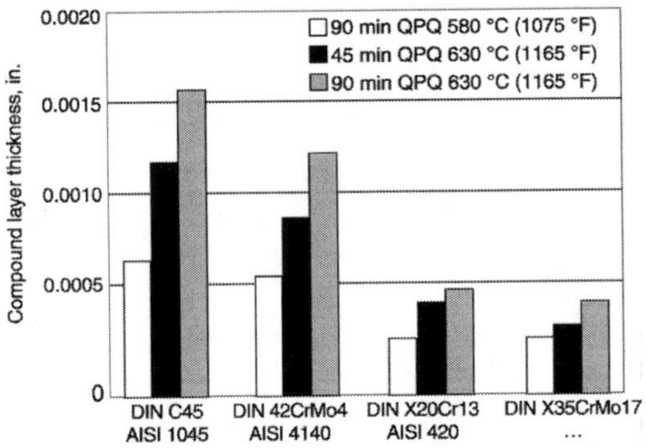

Fig. 13 Influence of temperature and treatment time on the compound layer thickness. QPQ, quench, polish, quench. Source: Ref 2

Salt Bath Nitrocarburizing plus Posttreatment

As an adjunct to conventional salt bath ferritic nitrocarburizing, a mechanical polish and post-salt-bath oxidative treatment are carried

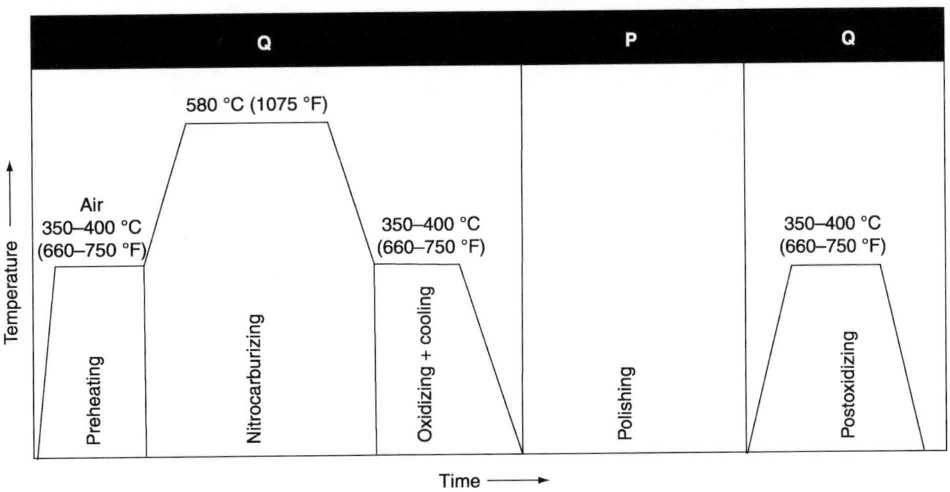

Fig. 14 Schematic of the QPQ nitrocarburizing treatment cycle. Source: Ref 2

out on the nitrocarburized surface. The quench-polish-quench (QPQ) process is based on a sequence of process steps that occur directly following the nitrocarburizing cycle. As shown in Fig. 14, the process begins with the treating cycle of the nitrocarburizing segment, that is, preheat, salt bath nitrocarburize, and salt bath quench, which produces a compound layer of ε iron nitride (Fig. 15). The next step is a mechanical polish of the nitride layer. This may be accomplished by vibratory polishing, lapping, centerless grinding, or by other similar means. Finally, to optimize the corrosion resistance, the component is reimmersed in the salt quench bath for 20 to 30 min, rinsed, and oil dipped.

The level of corrosion protection provided by salt bath nitrocarburizing and the QPQ variant is shown in Fig. 16. The results demonstrate that the QPQ process provides maximum corrosion resistance, as compared with chromium plating, nickel plating, and conventional salt bath nitrocarburizing. Another comparative evaluation of corrosion resistance based on the ASTM B 117 salt spray test is shown in Fig. 17. These results also demonstrate the superior protection provided by the QPQ treatment, even after 336 h exposure to the salt spray testing environment.

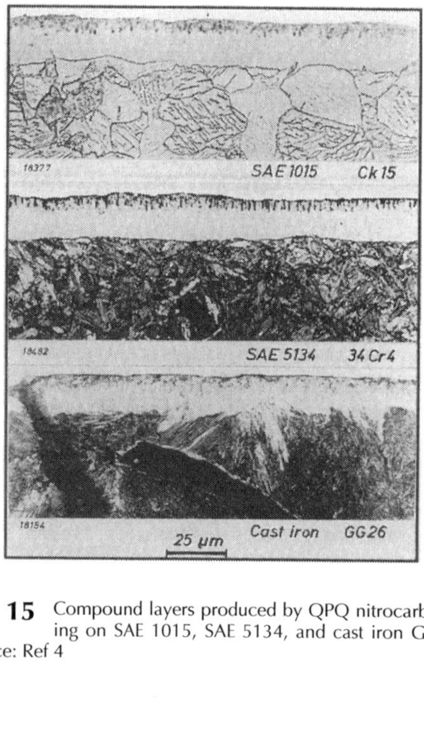

Fig. 15 Compound layers produced by QPQ nitrocarburizing on SAE 1015, SAE 5134, and cast iron GG26. Source: Ref 4

The QPQ treatment also improves wear and fatigue properties of steel parts.

Gas Ferritic Nitrocarburizing

As with the salt bath nitrocarburizing process, gas ferritic nitrocarburizing also involves the introduction of carbon and nitrogen into a steel in order to produce a thin layer of iron carbonitride and nitrides, the compound or white layer, with an underlying diffusion zone con-

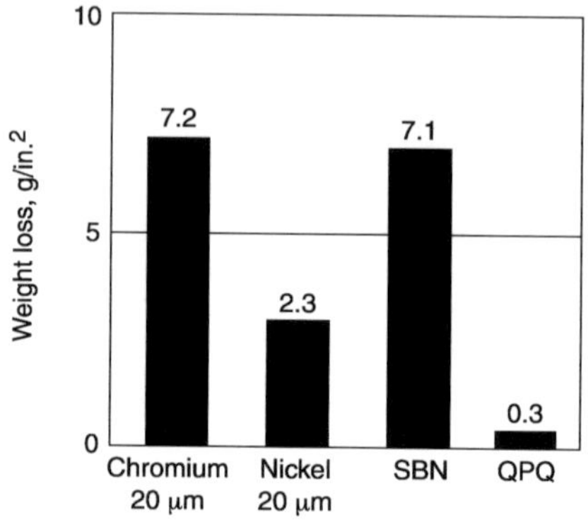

Fig. 16 Comparison of corrosion resistances of various surface treatments based on field immersion tests. Test conditions: full immersion for 24 h in 3% sodium chloride plus 3 g/L hydrogen peroxide. SBN, salt bath nitrocarburizing (no post-treatment). Source: Kolene Corporation

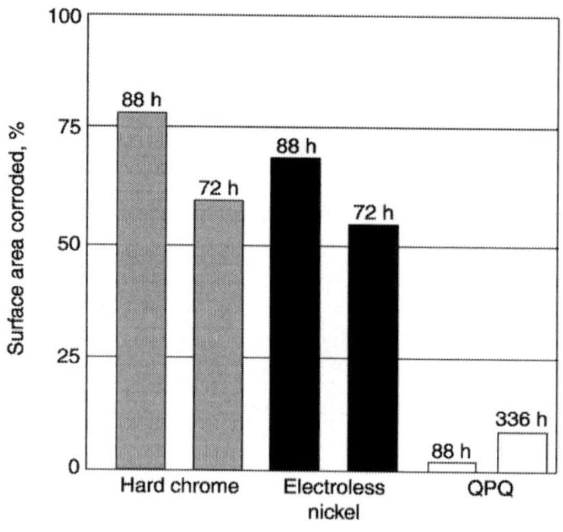

Fig. 17 Corrosion resistance evaluation of surface-treated steel spool shafts used in automotive steering columns based on the ASTM B 117 salt spray test. Source: Kolene Corporation

taining dissolved nitrogen and iron (or alloy) nitrides. The compound layer enhances surface resistance to galling/scuffing and wear, while the diffusion zone increases the fatigue endurance significantly, especially in carbon and low-alloy steel.

The compound-diffusion layer may contain varying amounts of γ′, ε phase, cementite, and various alloy carbides and nitrides. The exact composition is a function of the nitride-forming elements in the steel and the composition of the atmosphere.

Following thorough cleaning (vapor degreasing is adequate for most applications), parts are gas nitrocarburized near 570 °C (1060 °F), a temperature just below the austenite range for the iron-nitrogen system. Treatment times generally range from 1 to 3 h. Although there are a number of proprietary gas mixtures, most contain ammonia (NH_3) and an endothermic gas. Batch furnaces with integral oil quenches are ideally suited for performing gas nitrocarburizing. Figure 18 shows a typical microstructure of a gas nitrocarburized low-carbon steel.

Process Variations (Ref 5)

Initial Developments. In 1961, before the availability of detailed structural and chemical analyses of the compound layer on salt bath nitrocarburized materials, a patent had been applied for by Joseph Lucas (Industries) Ltd. for a type of gas nitrocarburizing (British patent 1,011,580). This treatment produced, on mild steel, a porous layer that was claimed to have good anti-frictional properties. The complete patent, when published, revealed that the gaseous atmosphere consisted of ammonia and hydrocarbon or other carbon-containing gases of unspecified proportions and that the treatment was undertaken in the temperature range of 450 to 590 °C (840 to 1095 °F). At that time, however, no detailed technical information on the property improvements achieved, or of the structures which were responsible, was published. During the 1960s, further research led to consideration of a large range of gas nitrocarburizing processes throughout the world. A wide variety of atmospheres were proposed and, indeed, employed in these processes. These included triethanolamine, ammonia/kerosene, and isopropanol/water/urea/ammonia. However, it was only in the early 1970s that gaseous nitrocarburizing received serious industrial attention with the introduction of a variety of gaseous techniques.

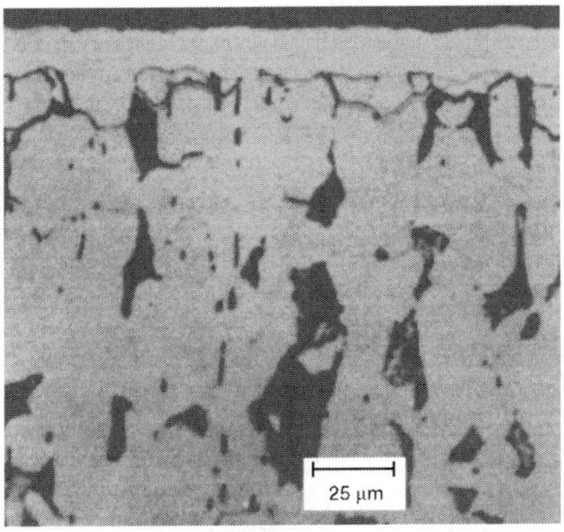

Fig. 18 Mild steel after 3 h gaseous nitrocarburizing in an ammonia/endothermic gas mixture at 570 °C (1060 °F) followed by oil quenching

The Nitemper process is usually carried out in sealed quench furnaces and uses an inert atmosphere consisting of 50% NH_3 and 50% endogas. The treatment temperature is 570 °C (1060 °F), and treatment times usually between 1 and 3 h are used, after which the components are either quenched into oil or cooled under recirculated protective gas. By 1975, the Nitemper process had been in use for several years, and furnaces performing the treatment were in operation in Germany, Sweden, the United States, Japan, and the United Kingdom for improving the scuffing and fatigue resistance of ferrous engineering components. The treatment is now used extensively throughout the world, and a two-stage Nitemper process has been developed. This involves the use of an atmosphere with a high carbon dioxide (CO_2) level in the initial stage to promote rapid compound layer formation. The influence of controlled additions of carbon dioxide to ammonia-based nitrocarburizing atmospheres under industrial conditions has been investigated for a wide range of alloy steels. It has been demonstrated that the proportion of the ε phase in the compound layer increased with increasing carbon dioxide content, that is, lower carbon activities, and that the ε phase more readily formed on alloy steels than on pure iron or plain carbon steels (Fig. 19). In the second stage of the modified Nitemper process, an atmosphere with a high carbon monoxide content is employed to increase the carbon content of the compound layer for enhanced wear resistance. In essence, therefore the process involves a combination of the Nitroc process, which uses unpurified exothermic gas as the carburizing medium (see subsequent description), and the Nitemper technology. A similar duplex treatment called Deganit has also been developed.

The Alnat-N process is a patented approach to nitrocarburizing whereby nitrous oxide is incorporated in the atmosphere to enhance, through the indirect presence of oxygen, the rate of formation of the compound layer. A further feature of the Alnat-N process is that the addition of a carburizing gas to the basic ammonia/nitrous oxide/nitrogen mixture is claimed to be unnecessary. Thus, the incorporation of carbon into the compound layer must be via diffusion from the matrix materials.

Control of Gaseous Nitrocarburizing Atmospheres. A possible limitation on the gas nitrocarburizing processes developed in the mid-1970s was that optimal processing conditions for all classes of material, including cast irons, tool steels, and stainless steels, could not be assured. A further and perhaps more serious limitation was that reproducibility could be impaired with variable loads and from furnace to furnace. These difficulties were, in part, overcome through the use of infrared monitoring and control systems. However, gas analysis of atmospheres containing both ammonia and carbon dioxide can be problematic, especially when high ammonia contents and high dew-points are concerned. If water is condensed in the sample gas piping, high amounts of ammonia dissolve, and the resulting strong solution can dissolve large amounts of carbon dioxide. If this solution becomes supersaturated, a mixture of ammonium carbonate and bicarbonate precipitate out in the form of a white powder. Experience has shown that when such precipitation has occurred, further measurement of ammonia and carbon dioxide is in error, and there is a distinct likelihood that the pipeline of the measurement system will be blocked. This problem can be largely overcome by suitable heating of the measurement instrumentation and the gas sampling pipeline.

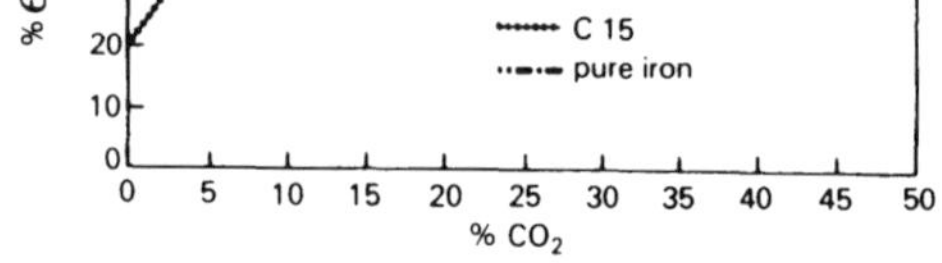

Fig. 19 The influence of CO_2 addition to ammonia on the structure of the compound layer, formed by nitrocarburizing at 580 °C (1075 °F) on pure iron, plain carbon steels, and low-alloy steels

Because of the limitations of the infrared gas analysis approach to the control of gas nitrocarburizing atmospheres, attention has been focused recently on the development of solid electrolyte gas sensors for the measurement and control of the nitrogen and oxygen potentials of nitriding and nitrocarburizing atmospheres. Such instruments are, in principle, similar to those widely used for carbon potential control of carburizing gas atmospheres.

Black Nitrocarburizing. Postnitrocarburizing oxidation treatments have been used on a commercial basis since 1976 to enhance the aesthetic properties of gas nitrocarburized components for the hydraulics industry. However, in 1982, Dawes and Tranter showed how such black nitrocarburizing treatments, including the Nitrotec process, could be used for the combined enhanced fatigue, wear, and corrosion resistance of mild steels (Ref 26). They showed that this could be achieved by specifically designing a range of cost-effective, aesthetically pleasing black oxidized electrical components for use in automobile manufacture.

Particular success has been achieved with vacuum degassed ultralow-carbon deep-drawing steels that have been stabilized with niobium and/or titanium additions. Such steels facilitate many complex thin-sectioned components to be manufactured by single-stage press operations. These steels have very low yield strengths of approximately 155 MPa (23 ksi), with elongation values of approximately 45%. To achieve optimal engineering properties in the final component, nitrocarburizing can be used. The influence of quenching temperature on the yield strength of these nitrocarburized special steels is shown in Fig. 20. It is clear that low-temperature quenching has resulted in nitride precipitation and loss of strength, and that if a high strength is to be achieved, a quenching temperature not less than 550 °C (1020 °F) is necessary. This is also the condition for optimal enhancement of the fatigue strength. Another essential feature of this treatment is that distortion of thin-sectioned material can be kept to a minimum by controlled quenching into an oil/water emulsion at a temperature of 70 to

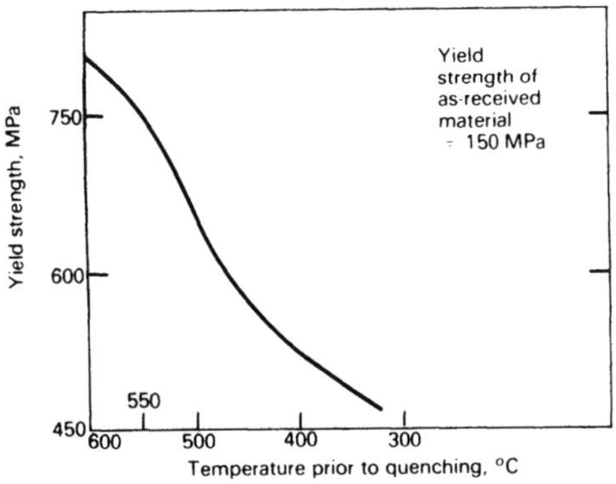

Fig. 20 The influence of quenching temperature on the yield strength of nitrocarburized deep-drawing steels

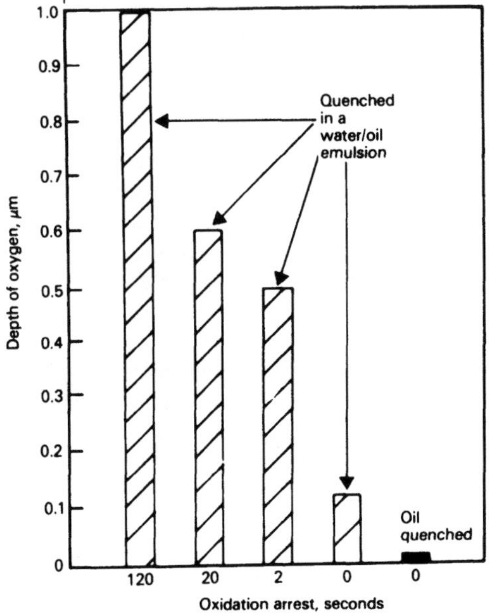

Fig. 21 Influence of depth of oxygen on surface coloration and corresponding oxidation arrest time for various quench media

80 °C (160 to 175 °F), and that the quench time involved is sufficient to produce an aesthetically pleasing black oxide film of Fe_3O_4, which needs to be less than 1.0 μm in thickness if exfoliation is to be avoided (Fig. 21).

The flash oxidation parameters of the basic Nitrotec process are designed to produce an oxide structure capable of both conferring a degree of corrosion resistance and acting as a carrier for an organic sealant. Investigations into the composition of organic sealants has resulted in the development of specific formulations that are based on either hydrocarbon-solvent-borne mixtures of metal soaps produced from rosin acids and oxidized petrolatums, or water-based mixtures of emulsified microcrystalline and synthetic hydrocarbon waxes with corrosion inhibitors.

The relative contributions of postnitrocarburizing oxidation and organic sealing to the overall corrosion resistance resulting from the Nitrotec process is shown in Fig. 22. It can be seen that the degreased nitrocarburized surface itself imparts little inherent corrosion resistance. The Nitrotec process, in conjunction with organic sealants, is widely used to treat a variety of automotive components, some of which were formerly zinc coated for corrosion resistance. Figure 23 compares salt spray test results for nitrocarburized/organically sealed low-carbon steels and zinc-coated steels.

An alternative black nitrocarburizing finish is the Ashland Nitro Black, which is a patented

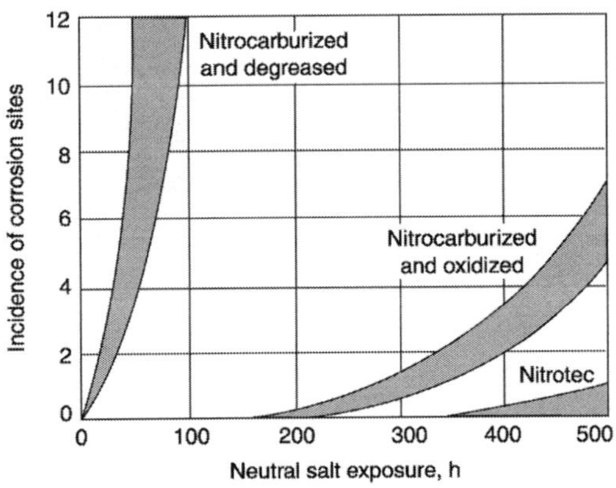

Fig. 22 Relationship between treatment sequence and salt corrosion resistance (ASTM B 117)

process using fluidized bed technology. The atmosphere used for the nitrocarburizing stage comprises a mixture of ammonia, natural gas, and nitrogen. After nitrocarburizing, the fluidized bed is purged with nitrogen for 2 min prior to the oxidation step, during which steam and air are injected via an integral coil system to impart a thin Fe_2O_3 layer on treated components. Fluidized bed quenching is then followed by coating with a proprietary polymeric emulsion sealant.

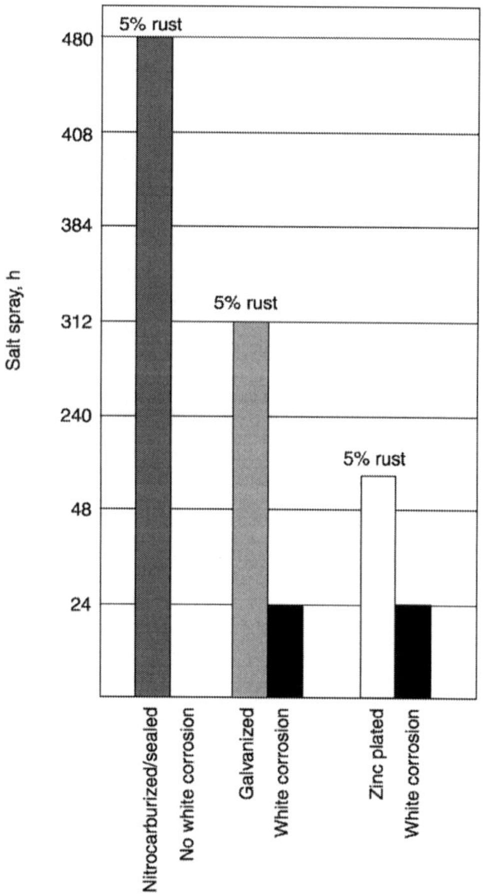

Fig. 23 Salt spray test (ASTM B 117) results for low-carbon steels that were gas nitrocarburized and organically sealed versus zinc-coated steel. The effectiveness of the nitrocarburizing/organic sealing process affords the use of less-expensive materials for corrosion-resistant applications (i.e., stainless steels can be replaced with low-carbon steel). Source: Erie Steel Treating, Inc.

Plasma Nitrocarburizing

Plasma nitrocarburizing is, in essence, a variant of the plasma (ion) nitriding method described in Chapter 6, "Nitriding." Advantages associated with plasma heat treatment technology include:

- No toxic fumes or waste produced
- No risks of explosion
- No significant dirt, noise, or heat pollution
- Reduced processing times
- Reduced energy consumption
- Reduced treatment gas consumption

Plasma nitrocarburizing is typically carried out at 570 °C (1060 °F) to produce a compound layer of >5 μm and a surface hardness of ≥350 HV. Plasma atmospheres consist of mixtures of hydrogen, nitrogen, and a carbon-bearing gas, such as methane, or carbon dioxide.

Physical Metallurgy of Plasma Nitrocarburizing. In respect to the tribological properties of nitrocarburized steels, evidence from gaseous and salt bath nitrocarburizing research indicates that the monophase ε structure is strongly preferred. However, plasma nitrocarburizing still faces the problem of controlling the quality and character of the compound layer structure to achieve the monophase ε carbonitride on a regular basis. Accordingly, with plasma nitrocarburizing, the compound layer usually consists of ε and γ′ phases for low-carbon-level atmospheres. Equilibrium thermodynamic considerations would indicate that increasing the carbon level in the atmosphere should produce the monophase ε structure. However, under the nonequilibrium thermodynamic conditions prevailing in the glow-discharge plasma, an increase in the carbon level does not automatically produce a 100% ε structure, and yet cementite does appear above a certain limit of the carbon level. Laboratory studies using methane as the source of carbon in the gaseous plasma have shown that some stabilization of the ε phase is possible, but above a certain limit (depending on the substrate materials), the cementite always appears, and soot formation is difficult to prevent. The use of controlled additions of oxygen-bearing gases to reduce the activity of carbon has shown some promise in stabilizing the ε phase, and the kinetics of compound layer growth are increased.

Laboratory experiments using 90% N_2/H_2 atmospheres with controlled additions of carbon dioxide (up to 2.5%) have been carried out at 570 °C (1060 °F) for 2 h. It was found that:

- With pure iron, increasing the carbon dioxide stabilized the ε phase, and an essentially monophase ε structure was formed at 1% CO_2 level (Fig. 24a and b). A further increase in carbon dioxide level to 2% led to the formation of surface oxides.
- With plain carbon steel, increasing the carbon dioxide level again stabilized the ε phase, but a mixture of the ε and γ′ phases was invariably present (Fig. 25a and b).
- With a low-alloy chromium-bearing steel, EN40B (0.20 to 0.28% C, 0.10 to 0.35% Si, 0.45 to 0.70% Mn, 3.0 to 3.5% Cr, 0.45 to 0.65% Mo), the γ′ phase was suppressed by even 0.5% CO_2, but cementite compounds were invariably formed (Fig. 26a and b).

These controlled laboratory experiments clearly illustrate the lack of tolerance of the

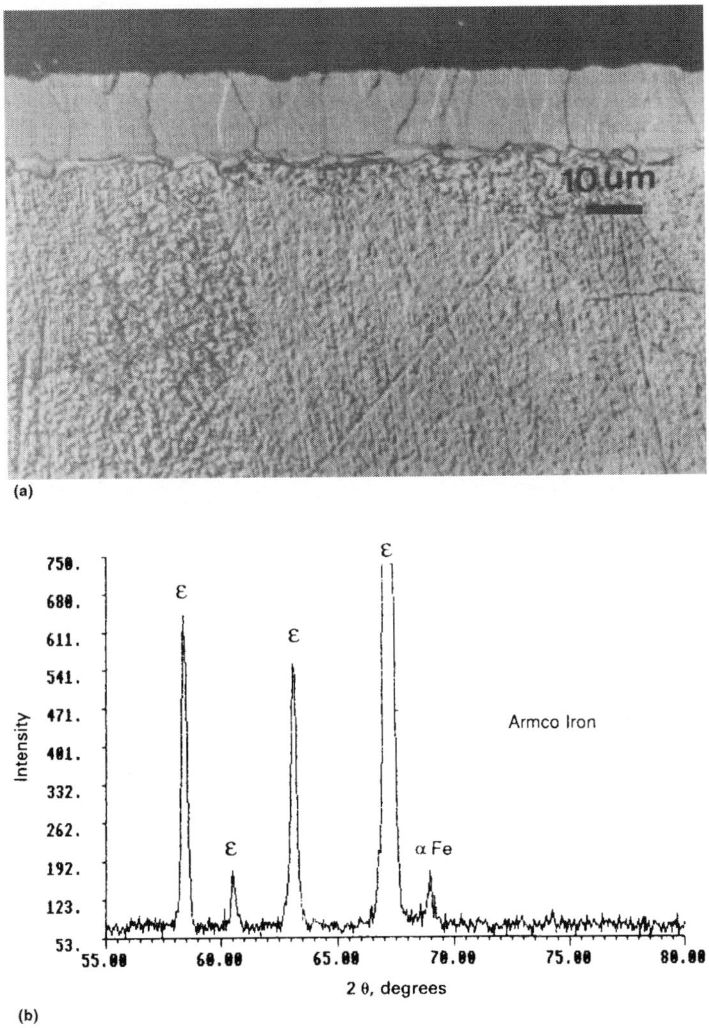

Fig. 24 (a) Microstructure of Armco iron plasma nitrocarburized at 570 °C (1060 °F) for 3 h at a gas pressure of 3.5 mbar. Gas mixture: 90 vol% N_2, 1 vol% CO_2, 9 vol% H_2. Etched in 1 mL mix of hydrochloric acid (HCl) and ethanol (1 part concentrated HCl + 10 parts ethanol) plus 99 mL 5% nital. (b) X-ray diffraction pattern of the compound layer of the sample

plasma nitrocarburizing process to minor variations in atmosphere condition.

Applications. The automotive industry is one of the large market areas for plasma nitrocarburizing. Plasma equipment can be easily integrated into high-production manufacturing lines. One typical application is the plasma nitrocarburizing of automotive seat slider rails. Loads of up to 3000 parts can be nitrocarburized automatically. Following nitrocarburizing, the workpieces are allowed to cool under controlled vacuum conditions.

Another growing area for plasma nitrocarburizing is in the powder metallurgy industry. This

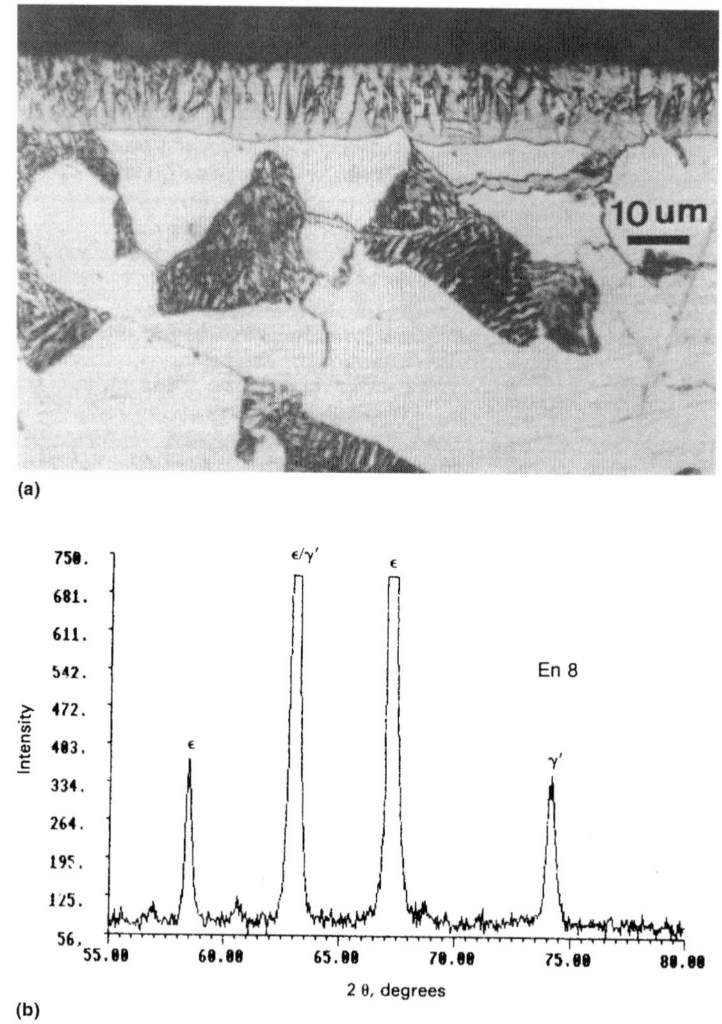

Fig. 25 (a) Microstructure of a plasma nitrocarburized EN8 steel sample with (b) the corresponding x-ray diffraction pattern. See Fig. 24 for processing details.

process is currently used in production of such parts as synchronizer hubs and cam lobes. The use of a plasma to deliver the nitrogen and carbon ions to the surface of the part allows for a more uniform control of surface concentrations and diffusion of the nitriding elements. It minimizes the nitridation of internal pore surfaces, thereby reducing the volume expansion that normally occurs during gas nitrocarburization.

The plasma nitrocarburizing process can be applied to most iron blends and prealloys with uniform formation of the ε iron-nitride hard (60 HRC) phase at the surface of parts with sinter densities exceeding 6.9 g/cm^3. Below this density, porosity variations in the part can lead to

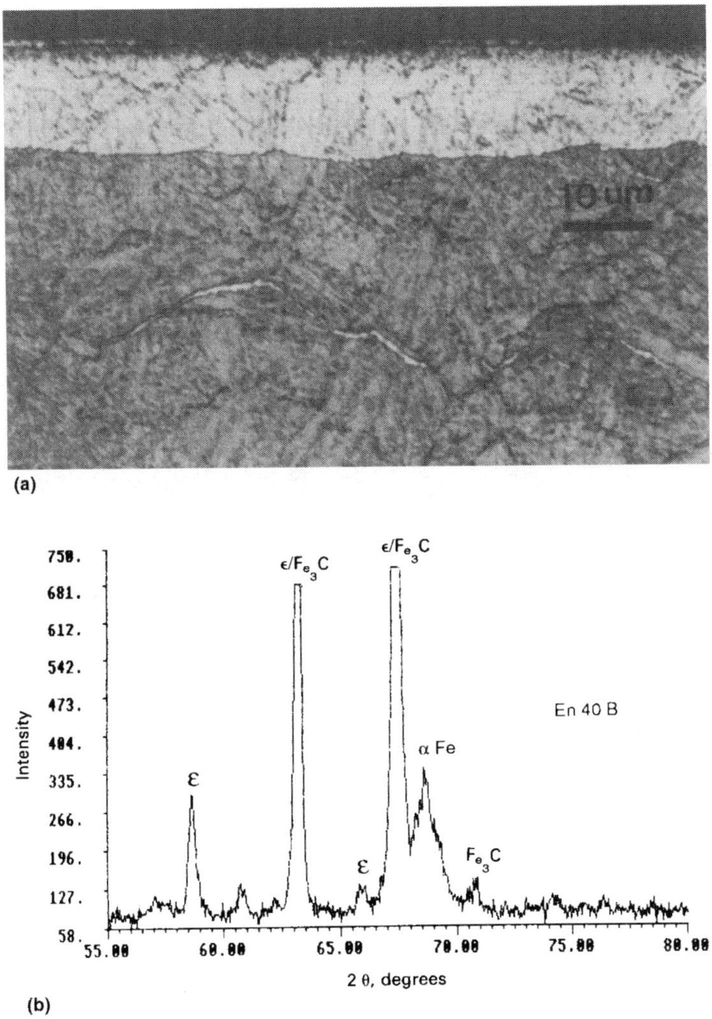

Fig. 26 (a) Microstructure of a plasma nitrocarburized EN40B steel sample with (b) the corresponding x-ray diffraction pattern. See Fig. 24 for processing details.

nonuniform dimensional changes occurring on sintered iron transverse rupture bars that have been nitrocarburized by various methods (Fig. 27).

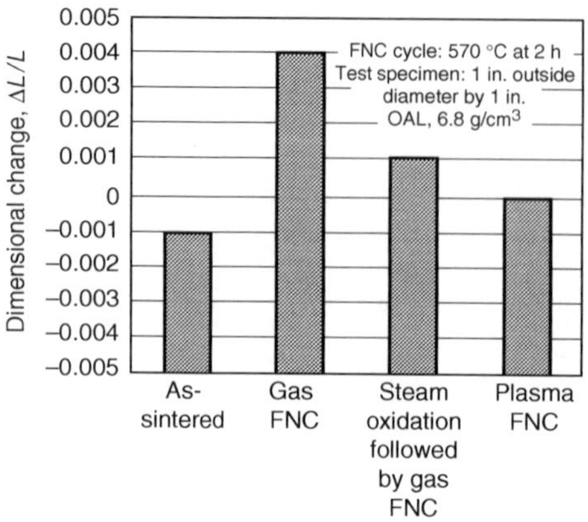

Fig. 27 Dimensional change in sintered iron after various ferritic nitrocarburizing (FNC) treatments.

REFERENCES

1. D.H. Herring, Comparing Carbonitriding and Nitrocarburizing, *Heat Treat. Prog.*, May 2002, p 17–19
2. R. Willing and C. Faulkner, New Ways to Use Salt-Bath Nitrocarburizing, *Ind. Heat.*, April 2001, p 33–36
3. S. Alwart and U. Baudis, Low-Temperature Nitrocarburizing, *Adv. Mater. Process.*, Sept 1998, p 41–43
4. G. Wahl, Nitrocarburizing for Wear, Corrosion, and Fatigue, *Adv. Mater. Process.*, April 1996, p 37–38
5. T. Bell, Gaseous and Plasma Nitrocarburizing, *Heat Treating*, Vol 4, *ASM Handbook*, ASM International, 1991, p 425–436.
6. C. Dawes and D.F. Tranter, Nitrotec Surface Treatment—Its Development and Application in the Design and Manufacture of Automobile Components, *Heat Treat. Met.*, Vol 4, 1982, p 85–90

CHAPTER 8

Boriding

BORIDING, also commonly referred to as boronizing, is a thermochemical surface hardening process that involves diffusion of boron into a well-cleaned base metal (steel) surface at high temperature. As a rule, the boriding process takes place at temperatures between approximately 850 and 950 °C (1560 and 1740 °F). The resulting metallic boride provides high hardness and wear resistance, high-temperature resistance, and corrosion resistance.

Boriding fills a gap between conventional surface treatments and the more exotic chemical and physical vapor deposition techniques. In a number of applications, boriding has replaced such processes as carburizing, nitriding, and nitrocarburizing (Ref 1). It has also replaced hard chromium plating in some cases, while achieving similar service life improvements.

Boron can be uniformly applied to irregular surfaces and can be applied to specific areas of a surface via paste boriding. It is also suitable for high-volume production applications, as first demonstrated by the European automotive industry (Ref 1).

This chapter describes:

- The advantages and limitations of the boriding process, with emphasis placed on pack boriding, the most commercially important boriding process. Pack boriding is very similar to pack carburizing (see Chapter 4, "Pack and Liquid Carburizing," in this book) except that a granulate rich in boron is used instead of a carbon-rich granulate as in pack carburizing.
- Boride layer characteristics
- Steel selection and effects of alloying elements
- Boriding processes, including multicomponent boriding
- Properties of borided steels
- Applications for pack boriding

Additional information on the boriding process may be found in the article "Boriding (Boronizing)" in *Heat Treating,* Volume 4 of the *ASM Handbook.*

Advantages and Disadvantages of Boriding

Advantages. Boride layers possess a number of characteristic features with special advantages over conventional case-hardened layers. One basic advantage is that iron boride layers have extremely high hardness values (between 1600 and 2000 HV). The typical surface hardness values of borided steels compared with other treatments and other hard materials are listed in Table 1. This clearly illustrates that the hardness of boride layers produced on carbon steels is much greater than that produced by any other conventional surface hardening treatments: it exceeds that of the hardened tool steel, hard chromium electroplate, and is equivalent to that of tungsten carbide.

The combination of a high surface hardness and a low surface coefficient of friction of the borided layer also makes a significant contribution in combating the main wear mechanisms: adhesion, tribooxidation, abrasion, and surface fatigue. This fact has enabled the mold makers to substitute easier-to-machine steels for the base metal and to still obtain wear resistance and antigalling properties superior to those of the original material.

Other advantages of boriding include:

- Hardness of the boride layer can be retained at higher temperatures than, for example, those for nitrided cases.
- A wide variety of steels, including through-hardenable steels, are compatible with the processes.
- Boriding can considerably enhance the corrosion-erosion resistance of ferrous materials in nonoxidizing dilute acids and alkali media,

and is increasingly used to this advantage in many industrial applications.
- Borided surfaces have moderate oxidation resistance (up to 850 °C, or 1550 °F) and are quite resistant to attack by molten metals.
- Borided parts have an increased fatigue life and service performance under oxidizing and corrosive environments.

Disadvantages of boriding treatments are:

- The techniques are inflexible and rather labor intensive, making the process less cost effective than other thermochemical surface hardening treatments such as gas carburizing and plasma nitriding. Both gas carburizing and plasma nitriding have the advantage over boriding, because those two processes are flexible systems, offer reduced operating and maintenance costs, require shorter processing times, and are relatively easy to operate.
- The growth (that is, the increase in volume) resulting from boriding is 5 to 25% of the layer thickness (for example, a 25 μm, or 1000 μin., layer would have a growth of 1.25 to 6.25 μm, or 50 to 250 μin.); its magnitude depends on the base material composition but remains consistent for a given combination of material and treatment cycle. However, it can be predicted for a given part geometry and boriding treatment. For treatment of precision parts, where little stock removal is permitted, an allowance of ~20 to 25% dimensional increase of the final boride layer thickness must be provided.
- Partial removal of the boride layer for closer tolerance requirements is made possible only by a subsequent diamond lapping, because conventional grinding causes fracture of the layer. Thus, precise boriding is mostly practiced for components with a large cross-sectional area.

Table 1 Typical surface hardness of borided steels compared with other treatments and hard materials

Material	Microhardness, kg/mm^2 or HV
Borided mild steel	1600
Borided AISI H13 die steel	1800
Borided AISI A2 steel	1900
Quenched steel	900
Hardened and tempered H13 die steel	540–600
Hardened and tempered A2 die steel	630–700
High-speed steel M42	900–910
Nitrided steels	650–1700
Carburized low-alloy steels	650–950
Hard chromium plating	1000–1200
Cemented carbides, WC + Co	1160–1820 (30 kg)
$Al_2O_3 + ZrO_2$ ceramic	1483 (30 kg)
$Al_2O_3 + TiC + ZrO_2$ ceramic	1738 (30 kg)
Sialon ceramic	1569 (30 kg)
TiN	2000
TiC	3500
SiC	4000
B_4C	5000
Diamond	>10,000

Boride Layer Characteristics

The boriding of steel alloys results in the formation of either a single-phase or double-phase layer of boride with definite compositions. The single-phase boride layer consists of Fe_2B, while the double-phase layer consists of an outer phase of FeB and an inner phase of Fe_2B. The FeB phase is brittle and forms a surface that is under high tensile stress. The Fe_2B phase is preferred because it is less brittle and forms a surface with a high compressive stress, the preferred stress state for a high-hardness, low-ductility case. Although small amounts of FeB are present in most boride layers, they are not detrimental if they are not continuous. However, a continuous layer of FeB can lead to crack formation at the FeB/Fe_2B interface of a double-phase layer. These cracks can lead to separation or spalling of a double-phase layer when a mechanical strain is applied or when the component is undergoing a thermal and/or mechanical shock (Fig. 1). Fortunately, continuous lay-

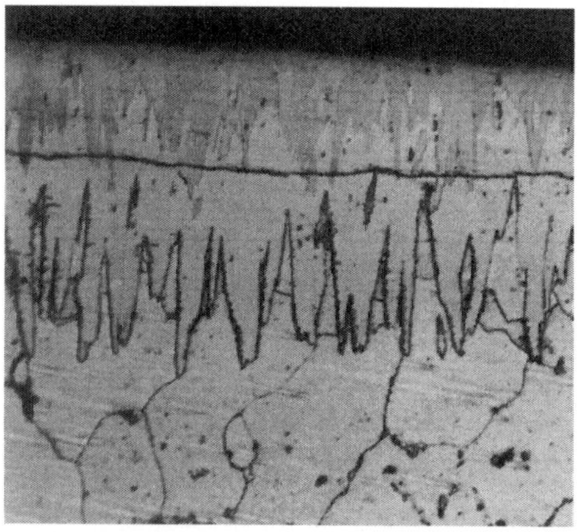

Fig. 1 Separation of two-phase boride layer on a carbon steel (borided at 900 °C, or 1650 °F, for 4 h) caused by grinding with a cutting-off disk. 200×.

ers of FeB can be minimized by diffusion annealing after boride formation. Also, boriding powders that minimize formation of FeB have been developed and are readily available.

As described in the following sections of this chapter, the morphology, growth, and phase composition of the boride layer can be influenced by the alloying elements in the base material. The preferred morphology is a "sawtooth" or "serrated" boride layer structure most easily obtained with carbon or low-alloy steels (Fig. 2). The microhardness of the borided layer also depends strongly on the composition and structure of the boride layer and the composition of the base material.

Steel Selection

Recommended Steels. Boriding can be carried out on plain carbon steels, through-hardening low-alloy steels, tool steels, stainless steels, and sintered steels. Carburized steels may also be borided ("carboborided"), then rehardened by post-boriding heat treatment. Most tool steels can be rehardened after boriding, provided that the austenitizing temperature is below 1100 °C (2000 °F). Above this temperature, the iron boride eutectic could melt. Some tool steels, such as high-speed steels, can be underhardened to develop sufficient core properties.

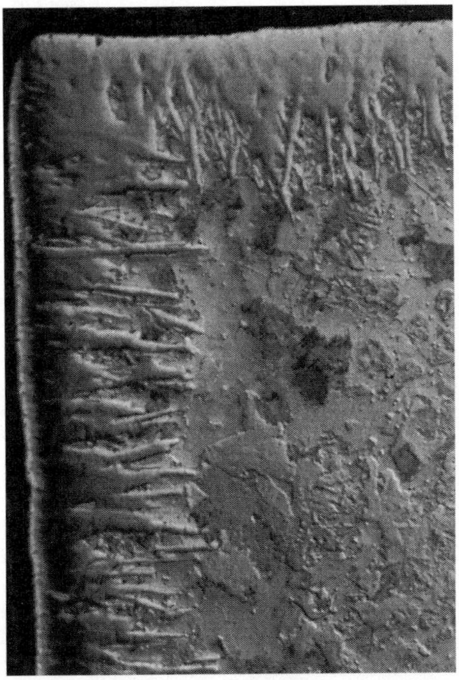

Fig. 2 Boride layer "sawtooth" structure formed on a plain carbon steel that contains both FeB and Fe_2B phases. The Fe_2B phase is preferred because it forms a surface under compressive stress. 30×. Source: Ref 1

Steels Not Recommended. Water-hardening steel grades are not borided because of the susceptibility of the boride layer to thermal shock. Resulfurized and leaded steels should not be used because they have a tendency toward case spalling and case cracking. Nitrided steels cannot be borided because nitrogen retards the diffusion of boron in steel, making nitrided steels sensitive to cracking.

Effects of Alloying Elements

The mechanical properties of the borided alloys depend strongly on the composition and structure of the boride layers. The characteristic "sawtooth" configuration of the boride layer is dominant with pure iron, unalloyed low-carbon steels, and low-alloy steels. As the alloying element and/or carbon content of the substrate steel is increased, the development of a jagged boride/substrate interface is suppressed, and for high-alloy steels, a smooth interface is formed (Fig. 3). Alloying elements mainly retard the boride layer thickness (or growth) caused by restricted diffusion of boron into the steel because of the formation of a diffusion barrier. Figure 4 shows the effect of alloying additions in steel on boride layer thickness. The effect of increasing alloying content and treatment time on boride layer thickness is shown in Fig. 5.

Carbon does not dissolve significantly in the boride layer and does not diffuse through the boride layer. During boriding, carbon is driven (or diffused away) from the boride layer to the matrix and forms, together with boron, borocementite $Fe_3(B,C)$ [or more appropriately, $Fe_3(B_{0.67}C_{0.33})$ in the case of Fe-0.08% C steel] as a separate layer between Fe_2B and the matrix.

Silicon and Aluminum. Like carbon, silicon and aluminum are not soluble in the boride layer, and these elements are pushed from the surface by boron and are displaced ahead of the boride layer into the substrate, forming iron silicoborides—$FeSi_{0.4}B_{0.6}$ and Fe_5SiB_2—underneath the Fe_2B layer. Steels containing high contents of these ferrite-forming elements should not be used for boriding because they reduce the wear resistance of the normal boride layer; they produce a substantially softer ferrite

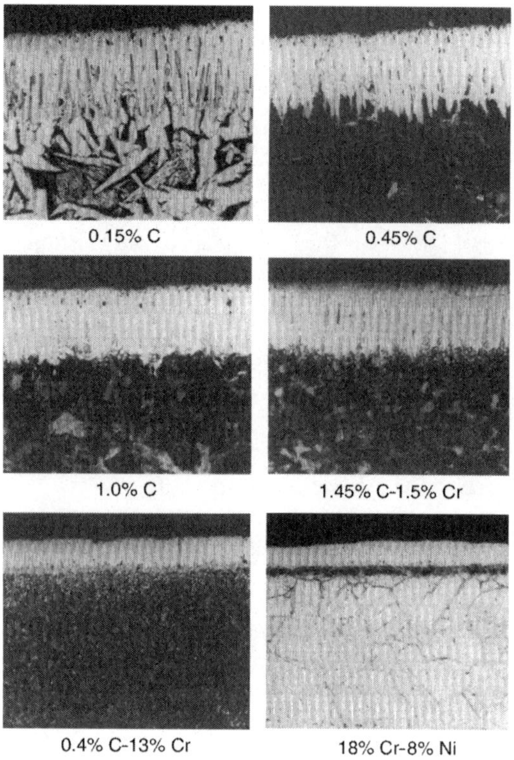

Fig. 3 Effect of steel composition on the morphology and thickness of the boride layer.

zone beneath the boride layer than that of the core. At higher surface pressure, this type of layer buildup results in the so-called "egg shell" effect, that is, at greater thicknesses, an extremely hard and brittle boride layer penetrates into the softer intermediate layer and is consequently destroyed.

Nickel. A reduction in the degree of both interlocking tooth structure and boride depth can occur with high-nickel-containing steels. Nickel has been found to concentrate below the boride layer; it enters the Fe_2B layer and in some instances promotes the precipitation of Ni_3B from the FeB layer. It also segregates strongly to the surface from the underlying zone corresponding to the Fe_2B layer. This is quite pronounced in austenitic stainless steels.

Chromium considerably modifies the structure and properties of iron borides. As the chromium content in the base material increases, the following effects are observed: formation of boron-rich reaction products, decrease in boride depth, and flattening or smoothing of the coating/substrate interface. A reduction of boride thickness has also been noticed in ternary Fe-12Cr-C steels with increasing carbon content.

Tungsten, molybdenum, and vanadium also reduce the boride layer thickness (Fig. 5) and flatten out the tooth-shaped morphology in carbon steel.

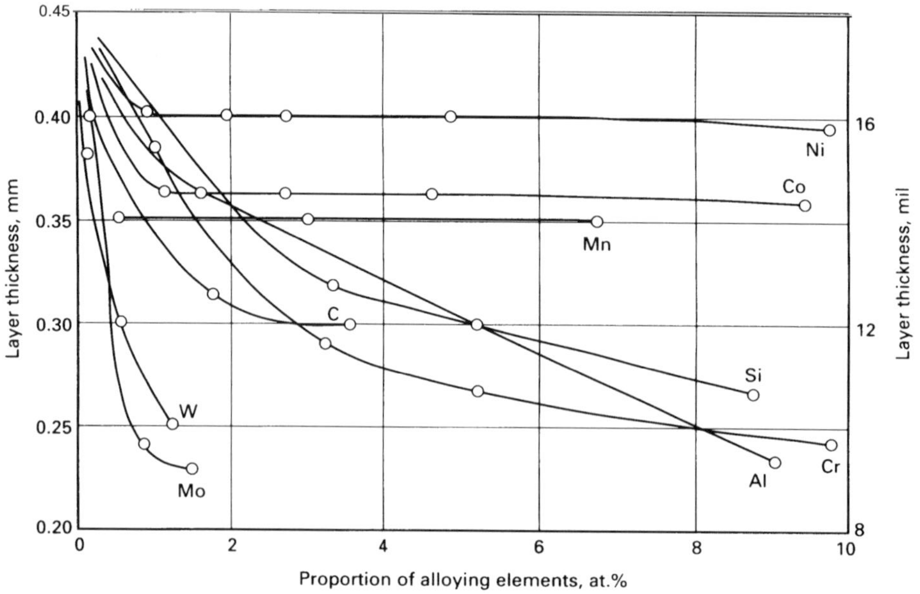

Fig. 4 Effect of alloying elements in steel on boride layer thickness

Boriding Processes

There are a variety of methods for diffusing boron into a steel surface. These include:

- Pack boriding, in which the boronaceous medium is a solid powder
- Paste boriding, in which the boronaceous medium is a boron-rich, water-based paste that is applied by dipping, brushing, or spraying
- Liquid boriding, in which the boronaceous medium is a salt bath
- Gas boriding, in which the boronaceous medium is a boron-rich gas, such as a (B_2H_6)-H_2 mixture
- Plasma boriding, which also uses boron-rich gases but is carried out at lower temperatures than gas boriding
- Fluidized bed boriding, which uses special boriding powders in conjunction with an oxygen-free gas such as a N_2-H_2 mixture

Of these various methods, only pack and paste boriding have reached commercial success, although work continues to be carried out on developing plasma boriding. Because of unsolved problems and serious technical deficiencies (e.g., toxicity problems), gas- and liquid-phase boriding have not become state-of-the-art and will not be discussed further in this section.

Pack Boriding (Ref 2, 3)

As stated earlier, pack boriding is the most common boriding method. With this process, parts are immersed in the boriding agent (powder), then placed in a sealed heat-resistant steel container. Parts are separated from each other with at least 10 mm (0.4 in.) of boriding agent, and covered with a layer of the material approximately 50 to 100 mm (2 to 4 in.) deep. This ensures uniform boriding and guarantees that both the formation and microstructure of the boride layer will be influenced only by the activity of the boriding agent, the treatment temperature, and the material being treated.

During subsequent furnace heating at 900 to 1000 °C (1650 to 1830 °F), boron diffuses into the metal and forms the boride layer. After a sufficient time at the boriding temperature, the box is removed from the furnace and allowed to cool at room temperature.

Some heat treating companies specializing in the boriding process suggest that to avoid com-

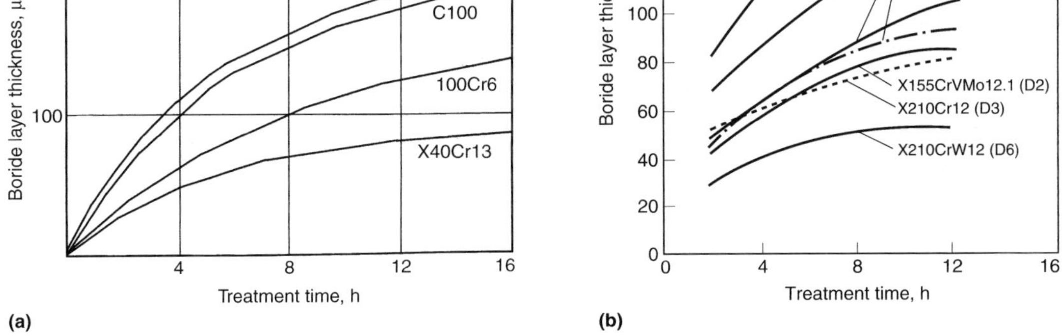

Fig. 5 Two graphs demonstrating how increasing levels of alloying elements such as chromium, vanadium, tungsten, molybdenum, and/or carbon will restrict the growth of the borided case and also reduce the degree of serration. (a) Boride layer thickness as a function of time for various steels. C45 and C100 are carbon steels approximately equivalent to AISI 1045 (0.43 to 0.50 C) and 1095 (0.90 to 1.03 C), while 100Cr6 is equivalent to the through-hardening low-alloy bearing steel AISI 52100 (0.98 to 1.10 C, 1.3 to 1.6 Cr). X40Cr13 is a heat resistant chromium stainless steel (0.35 to 0.42 C, 12.5 to 14.5 Cr). Source: Houghton International Inc., Valley Forge, PA. (b) Similar data for carbon, low-alloy, and various tool steels that were borided at 900 °C (1650 °F) in a boriding powder with a grain size <850 µm. AISI equivalents are in parentheses. Source: BorTec GmbH, Hürth Germany

plications, boriding should be performed in a protective-gas atmosphere. This is accomplished by packing the containers into a protective-gas retort and heat treating them in a chamber furnace, or else boriding directly in a retort furnace with the necessary protective-gas supply. The protective gas may be pure argon, pure nitrogen, a mixture of hydrogen and either argon or nitrogen, or, in special cases, pure hydrogen. It is important to note that oxygen-bearing compounds adversely affect boriding. For this reason, gases containing carbon monoxide should not be used. The parts to be borided are placed in the retort, which is then flushed with the protective gas in order to expel the oxygen. The flow of protective gas must be maintained after boriding until the retort has cooled to about 300 °C (570 °F).

Powder agents provide a boride layer of excellent quality and are particularly suitable for treating small tools and other parts. The fine granular powders also are ideal for treating intricately shaped parts. The very small granules ensure excellent contact with the metal surface. Note that powder agents may form a crust-like deposit on the parts during boriding. However, the crust can be easily broken off after parts have cooled.

These agents may be reused several times in normal boriding operations, which typically take 3 to 5 h. After each cycle, 30 to 40% fresh powder is mixed in with the recycled material. Complete replacement of the boriding agent is usually required only if the treatment time is especially long—about 20 h or more.

Boriding Agents. The boriding agent, or powder, is composed of an active source of boron (B_4C), an inert filler (SiC), and an activator. $NaBF_4$ KBF_4 $(NH_4)_3BF_4$, NH_4Cl, Na_2CO_3, BaF_2, and $Na_2B_4O_7$ are the boriding activators. There are special proprietary brands of boriding compounds, such as different grades of Ekabor (BorTec GmbH, Hürth, Germany), available on the market that can be used with confidence.

Typical compositions of commercial solid boriding powder mixtures are:

- 5% B_4C, 90% SiC, 5% KBF_4
- 50% B_4C, 45% SiC, 5% KBF_4
- 85% B_4C, 15% Na_2CO_3
- 95% B_4C, 5% $Na_2B_4O_7$
- 84% B_4C, 16% $Na_2B_4O_7$

Case Depth. The thickness of the boride layer depends on the substrate material being processed, boron potential of the boriding compound (Fig 6), boriding temperature, and time (Fig 7). In ferrous materials, the heating rate, especially between 700 °C (1300 °F) and the boriding temperature (800 to 1000 °C, or 1470 to 1830 °F), should be high in order to minimize the formation of FeB.

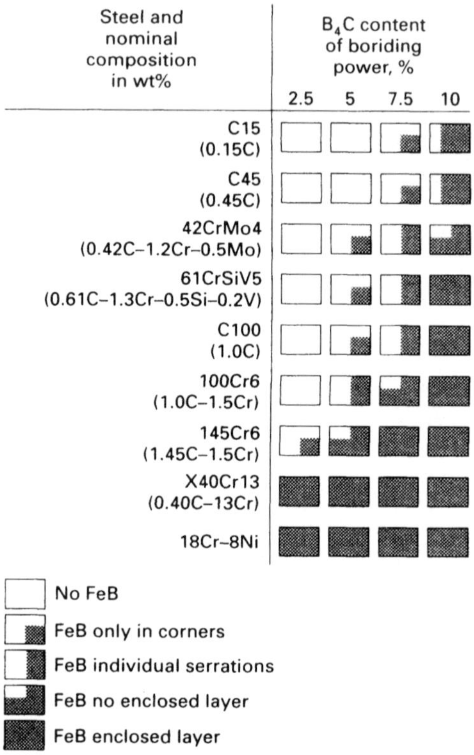

Fig. 6 Diagram showing the influence of the B_4C content of the boriding powder on the proportion of FeB phase in the boride layer of various steels borided with pack powder at 900 °C (1650 °F) for 5 h

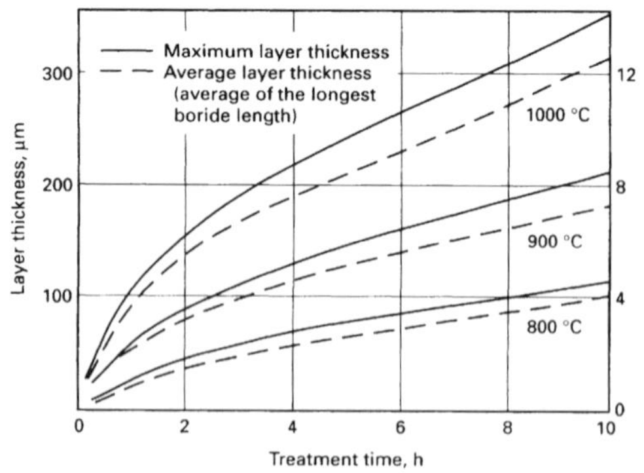

Fig. 7 Effect of pack boriding temperature and time on the boride layer thickness in a medium-carbon (Ck 45) steel

It is usual practice to match the case depth with the intended application and base material. As a rule, thin layers (for example, 15 to 20 μm, or 0.6 to 0.8 mil) are used for protection against adhesive wear (such as chipless shaping and metal-stamping dies and tools), whereas thick layers are recommended to combat abrasive wear (for example, extrusion tooling for plastics with abrasive fillers and pressing tools for the ceramic industry). The commonly produced case depths are 0.05 to 0.25 mm (0.002 to 0.01 in.) for low-alloy and low-carbon steels and 0.025 to 0.076 mm (0.001 to 0.003 in.) for high-alloy steels. However, case depths >0.089 mm (>0.0035 in.) are uneconomical for highly alloyed materials such as stainless steels and some tool steels.

Heat Treatment After Boriding. Borided parts may be heat treated to optimize core properties without loss of layer hardness. However, care must be taken to protect the boride layer from oxidation at temperatures above 650 °C (1200 °F). For this reason, vacuum furnaces designed for heat treating tool steels (A-2, D-2) are the best choice. Vacuum furnaces with internal oil quench systems may be used for hardening alloy steels. Fluidized bed furnaces equipped with an inert atmosphere such as argon also provide good results.

Endothermic and exothermic atmospheres are not suitable because these atmospheres cause the boride layer to oxidize, resulting in a loss of hardness. Plain carbon steels that require severe quenching (water) are not acceptable substrates because water quenching can fracture the boride layer.

Paste Boriding

Paste boriding was developed as a cost-effective means of boriding large components or those requiring partial, or selective, boriding. In this process, a paste of 45% B_4C (grain size 200 to 240 μm,) and 55% cryolite (Na_3AlF_6, flux additive), or conventional boronizing powder mixture (B_4C-SiC-KBF_4) in a good binding agent (such as nitrocellulose dissolved in butyl acetate, aqueous solution of methyl cellulose, or hydrolyzed ethyl silicate) is repeatedly applied (that is, dipped brushed, or sprayed) at intervals over the entire part or selected portion(s) of parts until, after drying, a layer about 1 to 2 mm (0.04 to 0.08 in.) thick is obtained. Subsequently, the ferrous materials are heated (say at 900 °C, or 1650 °F, for 4 h) inductively, resistively, or in a conventional furnace to 800 to 1000 °C (1470 to 1830 °F) for 5 h. Paste boriding necessitates the use of a protective atmosphere (for example, argon, cracked NH_3, or N_2). A layer in excess of 50 μm (2 mil) thickness may be obtained after inductively or resistively heating to 1000 °C (1830 °F) for 20 min (Fig 8). At the conclusion of the process, the paste may be removed by means of blast cleaning, brushing, or washing.

Plasma Boriding

Although still in its developmental stages, plasma boriding may be considered the key to increased commercial acceptance of the boriding process. Both mixtures of B_2H_6-H_2 and BCl_3-H_2-Ar may be used successfully in plasma boriding. However, the former gas mixture can be applied to produce a boride layer on various steels at relatively low temperatures, such as 600 °C (1100 °F), which is impossible with a pack boriding process. Plasma boriding in a mixture of BCl_3-H_2-Ar gases facilitates better control of BCl_3 concentration, reduction of the discharge voltage, and higher microhardness of the boride films.

The dual-phase layer is characterized by visible porosity, occasionally associated with a black boron deposit. This porosity, however, can be minimized by increasing the BCl_3 concentration. Boride layers up to 200 μm (8 mil) in thickness can be produced in steels after 6 h treatment at a temperature of 700 to 850 °C (1300 to 1560 °F) and a pressure of 270 to 800 Pa (2 to 6 torr).

Advantages of this process are:

- Control of composition and depth of the borided layer

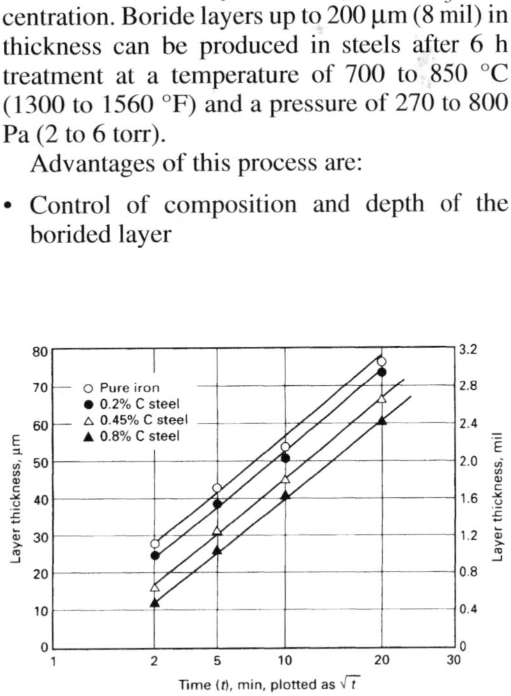

Fig. 8 A linear relationship between boride layer thickness and the square root of time for iron and steel boronized with B_4C-$Na_2B_4O_7$-Na_3AlF_6-based paste at 1000 °C (1830 °F)

- Increased boron potential compared to conventional pack boronizing
- Finer plasma-treated boride layers
- Reduction in temperature and duration of treatment
- Elimination of high-temperature furnaces and their accessories
- Savings in energy and gas consumption

The only disadvantage of the process is the extreme toxicity of the atmosphere employed. As a result, this process has not gained commercial acceptance. To avoid the above shortcoming, boriding from paste containing a mixture of amorphous boron and liquid borax in a glow discharge at the impregnating temperature has been developed, which is found to greatly increase the formation of the surface boride layer. Such paste mixtures vary from 30 to 60% amorphous boron to 40 to 70% borax, depending on the substrate material (e.g., carbon steel versus stainless steel).

Multicomponent Boriding (Ref 4)

Multicomponent boriding is a thermochemical treatment involving consecutive diffusion of boron and one or more metallic elements such as aluminum, silicon, chromium, vanadium, and titanium into the component surface. This process is carried out at 850 to 1050 °C (1560 to 1920 °F) and involves two steps:

1. Boriding by conventional methods—notably pack and paste methods. Here, the presence of FeB is tolerated, and, in some cases, may prove beneficial. Among these methods, much work has been done on pack method (Table 2), which produces a compact layer at least 30 μm (1 mil) thick
2. Diffusing metallic elements through the powder mixture or borax-based melt into the borided surface. If the pack method is used, sintering of particles can be avoided by passing argon or H_2 gas into the reaction chamber

There are six multicomponent boronizing methods: boroaluminizing, borosiliconizing, borochromizing, borochromtitanizing, borochromvanadizing, and borovanadizing.

Boroaluminizing When boroaluminizing involves boriding followed by aluminizing, the compact layer formed in steel parts provides good wear and corrosion resistance, especially in humid environments.

Borosiliconizing results in the formation of FeSi in the surface layer, which enhances the corrosion-fatigue strength of treated parts.

Borochromizing (involving chromizing after boriding) provides better oxidation resistance than boroaluminizing, the most uniform layer (probably comprising a solid-solution boride containing iron and chromium), improved wear resistance compared with traditionally borided steel, and enhanced corrosion-fatigue strength. In this case, a post-heat-treatment operation can be safely accomplished without a protective atmosphere.

Borochromtitanizing of structural alloy steel provides high resistance to abrasive wear and corrosion as well as extremely high surface hardness 5000 HV (15 g load). Figure 9 shows

Table 2 Multicomponent pack boriding treatments

Multicomponent boriding technique	Media composition(s), wt%	Process steps investigated(a)	Substrate(s) treated	Temperature, °C (°F)
Boroaluminizing	84% B_4C + 16% borax 97% ferroaluminium + 3% NH_4Cl	S B-Al Al-B	Plain carbon steels	1050 (1920)
Borochromizing	5% B_4C + 5% KBF_4 + 90% SiC (Ekabor II) 78% ferrochrome + 20% Al_2O_3 + 2% NH_4Cl	S B-Cr Cr-B	Plain carbon steels	Borided at 900 (1650) Chromized at 1000 (1830)
Borosiliconizing	5% B_4C + 5% KBF_4 + 90% SiC (Ekabor II) 100% Si	B-Si Si-B	0.4% C steel	900–1000 (1650–1830)
Borovanadizing	5% B_4C + 5% KBF_4 + 90% SiC (Ekabor II) 60% ferrovanadium + 37% Al_2O_3 + 3% NH_4Cl	B-V	1.0% C steel	Borided at 900 (1650) Vanadized at 1000 (1830)

(a) S, simultaneous boriding and metallizing: B-Si, borided and then siliconized; Al-B, aluminized and then borided. Note: Ekabor is a trademark of BorTec GmbH (Hürth, Germany).

the microstructure of the case of a borochromtitanized constructional alloy steel part exhibiting titanium boride in the outer layer and iron-chromium boride beneath it.

Borovanadizing and borochromvanadizing produce layers that are quite ductile with their hardnesses exceeding 3000 HV (15 g load). This reduces drastically the danger of spalling under impact loading conditions.

Wear Resistance of Multicomponent Coatings. Various methods have been used to gage the wear resistance of these coatings. The Faville test, for example, has been called on to assess their performance under conditions of metal-to-metal wear. Typical comparative test data are plotted in Fig. 10(a). In all cases, the substrate was C 45 steel (AISI 1043), and both members of the couple had the same coating.

Abrasive wear resistance was measured via a grinding disk test in which coated C 45 steel samples were run against silicon carbide. Weight loss versus time data for this test are shown in Fig 10(b).

Properties of Borided Steels

Boriding can impart a number of desirable properties to the surface, including improved wear resistance and corrosion resistance. Service lives have been shown to improve by a factor of three to ten due to the boriding process.

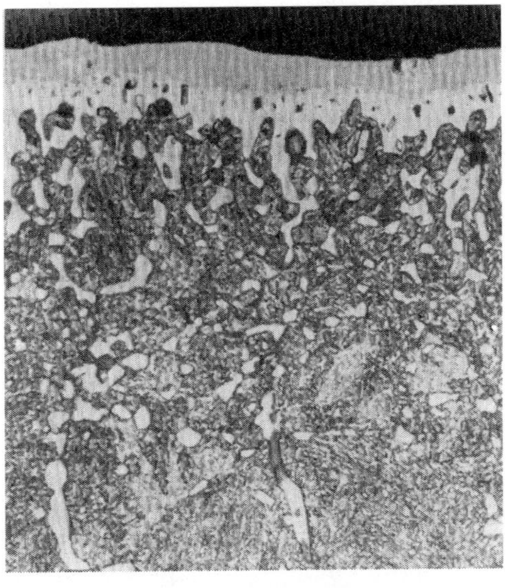

Fig. 9 Microstructure of the case of a borochromtitanized construction alloy steel. Source: Ref 4

Abrasive Wear Resistance. High hardness provides high wear resistance. The thickness of the boride layer can be tailored to the application. Layers between 50 and 150 μm (0.002 and 0.006 in.) thick are usually adequate to impart wear resistance to machine parts. Prevention of wear by abrasive materials calls for a case at least 200 μm (0.008 in.) deep.

Figure 11 shows the effect of boriding on abrasive wear resistance of a borided C 45 steel as a function of number of revolutions (or stressing period) based on the Faville test. Figure 12 shows the influence of steel composition on abrasive wear resistance.

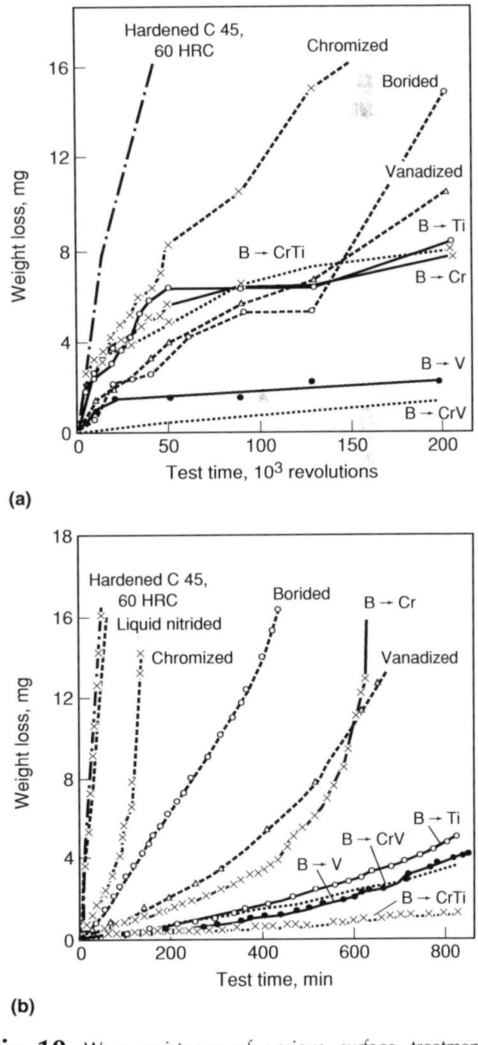

Fig. 10 Wear resistance of various surface treatments, including multicomponent coatings. (a) Metal-to-metal wear (Faville test). Substrate: Medium-carbon steel (C 45). (b) Abrasive wear (grinding disk test). Substrate: Medium-carbon steel (C 45). Source: Ref 4

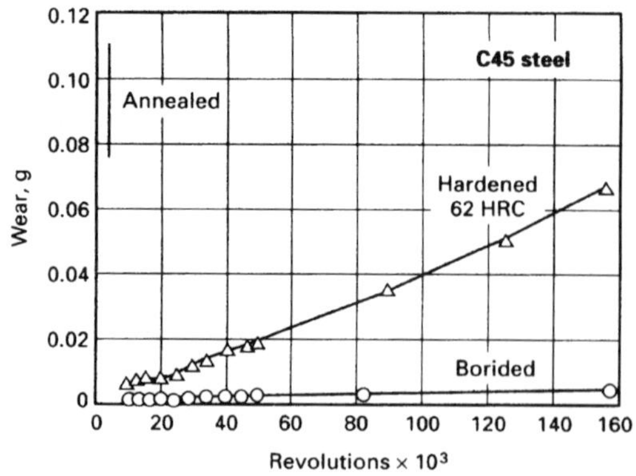

Fig. 11 Effect of boriding on the wear resistance (Faville test) of a 0.45% C (C45) steel borided at 900 °C (1650 °F) for 3 h

Adhesion Resistance. Tests have shown that borided surfaces show little tendency to cold weld (Ref 3). As a result, borided tools are used for the cold forming of metals such as aluminum and copper.

Toughness. Good bonding between the boride layer and the base metal ensures that the case will not flake or peel off under load. Because a borided component is actually a composite material, its toughness depends on case depth, cross section, and mechanical properties. In bend tests of borided samples having single-phase microstructures and average case depths of 150 to 200 μm (0.006 to 0.008 in.), elongations of up to about 4% were recorded without cracking. This means that borided parts can also survive a certain amount of post-treatment straightening without cracking.

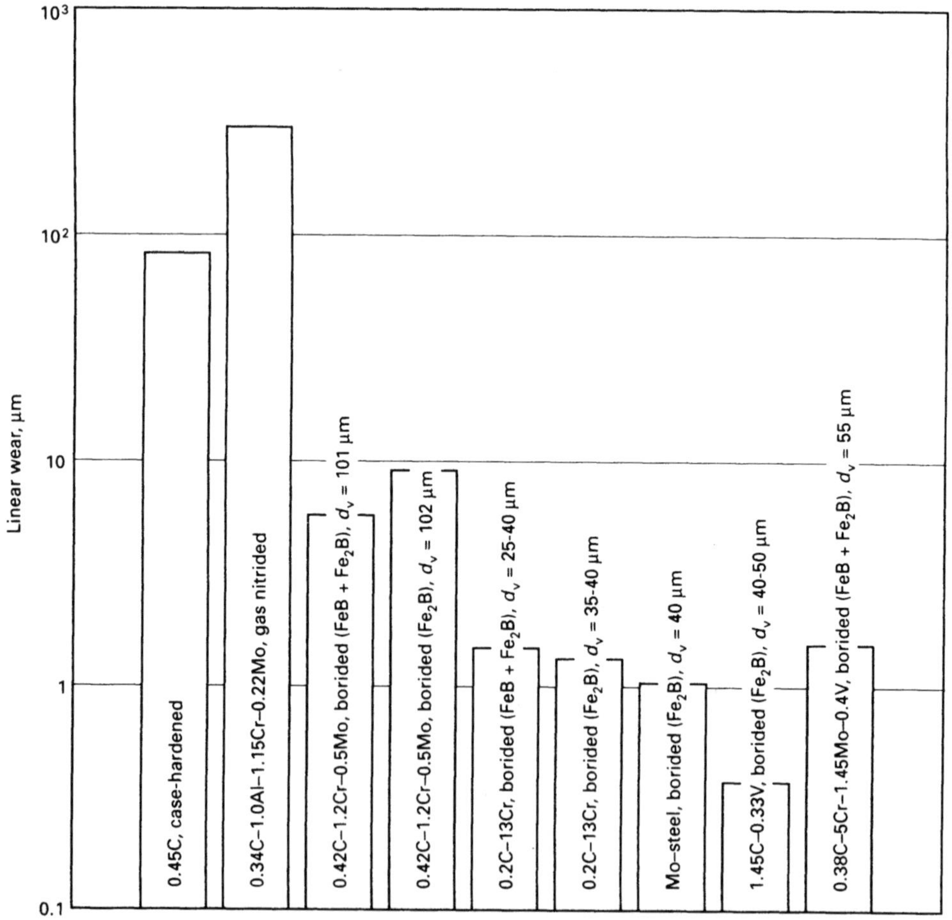

Fig. 12 Effect of steel composition (nominal values in wt%) on wear resistance under abrasive wear (d_v = thickness of the boride layer). Test conditions: DP-U grinding tester, SiC paper 220, testing time 6 min

Corrosion Resistance in Acids. Boriding increases the corrosion resistance of carbon and alloy steels to hydrochloric, sulfuric, and phosphoric acids, and improves the resistance of austenitic stainless steel to hydrochloric acid (Fig. 13). The process has been used to treat textile machinery components, ceramic extrusion dies, stamping and die casting dies, glass molds, material handling equipment, and various tools that were previously throwaway items. It should be noted, however, that the resistance of borided steels to oxidizing acids such as nitric acid is not as good as that in the aforementioned mineral acids.

Corrosion Resistance in Liquid Metals. As stated earlier in the section "Advantages and Disadvantages of Boriding" borided surfaces exhibit resistance to attack by molten metals. One study examined the degradation behavior of borided carbon (>0.2% C) and high-alloy steels (20% Cr and 1% Mo) in molten aluminum and zinc (Ref 5). The samples were pack borided at 900 °C (1650 °F) for 4 h and immersed in molten aluminum and zinc for periods ranging from 6 to 120 h and at temperatures of 630 °C (1165 °F) (aluminum melt) and 500 °C (930 °F) (zinc melt). Figure 14 shows the improved performance of the borided samples. Such tests demonstrate that borided steel components could find a wide range of applications in various industries handling molten metals including:

- Components to handle molten zinc in hot dip galvanizing industries
- Molten metal pumps for handling aluminum
- Systems for molten aluminum filtration and degassing

Applications for Pack Boriding

Borided parts have been used in a wide variety of industrial applications (Table 3) because of the numerous advantageous properties of boride layers. In sliding and adhesive wear situations, boriding is applied to:

- Spinning steel rings, steel rope, and steel thread guide bushings (made of DIN St 37 steel)
- Grooved gray cast iron drums (thread guides) for textile machinery
- Diesel engine oil-pump gears (made from borided, then hardened and tempered 4140 alloy steel)
- Four-holed feed water regulating valves (made from DIN 1.4571, or AISI 316 Ti steel)
- Drive, worm, and helically toothed steel gears in various high-performance vehicle and stationary engines

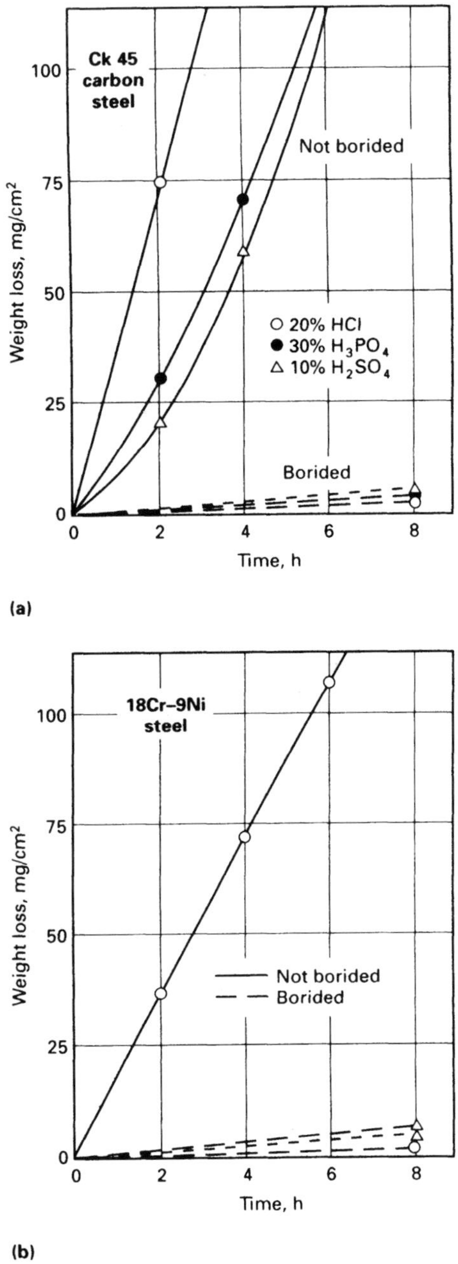

Fig. 13 Corrosive effect of hydrochloric and phosphoric acids on borided and nonborided steels at 55 °C (130 °F). (a) Carbon steel, 0.45% C (Ck 45). (b) Austenitic stainless steel (18Cr-9Ni)

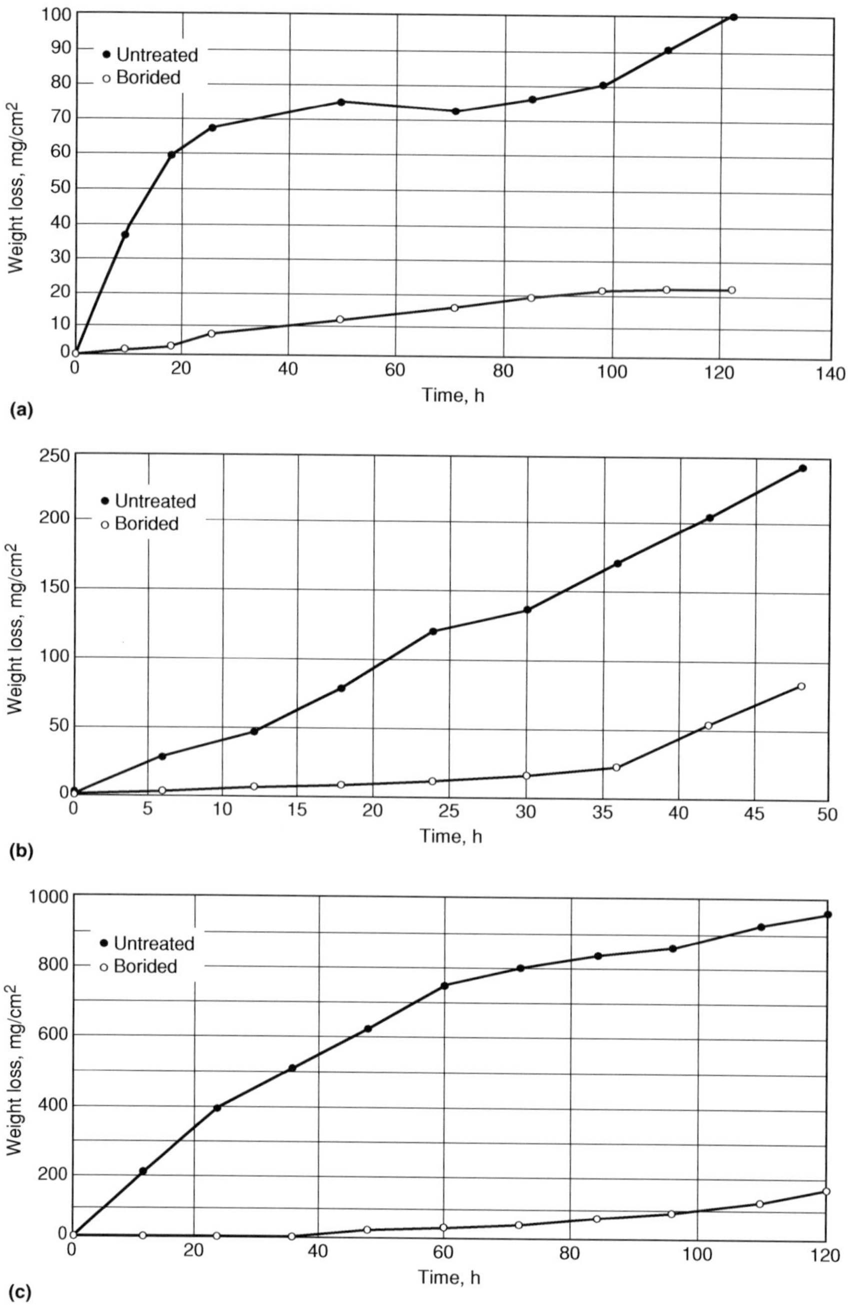

Fig. 14 Weight loss versus number of cycles for circular steel coupons (3 to 5.5 mm thickness and 9 to 20 mm radius) immersed in molten aluminum and zinc. (a) Carbon steel in aluminum. (b) Carbon steel in zinc. (c) High-alloy steel in zinc. Source: Ref 5

Table 3 Proven applications for borided steels

Substrate material			
AISI	BSI	DIN	Application
		St37	Bushes, bolts, nozzles, conveyer tubes, base plates, runners, blades, thread guides
1020	...	C15 (Ck15)	Gear drives, pump shafts
1043	...	C45	Pins, guide rings, grinding disks, bolts
		St50-1	Casting inserts, nozzles, handles
1138	...	45S20	Shaft protection sleeves, mandrels
1042	...	Ck45	Swirl elements, nozzles (for oil burners), rollers, bolts, gate plates
		C45W3	Gate plates
W1	...	C60W3	Clamping chucks, guide bars
D3	...	X210Cr12	Bushes, press tools, plates, mandrels, punches, dies
C2	...	115CrV3	Drawing dies, ejectors, guides, insert pins
		40CrMnMo7	Gate plates, bending dies
H11	BH11	X38CrMoV51	Plungers, injection cylinders, sprue
H13	...	X40CrMoV51	Orifices, ingot molds, upper and lower dies and matrices for hot forming, disks
H10	...	X32CrMoV33	Injection molding dies, fillers, upper and lower dies and matrices for hot forming
D2	...	X155CrVMo121	Threaded rollers, shaping and pressing rollers, pressing dies and matrices
		105WCr6	Engraving rollers
D6	...	X210CrW12	Straightening rollers
S1	~BS1	60WCrV7	Press and drawing matrices, mandrels, liners, dies, necking rings
D2	...	X165CrVMo12	Drawing dies, rollers for cold mills
L6	BS224	56NiCrMoV7	Extrusion dies, bolts, casting inserts, forging dies, drop forges
		X45NiCrMo4	Embossing dies, pressure pad and dies
O2	~BO2	90MnCrV8	Molds, bending dies, press tools, engraving rollers, bushes, drawing dies, guide bars, disks, piercing punches
E52100	...	100Cr6	Balls, rollers, guide bars, guides
		Ni36	Parts for nonferrous metal casting equipment
		X50CrMnNiV229	Parts for unmagnetizable tools (heat treatable)
4140	708A42 (En19C)	42CrMo4	Press tools and dies, extruder screws, rollers, extruder barrels, non-return valves
4150	~708A42 (CDS-15)	50CrMo4	Nozzle base plates
4317	...	17CrNiMo6	Bevel gears, screw and wheel gears, shafts, chain components
5115	...	16MnCr5	Helical gear wheels, guide bars, guiding columns
6152	...	50CrV4	Thrust plates, clamping devices, valve springs, spring contacts
302	302S25 (En58A)	X12CrNi188	Screw cases, bushes
316	~316S16 (En58J)	X5CrNiMo1810	Perforated or slotted hole screens, parts for the textile and rubber industries
		G-X10CrNiMo189	Valve plugs, parts for the textile and chemical industries
410	410S21 (En56A)	X10Cr13	Valve components, fittings
420	~420S45 (En56D)	X40Cr13	Valve components, plunger rods, fittings, guides, parts for chemical plants
		X35CrMo17	Shafts, spindles, valves

As abrasive wear-resistance materials, borided stainless steels are used for parts such as screw cases and bushings, perforated and slotted hole screens, rollers, valve components, fittings, guides, shafts, and spindles. Other abrasive wear applications for borided steels include:

- Burner nozzle used for oil firing and liquid waste disposal in the chemical industry (made from 1045 carbon steel)
- Screws, tips, non-return valves and cylinders for the extrusion of glass-filled plastics (made from 4140 alloy steel)
- Nozzles for handling prussic acid (made from type 316 stainless steel)
- Nozzles of bag filling equipment
- Extrusion screws, cylinders, nozzles, and reverse-current blocks in plastic production machinery (extruder and injection molding machinery)
- Bends and baffle plates for conveying equipment for mineral-filled plastic granules in the plastics industry
- Punching dies (for making perforations in accessory parts for cars), press and drawing matrices, and necking rings (made from S1 tool steel)
- Press dies, cutting templates, punched plate screens (made of DIN St 37 steel)
- Screw and wheel gears, bevel gears (from AISI 4317 steel)
- Steel molds (for the manufacture of ceramic bricks and crucibles in the ceramics industry), extruder barrels, plungers, and rings (from 4140 steel)
- Extruder tips, non-return valves and cylinders (for extrusion of abrasive minerals or glass fiber-filled plastics, from 4150 steel)
- Casting fillers for processing nonferrous metals (from AISI H11 steel)
- Transport belts for lignite coal briquettes

Borided parts also find applications in die-casting molds; bending blocks; wire draw blocks; pipe clips; pressing and shaping rollers, straightening rollers, engraving rollers, and rollers for cold mills; mandrels; press tools; bushings; guide bars; discs; casting inserts; various types of dies including cold heading, bending, extrusion, stamping, pressing, punching, thread rolling, hot forming, injection molding, hot forging, drawing, embossing, and so on in A2, A6, D2, D6, H10, H11, O2, and other tool steels.

Borided steel parts have also been used as transport pipe for molten nonferrous metals such as aluminum, zinc, and tin alloys (made from DIN St 37), corrosion-resistant transport pipe elbows for vinyl chloride monomer, grinding discs (made from DIN Ck 45), die-casting components, air foil erosion-resistant cladding, data printout components (for example, magnetic hammers and wire printers), and engine tappets.

Some examples of multicomponent boriding include: improving the wear resistance of austenitic steels (borochromizing), parts for plastics processing machines (borochromtitanizing), and dies used in the ceramics industry (borochromizing).

ACKNOWLEDGMENTS

Portions of this chapter were adapted from A.K. Sinha, Boriding (Boronizing), *Heat Treating,* Vol 4, *ASM Handbook,* ASM International, 1991, p 437–447

REFERENCES

1. K. Stewart, Boronizing Protects Metals Against Wear, *Adv. Mater. Process.* March 1997, p 23–25
2. C.H. Faulkner, Optimizing the Boriding Process, *Adv. Mater. Process.* April 1999, p H43–H45
3. "Boronizing," product literature from BorTec GmbH, Hürth, Germany, available on the Internet at http://www.bortec.de/boronizing.htm
4. R. Chatterjee-Fischer, Time to Take a Look at Multicomponent Boriding, *Met. Prog.* April 1986, p 24, 25, 37
5. D.N. Tsipas, G.K. Triantafyllidis, J.K. Kiplagat, and P. Psillaki, Degradation Behavior of Boronized Carbon and High Alloy Steels in Molten Aluminum and Zinc, *Mater. Lett.* Oct 1998, p 128–131

CHAPTER 9

Thermal Diffusion Process

THE THERMAL DIFFUSION (TD) PROCESS, which is also referred to as the thermoreactive deposition/diffusion (TRD) process and the Toyota diffusion process, is a high-temperature surface modification process that forms a hard, thin, wear-resistant layer of carbides, nitrides, or carbonitrides on steels as well as other carbon-containing materials such as nickel and cobalt alloys, cemented carbides, and steel-bonded carbides (TiC dispersed in a steel matrix). In the TD process, the carbon and nitrogen in the steel substrate diffuse into a deposited layer with a carbide-forming element such as vanadium, niobium, tantalum, chromium, molybdenum, or tungsten. The diffused carbon or nitrogen reacts with the carbide- and nitride-forming elements in the deposited coating so as to form a nonporous, metallurgically bonded coating at the substrate surface.

The TD process is unlike conventional case-hardening methods, where the specific elements (carbon and nitrogen) in a treating agent diffuse into the substrate for hardening. Unlike conventional diffusion methods, the TD method also results in an intentional buildup of a coating at the substrate surface (Fig. 1). These TD coatings, which have thicknesses of about 2 to 20 μm (0.08 to 0.8 mil), have applications similar to those of coating produced by chemical vapor deposition (CVD) or physical vapor deposition (PVD). In comparison, the thickness of typical CVD coatings (usually less than 25 μm) has about the same range as TD coatings. Figure 2 compares the processing characteristics of TD, CVD, and PVD processes. Process sequences for these coating methods are summarized in Fig. 3.

Process Characteristics

The TD process is performed by immersing parts in a fused salt bath (molten borax) kept at temperatures ranging from 800 to 1250 °C (1475 to 2285 °F) for 1 to 8 h. This temperature range is suitable for quench hardening many grades of steels that contain a carbon content of 0.3% or greater. This includes many grades of tool steels, including powder metallurgy (P/M) tool steels and low-alloy steels. Steels containing less than 0.3% C must be carburized prior to TD processing. The coated steels may be cooled and reheated for hardening, or the bath temperature may be selected to correspond to the steel austenitizing temperature (Fig. 2), permitting the steel to be quenched directly after cooling (Fig. 3).

Carbide layers commonly produced include vanadium carbide, which is the most commonly used coating for tooling applications, niobium carbide, and chromium carbide, depending on the carbide-forming elements used in the salt bath. Vanadium and niobium carbide layers exhibit superior peel strength and resistance to wear, corrosion, and oxidation when compared to other processes. Chromium carbide has light wear resistance and high resistance to oxidation.

Coating Procedure and Mechanism of Coating Formation. Before parts are TD processed, they are first preheated to minimize distortion and to lower the TD processing time. They are then TD processed at the austenitizing temperature for the particular grade of steel. After TD processing the parts are quenched in air, salt, or oil to produce a hardened substrate. After quenching, tempering is carried out. Figure 4 shows a schematic of a typical cycle. High-speed steels and other steels that have austenitizing temperatures greater than 1050 °C (1920 °F) may be post-TD heat treated in vacuum, gas, or protective salt to achieve full substrate hardness.

When substrate materials containing carbon and nitrogen are kept in contact with treating agents at appropriately elevated temperatures, carbon and nitrogen chemically combine with

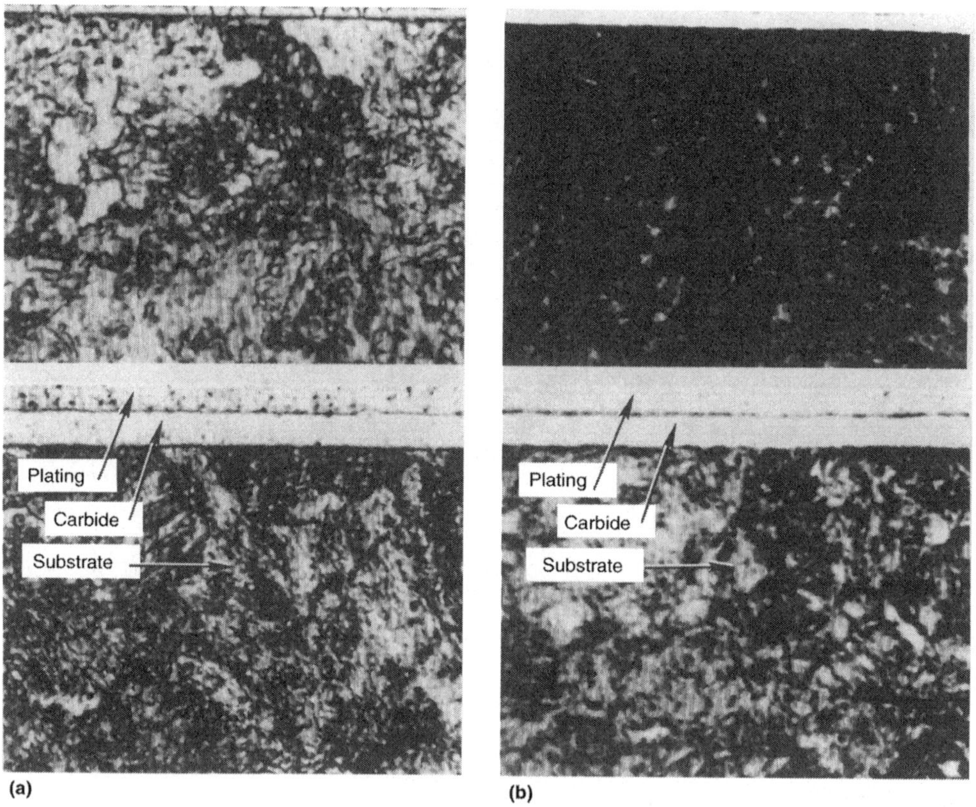

Fig. 1 Carbide coating grown during TD process. Substrate, W1 steel; temperature, 900 °C (1650 °F). Salt: borax, V_2O_5 and B_4C borax and chromium. (a) Vanadium carbide coating. Upper, 5 min; lower, 30 min. (b) Chromium carbide coating. Upper, 5 min; lower, 30 min

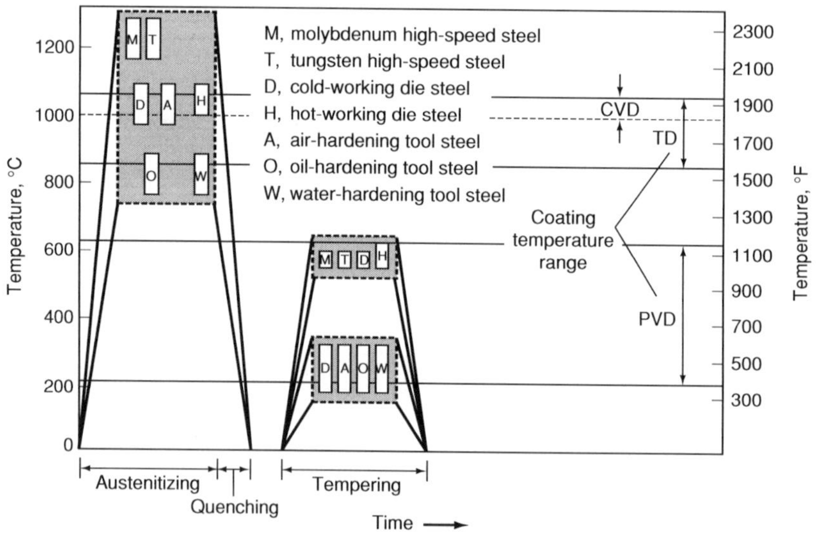

Fig. 2 Coating temperature comparison for applying thin coatings by thermal diffusion (TD), chemical vapor deposition (CVD), and physical vapor deposition (PVD). Note that some cold-working die steels (e.g., D2) are tempered at either high or low temperatures, depending on the application/required properties. Source: Ref 1

the carbide- and nitride-forming elements of the treating agent due to their small free energies for carbide and nitride formation. This formation of carbides, carbonitrides, and nitrides on the substrate results in the growth of a layer, as shown in Fig. 1 for vanadium carbide and chromium carbide coatings. Carbide layers are formed in the following steps:

- Carbide-forming elements dissolve into borax from added powders.
- Carbon in steel combines with the carbide-forming elements to produce a carbide layer on the surface.
- The carbide layer grows at the surface front through reaction between carbide-forming elements and carbon atoms successively supplied from the substrate.

Vanadium and chromium diffuse into the steel substrate to form iron-chromium or iron-vanadium solid-solution layers beneath the carbide layer.

Reagents Used. The carbide-forming elements (CFE) and the nitride-forming elements (NFE) must be in an active state to combine with carbon and nitrogen. Typical reagents have the CFE and NFE dissolved into molten salt during salt bath processing. Therefore, borax with additions of CFE and NFE contained in ferroalloy powder or with oxides of CFE and NFE and their reducing agents, such as boron

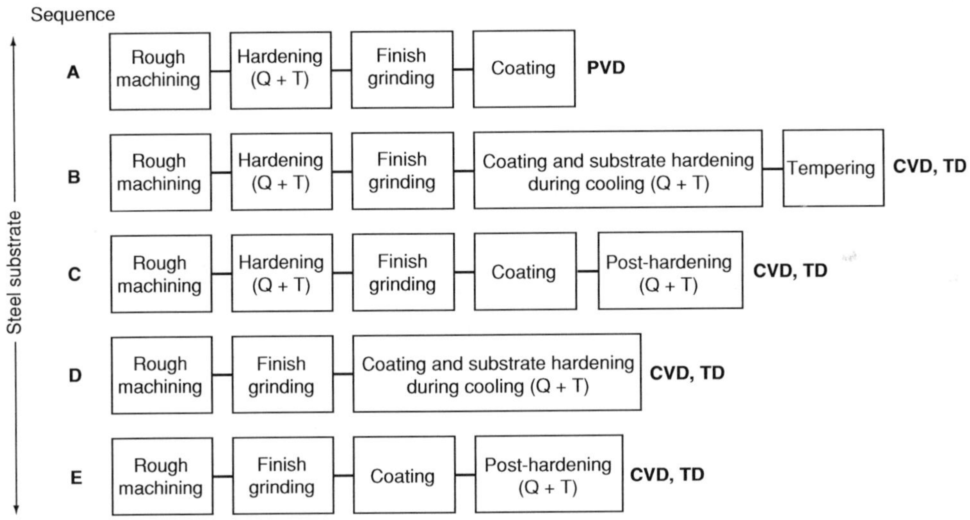

Fig. 3 Process sequences for physical vapor deposition (PVD), thermal diffusion (TD), and chemical vapor deposition. Hardening after coating is not required for PVD coatings (sequence A). For both TD and CVD coatings, hardening may occur during coating/substrate cooling (sequences B and D) or by post-hardening (sequences C and E). Q + T, quench and temper. Source: Ref 1

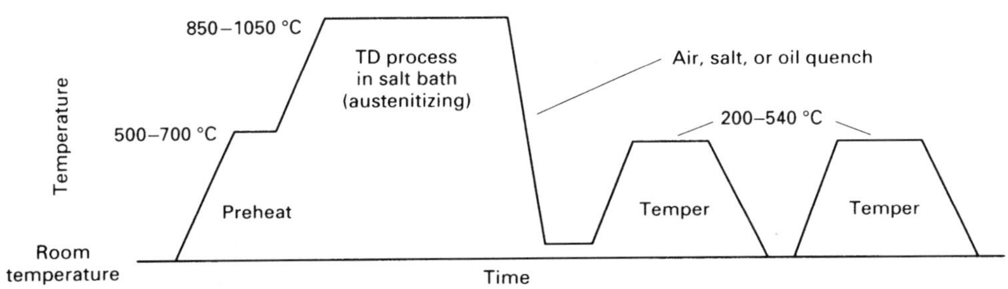

Fig. 4 Schematic of typical TD-processing cycle

carbide and aluminum, are successfully used as bath agents.

Effect of Treating Parameters. The coating growth rate is determined by the number of carbon atoms and nitrogen atoms that can be supplied to the coating from the substrate by diffusion, if the treating reagents can supply CFE and NFE in excess of the critical amount required to combine with the carbon and nitrogen supply from the substrate. Excess amounts of material containing CFE and NFE (for example, more than 10 wt% Fe-V, or 20 wt% V_2O_5 and 5 wt% B_4C in molten borax for vanadium carbide coating), are usually added to maintain this requirement. Therefore, the coating growth rate is determined by factors that affect only the amount of CFE and NFE required for coating: temperature, time, type of substrate, and type of coating.

As in many diffusion treatments, the effect of temperature and time on coating thickness (d) is expressed by the equation:

$$d^2/t = K = K_o \exp(-Q/RT)$$

where d is the thickness of coating (cm), t is time (s), K is the growth rate constant (cm^2/s), K_o is the constant term of K (cm^2/s), Q is the activation energy (KJ/mol), T is absolute temperature (K), and R is the gas constant. Figure 5 shows the relation between the thickness of the vanadium-carbide layer formed on W1 steel versus salt bath temperature and immersion time. The temperature is usually selected around the hardening temperature of steels, that is, 800 to 1250 °C (1475 to 2285 °F).

The carbon and nitrogen content in the substrate has a positive effect on the growth rate. However, the total content in the substrate does not have a direct effect. For example, in steels the carbon content in the austenite matrix, not the total carbon content, is nearly linear in relation to the thickness of the carbide coating. This is shown in Fig. 6. In the case of alloyed steels, an increase of temperature increases the carbon content in the matrix phase, as well as the diffusion rate of carbon in the carbide layer and in the substrate, resulting in a considerable increase of coating thickness. Figure 7 exemplifies the relation between bath temperature and immersion time needed for producing a 4 and 7 μm thick VC coating on four types of steel. The diffusion rate and its temperature dependence in relation to the carbon and nitrogen content are different between coatings. However, the difference in thickness among vanadium carbide (VC), niobium carbide (NbC), chromium carbide (Cr_7C_3, $Cr_{23}C_6$), and titanium carbide (TiC) is negligibly small.

Control of Distortion. (Ref 2). The possibility of distortion is present with the high-temperature process. Distortion entails dimensional change and deformation. Dimensional change is due to phase transitions in heat treatment of the base steel and to formation of the carbide layer. Deformation is a change in shape.

Thermal diffusion processing usually hardens a material. Therefore, to minimize dimensional change, it is best to start with a part that

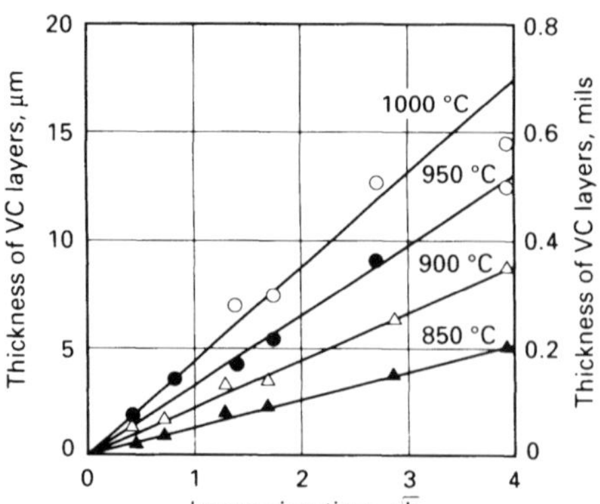

Fig. 5 Effect of temperature and time on thickness of vanadium carbide layer in a borax bath containing 20 wt% Fe-V powder

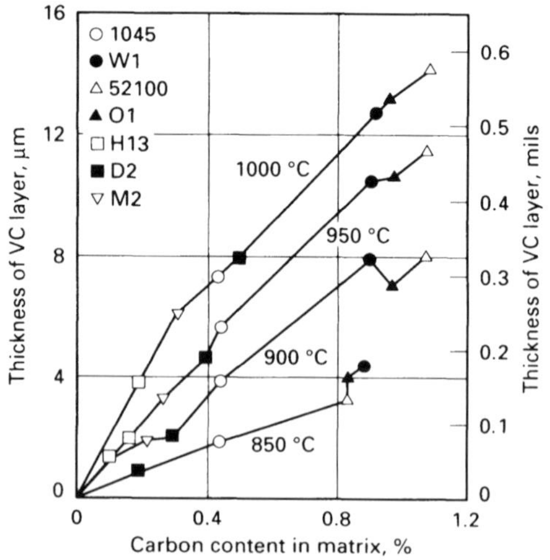

Fig. 6 Effect of carbon content in matrix phase on thickness of vanadium carbide layer in a borax bath containing 20 wt% Fe-V powder. Immersion time, 4 h

has been hardened and finish ground. Even then, there will be some dimensional change due to differences in the amount of retained austenite. Cemented carbide is not hardened in the process, therefore it has very little dimensional change.

The amount of retained austenite before TD processing should equal the amount after processing. The easiest method of controlling retained austenite is to reduce it to 0% before and after the TD process. This can be achieved in D2 tool steel by tempering at 520 to 535 °C (975 to 1000 °F) to decompose the retained austenite. Subzero treatment is another method of decomposing retained austenite.

Deformation is caused by thermal stresses, transformation stresses, creep during heating, anisotropy of the substrate structure, and residual stresses. The following steps can be taken to minimize deformation:

- Minimize variations in cross-sectional area.
- Use air-hardening grades of tool steel, which can be slow cooled.
- Machine tools so that critical dimensions are transverse of the rolling direction of the raw material.
- Use P/M steels for tooling that requires minimum out-of-roundness, because of their very uniform distribution of fine carbide particles.

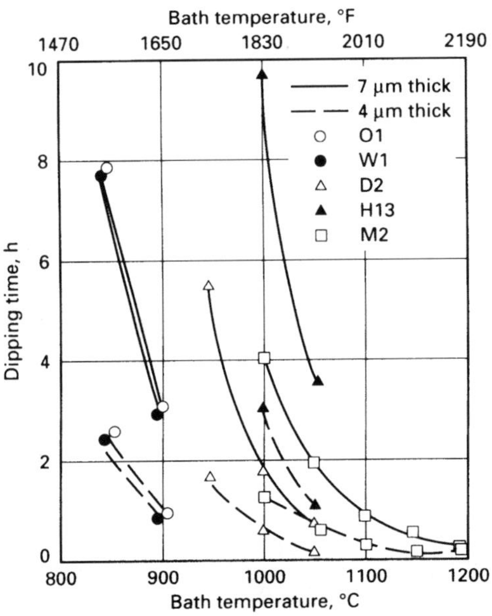

Fig. 7 Effect of bath temperature and substrate steel on the immersion time required to form a 7 and 4 μm thick vanadium carbide layer in a borax bath

- Relieve residual stresses caused by machining and grinding.

Parts made from air-hardened steels requiring tight tolerances should be double high-tempered before using the TD process. In making new tooling, it is recommended to leave stock on nonworking surfaces and finish only the working surfaces. The nonworking surfaces may then be finished after TD processing.

Edge preparation of cutting and piercing tools is important. An edge that is too sharp or that contains burrs will break. The cutting edge should be rounded to a radius of 0.05 to 0.25 mm (0.002 to 0.010 in.) with a stone or emery paper. A worn cutting edge may be resharpened. This is not detrimental because performance is governed by the carbide layer on the side surface of the cutting edge.

The surface finish and polishing direction of a forming die prior to TD processing is very important. Due to the high-hardness carbide layer, a TD-processed tool that has a rough surface finish will perform worse than a regular uncoated tool. This is shown in Fig. 8. The surface should be finished to a maximum peak-to-valley roughness height (R_{max}) of 3 μm (0.1 mil). All large scratches and machining marks should be removed. When plated steel, stainless steel, and high-strength steels are the materials being processed, a finish of 0.5 to 1 μm (0.02 to 0.04 mil) for R_{max} is recommended on the tool being used. The polishing lines should be parallel to the metal flow. The characteristic white layer that is produced in electrical discharge

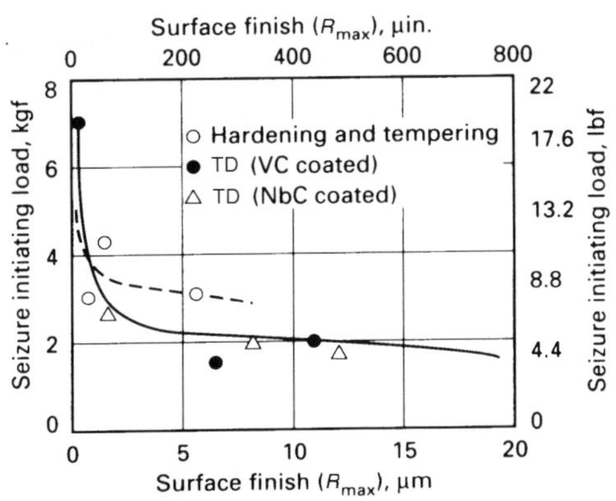

Fig. 8 Influence of tool surface finish on seizure-initiating load for a TD-coated tool and uncoated tool. Mating material, type 304 stainless steel; speed, 2.6 m/s (8.5 ft/s); lubricant, none

machining should be removed before TRD processing.

Re-treating. Tools processed by TRD may be re-treated by TRD. Some tools have been re-treated eight times. After the worn areas are refinished, tools can be re-treated without removing the sound carbide. The difference in layer thicknesses will be insignificant due to the slower growth rate of the carbide layer on the previously coated areas.

Characteristics of TD-Processed Materials

Hardness. Vanadium TD-treated materials show surface hardness in the range of 3200 to 3800 on the Vickers hardness scale (Fig. 9). Vanadium carbide retains exceptional hardness of 1000 HV at temperatures as high as 800 °C (1470 °F). Furthermore, hardness will be returned to previous levels once the layer is cooled to room temperature after exposure to high temperature.

Wear Resistance. Carbide layers from the TD process exhibit high wear resistance against materials such as steel, nonferrous alloys, plastics, and rubber. Figure 10 shows results obtained by measuring the abrasion of hardened-and-tempered, chromium-plated, and VC-coated tool steel dies after continuous coining of cold-rolled mild steel workpieces. Hardened-and-tempered dies suffered considerable wear losses, whereas the TD-processed dies exhibited little abrasion.

Seizure Resistance. Vanadium-carbide-coated steel from the TD process resists seizing at any temperature. In the case where the mating material is stainless steel, the seizure resistance of a TD-treated VC layer is considerably better than that of cemented carbide. Vanadium carbide also shows superior score resistance, regardless of mating materials. Figure 8 shows the influence of surface finish on seizure resistance.

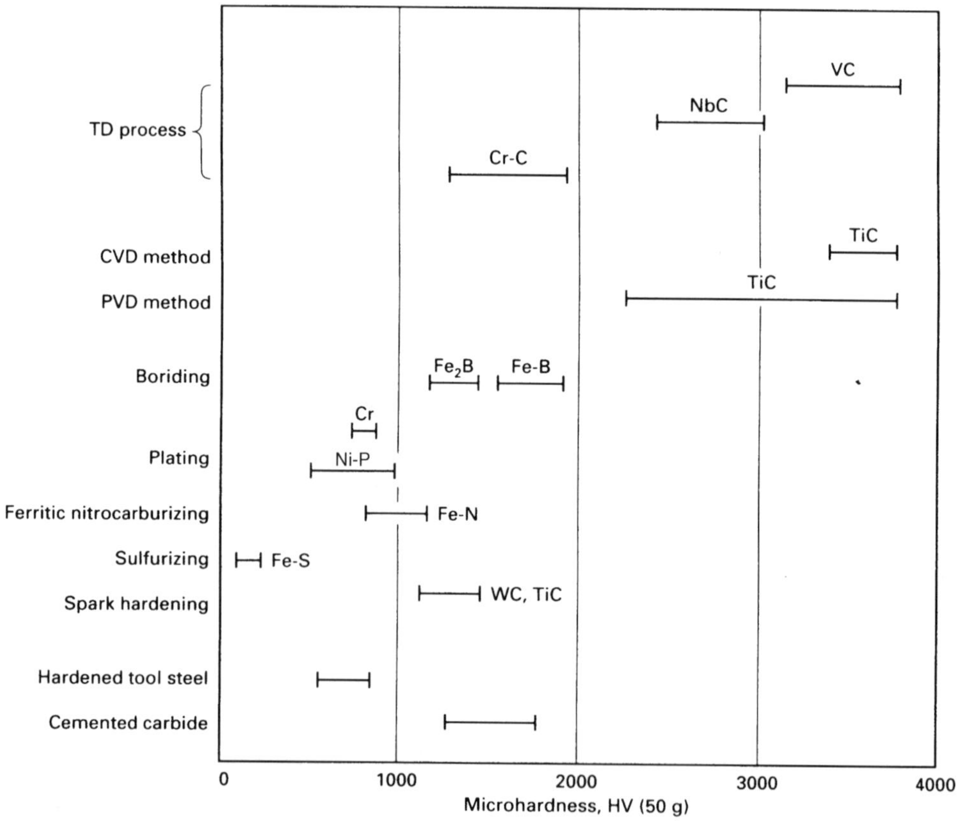

Fig. 9 Surface hardness of carbide layers by TD process in relation to other surface hardening processes

Impact Resistance. In the Izod impact test, TD-treated steels are equivalent in impact values to hardened-and-tempered steels, regardless of the substrate. Therefore, if a material having high impact resistance is selected for the substrate, it will be effective against breaking and chipping after TD treatment.

Corrosion and Oxidation Resistance. Thermal diffusion-processed steels exhibit little or no corrosion when immersed in 36% hydrochloric acid (HCl). Figure 11 compares weight loss data for uncoated, chromium-plated, and TD-, CVD-, and PVD-processed steels in HCl.

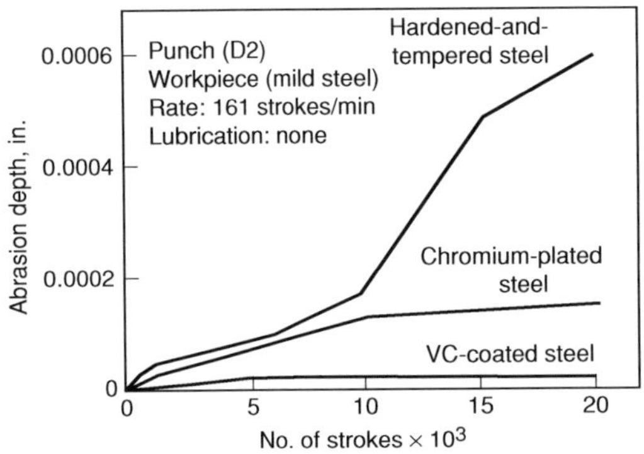

Fig. 10 Wear resistance of a TD-processed punch during a coining operation. Source: Ref 2

When oxidation resistance is required, chromium-carbide-coated steels are recommended. As shown in Fig. 12, chromium-carbide-coated steels are resistant to oxidation at temperatures as high as 900 °C (1650 °F).

Peeling Resistance. Unlike plating, the treated layer produced by the TD process will not easily peel off. The VC layer is metallurgically bonded versus deposited or mechanically bonded. In tests, various surfaces were repeatedly struck on the same spot with an acuminated hammer (Ref 2). A chromium plated layer was cracked after a small number of strikes and peeled off after about 50,000 strikes. The TiC layer produced by the CVD method or PVD method cracked after 50,000 strikes and peeled off after 100,000 strikes. The TD-treated VC layer suffered neither cracks nor peeling after 200,000 strikes.

Applications

Tooling Applications. The TD process has been used on tooling and dies for the following industries and components:

- Sheet metal
- Cold-forming dies
- Hot-forming dies

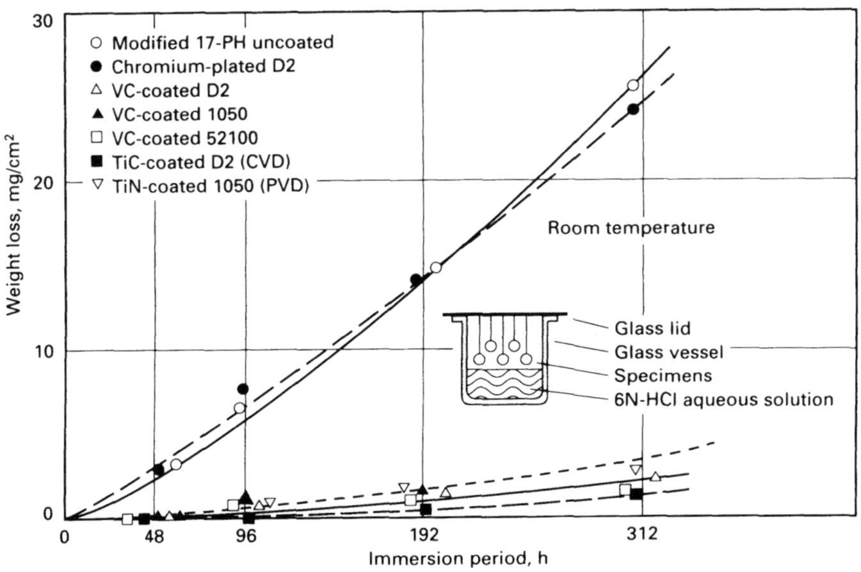

Fig. 11 Comparative weight loss by corrosion in hydrochloric acid vapor

- Powdered metal production
- Glass
- Textiles
- Pump parts
- Machine parts
- Engine parts
- Wire and tube production

Applications for TD-processed tooling are summarized in Table 1. The range in size for parts treated has been 1.2 mm (0.047 in.) diam punches to 160 kg (350 lb) rolls for forming. In many cases, tool life improvements of 30 to 50 times have been achieved after TD treatment.

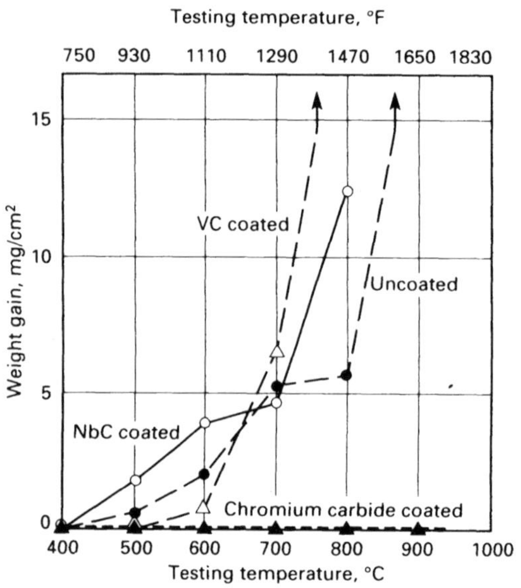

Fig. 12 Comparative weight gain in a high-temperature oxidation test. Substrate, D2; testing period, 40 h

Substrate Requirements for Tooling. The substrate hardness may be the same or lower than normal in some applications. In applications where tool chipping or breakage is the problem, a lower substrate hardness with increased toughness can be used. The hard carbide coating provides the surface wear resistance. Underhardened high-speed steel could be used to provide needed substrate toughness.

In applications with high surface pressures, such as extrude dies and cold-forging dies, the carbide layer has to be supported by a hard substrate. High-speed steels should be post-TRD hardened. Some powdered high-speed steels that contain cobalt can be treated at the maximum TRD processing temperature of 1050 °C (1920 °F) to give hardnesses of 60 to 65 HRC. The hardest substrate available is cemented carbide, which can be TD-treated very successfully.

Non-tooling applications requiring wear and corrosion resistance include:

- *Components used in high-performance machines:* roller chain for racing bicycles, motorcycles, and automobiles; traveller rings used under extremely high-velocity spinning; and pump plungers used under extremely high pressure
- *Components used in corrosive or adverse operating conditions:* vanes in vane pumps, spraying nozzles that work with corrosive liquids, and liquids in which abrasive particles exist; link components in glass-molding machines; and automobile components that are susceptible to oxidation and corrosion by exhaust gas

Structural steels such as 10*xx* series carbon steel, and 41*xx* series low-alloyed steel are

Table 1 Applications of TD-processed tooling

Application	Tool
Sheet metal working	Draw die, bending die, pierce punch, form roll, embossing punch, coining punch, shave punch, seam roll, shear blade, stripper guide pin and bushing, pilot pin, and so on
Pipe and tube manufacturing	Draw die, squeeze roll, breakdown roll, idler roll, guide roll, and so on
Pipe and tube working	Bending die, pressure die, mandrel, expand punch, swaging die, shear blade, feed guide, and so on
Wire manufacturing	Draw die, straightening roll, descaling roll, feed roll, guide roll, cutting blade
Wire working	Bending die, guide plate, guide roll, feed roll, shear blade
Cold forging and warm forging	Extrusion punch and die, draw die, upsetting punch and die, coining punch and die, rolling die, quill cutter, and so on
Hot forging	Press-forging die, rolling die, upsetting die, rotary swaging die, closed-forging die, and so on
Casting (aluminum, zinc)	Gravity-casting core pin, die-casting core pin, core, sleeve, and so on
Rubber forming	Form die, extrusion die, extrusion screw, torpedo, cylinder sleeve, piston, nozzle, and so on
Plastic forming	Form die, injection screw, sleeve, plunger, cylinder, nozzle, gate, and so on
Glass forming	Form die, plunger, blast nozzle, machine parts, and so on
Powder compacting	Form die, core rod, extrusion die, screw, and so on
Cutting and grinding	Cutting tool, cutting knife, drill, tap, gage pin, tool holder, guide plate, and so on

widely used for these applications. Low-carbon steels are often carburized prior to TD processing. Substrate hardening is done during cooling in TD treatment or by reaustenitizing hardening, if it is necessary. Attention should be paid to surface finishing and edge preparation for components used in severe conditions. Barrel finishing is often used for surface finishing of small components in large volume.

ACKNOWLEDGMENT

Parts of this chapter were adapted from T. Arai and S. Harper, Thermoreactive Deposition/Diffusion Process, *Heat Treating,* Volume 4, *ASM Handbook,* ASM International, 1991, p 448–453.

REFERENCES

1. T. Arai and H.M. Glaser, Substrate Selection for Tools Used with Thin Film Coatings, *Met. Form.,* June 1998, p 31–40
2. H.M. Glaser, The Thermal Diffusion (TD) Process: Technology and Case Studies, *Worldclass Productivity Conf.,* (Chicago, Il), 10–13 March, Vol 1 (No. 2), Precision Metalforming Association, 1991, p 507–527

SELECTED REFERENCES

- T. Arai and S. Harper, Diffusion Carbide Coating for Distortion Control, *Die Casting Engineer,* March/April 2000, p 84–94
- T. Arai and N. Komatsu, Carbide Coating Process by Use of Salt Bath and its Application to Metal Forming Dies, *Proc. 18th International Machine Tool Design and Research Conf.* 14–16 Sept 1977, p 225–231
- T. Arai, Carbide Coating Process by Use of Molten Borax Bath in Japan, *J. Heat Treat.,* Vol 18 (No. 2), 1979, p. 15–22
- T. Arai, H. Fujita, Y. Sugimoto, and Y. Ohta, Vanadium Carbonitride Coating by Immersing into Low Temperature Salt Bath, *Heat Treatment and Surface Engineering.* G. Krauss, Ed., ASM International, 1988, p 49–53

CHAPTER 10

Surface Hardening by Applied Energy

APPLIED-ENERGY SURFACE HARDENING PROCESSES are those in which thermal energy is selectively applied to specific areas of a part for the purposes of altering such properties as hardness, wear resistance, strength, and torsional and bending fatigue resistance. Also referred to as selective surface hardening, these processes include flame hardening, induction hardening, laser-beam hardening, and electron-beam hardening.

Flame and induction hardening are methods for hardening the surfaces of components, usually in selected areas, by the short-time application of high-intensity heating followed by quenching. The heating and hardening effects are localized, and the depth of hardening is controllable. Unlike thermochemical case-hardening treatments (carburizing, nitriding, carbonitriding, etc.) applied to low-carbon steels, flame and induction hardening do not promote chemical enrichment of the surface with carbon or nitrogen but rely on the presence of an adequate carbon content already in the material to achieve the hardness level required. The properties of the core remain unaffected and depend on material composition and prior heat treatment.

Surface heating by high-energy laser and electron beams are being increasingly applied for localized hardening of steels. Compared to flame and induction thermal processing, laser and electron beams concentrate their energy very close to the surface of a part, and, therefore, relatively shallow hardening develops. The intense concentration of energy, however, results in very rapid heating and cooling by thermal conduction into the substrate, and thus very fine martensitic microstructures may develop. Like flame and induction hardening, there is no chemistry change produced by laser- or electron-beam hardening treatments.

Flame Hardening

Flame hardening involves the direct impingement of oxyfuel gas flames from suitably designed and positioned burners onto the surface area to be hardened, followed by quenching. The result is a hard surface layer of martensite over a softer interior core. There is no change in composition, and, therefore, the flame-hardened steel must have adequate carbon content for the desired surface hardness. The rate of heating and the conduction of heat into the interior appear to be more important in establishing case depth than the use of a steel of high hardenability.

Flame-heating equipment may be a single torch with a specially designed head or an elaborate apparatus that automatically indexes, heats, and quenches parts. Large parts such as gears and machine tool ways, with sizes and shapes that would make furnace heat treatment impractical, are easily flame hardened. With improvements in fuel- and oxygen-flow control, infrared temperature measurement and control, and burner design, along with computerized monitoring of the process, flame hardening has been accepted as a reliable heat-treating process that is adaptable to general or localized surface hardening for small and medium-to-high production requirements.

Hardening by flame differs from true case hardening because the hardenability necessary to attain high levels of hardness is already contained in the steel, and hardening is obtained by localized heating. Although flame hardening is mainly used to develop high levels of hardness for wear resistance, the process also improves bending and torsional strength and fatigue life. One of the major advantages of flame hardening is the ability to satisfy stringent engineering requirements with carbon steels.

Scope and Application

Flame hardening is applied to a wide diversity of workpieces and ferrous materials for one or more reasons. This process is used because:

- Parts are so large that conventional furnace heating and quenching are impracticable or uneconomical. Typical examples include large gears, machineways, dies, and rolls.
- Only a small segment, section, or area of a part requires heat treatment, or because heat treating all over would be detrimental to the function of the part. Typical examples include the ends of valve stems and pushrods and the wearing surfaces of cams and levers.
- The dimensional accuracy of a part is impracticable or difficult to attain or control by furnace heating and quenching. A typical example is a large gear of complex design for which flame hardening of the teeth would not disturb the dimensions of the gear.
- The use of flame hardening permits a part to be made from a less costly material, thereby effecting an overall cost saving in comparison with other technically acceptable methods. The process gives inexpensive steels the wear properties of alloyed steels, and parts can be hardened without scaling or decarburization, thereby eliminating costly cleaning operations. For example, a large steel part might be made at a lower cost if produced from a flame-hardened plain carbon steel rather than from a carburized low-carbon alloy steel.

Table 1 compares the benefits of flame hardening with those of other commonly used surface hardening methods.

Methods of Flame Hardening

Flame hardening can be carried out by a number of methods including:

- Spot hardening, in which the flame is directed to the spot to be heated and hardened. After being heated, the parts are usually immersion quenched.
- Spin hardening, in which the workpiece is rotated in contact with the flame. After the surface has been heated to the desired temperature, the flame is extinguished or withdrawn and the part is quenched by immersion or spray or a combination of both.
- Progressive hardening, in which the torch and quenching medium move across the workpiece (Fig. 1). The progressive method is used to harden large areas that are beyond the scope of the spot method.

Flame-Hardenable Steels

Selection Criteria. Maximum hardness is not the sole criterion used in selecting flame hardening as a heat treatment. Proper steel selection is essential to minimize distortion, for example. Plain carbon steels should be used, if possible, instead of steels whose deep-hardening characteristics are more likely to incur higher internal stresses.

Some flame hardeners feel it is important to stress relieve all alloy steels and other steels with more than 0.40% carbon at 175 to 245 °C (350 to 475 °F), depending on customer specifications. This low-temperature tempering decreases hardness, but it also removes internal stress and restores toughness and ductility.

Table 1 Relative benefits of five surface hardening processes

Carburizing	Hard, highly wear-resistant surface (medium case depths); excellent capacity for contact load; good bending fatigue strength; good resistance to seizure; excellent freedom from quench cracking; low-to-medium-cost steels required; high capital investment required
Carbonitriding	Hard, highly wear-resistant surface (shallow case depths); fair capacity for contact load; good bending fatigue strength; good resistance to seizure; good dimensional control possible; excellent freedom from quench cracking; low-cost steels usually satisfactory; medium capital investment required
Nitriding	Hard, highly wear-resistant surface (shallow case depths); fair capacity for contact load; good bending fatigue strength; excellent resistance to seizure; excellent dimensional control possible; good freedom from quench cracking (during pretreatment); medium-to-high-cost steels required; medium capital investment required
Induction hardening	Hard, highly wear-resistant surface (deep case depths); good capacity for contact load; good bending fatigue strength; fair resistance to seizure; fair dimensional control possible; fair freedom from quench cracking; low-cost steels usually satisfactory; medium capital investment required
Flame hardening	Hard, highly wear-resistant surface (deep case depths); good capacity for contact load; good bending fatigue strength; fair resistance to seizure; fair dimensional control possible; fair freedom from quench cracking; low-cost steels usually satisfactory; low capital investment required

Source: Ref 1

Selective heating has the disadvantage of developing residual tensile stresses in the surface. As one area of a piece of metal is heated while the remainder stays cold, the hot metal expands; if restraint is sufficient, the heated metal will upset itself. On cooling, this upset metal becomes short. As it cools to room temperature, it often stabilizes in a state of tension, which can be high enough to crack the part.

When a part is to be induction or flame hardened, the materials engineer should work closely with the designer to keep the level of hardness and the necessary carbon as low as possible, while still meeting engineering requirements. Carbon content is the most important factor determining the level of hardness that can be attained in steels by induction or flame heating. It controls hardness level, the tendency of the part to crack, the magnitude of the part to crack, and the magnitude of residual surface stresses.

The practical level of minimum surface hardnesses attainable with water quenching for various carbon contents is shown in Fig. 2. The curve is applicable for induction hardening as well as for flame hardening. It applies also for alloy steels, except those containing stable carbide formers such as chromium and vanadium.

For best results, steels to be induction or flame hardened should be as-rolled, normalized (particularly from a high temperature), air-blast quenched, or quenched and tempered. These preferred heat treatments result in microstructures conducive to rapid and complete austenitization and full hardening. In selecting steels for either induction or flame hardening, it is important that the necessary steps be taken to ensure that the areas to be hardened are free of decarburization. Depending on stock size, steel grade, producing mill, and several other factors, the depth of decarburization for as-rolled bar may run from near 0 to 3.2 mm (0.125 in.). It should not be assumed that turned and polished bar is free of decarburization unless it is specifically ordered with this requirement. Carbon-restored and cold-finished bar is available from mills in various carbon and alloy grades.

When maximum resistance to fatigue is desired, the hardened surface should contain residual compressive stresses; a recommended level is 172 MPa (25 ksi). Because surfaces hardened to depths of less than 1.9 mm (0.075 in.) are commonly residually stressed in tension, it is suggested that depth of hardening be at least 2.7 mm (0.105 in.) to ensure that residual stresses are compressive. This depth is particularly appropriate for manufacturers not equipped with residual stress-measuring equipment. Further, microstructure should be at least 90% martensite, with no ferrite visible at a magnification of 500×.

Carbon Steels. Plain carbon steels in the range of 0.37 to 0.55% C are the most widely used for flame-hardening applications. They can be through hardened in sections up to 13 mm (0.5 in.). This response permits the use of carbon steel for selectively flame-hardened small gears, shafts, and other parts of small cross section in which uniform properties are needed throughout the section. These same

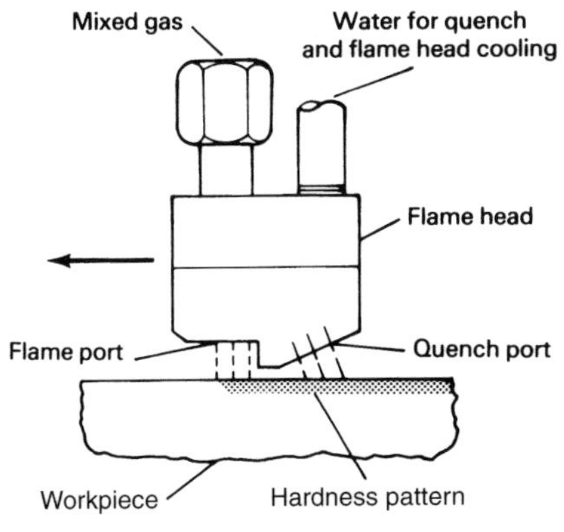

Fig. 1 The progressive flame hardening method

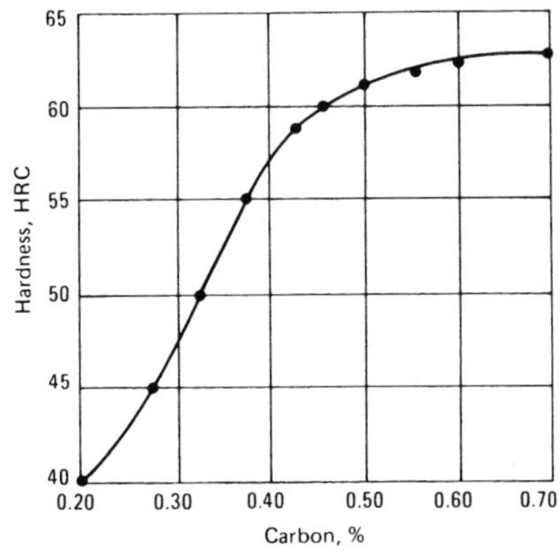

Fig. 2 Relationship of carbon content to minimum surface hardness attainable by flame or induction heating and water quenching. Practical minimum carbon contents can be determined from this curve.

steels can be used for larger parts in which hardness is necessary only to shallow depths from 0.8 to 6.4 mm (1/32 to 1/4 in.).

Carbon steels 1035 to 1053 are suitable for flame hardening; 1042 and 1045 are the most widely available and are recommended for all flame-hardening applications except when they would be incapable of meeting requirements, for example:

- Failure of a 1045 steel part to harden with a given quench would necessitate the use of a steel of higher hardenability, for example, one with higher carbon or manganese or both, or possibly an alloy steel.
- If increased depth of hardening is required, 1042 or 1045 may be inadequate where heavy sections are progressively hardened; therefore, the substitution of 1541, 1552, or an alloy steel would be necessary.
- In applications in which wear resistance is of prime importance, it might be advisable to use a steel with 0.60% C or more to produce maximum surface hardness. Steels this high in carbon content are often quenched in oil or simulated oil to avoid the possibility of cracking due to water quenching. Thus, greater hardenability may be needed with the higher carbon content.
- When a severe quench in brine or caustic is required for hardening 1042 or 1045 steel and such quenching causes cracking, a steel of higher hardenability—either carbon or alloy, which can be hardened by a less severe quench—should be selected.

Alloy Steels. The use of alloy steels for flame-hardening applications is justified only when:

- High core strength is required (through heat treatment before flame hardening), and carbon steels are inadequate to achieve this strength in the section sizes involved.
- The mass and shape of a part, restrictions on distortion, or the hazard of cracking preclude the use of carbon steel quenched in water.
- Certain alloy grades may be more readily obtainable than carbon grades (particularly the higher-manganese carbon grades) appropriate for the application. Steels such as 4135H, 4140H, 6150H, 8640H, 8642H, and 4340H are typical of the more readily obtainable alloy steels.

Steel Castings. Carbon and alloy steel castings are widely used for flame-hardening applications. The selection of a specific composition or grade is made on much the same basis as for wrought carbon and alloy steels.

Attainable Hardness Levels and Depth of Hardness

Hardness of the case in flame hardening is a function of the carbon content of the steel and will range up to 65 HRC (Table 2). Medium-carbon steels with 0.40 to 0.50% carbon are ideal for flame hardening, but steels with carbon contents as high as 1.50% also can be flame hardened with special care. Normally, hardening depth ranges from 1.3 to 6.4 mm (0.05 to 0.25 in.). Heavier sections, such as large rolls and wheels, can have case depths of up to 13 mm (0.5 in.). Manganese-bearing alloys aid in the depth of hardening by decreasing the critical

Table 2 Response of steels to flame hardening

	Typical hardness, HRC, as affected by quenchant		
Material	Air(a)	Oil(b)	Water(b)
Plain carbon steels			
1025–1035	. . .	. . .	33–50
1040–1050	. . .	52–58	55–60
1055–1075	50–60	58–62	60–63
1080–1095	55–62	58–62	62–65
1125–1137	. . .	. . .	45–55
1138–1144	45–55	52–57(c)	55–62
1146–1151	50–55	55–60	58–64
Carburized grades of plain carbon steels(d)			
1010–1020	50–60	58–62	62–65
1108–1120	50–60	60–63	62–65
Alloy steels			
1340–1345	45–55	52–57(c)	55–62
3140–3145	50–60	55–60	60–64
3350	55–60	58–62	63–65
4063	55–60	61–63	63–65
4130–4135	. . .	50–55	55–60
4140–4145	52–56	52–56	55–60
4147–4150	58–62	58–62	62–65
4337–4340	53–57	53–57	60–63
4347	56–60	56–60	62–65
4640	52–56	52–56	60–63
52100	55–60	55–60	62–64
6150	. . .	52–60	55–60
8630–8640	48–53	52–57	58–62
8642–8660	55–63	55–63	62–64
Carburized grades of alloy steels(d)			
3310	55–60	58–62	63–65
4615–4620	58–62	62–65	64–66
8615–8620	. . .	58–62	62–65

(a) To obtain the hardness results indicated, those areas not directly heated must be kept relatively cool during the heating process. (b) Thin sections are susceptible to cracking when quenched with oil or water. (c) Hardness is slightly lower for material heated by spinning or combination progressive-spinning methods than it is for material heated by progressive or spot methods. (d) Hardness values of carburized cases containing 0.90 to 1.10% C cooling rate, which contributes to deep hardening. Therefore, manganese and free-machining grades of steel are considered excellent for flame hardening.

When hardening depths are required beyond the capabilities of ordinary carbon steels (0.60 to 0.90% Mn), elevated manganese ranges such as 0.80 to 1.10%, 1.00 to 1.30%, or 1.10 to 1.40% can be used efficiently. Wear resistance in many cases is not the only critical design criterion. Under high compressive loading, the hardened layer must be deep enough not only to provide the required wear life of the part, but also to contribute to the support of heavy contact loads. The case must be fully martensitic, and the material supporting the hardened layer must be of sufficient strength. However, increased hardenability may lead to cracking problems, at least with water quenching.

Tempering of Flame-Hardened Parts

It is usually desirable to temper parts that have been flame hardened; the need for tempering martensite is the same regardless of the heat-treating method used to produce it. Flame-hardened steel will respond to a tempering treatment in the same manner as it would if hardened to the same degree by any other method. Standard procedures, equipment, and temperatures can be used. However, for work that is flame hardened because it is too large to be heated in a furnace, flame tempering may be the only feasible method of tempering available.

Surface Condition Effects

For wrought and cast steel parts, the surface conditions likely to be detrimental to successful flame hardening are, in general, those that interfere with heating or quenching, cause localized overheating, initiate cracking, or result in the presence of a soft surface skin after proper heating and quenching.

Table 3 summarizes the more common defects or conditions, their origins, and the detrimental effects to be expected when they are present on flame-hardened areas. The extent of these defects determines the amount of difficulty they may cause.

Causes of Distortion

Distortion can occur in flame hardening due to the following causes:

- Shape of part or relationship of portion to be hardened to remainder of section not well adapted to flame hardening

Table 3 Surface conditions detrimental to flame hardening of steel parts

Defect or condition	Probable origin of condition	Detrimental effects to be expected on flame-hardened areas
Laps, seams, folds, fins (wrought parts)	Rolling mill or forging operations	Localized overheating (or, at worst, surface melting), with consequent grain growth, brittleness, and greater hazard of cracking
Scale (adherent)(a)	Rolling or forging: prior heat treatment; flame cutting	Insulating action against heating, with resulting underheated areas and soft spots Localized retardation of quench, causing soft spots
Rust, dirt(a)	Storage and handling of material or parts	Similar to scale condition as noted above left Severe rusting may result in surface pitting that will remain after hardening.
Decarburization	Present in as-received steel bar stock; heating for forging or prior heat treatment of parts or stock	In severely decarburized work, no hardening response will be found when parts are tested by file or other superficial means(b).
Pinholes, shrinkage (castings)	Casting defects	Localized overheating (or, at worst, surface melting), with consequent grain growth, brittleness, and greater hazard of cracking
Coarse-grain gate areas (castings)	Casting gates located in areas to be flame hardened (avoid, if possible)	Increased cracking hazard during quenching, compared with nongated areas; shrinkage defects also likely in these areas
Improper welds	Parts welded with an alloy dissimilar to base metal	Weld zone reaction dissimilar to base-metal reaction. Weld may separate, requiring rewelding or scrapping of the part(c).

(a) In addition to having detrimental effects on flame-hardened surfaces, scale, rust, or dirt in the path of the flame may become dislodged and cause fouling of oxy-fuel gas burners or react chemically with ceramic air-fuel gas burner parts (causing rapid deterioration). When such materials enter a closed quenching system, they may clog strainers, plug quench orifices, and cause excessive wear of pumps. (b) Partial decarburization lowers surface hardness as a direct function of actual carbon content of stock lost at surface, provided that steel was adequately heated and quenched. (c) To avoid these and other problems, it is mandatory that the flame hardener be given accurate and complete information on any changes in composition and past processing of the part. For example, previously hardened parts should never be flame hardened unless they have been annealed; otherwise, cracking is inevitable Source: Ref 2

- Metallurgically unsuitable prior structure
- Heating cycle too long
- Nonuniform heating
- Nonuniform quenching
- Excessive rate of quenching
- Material hardenability excessive

Induction Surface Hardening

Electromagnetic induction is one method of generating heat within a part for hardening or tempering a steel or cast iron part. Any electrical conductor can be heated by electromagnetic induction. As alternating current (ac) from the converter flows through the inductor, or work coil, a highly concentrated, rapidly alternating magnetic field is established within the coil. The strength of this field depends primarily on the magnitude of the current flowing in the coil. The magnetic field thus established induces an electric potential in the part to be heated, and because the part represents a closed circuit, the induced voltage causes the flow of current. The resistance of the part to the flow of the induced current causes heating.

The pattern of heating obtained by induction is determined by the shape of the induction coil producing the magnetic field, number of turns in the coil, operating frequency, ac power input, and nature of the workpiece. Four examples of magnetic fields and induced currents produced by induction coils are shown in Fig. 3.

The rate of heating obtained with induction coils depends on the strength of the magnetic field to which the part is exposed. In the workpiece, this becomes a function of the induced currents and of the resistance to their flow.

The depth of current penetration depends on workpiece permeability, resistivity, and the ac frequency. Because the first two factors vary comparatively little, the greatest variable is frequency. Depth of current penetration decreases as frequency increases. High-frequency current is generally used when shallow heating (thin case) is desired; intermediate and low frequencies are used in applications requiring deeper heating.

Most induction surface-hardening applications require comparatively high power densities and short heating cycles in order to restrict heating to the surface area. The principal metallurgical advantages that may be obtained by surface hardening with induction include increased wear resistance and improved fatigue strength. Another advantage of induction hardening is that localized heating can be used to strengthen components at critical points while leaving other areas soft, without the need for stopoff procedures required in thermochemical treatments such as carburizing (see, for example, the discussion of selective carburizing in Chapter 2, "Gas Carburizing"). Figure 4 shows a constant-velocity automotive front-wheel drive component that has been sectioned and etched to show the pattern obtained by induction heat treating. This component requires two areas of hardness with different strength, load, and wear requirements. The "stem" needs torsional strength as well as a hard outer surface, whereas the soft core must be ductile and therefore able to handle the mechanical shock from constant pulsing. The inner surface of the "bell" needs hardness for wear purposes, because ball bearings ride in the track or raceways.

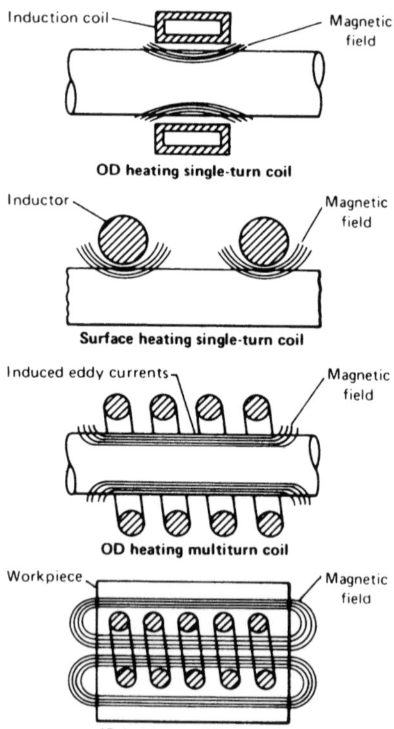

Fig. 3 Magnetic fields and induced currents produced by various induction coils. OD, outside diameter; ID, inside diameter

Induction Heat-Treating Equipment

An induction heat treating system typically consists of a power supply, a workstation, an inductor (heating) coil, workpiece handling equipment, and a quench system. A variety of manipulation procedures can be employed to suit the geometry of the component including "single-shot hardening," in which the entire area to be hardened is heated on a timed basis in one operation and then quenched, and "scan hardening," which involves relative movement between the heating coil, quench head, and the workpiece.

Power Supplies

Most induction power supplies sold for heat treating today are either some type of solid-state or oscillator (vacuum) tube. Many types and models of induction power supplies are made to meet the diverse requirements for different frequencies and output power requirements for induction heat treating (Fig. 5). Regardless of the electronic technology, the power supplies perform a common function. Figure 6 shows a block diagram of modern high-frequency power supplies performing line frequency conversion into high frequency. The power supplies are basically frequency changers that change the 60 Hz (U.S.), three-phase current furnished by the electric utility into a higher-frequency, single-phase current for induction heating. These power supplies are often referred to as converters, inverters, or oscillators, depending on the circuits and electronic devices used, with many possible combinations of conversion techniques.

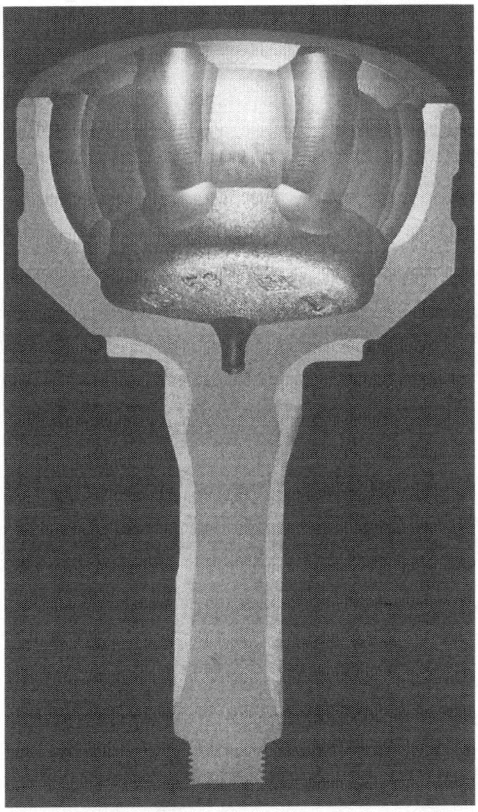

Fig. 4 Induction-hardened constant-velocity automotive front-wheel drive component. The part is sectioned to show two separate heat treat patterns. Source: Inductoheat, Inc.

Solid-state power supplies convert the line alternating voltage (ac) to produce single-phase, direct-current (dc) voltage. Inversion is then accomplished through use of thyristors (silicon controlled rectifiers, or SCRs), or transistors such as isolated gate bipolar transistors (IGBTs) or metal-silicon-dioxide field-effect transistors (MOS FETs), to produce dc pulses that are then made sinusoidal to form high-frequency ac. (Some current source power supplies do this in one step.) Radio frequency (RF) (oscillator or vacuum tube) power supplies use a transformer to change the input voltage to high voltage before conversion to dc. The oscillator tube is used to produce dc pulses that are likewise changed into the high-frequency ac current. The higher output voltage of the RF tube power supplies is one of their more distinguishable features.

Figure 7 shows the use of SCRs, transistors (IGBTs and MOS FETs), and vacuum tube oscillators currently in use. At the lower end of the frequency range, up to 10 kHz, SCRs are in wide use for the switching devices because of device cost versus current carrying capabilities. In the medium frequencies, IGBT transistors are used, and in the higher frequencies, MOS FET transistors are used. In the future, as the current-carrying ability of transistors is increased and the cost is decreased, transistors are expected to come into wider use over the full frequency range. Solid-state power supplies and the RF oscillator tube power supplies have considerable differences in efficiency, as shown in Table 4. The lower frequency, solid-state power supplies are more efficient in energy conversion. The RF oscillator tube has a filament that consumes energy being heated all the time, and the switching losses in oscillator tubes are high.

Solid-State Advantages. Solid-state power supplies are preferred when the workpieces are large enough to permit cost-efficient frequency selection. High power units are less expensive and smaller in size than oscillator tube units, while having higher efficiency in conversion from line frequency to terminal output. Solid-state power supplies require no warm up, and they have a high degree of reliability. Finally, solid-state power supplies inherently have better power regulation with the ability to produce full power over an entire heating cycle. At the higher frequencies, such as above 50 kHz, the smaller MOS FET transistors are used. Higher frequencies cause more switching losses, resulting in reduction of the output power rating. With the higher frequencies, such as above 300 kHz, vacuum tube oscillators are still widely used.

Oscillator tube units operate in the 200 kHz up through 2 MHz frequency range and tend to have higher cost per kW of power sold. Older power supplies used rectifier tubes to complete the rectification to dc, while modern units use solid-state diodes. (The only tube in a modern power supply is the oscillator tube.) The output power of an RF oscillator decreases when magnetic steel parts are heated through the Curie temperature, so it is harder to maintain full power output. However, RF power supplies have been around for many years and have more versatility in impedance matching and tuning than solid-state power supplies. Radio fre-

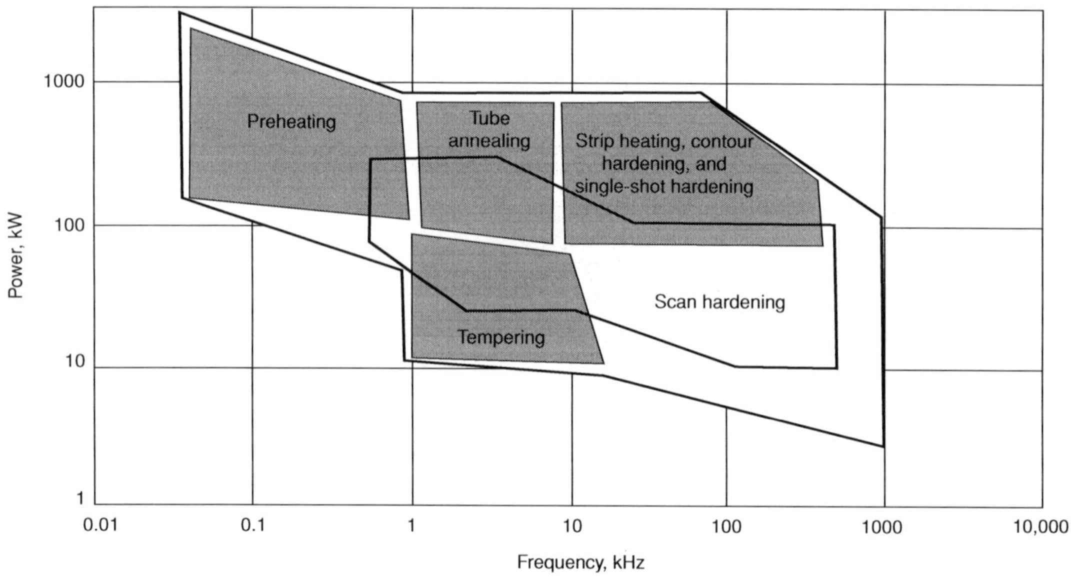

Fig. 5 Power level and frequency combinations for various induction heating applications. Source: Ref 3

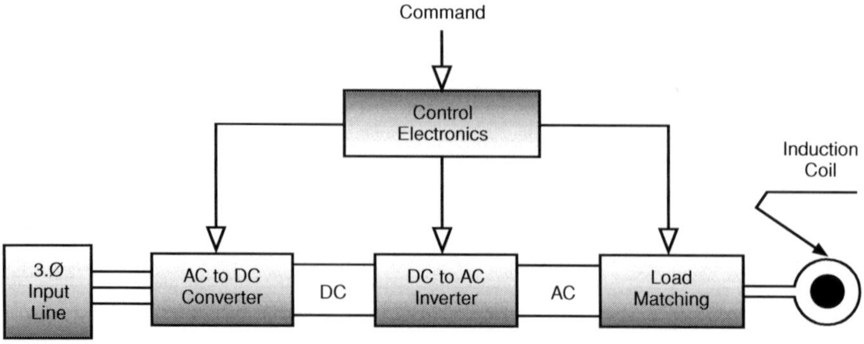

Fig. 6 Induction heat treat power supply basic diagram. ac, alternating current; dc, direct current. Source: Ref 3

quency units are easy to tune, and when there is a component failure, they are easy to troubleshoot. Radio frequency tube power supplies have been in wide use for 50 years and have a good history of operation. Although oscillator tubes have 1,000 h warranties, tube life up to 25,000 h or more is not unusual.

Selection of Frequency, Power, and Duration of Heating

The distribution of induced current in a part is maximum on the surface and decreases rapidly within the part; the effective penetration of current increases with a decrease in the frequency. The distribution of induced current is influenced also by the magnetic and electrical characteristics of the part being heated; and because these properties change with temperature, the current distribution will change as the work is heated.

Because the heat rapidly progresses to the interior by conduction as soon as the surface is heated, the actual depth of heating is determined by the duration of heating and the power density (kilowatts per square inch of surface exposed to the inductor), as well as by the frequency. Maximum power density, minimum duration of heating, and high frequency produce a minimum depth of heating.

Selection of Frequency. In analyzing the frequency and power required for a specific application, it is desirable to consider the frequency first. Primary considerations are the depth of heating and the size of the part. Table 5

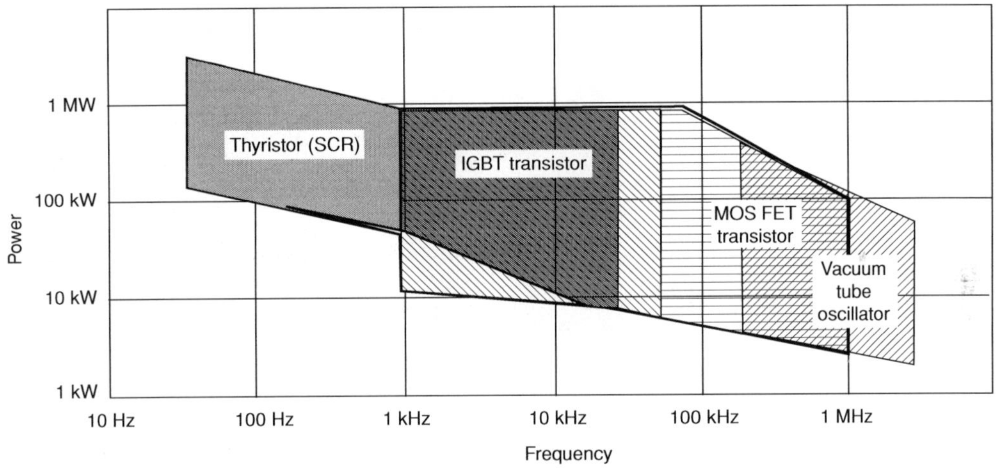

Fig. 7 Modern inverter types for induction heat treatment. Source: Ref 3

Table 4 Comparative efficiencies of various power sources

Power source	Frequency	Terminal efficiency, %	Coil efficiency, %	System efficiency, %
Supply system	50 to 60 Hz	93 to 97	50 to 90	45 to 85
Frequency multiplier	50 to 180 Hz	85 to 90	50 to 90	40 to 80
	150 to 540 Hz	93 to 95	60 to 92	55 to 85
Motor-generator	1 kHz	85 to 90	67 to 93	55 to 80
	3 kHz	83 to 88	70 to 95	55 to 80
	10 kHz	75 to 83	75 to 96	55 to 80
Static inverter	500 Hz	92 to 96	60 to 92	55 to 85
	1 kHz	91 to 95	70 to 93	60 to 85
	3 kHz	90 to 93	70 to 95	60 to 85
	10 kHz	87 to 90	76 to 96	60 to 85
Radio-frequency generator	200 to 500 kHz	55 to 65	92 to 96	50 to 60

Source: Ref 4

lists the frequencies most commonly used In induction hardening. As shown in this tabulation, the lower frequencies are more suitable as the size of the part and the case depth increase. Use of the wrong frequency will result in a decrease in electrical efficiency, sometimes in failure to maintain a minimum case depth where shallow cases are required and sometimes in failure to heat uniformly throughout the piece where through hardening is required.

Selection of Power. The size of the converter or the power required should be determined on the basis of power density, section size, heating method, and production requirements. In surface hardening, the area heated at one time multiplied by the power density indicates the total power input (kilowatts). This area is obtained by multiplying the perimeter of the part by the length of the inductor. Typical power ratings for surface hardening of steel are listed in Table 6.

Selection of Duration of Heating. When the frequency and power density have been selected, the duration of the heating cycle becomes a fixed value for a specific set of conditions. To calculate duration of heating for surface hardening by the single-shot method, divide the value for kilowatt seconds per square inch by power density (kilowatts per square inch). The value of kilowatt seconds per square inch is affected by case-depth requirements, type of steel, and prior structure and may be derived by experiment or be based on previous experience. To calculate heating time for surface hardening by the scanning method, divide kilowatt seconds per square inch by power density and inductor length.

Table 5 Choice of induction hardening frequency for a minimum hardness of 50 HRC

Hardening depth		Part diameter		Frequency, kHz(a)(b)			
mm	in.	mm	in.	1	3	10	100
0.3–1.2	0.01–0.05	6–25	0.2–1	...	...	...	1
1.2–2.5	0.05–0.1	11–56	0.4–2.2	...	...	2	1
		16–25	0.6–1	...	...	1	1
		25–50	1–2	...	2	1	2
		50	2	2	1	1	...
2.5–5	0.1–0.2	19–50	0.7–2	...	1	1	2
		50–100	2–4	1	1	2	...
		100	4	1	2	3	...

(a) 1, best; 2, satisfactory; 3, acceptable. (b) Note that a frequency of 10 kHz covers a wide range of applications. Source: Inductoheat, Inc.

Selection of Coil Design

Coil design is influenced by a number of factors, including the dimensions and configuration of the part to be heated, the heat pattern desired, whether the part is heated throughout its length at the same time or progressively, the number of parts to be heated, and the frequency and power of the induction heater.

Basic Designs. Five basic designs of work coils for use with high-frequency (over 200 kHz) units and the heat patterns developed by

Table 6 Power densities required for surface hardening of steel

			Input(b)(c)					
	Depth of hardening(a)		Low(d)		Optimum(e)		High(f)	
Frequency, kHz	mm	in.	kW/cm2	kW/in.^{2}	kW/cm^2	kW/in.^{2}	kW/cm^2	kW/in.^{2}
500	0.381–1.143	0.015–0.045	1.08	7	1.55	10	1.86	12
	1.143–2.286	0.045–0.090	0.46	3	0.78	5	1.24	8
10	1.524–2.286	0.060–0.090	1.24	8	1.55	10	2.48	16
	2.286–3.048	0.090–0.120	0.78	5	1.55	10	2.33	15
	3.048–4.064	0.120–0.160	0.78	5	1.55	10	2.17	14
3	2.286–3.048	0.090–0.120	1.55	10	2.33	15	2.64	17
	3.048–4.064	0.120–0.160	0.78	5	2.17	14	2.48	16
	4.064–5.080	0.160–0.200	0.78	5	1.55	10	2.17	14
1	5.080–7.112	0.200–0.280	0.78	5	1.55	10	1.86	12
	7.112–8.890	0.280–0.350	0.78	5	1.55	10	1.86	12

(a) For greater depths of hardening, lower kilowatt inputs are used. (b) These values are based on use of proper frequency and normal overall operating efficiency of equipment. These values may be used for both single-shot and scanning methods of heating; however, for some applications, higher inputs can be used for scan hardening. (c), Kilowattage is read as maximum during heat cycle. (d) Low kilowatt input may be used when generator capacity is limited. These kilowatt values may be used to calculate largest part hardened (single-shot method) with a given generator. (e) For best metallurgical results. (f) For higher production when generator capacity is available.

each are shown in Fig. 8(a) through (e). These basic shapes are (a) a simple solenoid for external heating; (b) a coil to be used internally for heating bores; (c) a pie-plate type of coil designed to provide high current densities in a narrow band for scanning applications; (d) a single-turn coil for scanning a rotating surface, provided with a contoured half-turn that will aid in heating the fillet; and (e) a pancake coil for spot heating. Solenoid coils for external heating are most efficient and should be used whenever possible.

The same designs are used for lower frequencies, although the higher powers may require milled copper coil construction. This type of coil construction involves milling or drilling out of holes, followed by brazing in of inserts, to form the cooling passages.

Ferrite concentrators can be used on coils to increase coil efficiency. Laminated iron concentrators can be used at 1 to 10 kHz to increase coil efficiency.

It usually is important to keep coil lead lengths as short as possible. If the lead lengths provide excessive power drops, they should be made wider or brought closer together (or both). The number of turns in a coil depends on the requirements of the area to be heated and on the ability to match the impedance of the power supply.

Commercial copper tubing may be used for coils. The tubing must be large enough to permit an adequate flow of water for cooling.

Coil Coolants. Water is commonly used for cooling inductors, although in some applications oil, modified water, or a polymer quench may be employed to serve the dual purpose of cooling the inductor and quenching the workpiece in a continuous heating and quenching operation. Generally, the water should have a hardness of less than 10 grains/gal. If the water-cooling passages are small relative to the current load carried by the inductor, it may be necessary to use distilled or deionized water to avoid a deposit buildup that could eventually stop circulation. Preferably, the water should be filtered to remove foreign particles that might clog small passageways, especially when intricately designed inductors are being used. The water should have an inlet temperature below 35 °C (95 °F), and flow should be sufficient to prevent the outlet temperature from rising above 66 °C (150 °F).

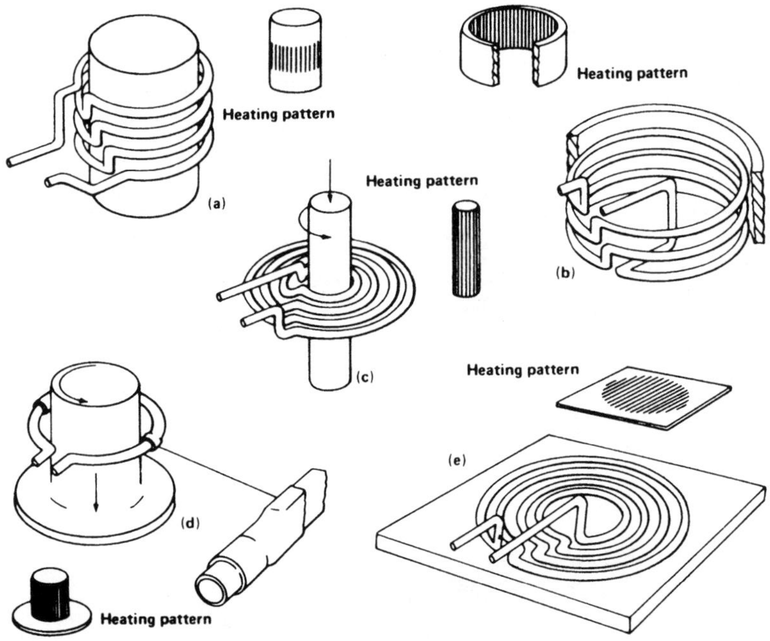

Fig. 8 Typical work coils for high-frequency units. See text for details.

Quenching

The type of quench used will depend on metallurgical considerations. A great many induction hardening applications employ water as the quenching medium. Other media, such as conventional quenching oil, water modified by organic polymer, and compressed air, are occasionally used. Water is easiest to handle, simple to install and maintain, and generally less hazardous than other media. Oil quenching produces the least distortion and provides the smallest tendency toward cracking. The modified-water compounds are compounds with organic polymers that are soluble in water. The temperature and concentration determine the quenching rate. Compressed air is used in shallow-case applications where the air and the massive heat sink of the workpiece are used to produce the required cooling rate.

Basic Systems for Quenching. Eleven basic arrangements for quenching induction hardened parts are shown schematically in Fig. 9(a) through (k). In correlation with the lettering there, these arrangements are briefly described as follows:

- (a) Heat in coil; manually lift part out of coil. Submerge part in tank of agitated quench medium. Used where limited production does not warrant the cost of an automated quench
- (b) Heat and quench in one position; quench by means of integral quench chamber in inductor. Called single-shot method
- (c) Heat in coil with part stationary; quench ring moves in place. Single-shot adaptation of scanning method
- (d) Part is hydraulically lowered into quench tank after single-shot heating. Quench medium is agitated by submerged spray ring or propeller.
- (e) Vertical or horizontal scanning with integral spray quench. Single-turn inductor. Used for shallow hardening
- (f) Vertical or horizontal scanning with multiturn coil and separate multirow quench ring. Used for deep-case or through hardening
- (g) Coil scans and heats workpiece; self-quench or compressed air quench. Used in special applications with high-hardenability steels
- (h) Horizontal cam-fed parts are pushed through coil, then dropped onto submerged quench conveyor.
- (i) Vertical scanning with single-turn inductor in combination with integral dual quench. One quench ring for scan hardening; the second for stationary quenching when the scanning travel stops. Used for parts having a diameter or a flange section too large to travel through the inductor, wherein it is desired to harden up to the shoulder or flange
- (j) Vertical scanning with single-turn inductor with integral spray quench and submerged quench in tank
- (k) Split inductor and integral split quench ring. Used for hardening crankshaft bearing surfaces

Induction Surface Hardening Metallurgy (Ref 5)

The rapid heating, short austenitizing times, and rapid cooling (quenching) characteristics of induction hardening produce structures that differ from those associated with conventional furnace hardening. However, the physical metallurgy involved is the same. This section compares the induction and conventional methods, with particular regard to austenitizing and subsequent cooling.

Austenitizing Heat Treatment. The precursor step to hardening is to austenitize the part. For conventional heat treatment, this involves heating the steel to the desired austenitizing temperature in the austenite (γ) phase field (Fig. 10) and holding for sufficient time to obtain a chemically homogeneous, single-phase austenite structure. Austenite forms by a nucleation and growth process, and the rate depends on the beginning microstructure (pearlite or bainite, for example).

The formation of austenite can be studied on isothermal transformation, but it is more useful to examine its formation on continuous heating, since this simulates actual heat treating practice. A time-temperature-transformation (TTT) diagram for the formation of austenite in a steel on continuous heating is shown in Fig. 11. Included is a typical heating curve for a part placed in an air furnace at 900 °C (1650 °F) (C, conventional heating). Note that homogeneous austenite is obtained after about 1 h (3600 s).

Also included is a curve labeled R for a part rapidly heated using a high-heat-input method such as induction. Note that a higher temperature is required to achieve homogeneous austenite. If a lower maximum austenitizing temperature is used, the structure may contain chemical gradients or undissolved carbides. This reduces hardenability, which means that more rapid cooling will be required to form martensite than if the structure were homogeneous austenite.

Hardening Methods Compared. Conventional and high-heat-flux surface hardening processes are compared in Fig. 12. In the conventional method (Fig. 12a), the part is heated in the furnace to the austenitizing temperature, held sufficiently long to ensure the formation of

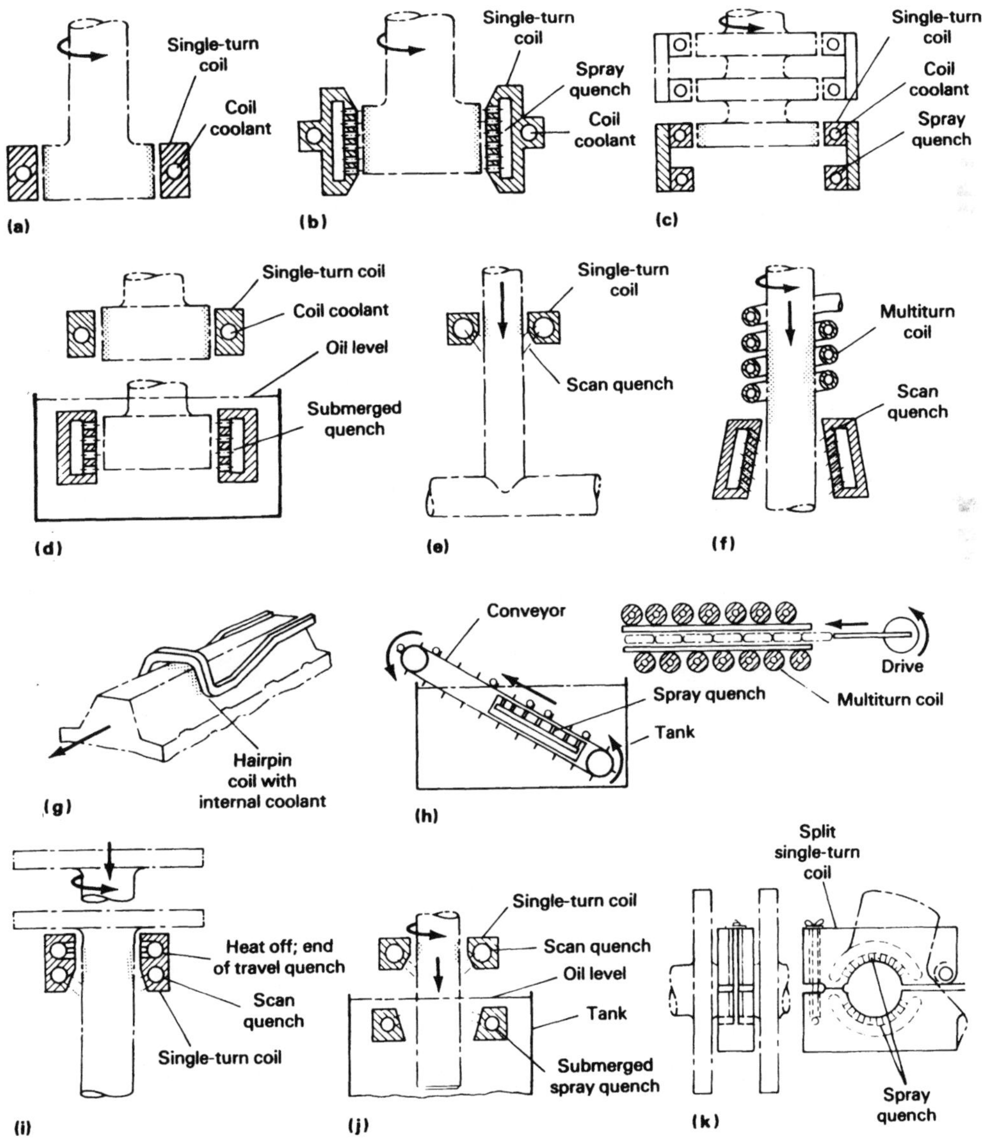

Fig. 9 Basic arrangements for quenching induction hardened parts. See text for details.

austenite (1 h, for example), and then quenched into a medium such as water or oil that has been selected to produce the desired hardness distribution. This heat treatment is then frequently followed by tempering (heating below the eutectoid temperature). Any martensite present is converted to a fine mixture of carbides in ferrite (called tempered martensite) having improved toughness.

In contrast, induction surface hardening (Fig. 12b) involves rapid heating to the austenite region, followed by rapid cooling. A high heat flux is applied to the surface for a time sufficient to austenitize the surface of the part but not its interior. When the surface has reached the proper austenitizing temperature for the desired time, the energy source is removed and the hot surface region cools by conduction of heat into the colder interior and by cooling from the surface inward (by water spraying, for example). Note that if the heated layer is small compared with the bulk of the part, then cooling of the surface by conduction into the unheated interior can be very rapid.

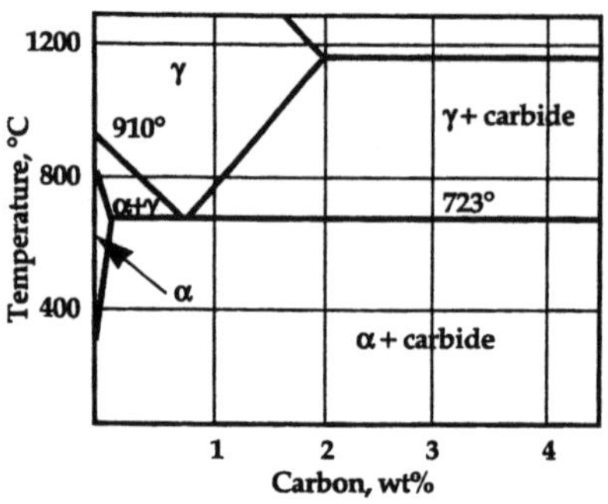

Fig. 10 A portion of the iron-carbon phase diagram. Source: Ref 6

Effects on Microstructure. The heating and cooling processes for induction hardening are shown schematically in Fig. 13. Heating times will be relatively short (on the order of seconds), as will cooling times.

The surface is in the austenite region longest and at higher temperatures, and hence the austenite grains are subject to growth after they form. Subsurface regions, which also are heated into the austenite range, may have smaller austenite grains (curve 2 in Fig. 13). In addition, there are locations which contain mixtures of

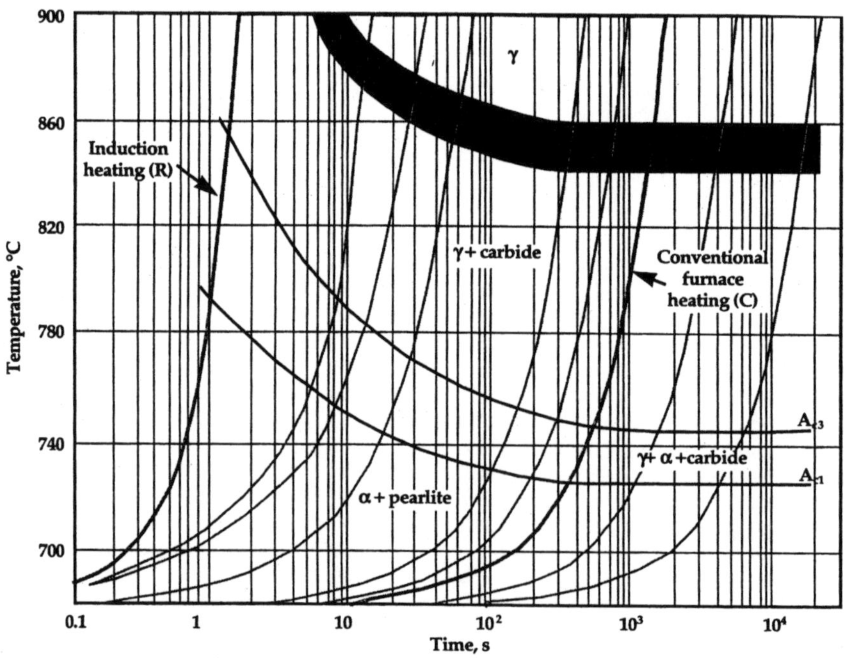

Fig. 11 A typical TTT diagram for the formation of austenite. Also shown are typical curves for conventional furnace heating, C, and rapid heating by induction, R. Source: Ref 7

austenite and the starting microconstituents (austenite and primary ferrite, for example).

Because cooling at the surface is usually very rapid, lasting only a few seconds, martensite usually forms, as shown schematically in Fig. 14. As the interior is approached, a region is reached that at high temperature is only partially austenite, which then forms on cooling a mixture of martensite and the unaffected structure.

Hardness Values. Hardness as a function of depth depends on the underlying microstructure and is a useful way to illustrate these effects. Figure 15 shows the hardness distribution after induction heating of a 0.8% C steel having an initial microstructure of all pearlite. Curves are given for various maximum temperatures reached at the surface of the steel.

Heating to 700 °C (1290 °F) produced no hardness variation, because this temperature is below the eutectoid (723 °C, or 1333 °F); no austenite formed as a result (see Fig. 10). Heating the 0.8% C steel to 800 °C (1470 °F) produced an all-austenite surface, which transformed on cooling to almost all martensite with a hardness of about 780 HK (~63 HRC). For maximum surface temperatures higher than this, the hardness increased slightly, to about 850 HK (65 HRC). Also, heating to a higher surface temperature produced austenite, and hence martensite, to a greater depth.

Use of Jominy Data. A convenient method of examining the influence of rapid heating on the transformation of austenite is to surface heat a Jominy bar. The bar is placed in the induction coil, heated for the desired time, and then end-quenched in a water spray. A small flat is ground along the length of the bar, and the hardness is measured. Hardness values are plotted versus distance from the quenched end. The depth of the flat is sufficiently small so that the hardness measured is essentially that at the surface that was induction heated.

Jominy bar data for three steels—AISI 1050, 4150, and 4340—both conventional furnace heated and induction heated are plotted in Fig. 16.

The AISI 1050 carbon steel (Fig. 16a), was induction heated for about 20 s and end-quenched when the surface reached 870 °C (1600 °F); that is, the bar was held for 0 s at 870 °C (1600 °F). Its conventionally hardened counterpart was heated 1 h at 870 °C (1600 °F) in an air furnace and then end-quenched. The shapes of the Jominy curves are similar, showing that the steel is of low hardenability, with an all-martensite structure forming only very near the quenched end.

Data for the two low-alloy steels, Fig. 16(b) and (c), show how the addition of alloying elements improves hardenability, as dramatized by the curves for the furnace heated steels. The curves for the induction heated steels are markedly lower, reflecting the fact that rapid heating reduces hardenability. The effect is

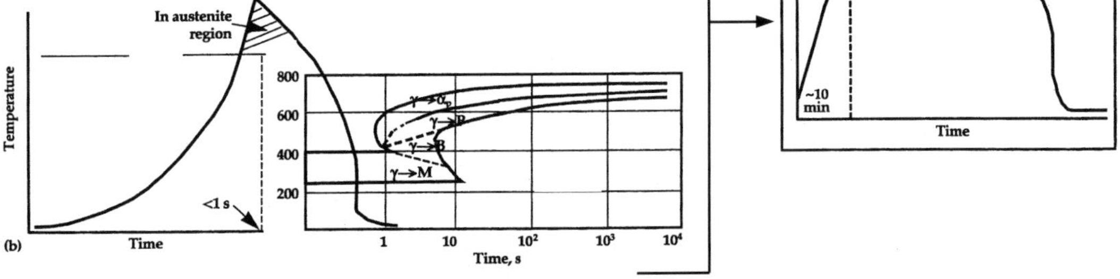

Fig. 12 (a) Conventional furnace hardening of steel is compared with (b) surface hardening using a high-heating-rate method such as induction. γ, austenite; α_p, primary ferrite; P, pearlite; B, bainite; M, martensite. The tempering step is the same for both methods.

associated with undissolved carbides and fine austenite grains in the microstructure of the rapidly heated steels. However, note that the hardness at the quenched end is essentially the same as that for steels conventionally austenitized. Although this region consists of austenite and undissolved carbides at the austenitizing temperature—a low hardenability structure—the very rapid cooling at the quenched end still produced a high-hardness martensite containing fine undissolved carbides.

The effects of induction heating time and maximum surface temperature on the Jominy curve for AISI 4150 are shown in Fig. 17. Increasing either of these parameters produces more homogeneous austenite and larger austenite grains, which increase hardenability and raise the Jominy curve. However, note that the hardness at the quenched end remains essentially the same, even though the hardenability of the steel may be lower. The rapid cooling at the end of the bar allowed a structure of martensite with some undissolved carbides to form, retaining high hardness.

This explains why high surface hardness can be obtained by rapid heating followed by rapid

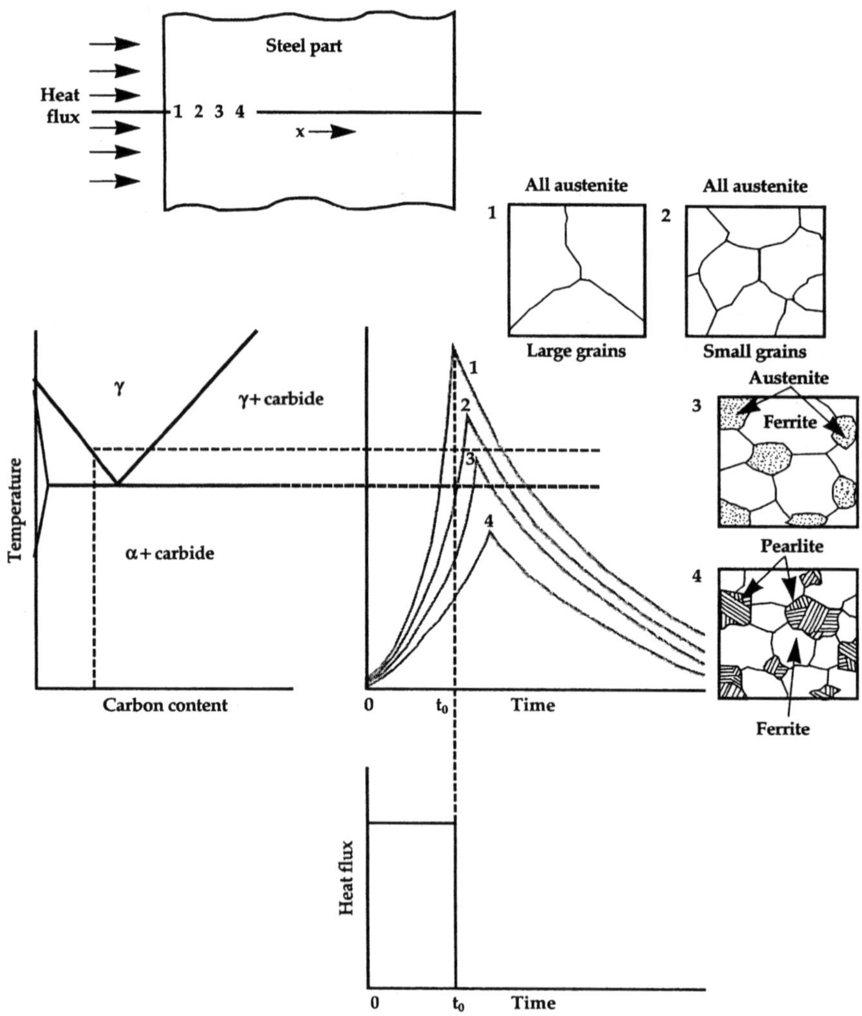

Fig. 13 Temperature-time curves as a function of depth for induction surface hardening of a 0.4% C steel. Schematic microstructures show austenite formation. Austenite grains at the surface are subject to growth after they form. Key: γ, austenite; α, ferrite. Source: Ref 8

cooling. In induction hardening, austenite only has time to form in the surface region; the interior of the part stays at approximately room temperature. Thus, when the energy source is removed, the hot surface quickly cools by conduction into the cold center. This is equivalent to a severity of quench approaching infinity, which is similar to a high-velocity water spray such as that used in the Jominy end-quench test. The hardenability at the surface may be low, but the cooling rate is high enough to promote the formation of a very hard structure.

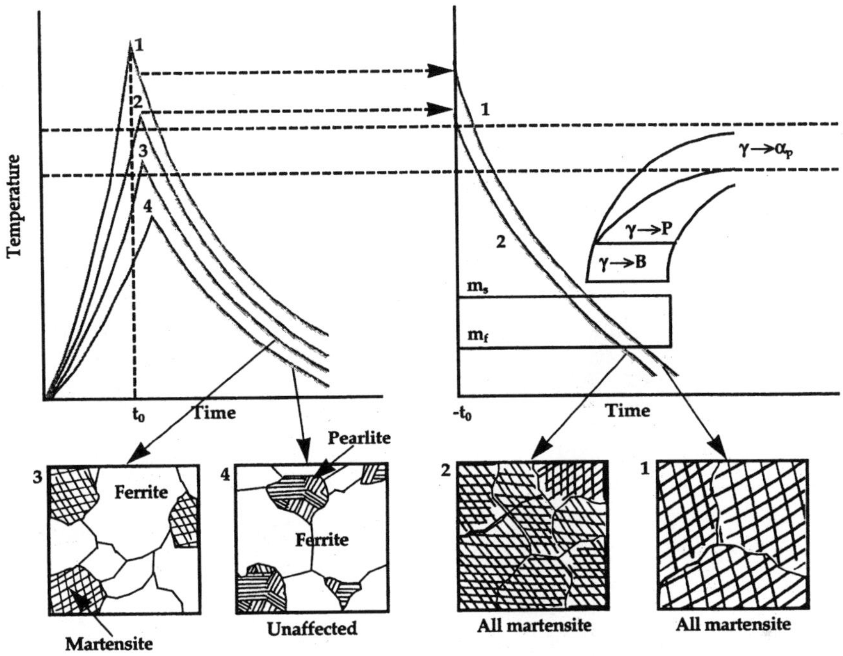

Fig. 14 Transformation of austenite upon cooling of a surface-heated 0.4% C steel. Numbers 1 to 4 refer to depth locations in the steel part (see Fig. 13). Martensite usually forms at the surface. Key: γ, austenite; α_p, primary ferrite; P, pearlite; B, bainite. Source: Ref 8

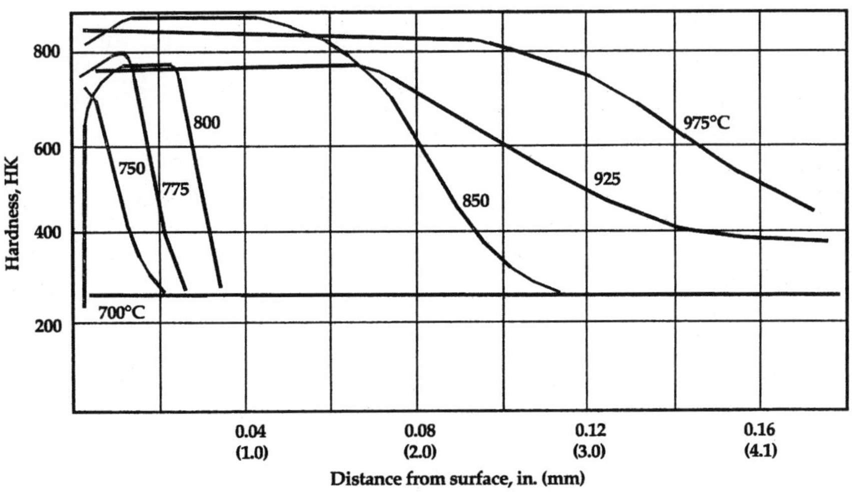

Fig. 15 Hardness profiles for an induction hardened 0.8% C steel for various maximum surface temperatures. The initial microstructure of the steel was all pearlite. Source: Ref 9

Summary. The most important difference between the hardening obtained by rapid surface heating and conventional heat treatment is that the former may produce inhomogeneous austenite. Undissolved carbides may be present, and there may be concentration gradients of carbon and alloying elements in the austenite. In alloy steels, these carbides may be relatively high in alloy content and hence dissolve more slowly than iron carbide. Also, substitutional alloying elements such as manganese, chromium, nickel, and molybdenum diffuse slowly. As a result, more time and higher temperatures are needed to form homogeneous austenite. In spite of the low hardenability of the surface region, the cooling rate is usually high enough to ensure a martensite-rich structure and, hence, high hardness.

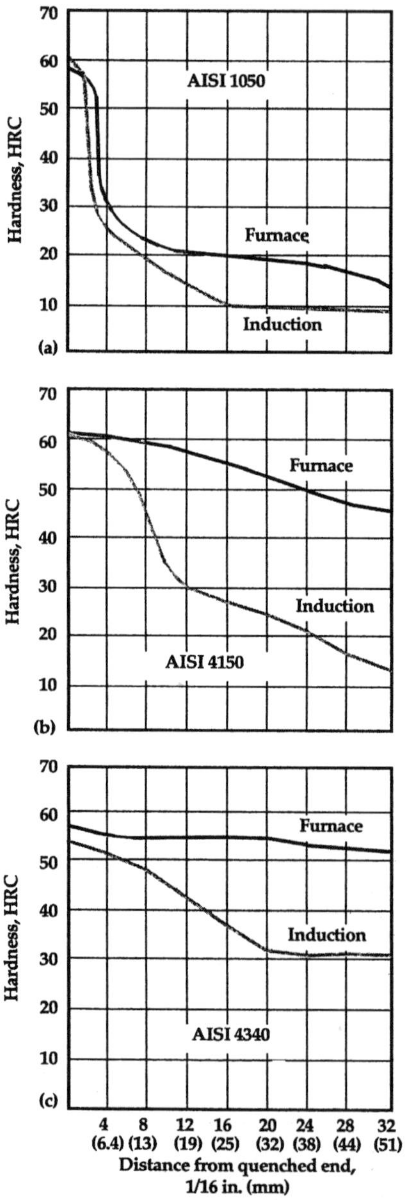

Fig. 16 Jominy curves for end-quenched bars of (a) AISI 1050, (b) 4150, and (c) 4340 steels, austenitized conventionally and by short-time induction heating. Source: Ref 10

Induction-Hardenable Steels, Case Depths, and Hardness Patterns

Induction-Hardenable Steels. Induction surface hardening is applied mostly to hardenable grades of steel, although some carburized parts are often reheated in selected areas by induction heating. Examples of steels surface hardened by induction include:

- Low-carbon steels are used when toughness rather than high hardness is required. These include AISI 1020–1035.
- Medium-carbon steels (AISI 1035–1050) are the most common induction-hardened steels. These steels are often used in automotive components such as front wheel drive components (Fig. 4), gears, and drive shafts.
- High-carbon steels (AISI 1050–1080) are used primarily for tools such as drill and rock bits and other tools due to their ability to achieve high hardness.
- Alloy steels are used for such things as bearings (AISI 52100) and automotive components and machine-tool components (AISI 4130, 4140, and 4340).
- More highly alloyed tool steels (O1, D2, D3, A1, and S1) and some martensitic stainless steels (AISI 416, 420, and 440C) are also sometimes induction hardened.

Table 7 lists steels suitable for induction hardening as well as their induction hardening austenitizing temperatures.

Case Depth. As described earlier in this chapter, frequency and power selection influence the case depth of induction hardened parts. A shallow, fully hardened case ranging in depth from 0.25 to 1.5 mm (0.010 to 0.060 in.) provides a part with good wear resistance for applications involving light to moderate loading. For this kind of shallow hardening, the depth of austenitizing may be controlled by using frequencies on the order of 10 kHz to 2 MHz, power densities to the coil of 800 to 8000 W/cm^2 (5 to 50 kW/in.^{2}), and heating time of not more than a very few seconds. Pump shafts, rocker arm shafts, and sucker rods are typical parts that benefit from a shallow-hardened case for wear resistance.

Where high loading stresses penetrate well below the surface, whether it be bending, torsion, or brinnelling, the metal needs to be strengthened so at any depth, its yield strength exceeds the maximum applied stress at that depth. Because loading stresses drop off exponentially from the surface to the center of a shaft, it is obvious a deep case with high hardness can be effective in strengthening below the surface. Consequently, parts subjected to heavy

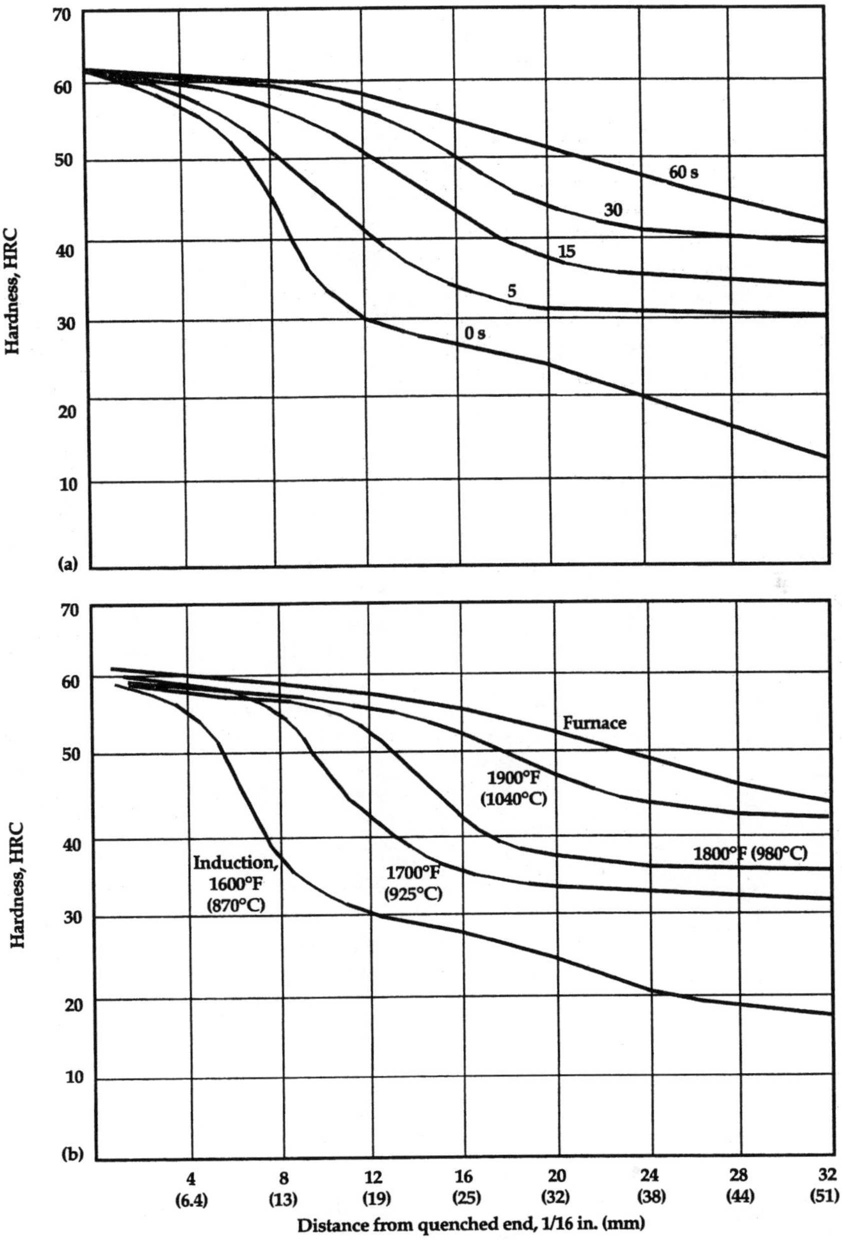

Fig. 17 Effect of time at (a) an 870 °C (1600 °F) austenitizing temperature and (b) maximum surface temperature on the Jominy curves for induction hardened AISI 4150 steel. The curve for conventional furnace heated 4150 is also shown in (b). Source: Ref 10

loads, particularly cyclic bending, torsion, or brinnelling, may require a thicker case depth (that is, deeper hardness). The hardened depth might then be increased to 1.5 to 6.4 mm (0.60 to 0.250 in.), which would require:

- Frequencies ranging from 10 kHz down to 1 kHz
- Power densities on the order of 80 to 1550 W/cm^2 (1/2 to 10 kW/in.^{2})
- Heating times of several seconds

Heavy duty gears, drive axles, wheel spindles, and heavily loaded bearings are typical parts to which this kind of strengthening surface heat treatment is most applicable.

Required hardness patterns can be determined from stress calculations, because hardness values can be translated to yield strength. The required case depth also depends on the distribution of the residual compressive stresses (induced by the transformation hardening of the surface region) and the loading stresses within the body of the part. Where a transformation hardened case ends, either in depth or at the termination of a hardened surface pattern, a stress reversal will most likely occur. This condition should be avoided in any region of the part that carries any significant portion of the load.

For example, the hardness pattern on a load-carrying gear should not terminate in the root when bending stresses tend to concentrate. On the other hand, fly wheel ring gears and some sprockets are just hardened on the tooth flanks only to resist wear. The discontinuous pattern reduces distortion because there is no hoop stress from hardening a continuous ring. If a spline or a keyway is in the torsional load transmitting part of a shaft, it should be hardened below the root or notch.

Table 7 Induction-hardenable steels and their approximate induction austenitizing temperatures

Steel	Carbon, %	Austenitizing temperature °C	Austenitizing temperature °F
1022	0.18/0.23	900	1650
1030	0.28/0.34	875	1600
10B35	0.32/0.38	855	1575
1040	0.37/0.44	855	1575
1045	0.43/0.50	845	1550
1050	0.48/0.55	845	1550
1141	0.37/0.45	845	1550
1144	0.40/0.48	845	1550
1541	0.36/0.44	845	1550
4130	0.28/0.33	870	1600
4140	0.38/0.43	875	1600
4150	0.48/0.53	845	1550
4340	0.38/0.43	845	1550
5160	0.56/0.64	845	1550
52100	0.98/1.1	800	1475
8620	0.18/0.23	875	1600
1018 Carb.	0.9 nom	815	1500
1118 Carb.	0.9 nom	815	1500
8620 Carb.	0.9 nom	815	1500
5120 Carb.	0.9 nom	815	1500
416 SS	<0.15	1065	1950
420 SS	>0.15	1065	1950
440C SS	0.95/1.2	1065	1950
O1	0.9	815	1500
D2	1.5	1020	1875
D3	2.25	980	1800
A1	1	980	1800
S1	0.5	955	1750

The induction austenitizing temperature can be up to 200 °F (110 °C) higher depending upon the prior microstructure and the rate of heating. Source: Ref 11

Revealing Hardness Patterns and Depth of Hardening by Macroetching (Ref 12)

Hardness patterns and depths of hardening must be evaluated in certain situations such as during the design of induction coils and fixtures, production setup, quality control, and failure analysis. This section describes methods of sample preparation and macroetching for visual examination of induction hardened parts. Three macroetchants found particularly useful for revealing hardness patterns and depth of hardening are described.

Sample Preparation: Cleaning, Sectioning, and Grinding. Parts must be cleaned of dirt, grease, and oil before etching. Rinse with acetone or other solvent to remove water, and dry with a clean air source. Shop air is not recommended, as it usually has oil added to lubricate the pneumatic tools connected to it. Do not touch the cleaned part with bare hands—skin oils may cause smudges.

If the part is to be sectioned before etching, extreme care must be used when cutting. Most induction hardened parts have residual compressive stresses that can cause burns on section faces or can clamp onto and break the cutoff blade. An abrasive cutoff machine with an oscillating or linear movement between wheel and workpiece is recommended. It is also important to select the proper cutoff wheel. Consult a metallographic consumables supplier for recommendations.

Some grinding after sectioning is recommended. Larger sections may need an initial grinding step on 120-grit abrasive to create a flat surface. After that, or with smaller sections, one grinding step with 180-grit abrasive is all that is needed. A belt grinder with zirconium oxide belts and a water coolant was used to prepare the samples discussed subsequently.

Macroetchant 1. This etchant consists of equal volumes of hydrochloric acid and water. It works best on larger sections. Mix enough to completely immerse the sample. Heat the solution to around 50 °C (125 °F) before use. Many sources recommend heating to 70 to 80 °C (160 to 180 °F), but this is primarily for showing grain flow. Heating to the lower temperature is sufficient for revealing induction hardened patterns.

Immerse the sample for approximately 30 min. Rinse the sample in hot running water, and scrub with a soft bristle brush (a clean, used toothbrush works well). Rinse with ethanol or methanol, and blow dry the surface with a clean air source. Re-etch if necessary to bring out the pattern.

If the sample is to be stored after examination for future reference, coat the surface with a thin layer of light oil or a clear water soluble spray (some hair sprays work, but try them on scrap parts first).

This etchant can show more than just the induction hardened pattern. An etched section through a heavy-duty axle shaft is shown in Fig. 18. The friction weld between the bell end and body and the forged grain flow of the bell end also are revealed.

Macroetchant 2. This is a two-step, room temperature process that uses 40 vol% nitric acid in water to etch and 4 to 10 vol% hydrochloric acid in ethanol to clean. It works on both cut sections and intact (unsectioned) parts.

Immerse the sample in the nitric acid etch or apply the solution with a squeeze bottle (immersion is recommended, particularly for larger samples). Etch for 10 to 20 s, allowing the surface to darken. Rinse in hot running water, follow with a rinse in ethanol, and then dry.

Apply the hydrochloric acid solution by either immersion or squeeze bottle. Allow 15 to 30 s for the solution to react and then gently scrub the surface with a soft bristle brush to remove the black "smut." Rinse in hot running water, rinse in ethanol, and dry.

The cross section in Fig. 19 shows the runout of the induction hardened pattern in the end of a shaft. Centerline segregation in the continuously cast shaft is also revealed. The cleaning solution contained 10 vol% hydrochloric acid in ethanol.

Circular shafts are often rotated inside the heating coil to make the hardened pattern around the diameter more even and are also slowly pulled through the coil (scanned) so that the coil does not have to be as long as the shaft. The shaft in Fig. 20 shows the "barber pole" hardened pattern caused by pulling the rotating shaft through the coil too quickly. Concentra-

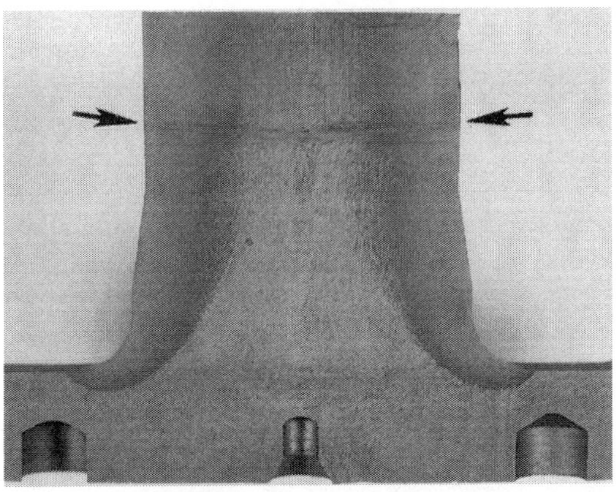

Fig. 18 Macroetching can reveal features of induction hardened steel parts. Visible in this cross section of an AISI 15B41 axle shaft are the hardened pattern on the shaft diameter (at top), as well as the friction weld between shaft and flange (arrows) and grain flow in the forged flange (bottom section). Macroetchant 1: 50 vol% hydrochloric acid, 50 vol% water. Source: Ref 12

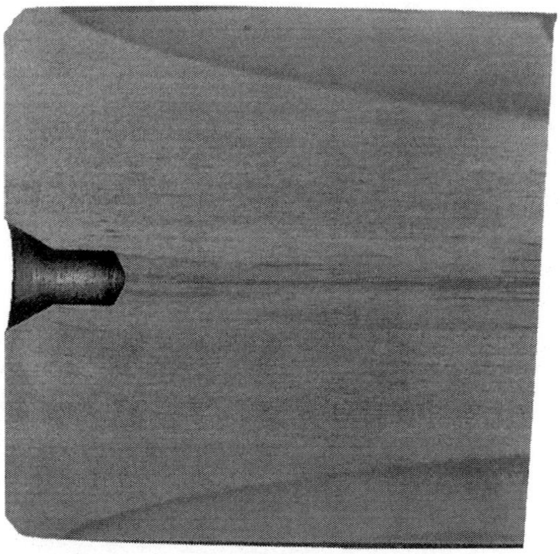

Fig. 19 The induction hardened pattern of a shaft typically runs out near the end, to reduce residual stresses at the end face. This macroetched section also shows centerline segregation in the continuously cast, AISI 15B41 bar product. Macroetchant 2: 40 vol% nitric acid in water; cleaner, 10 vol% hydrochloric acid in ethanol. Source: Ref 12

tion of the hydrochloric acid cleaner in this case was 4 vol%. The effects of this lower-concentration solution are more subtle.

Macroetchant 3 also is a two-step procedure used at room temperature. Mixing instructions for 200 mL of each solution (A and B) to produce macroetchant 3 are given in Table 8. Use multiple quantities of the ingredients if more is needed.

Both cut sections and intact parts can be evaluated.

Immerse the sample in solution A for approximately 10 s. Rinse in warm running water, scrub with a soft bristle brush, rinse with methanol, and dry.

Then immerse the sample in solution B for approximately 10 s. Rinse thoroughly with methanol, and follow immediately with a rinse in hot running water. Finally, rinse with methanol and dry.

An etched section through a notched shift bar is shown in Fig. 21. Fully hardened areas on the

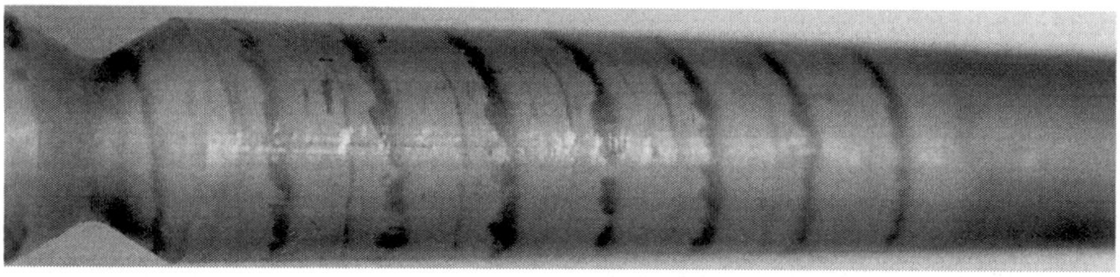

Fig. 20 The helical "barber pole" pattern on this shaft was caused by incomplete transformation during heating. The AISI 1045 steel part was rotated and pulled—too quickly—through the induction coil during scan hardening. Macroetchant, 2: 40 vol% nitric acid in water; cleaner, 4 vol% hydrochloric acid in ethanol. (Compare with Fig. 19, where 10 vol% HCl was used.). Source: Ref 12

Table 8 Mixing instructions for macroetchant 3 described in text

Solution	Recipe	Special instructions
A: Strong acid	40 mL sulfuric acid 60 mL nitric acid 100 mL water	1. Add nitric acid to water first. 2. Add sulfuric acid to solution and stir. 3. Allow solution to cool to room temperature before using or storing.
B: Weak acid	3.5 g picric acid 11 mL hydrochloric acid 189 mL methanol	1. Add picric acid to methanol first. Stir until dissolved. 2. Slowly add hydrochloric acid to the solution and stir.

Source: Ref 12

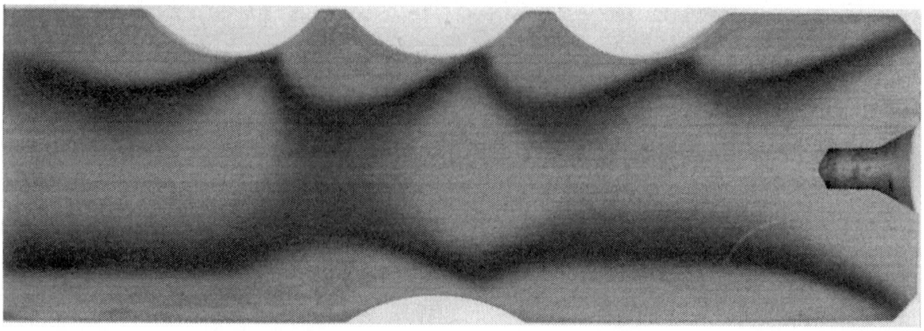

Fig. 21 Any change in distance between coil and workpiece, such as that caused by the notches in this AISI 1045 shift bar, can have a dramatic effect on case depth. Macroetchant 3 was used (see Table 8). Source: Ref 12

outside diameter have a golden yellow tint when photographed in color, transition zones are black, and unhardened areas in the center are light gray.

Note the change in the hardened depth at the notch. This dramatically illustrates the effect of the distance between coil and workpiece. The bottom of the notch was farther from the coil than the outside diameter of the bar during heating. Since the induction coil did not follow the contour of the notch, the hardened depth was drastically reduced.

Figure 22 shows an intact shift bar. The area around the drilled blind hole is not hardened, so it has a light gray color compared with the darker hardened portions of the part.

Distortion of Induction-Hardened Steels

Steel parts that have been surface hardened by induction generally exhibit less total distortion or distortion more readily controllable than that for the same parts quenched from a furnace. The decrease in distortion is a result of the support given by the rigid, unheated core metal, and of uniform, individual handling during heating and quenching. In scanning, distortion is controlled further by heating and quenching only a narrow band of the steel at one time. Unless a part through hardened by induction is scanned, the distortion encountered will approach the distortion that is experienced in furnace hardening.

As in furnace heat treating, the distortion from induction hardening arises during austenitizing or quenching. Distortion during austenitizing usually results from relief of residual stresses introduced during forging, machining, and so forth, or from nonuniform heating. When the part is only surface austenitized and hardened, the cool metal in the core of the workpiece minimizes distortion. Small amounts of distortion in induction surface hardened parts with shallow cases are often eliminated by means of a subsequent mechanical sizing (for example, straightening) operation. Furthermore, the use of induction scanning, in which only a small portion of the workpiece is heated at any one time, is helpful in preventing problems of this type. Scanning is also helpful in keeping distortion levels low in through-hardening applications. In these instances, rotation of the part, provided that it is symmetrical, enhances the uniformity of heating and decreases the likelihood of nonuniformities in the final shape.

Distortion resulting from quenching is largely a function of the austenitizing temperature, the uniformity of the quench, and the quench medium. Higher austenitizing temperatures, which give rise to higher residual stresses, increase the amount of nonuniform contraction during cooling. Severe quenches such as water or brine, which also tend to produce high residual stresses, can lead to severe distortions as well. This problem can be especially troublesome when alloy steels are quenched in water. However, these steels usually have sufficient hardenability such that oil can often be employed instead.

In extreme cases, distortion may lead to cracking. This cracking is intimately related to part design, as well as to the residual stresses which are developed. Components with large discontinuities in cross section are particularly difficult to heat treat for this reason. In addition, there often is a limiting case depth beyond which cracking will occur; in these instances, tensile stresses near the surface of the induction hardened part, which balance the compressive residual stresses generated, can be blamed for the cracking problem.

Steel composition also plays a role in the tendency toward cracking in induction hardening applications. This tendency increases as the carbon or manganese content is increased. This is

Fig. 22 Parts can be selectively hardened using induction heating technology, leaving sections unhardened for subsequent machining. The area around the drilled blind hole on this AISI 1045 steel shifter bar is not hardened. Macroetchant 3 was used (see Table 8.) Source: Ref 12

not to say, however, that critical levels of either element can be specified, because other factors such as case depth (in surface hardening applications), part design, and quench medium are also important. The effect of carbon content on the tendency toward quench cracking is greatest in through-hardened parts and arises because of its influence on the depression of the martensite-start (M_s) temperature and the hardness of the martensite.

Surface-Hardening Applications

This section describes five very common applications for induction surface hardening. It should be noted that there are many other applications associated with induction surface hardening, and many of these are described in Ref 4 and 11 and on internet sites dealing with induction heating and hardening.

Crankshafts for internal combustion engines were probably the first parts to which induction hardening techniques were applied. Because the explosive forces of the engine must pass through the crankshaft, severe demands in terms of strength and wear resistance are placed on the steel used in manufacturing the crankshaft. These demands are ever increasing with the rising horsepower ratings of engines used in automobiles, tractors, and other vehicles.

The most stringent demands are placed on the journal and bearing surface. Journals are the parts of the rotating shaft that turn within the bearings. Before the advent of induction heating, methods such as furnace hardening, flame hardening, and liquid nitriding were used. However, each of these processes presented problems such as inadequate or nonuniform hardening and distortion. Induction hardening overcomes many of these problems. Through proper selection of frequency, power, and the particular induction process, low-distortion case hardening can be done. In one of the most common steels used for crankshafts, 1045, case hardnesses over 55 HRC are readily obtained. Other advantages of the induction process for crankshafts include:

- Only the portions that need to be hardened are heated, leaving the remainder of the crankshaft relatively soft for easy machining and balancing.
- Induction hardening results in minimum distortion and scaling of the steel. The rapid heating associated with induction heat treating is advantageous in avoiding heavy scaling in other applications as well.
- Because induction heat treating processes can be automated, an induction tempering operation immediately following the hardening treatment can be done in manufacturing cells.
- The properties of induction-hardened crankshafts have been found to be superior to those of crankshafts produced by other techniques. These properties include strength and torsional and bending fatigue resistance.

Presently crankshafts are being made from steel forgings as well as from cast iron. In the latter case, surface hardness levels of higher than 50 HRC are easily obtainable after induction heating and air quenching. The resultant microstructure is a mixture of bainite and martensite, avoiding 100% martensite to minimize the danger of crack formation at holes and eliminating the need for chamfering and polishing in these regions. The air quench allows the initial formation of bainite during cooling. After a prescribed period of time, the air quench is followed by a water quench during which the martensite phase is produced from the remaining austenite. Sufficient residual heat is left in the part to self-temper the martensite.

Axle shafts used in cars, trucks, and farm vehicles are, with few exceptions, surface hardened by induction. Although in some axles a portion of the hardened surface is used as a bearing, the primary purpose of induction hardening is to put the surface under a state of compressive residual stress. By this means, the bending and torsional fatigue life of an axle may be increased by as much as 200% over that for parts conventionally heat treated (Fig. 23). Induction hardened axles consist of a hard, high-strength, and

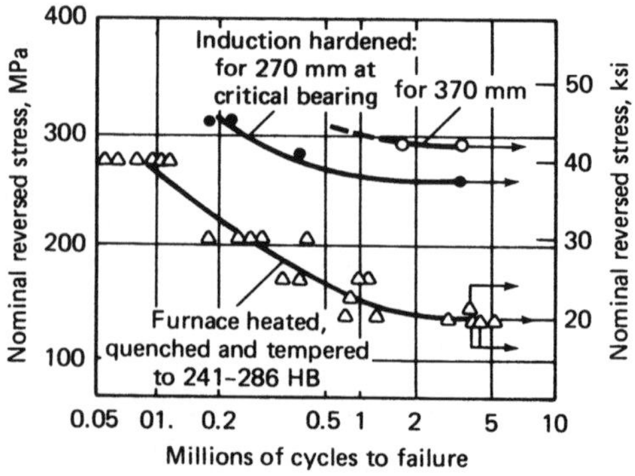

Fig. 23 Bending fatigue response of furnace-hardened and induction-hardened medium-carbon steel tractor axles. Shaft diameter: 70 mm (2.75 in.). Fillet radius: 1.6 mm (0.063 in.) Source: Ref 4

tough outer case with good torsional strength and a tough, ductile core. Many axles also have a region in which the case depth is kept very shallow so that the part can be readily straightened following heat treatment. In addition to substantially improving strength, induction hardening is also very cost-effective. This is because most shafts are made in inexpensive, unalloyed medium-carbon steel that is surface hardened to case depths of 2.5 to 8 mm (0.10 to 0.30 in.), depending on the cross-sectional size. As with crankshafts, typical hardness (after tempering) is around 50 HRC. Such hard, deep cases improve yield strength considerably as well.

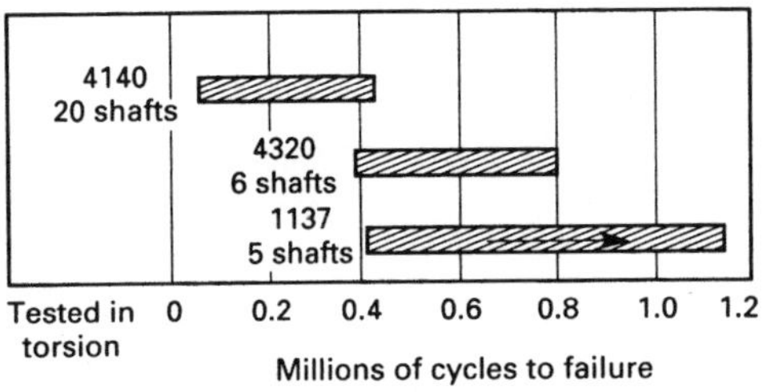

Steel	Surface hardness, HRC	Method of hardening
4140	36–42	Through-hardened
4320	40–46	Carburized to 1.0–1.3 mm (0.040–0.050 in.)
1137	42–48	Induction hardened to 3.0 mm (0.120 in.) min effective depth and 40 HRC

Fig. 24 Comparison of fatigue life of induction surface hardened transmission shafts with that of through-hardened and carburized shafts. Arrow in lower bar (induction-hardened shafts) indicates that one shaft had not failed after testing for the maximum number of cycles shown. Source: Ref 4

Modern transmission shafts—particularly those for cars with automatic transmissions—are required to have excellent bending and torsional strength, as well as surface hardness for wear resistance. Under well-controlled conditions, induction hardening processes are most able to satisfy these needs, as shown by the data in Fig. 24, which compares the fatigue resistance of through-hardened, case carburized, and surface induction hardened axles. The induction hardening methods employed are quite varied and include both single-shot and scanning techniques.

Induction hardening of crankshafts, axles, and transmission shafts is becoming an increasingly automated process. Often parts are induction hardened and tempered in-line. One such line for heat treating of automotive parts is depicted schematically in Fig. 25. It includes an automatic handling system, programmable controls, and fiber-optic sensors. Mechanically, parts are fed by a quadruple-head, skewed-drive roller system (QHD) after being delivered to the heat treatment area by a conveyor system. The roller drives, in conjunction with the check guides, impart both rotational and linear forward movement of the workpiece through the coil. Once a part enters the "ready position," the fiber-optic sensor senses its position and initiates the heating cycle for austenitization, subsequent in-line quenching, and then induction tempering. The workpieces are round bars that are fed end-to-end continuously.

In the hardening cycle of the QHD system, the induction power supply frequency is general either in the radio frequency range (approximately 500 kHz) for shallow cases or in the

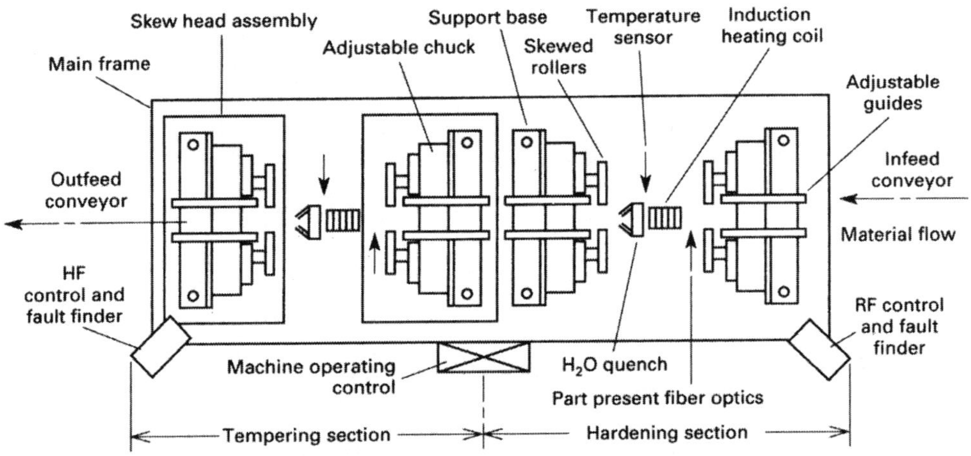

Fig. 25 Automated, quadruple-head, skewed drive roller system used for in-line induction hardening and tempering of automotive parts. RF, radio frequency; HF, high frequency. Source: Ref 4

range for 3 to 10 kHz if deeper cases are needed. For rejection purposes, a temperature monitor senses if the workpiece has been either underheated or overheated. Assuming that the workpiece has been heated properly, it then passes through a quench ring. After quenching, the workpiece is moved into the induction-tempering part of the heat treating line. Again, a fiber-optic sensor senses the presence of the workpiece and begins the heating cycle, generally using a lower frequency power supply (lower frequency can be used because the workpiece is still magnetic during tempering and accordingly has a shallower reference depth). Depending on the surface hardness as-quenched and the desired final hardness, the desired tempering temperature can be as high as approximately 400 °C (750 °F). Induction tempering requires a higher tempering temperature than furnace tempering because of the short heat cycle. When the tempering is complete, the workpiece is moved onto a conveyor for transportation to grinding.

The control system of this line is designed to allow decision making by a programmable controller. Thus, all aspects of the heat treating process and mechanical operations are preprogrammed and may be changed easily to accommodate different part sizes and heat treating parameters. With such a process, users have been able to increase production rates more than threefold over those obtainable with conventional heat treating lines.

Gears. Reliability and high dimensional accuracy (to ensure good fit) are among the requirements for gears. Keeping distortion as low as possible during heat treatment is very important. Induction heat treating is one of the very important processes used for heat treatment of gears. Gears, because of the wide varieties, sizes, and differences in tooth profiles, represent unique applications. External spur and helical gears, bevel and worm gears, and internal gears, racks, and sprockets are good examples of the kinds of gears in which the size can range from less than 6 mm (0.25 in.) to greater than 3 m (12 ft). The hardened pattern for gears, as for shafts, may be through the cross section, as with small-armature shafts, or be limited to single-tooth case hardening, as is done with large gears. A wide variety of frequencies and induction processes are used, because of the way the induced currents are produced in gear teeth with different profiles, sizes, and pitches. The heat treating processes use single-shot heating techniques and a variety of scanning techniques. A wide number of different frequencies are used to accommodate the different patterns and tooth profiles.

The size of gear, the hardening requirements, and the production requirement influence the type of induction-hardening process used. High-quantity production lots can be single-shot induction hardened, whereas small quantities of large gears need to be run one tooth at a time to keep the capital equipment costs low.

Single-shot, through hardening of the ends of small armature shafts has been done since the 1950s. In addition, there is a wide variety of different types of case patterns that are produced on gear teeth. Figure 26 shows eight different induction patterns that can be produced with induction. Patterns A, B, and C are similar in that a portion of the tooth is hardened, but not the root. These patterns were originally used on gears with large pitch teeth. Pattern A used single-shot, channel-type coils heating the entire tooth at one time or scanning. If there is no maximum case depth specified, small gears may be through-hardened. Use of a frequency high enough that root penetration does not occur produces patterns B and C. Figure 27 shows how high frequencies tend to heat the tips of teeth, while low frequencies tend to heat the root. Pattern D in Fig. 26 shows the root heating effect of a frequency that is too low. Patterns like this are not acceptable because the upper portion of the tooth is not hard and will be subject to wear.

Gear manufacturers have found that the greatest stress on a gear is from the pitch diameter through the fillet of the root. Failure is most likely to occur at these points. Therefore, it is highly desirable that the wear surface and the root of gearing be hard. Patterns E, F, G, and H

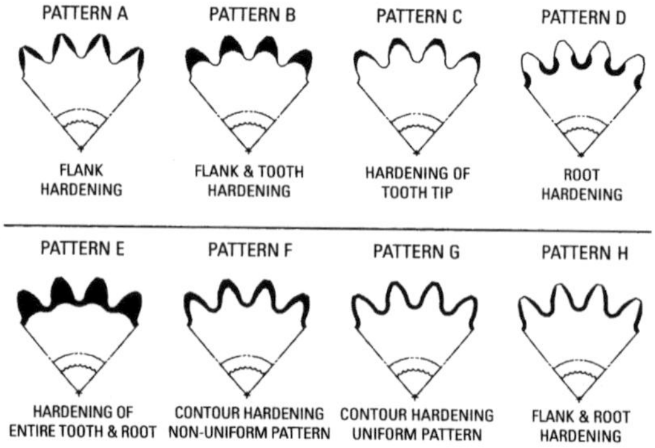

Fig. 26 Induction-hardening patterns for gears. Source: Ref 11

show patterns that meet these criteria. Pattern E represents one of the most common patterns produced by induction and is produced by either single-shot heating or scanning. The specifications commonly call for the gear to be induction hardened to a minimum hardness below the root. Figure 28 shows the frequencies versus gear pitch that are used to produce this type of pattern. A frequency of 450 kHz has difficulty in producing case depths below 1.5 mm (0.060 in.) on even the fine-pitch gears. Single-shot hardening is limited by the power available on the power supply. Larger gears can be scanned to keep the power requirements reasonable. When distortion is excessive, the pattern requirements may be changed to that shown in patterns F and G. With pattern F, the frequency is lowered, and the power density is increased. The case depth at the root is 30 to 40% of the case depth at the outer portion of the tooth. Pattern F attempts to produce a near contour pattern, while pattern G attempts to produce a uniform contour. Pattern F uses pulsed or dual-frequency heating techniques. These patterns attempt to produce gears that not only quench to net shape without distortion but also have the surface in compression so that the overall properties are increased.

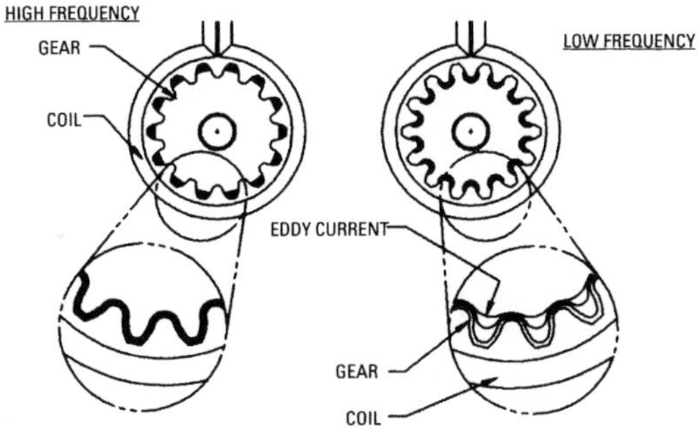

Fig. 27 Frequency influence on hardness profile with an encircling induction coil. Source: Ref 11

Rolling-Mill Rolls. During service, roll life is limited by abrasive wear. As the diameter is reduced by wear, adjustments are made to bring the rolls closer together in order to maintain a given rolling reduction. These adjustments are sufficient until the rolls have worn approximately 40 mm (1.5 in.); once this amount of wear is exceeded, the rolls must be replaced. The objective of induction heat treatment is, therefore, to produce a deep hardened case approximately 20 to 40 mm (0.75 to 1.5 in.) deep. This is done employing a low-frequency (60 Hz) power supply.

In the scanning method of induction hardening, the roll, hanging vertically, is lowered into the induction coil, in which its surface temperature is gradually raised to 955 °C (1750 °F). By controlling the power input and feed rate, a temperature profile is developed such that the temperature ranges from 900 °C (1650 °F) at 40 mm (1.5 in.) below the surface to less than 260 °C (500 °F) at 50 mm (2 in.) below the surface. Following heating, the roll is quenched using water precooled to 5 °C (40 °F). Because roll steels usually contain 0.8 to 0.9% C and substantial amounts of nickel, chromium, molybdenum, and vanadium, they have high hardenability and develop high hardness to the entire depth to which the steel was austenitized. A typical hardness profile is shown in Fig. 29. Here, the drop

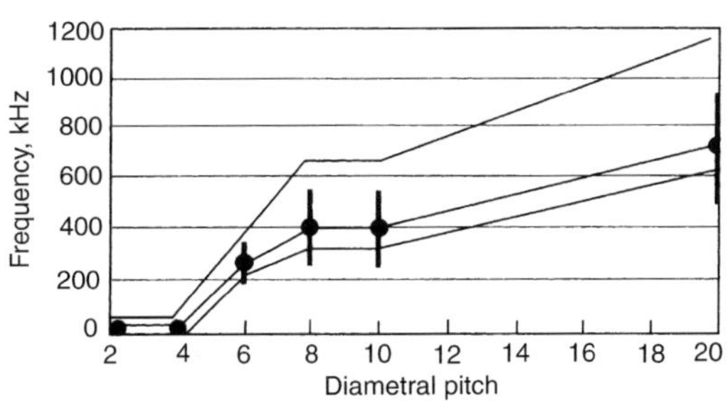

Fig. 28 Proper frequency selection is needed to accomplish even heating. Too low a frequency will result in field cancellation and inadequate heating; too high a frequency could overheat the surface. Nominal SAF 1050 steel; nominal case depth accomplished is a function of frequency. Case depth will be 1.0–1.5 times the reference depth of each frequency. Nominal power density of 10 kW/in.^{2} of surface area. Source: Ref 11

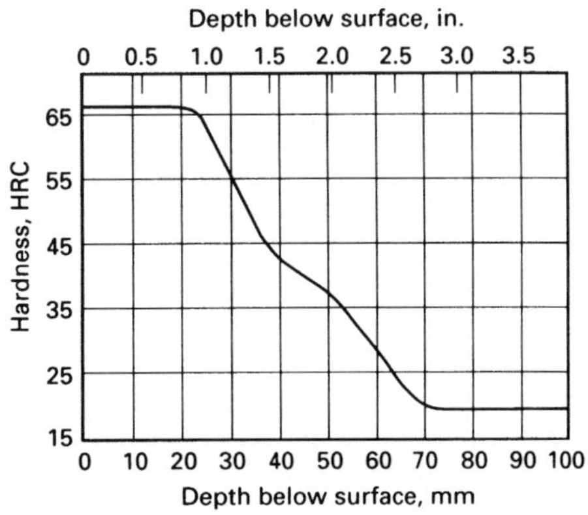

Fig. 29 Hardness pattern developed in rolling mill rolls induction hardened using a 60 Hz generator

in hardness beyond about 25 mm (1 in.) can be attributed to heat losses due to conduction, which could have resulted in the formation of pearlite or bainite prior to quenching, at which time the remaining austenite would have transferred to martensite.

Laser Surface Hardening

In conventional methods of heat treatment, the component is heated to the required temperature and then quenched in oil or water to achieve the desired hardness at the surface. In most industrial applications, wear occurs only in selected areas of the component, hence it is sufficient to harden these areas to enhance the performance of the component. Rapid advances in laser technology in the past decade have made it possible to perform various operations such as heat treating, glazing, alloying, and cladding on surfaces of materials, resulting in better physical properties of the surface and improved performance in a given environment. Because a laser is an expensive source of energy, it is used only in cases where it offers some technical and/or economic benefits compared to conventional methods. The advantage of using a laser for surface processing results from its highly directional nature and from the ability to deliver controlled amounts of energy to desired regions. In laser heat treatment, which involves using a laser as a heat source, the beam energy is applied to harden a surface with the rest of the component acting as a heat sink. Because ferrous materials are very good heat conductors, the high heat fluxes generated by lasers are most suitable to heat the surface layer to austenitization levels without affecting the bulk temperature of the sample. The ensuing self-quenching is rapid enough to eliminate the need for external quenching to produce the hard martensite in the heated surface. Thus a highly wear resistant surface with the desired core properties of the component can be obtained. This process is known as *laser surface transformation hardening.* Laser surface transformation hardening not only increases the wear resistance of materials but under certain conditions also increases fatigue strength due to the compressive stresses induced on the surface of the component. Components that have undergone laser surface hardening treatments include such highly stressed machine parts as gears and gear teeth, camshafts, gear housing shafts, cylinder liners, axles, and exhaust valves and valves guides. Many of these applications are in the automotive industry, which was among the first mass-production industries to exploit lasers for surface treatment.

This section briefly describes some processing parameters important to the success of the laser surface transformation hardening process, advantages and drawbacks of the process, and the types of steels that are amenable to laser processing. More detailed information on the fundamentals of laser surface hardening, the equipment used, and descriptions of various experimental studies carried out on different types of laser-processed steels can be found in several excellent reviews on laser heat treating (Ref 13–15).

Lasers for Surface Hardening

The majority of laser models used for metalworking are either the neodymium yttrium-aluminum garnet (Nd:YAG) solid-state type or the carbon dioxide (CO_2) gas type. These layers may have pulsed or continuous output power. Both types, whether pulsed or continuous wave, can be used for transformation surface hardening. The design and operating principles of these industial laser types is discussed in Ref 13.

Processing Parameters

When a laser beam impinges on a surface, part of its energy is absorbed as heat at the surface. If the power density of the laser beam (usually given in watts per square centimeter) is sufficiently high, heat will be generated at the surface at a rate higher than heat conduction to the interior can remove it, and the temperature in the surface layer will increase rapidly. In a very short time, a thin surface layer will have reached austenitizing temperatures, whereas the interior of the workpiece is still cool. Even with a relatively moderate power density of 500 W/cm^2 (3200 W/in.^{2}), temperature gradients of 500 °C/mm (25 °F/mil) can be obtained. By moving the laser beam over the workpiece surface (see Fig. 30), a point on the surface within the path of the beam is rapidly heated as the beam passes. This area is subsequently cooled rapidly by heat conduction to the interior after the beam has passed. By selecting the correct power density and speed of the laser spot, the material will harden to the desired depth.

Power Density. A relatively broad area beam, often in the shape of a square or a rectangle, is used in the laser hardening process. The power density of a focused laser beam used for hardening is much lower than the power density of the small, intense focused spots used for welding and cutting. The power density is typically in the 1,000 to 2,000 W/cm^2 (6,400 to 13,000 W/in.^{2}) range, occasionally as high as 5,000 or as low as 500 W/cm^2 (32,000 to 3,200 W/in.^{2}). Figure 31 compares the power densities associated with laser processing methods, including transformation hardening, welding, cutting, drilling, and surface modification by laser melting, alloying, and cladding. Additional information on laser alloying and cladding can be found in Chapter 11, "Surface Hardening by Coating or Surface Modification."

Depth of Hardening. The resulting depth of hardening will depend on the hardening response of the material, but it will rarely be more than 2.5 mm (0.1 in.). For steel with low hardenability, such as low- and medium-carbon steel, the depth of hardening obtainable is much smaller, varying from perhaps 0.25 mm (0.01 in.) in mild steels to 1.3 mm (0.05 in.) in a medium-carbon steel. Because of the very high heating and cooling rates obtainable, it is possible to harden steels not normally considered hardenable, such as SAE 1018. For the same reason, the hardness obtainable by the laser hardening process can, in some instances, be slightly higher than that considered possible with conventional methods.

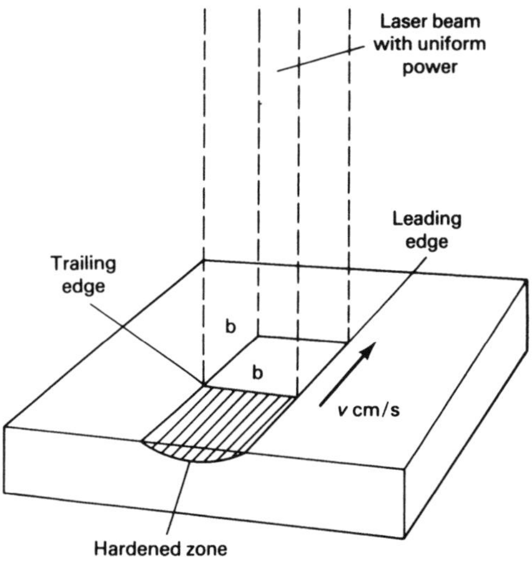

Fig. 30 Square laser beam with uniform power density on a flat plate

General guidelines for processing conditions are as follows:

- Usable power densities in laser surface hardening are in the 500 to 5,000 W/cm^2 (3,200 to 32,000 W/in.^{2}) range. Corresponding dwell times are in the range 0.1 to 10 s. For carbon steels, the power density is usually from 1,000 to 1,500 W/cm^2 (6,400 to 9,700 W/in.^{2}), and the dwell time 1 to 2 s.
- Materials with high hardenability can be processed at low power density and high dwell time (low speed), whereas materials with low hardenability should be processed at high power density and low dwell times.
- Rectangular, square, or sometimes round laser spots with uniform power density are suitable in obtaining uniform hardened case.
- High power density and low dwell time give shallow case but high cooling rates. The reverse is true for low power densities.
- Maximum surface temperature is approximately proportional to the square root of the processing speed. Hence, a doubling of the power density requires a quadrupling of the speed to obtain equivalent maximum surface temperatures.
- Increasing the power density results in lower total energy input for the same maximum surface temperature.
- Steel with normalized, annealed, or spheroidized structures; steel with proeutectoid cementite; cast irons and steels with stable alloy carbides require longer dwell times than steels that have been hardened and tempered.
- Small workpieces will require higher power densities and lower dwell times than large pieces, unless external quenching media are used.

Use of High-Absorptivity Coatings

Laser heat treating involves solid-state transformations, so the surface of the metal is not melted. The fraction of the beam power absorbed by the material is controlled by the absorptivity of the material surface. Because steels are not good absorbers of infrared and far-infrared electromagnetic radiation, special high-absorptivity coatings must be applied to their surfaces to allow efficient use of the laser energy. Chemical coatings, such as manganese phosphate and paints of graphite, silicon, and carbon, have all been used successfully. Some of these coatings may burn off during the heating process, and some may leave a residue that in itself can be an indicator of the maximum surface temperature reached. In any event, the absorptivity of these coatings at the beginning of the heating cycle is high (90% or better) and continues to be higher than that of the bare material throughout the temperature excursion. The overall absorptivity of these coatings, applied like a spray paint, is typically about 80% (pure iron has an absorptivity of ~4% at room temperature).

Advantages and Limitations of Laser Hardening

The major advantages of laser surface hardening include: close control of the power input with metal-working lasers; the laser can provide high power density in selected areas, which, in turn, minimizes the total energy input and thereby dimensional distortion; and the ability of the laser to reach normally inaccessible areas on the workpiece surface. Because no vacuum or protective atmosphere enclosure is needed and the distance from the workpiece to the last optical element of the laser system can be quite long, it is possible to process very large or irregular-shaped workpieces. The laser beam can also be optically shaped or split to accommodate different geometries.

On the negative side, the depth of case obtainable is limited to approximately 2.5 mm (0.1 in.), usually less than half of this, and the capital cost of the equipment may be high. Therefore, careful analysis of a potential application for laser hardening is needed to ascertain the cost-effectiveness of the process. One example of a cost analysis compared laser hardening and selective carburizing (Ref 16). In an environment where the laser is busy two shifts per day, laser treatment was shown to be cost effective for large gears where a limited area was to be treated. Reasons why laser hardening replaced gas carburizing included:

- Reduced hardening time
- Reduced scrap rate
- Elimination of complex quenching, plating, masking, stripping, and cleaning steps
- Reduced work-in-progress inventory

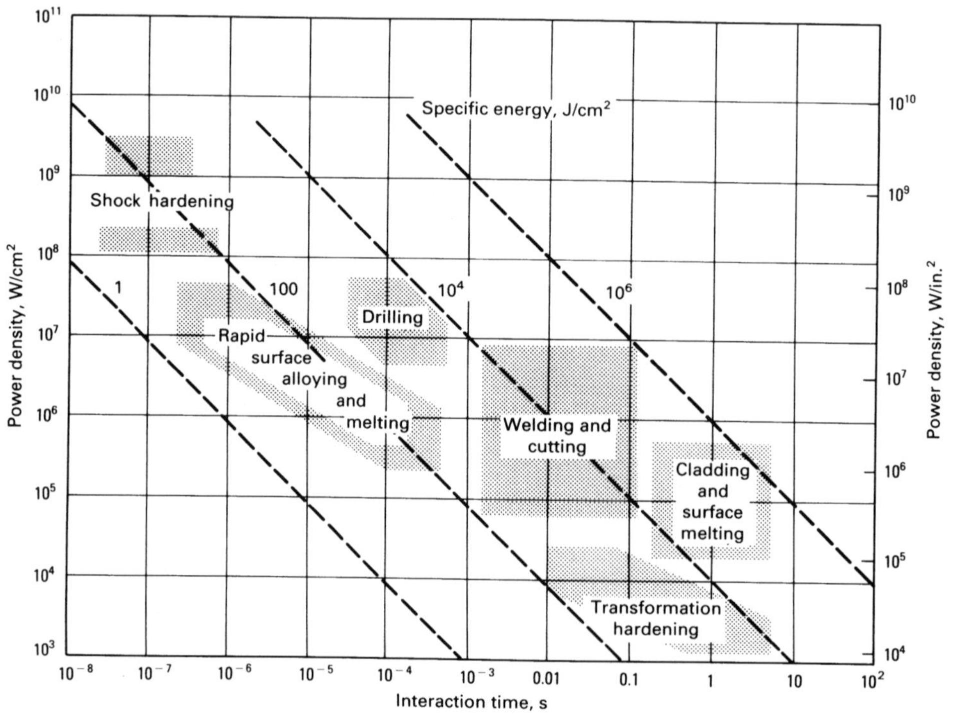

Fig. 31 Interaction times and power densities necessary for various laser processing methods

- Quicker turnaround, less material handling
- Reduced floor space requirements
- Reduced pollution by elimination of copper plating used for selective carburizing
- Reduced energy use

Residual Stresses in Laser Heat Treatment (Ref 13)

Residual stresses form in the laser-treated surfaces because of rapid thermal heating and constrained cooling due to clamping of the workpieces. The nature and magnitude of these stresses on the surfaces formed during laser processing depend on the type of processing, temperature gradients, and phase-change kinetics. This in turn may or may not give rise to cracking tendency after processing, depending on the level of stress, the distribution and nature of the type of stress distribution, and the mechanical strength of the phases present in the laser-treated microstructures. Residual compressive stresses are beneficial to enhance the fatigue resistance because they will help to retard the crack growth. On the other hand, residual tensile stresses are deleterious for the fatigue resistance due to the enhancement of crack propagation rates. Hence it is generally recommended that a simple postprocessing step such as annealing or a pretreatment step such as preheating of the base material before laser processing be carried out to minimize the chances of cracking tendency.

Laser-Hardenable Steels

Steels suitable for laser surface transformation hardening include a wide range of carbon, alloy, and tool steels. Some representative results including the processing parameters and microstructural and hardness characteristics for laser-hardened steels are listed in Table 9. Select microhardness data are shown in Fig. 32, which illustrates the typical final hardness values of ferrous materials after laser surface hardening. The hardness values depend on the laser heat treatment parameters as well as the microstructural and alloy composition of the materials being treated. The published research data on hardness characteristics of the laser-hardened alloys include hardness variation along the depth and a single average value or range (see Table 9). Because of the unavailability of sufficient data of the processing parameters with specific hardness values as a function of laser heat treatment conditions (e.g., laser power, spot size, and process speed), the exact values of laser power density and process speed are not included in Fig. 32. This figure and Table 9 serve as references for the readers to estimate average surface hardness available for various alloys. The general trend of the hardness data indicates that most of the high-carbon ferrous alloys such as bearing steel (e.g., 52100 steel), tool steels (No. 18, 19, 20, and 21 in Fig. 32), Fe-Cr-Mn-C, and Fe-W-Cr-V-C hardfacing alloy steels have relatively higher hardnesses compared to the low- and medium-carbon and alloy steels.

Electron-Beam Surface Hardening

Electron-beam heat treating is a selective hardening process in which the surface of a hardenable ferrous alloy is heated rapidly above the transformation temperature of the alloy by direct bombardment or impingement of an accelerated stream of electrons. The stream of electrons must have line-of-sight access to the area requiring heat treatment and a beam-impingement angle of at least 25 degrees. Guidelines for acceptable part configurations are shown in Fig. 33.

At the end of a heating cycle of 0.5 to 2.5 s, the flow of electrons is stopped abruptly to allow the part or workpiece being processed to self-quench and to form a martensitic structure with a compressive stress on the surface of the hardened area. Typical hardening depths obtained by electron-beam hardening range from 0.1 to 1.5 mm (0.004 to 0.006 in.).

The electron-beam hardening process is normally applied to finish-machined or ground surfaces. Because the buildup of energy is rapid and well controlled, postheat treatment operations such as grinding or straightening usually are not needed.

Despite the advantages of extremely low hardening distortion and relatively low energy consumption, the electron-beam hardening process has found limited application in the metals industry. This is primarily due to the very high capital costs associated with the process equipment. However, for certain specialized applications, electron-beam hardening is competitive with both case hardening and induction hardening processes in the heat-treating marketplace. Economic benefits can also be derived in facili-

Table 9 Laser heat treatment data for various steels

Base material (wt%)	Coating	Process parameters: (a) Laser type (b) Power, kW (c) Beam diam., mm	(d) Transverse mode (e) Process speed, mm/s (f) Interaction time, s	(g) Power density, kW/cm² (h) Specific energy, J/cm² (i) Shielding gas	Microstructural characteristics and properties	Depth/width of hardening	Hardness, kgf/mm²
AISI 1018 steel		(a) CO_2 (b) 9 (c) 12 × 12 (square)	(d) . . . (e) 63.5–169 (f) 0.071–0.20	(g) 6.25 (h) 444–1181 (i) Air	Low-carbon martensite	HAZ = 254 μm	446
AISI 1018, and 1045 steels		(a) CO_2 (b) 1.0 (c) 2.54	(d) . . . (e) 25.4 (f) 0.1–0.5	(g) 0.789–19.7 (h) 315–600 (i) . . .	Ferrite + martensite; proportion of martensite in 1045 much higher than in 1018	Depth = 254 μm Width = 1.65 mm	AISI 1045 Base: 354 HAZ: 720
En 8 (0.36% C steel)	Graphite	(a) CO_2 (b) 1.2–2.0 (c) 1.6–5.8	(d) Gaussian (e) 25–400 (f) 0.004–0.232	(g) 4.54–99.5 (h) 51.7–5000 (i) . . .	Martensite + proeutectoid ferrite	HAZ depth = 700 μm	500–680
AISI 1045	Black paint	(a) CO_2 (b) 2.5–4.15 (c) 18 × 18 (square)	(d) . . . (e) 8.5–12.7 (f) 1.42–2.12	(g) 0.772–1.28 (h) 1093–2712 (i) . . .	Martensite	HAZ = 520–1240 μm	446
AISI 1045	Black paint	(a) CO_2 (b) 8.8 (c) 12.5 × 25.4 (dual beam)	(d) . . . (e) 8.33 (f) . . .	(g) 1.38 (h) . . . (i) . . .	Martensite	HAZ = 1.2–1.3 mm	674–697
Hypoeutectic, eutectic, hypereutectic, and ledeburitic steels	Specimens oxidized to improve absorptivity	(a) CO_2 (b) 1.25 (c) 9	(d) . . . (e) 5 (f) 1.8	(g) 1.96 (h) 2778 (i) . . .	Inhomogeneous structures for the hypereutectic and ledeburitic steels; ferrite-cementite, austenite, and martensite in the HAZ		640–870
SK5 tool steel	Manganese phosphate	(a) CO_2 (b) 1.3 (c) 12 × 12 (square)	(d) TEM_{00} + TEM_{11} (e) 10 (f) 1.20	(g) 0.903 (h) 1083 (i) . . .	Martensite	HAZ depth = 920 μm width = 6.8 mm	850
Tool steels A6-air hardened O1-oil hardened		(a) CO_2 (b) 1.0 (c) 2.54–12.7	(d) . . . (e) 38.1 (f) 0.0667–0.333	(g) 0.789–19.7 (h) 1462 (i) . . .	Martensite	HAZ = 305 μm	A6: 653–746 O1: 800–865
Tool steel (0.95C-1.7Mn-0.25Cr-0.25V)		(a) CO_2 (b) 2.8 (c) 5	(d) . . . (e) 23.3 (f) 0.215	(g) 14.3 (h) 2400 (i) . . .	Martensite		800
Fe-3.11 Cr-1.98 Mn-0.5 Mo-0.26 C (3 Cr) and Fe-9.85 Cr-1.0 Mn-0.5 A1-0.2 C (10Cr)		(a) CO_2 (b) 1.25 (c) Narrow elliptical spot	(d) . . . (e) 4.23–8.47 (f) . . .	(g) . . . (h) . . . (i) He	Packet martensite (retained austenite films surrounding dislocated and fine twinned laths); martensitic laths oriented in resolidified and unmelted parts of grains; in melted zone, grains large and fine cellular; just below melt zone, grains coarse and irregular; in HAZ, grains fine martensitic and irregular; presence of 8-ferrite		600–700

Source: Ref 13

(continued)

Table 9 (continued)

AISI 4140 steel	Black paint	(a) CO_2 (b) 3.5 at tip, 7.5 at cylindrical portion (c) 5	(d) Annular (ring) (e) 1300 rpm 3.7 mm/s (f) . . .	(g) . . . (h) . . . (i) He	Martensite	HAZ depth = 1.65–2 mm	653–674
(a) AISI 4340 (b) AISI 8620 (c) AISI 52100		(a) CO_2 (b) 1.0 (c) 2.54–12.7	(d) . . . (e) 19–25.4 (f) 0.1–0.0667	(g) 0.789–19.7 (h) 310–2072 (i) . . .	In all cases very fine martensite: negligible distortion	HAZ depth/width, μm (a) 406/2500 (b) 356/2300 (c) 178/1350	(a) 633–674 (b) 513 (c) 697–800
AISI 4340 and 300-M		(a) CO_2 (b) 1.2 (c) 2.5	(d) . . . (e) 4.2 (f) 0.595	(g) 24.5 (h) 11429 (i) . . .	In 4340, a mixture of dislocated and twinned martensites with presence of twins in lath martensite; retained austenite; homogeneous dispersion of self-tempered cementite particles (both at the lath boundaries and along the internal twins) within the martensite; in both systems, substructure of grain boundary is blocky martensite (without twins or carbides) or massively transformed ferrite	HAZ depth = 1.1 mm Width = 3.6 mm	AISI 4340 Base: 354 HAZ: 720
12% Cr Steel (tempered martensitic)	Graphite	(a) CO_2 (b) . . . (c) . . .	(d) . . . (e) . . . (f) 0.63–1.62	(g) 1.8–2.9 (h) . . . (i) . . .	Very fine lamellar martensite with extremely high dislocation density	HAZ depth = 0.7–1.4 mm	HAZ: 500–600 Base: 300
Fe-.6C-.09Si-0.99Mn-0.24Cr and Fe-18W-4.25Cr-0.75C-1.05V and Fe-11.5Cr-2.05C-0.7W		(a) CO_2 (b) 1.3 (c) 19	(d) TEM_{00} (e) 10–400 (f) 0.0425–1.9	(g) 0.459 (h) 17.1–684 (i) . . .	8-ferrite, austenite, and martensite Softened tone = 200–500 μm	Fe.6C HAZ width = 550–800 μm Fe-W: 800–1000 Fe-Cr: 460–530	Fe-C: 800–1000

Source: Ref 13

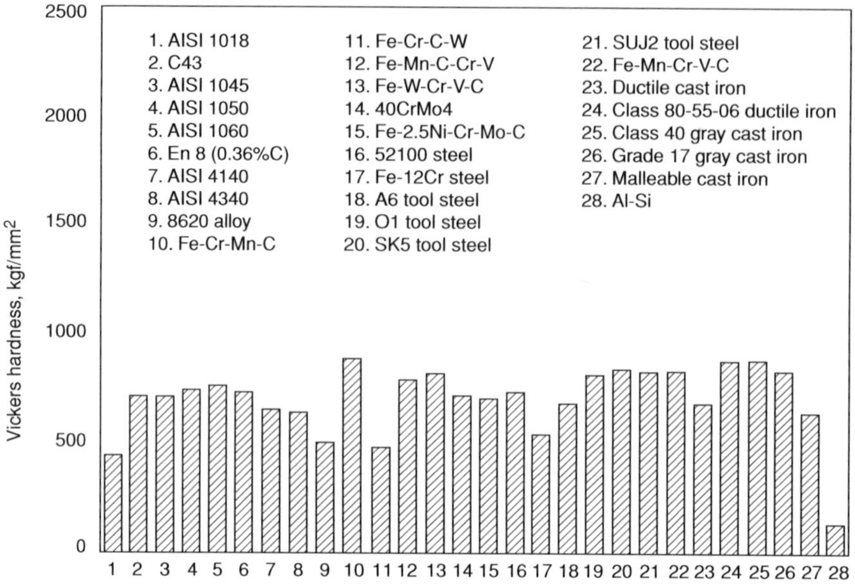

Fig. 32 Hardness data for various laser-hardened materials. Source: Ref 13

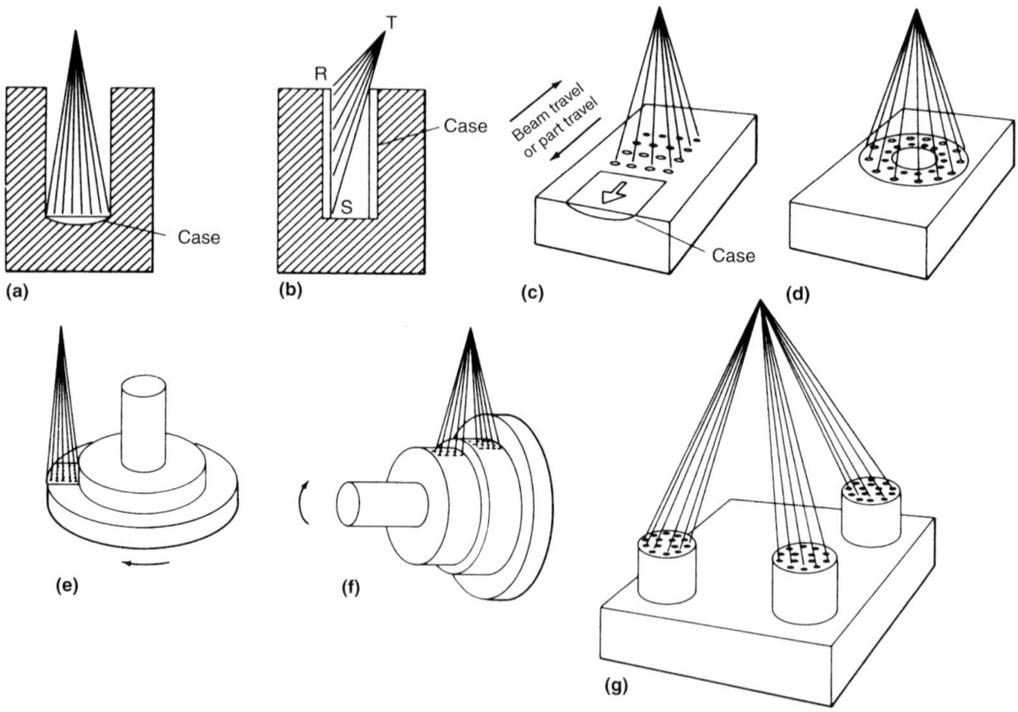

Fig. 33 Workpiece configurations and heating patterns for electron-beam heat treating. (a) Display static pattern within cavity in workpiece. (b) Maintain angle of workpiece rotation, RST, at 25° minimum. (c) Display static pattern and move the pattern or the workpiece to heat treat large areas. (d) Display static pattern; this annular pattern has well-defined inside and outside diameters. (e) Display static pattern and rotate workpiece. (f) Display more than one pattern and rotate workpiece. (g) Display multiple patterns on one workpiece or on a small group of workpieces for simultaneous hardening; patterns may be similar or dissimilar in geometric shape.

ties that can use the electron beam system for multiple tasks, for example, welding, machining, and heat treating.

Electron-Beam Equipment

In electron-beam heat treating, a highly concentrated beam of high-velocity electrons is used to heat selective surface areas. These electrons are accelerated and collimated into a dense, extremely energetic beam by the accelerating potential between the cathode and the anode. The high-energy beam thus formed passes through a small-diameter hole in the anode. Because of the mutual repulsion among neighboring electrons, the beam requires further collimation below the anode. This additional collimation is controlled with a focus coil that allows variation of the distance from the gun to the workpiece. A deflection coil deflects the reconverging beam to a designated location on the workpiece.

A high vacuum is needed in the region where the electrons are emitted and accelerated, both to protect the emitter from oxidation and to prevent interference with the electrons while they are still at low velocity. Therefore, the electron-gun housing is pumped and maintained at a vacuum of 10^{-5} torr. The workpieces are contained in an enclosure under a vacuum of approximately 5×10^{-2} torr. An intermediate vacuum level provides short evacuation times and higher production rates. Treating at one atmosphere does not require any evacuation time.

In electron-beam heat treating, the energy exchange is simply a matter of the electrons in the beam transferring their kinetic energy to the atomic structure of the target material in the form of heat. The electron beam, when sharply focused for welding, is capable of impingement power densities on the order of 10 MW/cm^2 (65 MW/in.^{2}). Because this powerful concentration of energy is easily controllable in power magnitude, power density, and beam position, it is well suited for surface hardening as well. These power densities are much too high for nondestructive heat treating, however. Destructive heat treating in this context refers to controlled remelting of ferrous and nonferrous materials.

An energy concentration of 3.1 kW/cm^2 (20 kW/in.^{2}) is more suitable for selective heat treating. To reduce the beam energy to this level, a single electron beam is programmed through a group of discrete beam positions referred to as a raster pattern.

Electron-Beam Applications

From an engineering aspect, practically all parts with surfaces that are accessible to the beam and with a thermal capacity that is sufficient for self-quenching can be considered candidates for electron-beam surface hardening. Table 10 lists steels commonly processed by electron-beam hardening. In terms of material specifications, the workpiece thickness that is in direct thermal contact with the hardened layer should be at least five to ten times the hardening depth. Another material consideration is the minimum temperature required for martensite formation. A very straightforward process regime is obtained with throughgoing, interrupted plane, or cylindrically hardened surfaces. For example, the circumferential surfaces of bores are hardenable with diameter-to-depth ratios up to unity. The flank-profile hardening of gears and racks generally calls for specialized beam guidance techniques. Occasionally, crucial problems may arise when the hardening of irregular and spherically bent surfaces is attempted.

Technologically, this process is preferred for hardening depths in a range of 0.3 to 1 mm (0.01 to 0.04 in.). Depending on the material being used, however, it can also produce a maximum hardening depth of approximately 2 mm (0.08 in.). In some cases, the maximum hardening depth is also fixed by the coarse-grain growth and the bainite or pearlite formation caused by a low cooling rate. Hardness penetrations below 0.3 mm (0.01 in.) (minimums of 10 μm, or 400 μin., on the order of the electron range are possible) can be implemented without difficulty but are not yet demanded in practice. It is especially these thin hardened layers where the most outstanding materials properties (for example, extreme grain fineness and very high hardness values) can be attained.

Electron beam hardening offers the following technological benefits:

- Precise control and reproducibility of the energy input with respect to location and time
- Constant hardening depth for both areal and laterally patterned hardening up to a track width of 50 to 100 mm (2 to 4 in.)
- Low thermal stress is imposed on the workpiece to minimize warpage.
- No scaling or oxidation of component surfaces
- No component-dependent means of energy transfer

Table 10 Steels commonly used in electron beam hardening applications

Material			Composition, wt%											
AISI	UNS No.	DIN(a)	C	Si	Mn	P	S	Cr	Mo	Ni	V	Al	Cu	Ti
Carbon and low alloy														
4140	G41400	42 CrMo 4	0.38–0.45	0.17–0.37	0.50–0.80	0.035 max	0.035 max	0.90–1.20	0.15–0.25	0.30 max	0.06 max	...	...	...
1340	G13400	42 MnV 7	0.38–0.45	0.17–0.37	1.60–1.90	0.035 max	0.035 max	0.30 max	0.10 max	0.30 max	0.07–0.12	...	...	...
E52100	G52986	100 Cr 6	0.95–1.05	0.17–0.37	0.20–0.45	0.027 max	0.020 max	1.30–1.65	...	0.30 max	...	...	0.25	...
1015	G10150	C 15	0.12–0.19	0.17–0.37	0.35–0.65	0.040 max	0.040 max	0.50 max	0.10 max	0.30 max	...	...	...	...
1045	G10450	C 45	0.42–0.50	0.17–0.37	0.50–0.80	0.040 max	0.040 max	0.50 max	0.10 max	0.30 max	...	...	...	...
1070	G10700	Ck 67	0.65–0.72	0.25–0.50	0.60–0.80	0.035 max	0.035 max	0.35 max	...	0.35 max	...	...	0.35	...
...	...	55 Cr 1	0.52–0.60	0.17–0.37	0.5–0.8	0.035 max	...	0.2–0.5	...	0.3 max	...	0.02–0.05	0.3 max	0.015
...	...	50 CrV 4	0.47–0.55	0.4 max	0.7–1.1	0.035	0.03 max	0.9–1.2	...	...	...	...	...	0.1–0.2
Tool steels														
O2	T31502	90 MnV 8	0.85–0.95	0.15–0.35	1.80–2.00	0.030 max	0.030 max	...	...	...	0.07–0.12	...	...	...
W1	T72301	C 100 W1	0.95–1.04	0.15–0.30	0.15–0.25	0.020 max	0.020 max	0.20 max	...	0.20 max(b)	...	...	...	...

(a) Deutsche Industrie-Normen. (b) 0.25 max Cu

- No preparation of surfaces to be hardened or of regions that have to be left untreated
- Computer numerical control (CNC) or computer-controlled processing similar to that used for machine tools, which are CAD/CAM (computer-aided design/computer-aided manufacturing) compatible and easy to integrate into mechanical flow lines
- Plant operation requires only electric power and low quantities of cooling water (usually in closed circulation systems) but neither transport media such as a working gas or an inert gas nor hardening salts or quenching oils are required.
- High energy efficiency
- High process productivity with available beam power ranging from 20 to 50 kW
- No waste products generated
- Technological equipment suitable for several processes such as deep welding, hardening, and fusion treatment of surfaces

Electron Beam Hardening versus Laser Beam Hardening

In recent years, heat treating engineers have focused their efforts on pinpointing the properties that differentiate electron beam hardening from laser beam hardening. The qualitative results of both hardening techniques are almost identical. However, differences do arise from the varying properties of both beam types and the entirely different methods that generate the beams. Comparing electron-beam hardening and laser-beam hardening to each other and to conventional hardening methods is the most accurate gage of their performance. Both techniques have common advantages and drawbacks compared to conventional hardening techniques, as well as specific positive and negative qualities with respect to each other.

Both beam techniques overshadow conventional hardening methods because of the following primary factors:

- Locally well-defined and reproducible energy transfer to the workpiece regions
- Low thermal stress imposed on the component

Secondary benefits include the capability to adapt to a CNC-control process similar to that used in mechanical processing lines and the omission of cooling media for quenching. However, the rather high capital investment costs of high-energy beam hardening facilities may be prohibitive to some customers.

Advantages for Laser-Beam Processing. Laser-beam hardening has the advantage over electron-beam hardening whenever the following factors are significant:

- Relatively large distance between beam source (laser oscillator) and process site
- Beam guidance aided by mirrors (that is, in robot levers and multistation machining)
- Comparatively low costs with low beam power setting (2 to 3 kW) sufficient for hardening
- Parts whose bulk and configuration prevent them from being placed in a vacuum and therefore processing at atmospheric pressure

Advantages for Electron-Beam Processing. The following factors favor electron beam hardening as a viable option:

- Inert environment characteristic of a low- or high-vacuum atmosphere is required.
- No working or protective gases are required.
- Energy absorption properties of the component surface are independent of the optical surface properties (no application of absorption layers for energy coupling).
- High overall energy efficiency of the installation increases with the beam power.
- Easy generation of limited energy transfer fields to a maximum of >10 cm^2 (>1.6 $in.^2$) and of any desired power density distribution via cyclic rf beam deflection programs (usability of the energy absorption layer as heat accumulator)
- High-volume productivity at beam powers of 10 kW and above

REFERENCES

1. R.F. Kern, Selecting Steels for Heat-Treated Parts, Part II: Case Hardenable Grades, *Met. Prog.,* Dec 1968, p 71–81
2. N.J. Fulco, Flame Hardening, *Heat Treat.,* Aug 1974, p 14–17
3. D.L. Loveless, R.L. Cook, and V.I. Rudnev, Chapter 11B, Induction Heat Treatment: Modern Power Supplies, Load Matching, Process Control, and Monitoring, *Steel Heat Treatment Handbook,* G.E. Totten and M.A.H. Howes, Ed., Marcel Dekker, Inc., 1997, p 873–911

4. S.L. Semiatin and D.E. Stutz, *Induction Heat Treatment of Steel,* American Society for Metals, 1986
5. C.R. Brooks, The Metallurgy of Induction Surface Hardening, *Heat Treat. Prog.,* Dec 2000, p H19–H23
6. *Metals Handbook,* 8th ed., Vol 8, *Metallography, Structures and Phase Diagrams:* American Society for Metals, Metals Park, Ohio, 1973
7. K.-E. Thelning, *Steel and Its Heat Treatment,* 2nd ed., Butterworths, London, U.K., 1975
8. C.R. Brooks, *Principles of the Surface Treatment of Steels,* Technomic Publishing Co., Lancaster, Pa., 1992
9. D.L. Martin and W.G. Van Note, Induction Hardening and Austenitizing Characteristics of Several Medium Carbon Steels, *Trans. ASM,* Vol 36, 1946, p 210
10. Joseph F. Libsch, Wen-Pin Chuang, and William J. Murphy, The Effect of Alloying Elements on the Transformation Characteristics of Induction-Heated Steels, *Trans. ASM,* Vol 42, 1950, p 121
11. R.E. Haimbaugh, *Practical Induction Heat Treating,* ASM International, 2001
12. R.M. Wood, Macroetching Induction Hardened Steel Parts, *Heat Treat. Prog.,* Dec 2000, p H26–H28
13. Surface Treatment: Heat Treating, *LIA Handbook of Laser Materials Processing,* J.F. Ready and D.F. Farson, Ed., Laser Institute of America/Magnolia Publishing Inc., 2001, p 223–261
14. K. Sridhar and A.S. Khanna, Laser Surface Heat Treatment, *Lasers in Surface Engineering,* Vol 1, *Surface Engineering Series,* N.B. Dahotre, Ed., ASM International, 1998, p 69–119
15. O.A. Sandven, Laser Surface Hardening, *Heat Treating,* Vol 4, *ASM Handbook,* ASM International, 1991, p 286–296
16. M.A. Howes, Lasers Can Replace Selective Carburization Economically, *Laser Surface Modification,* Proceedings from the 1988 Conference, (New Orleans, 14–15 April 1988), American Welding Society, Miami, p 43–69

CHAPTER 11

Surface Hardening by Coating or Surface Modification

THE SURFACE-ENGINEERING METHODS described in this chapter are those that:

- Involve an intentional buildup or the addition of a new layer on a steel substrate—that is, the application of a coating—to enhance surface hardness. Examples of coating processes commonly employed include electroplating, electroless plating, weld overlays (hardfacing), thermal spraying, chemical vapor deposition (CVD), and physical vapor deposition (PVD).
- Alter the steel surface composition or structure by the use of high-energy or particle beams—that is, surface modification—to enhance surface hardness. Examples of surface-modification methods commonly employed include laser surface processing and ion implantation.

The primary purpose of these surface treatments is to improve wear resistance of steel components, although increased corrosion resistance and/or improved appearance of the treated component may also result.

Figure 1 shows the surface thickness ranges associated with four types of surface processing. These include the thermochemical (diffusion) processes discussed in Chapters 2 through 9, plating and coating processes, surface modification processes, and thermal hardening processes (flame and induction hardening). As this figure shows, surface thickness or depth of hardening can range from thin films produced by PVD and CVD (1 μm, or 0.4 mil or less) to very thick coatings applied by various welding processes (>10 mm, or 0.4 in.). The key to proper selection of the surface treatments shown in Fig. 1 is in the identification of the performance requirements for a given surface-engineered material system in a given application. Not only must the properties of the surface be considered but also the properties of the substrate and the interface between the surface and substrate. In some systems there is a gradual change in properties between the surface and interior, as for example in nitrided and carburized components, while in others there is an abrupt change, as for example for parts where a vapor-deposited coating of titanium nitride has been deposited on steel. Such interface characteristics may significantly influence the performance of a surface-engineered system.

The performance requirements of surface-modified systems may vary widely. For example, heavily loaded systems such as bearings and gears require deep cases to resist rolling contact and bending stresses that result in fatigue damage. Other applications may require only very thin surface modification to resist corrosion, to resist near surface abrasion or scuffing, or to reduce friction between moving surfaces. Many of these requirements are based on complex interactions between applied static and cyclic stress states and gradients in structures and properties of the surface-engineered systems. The identification of the mechanisms of these interactions is not well understood and is an active area of research. See, for example, the studies carried out on microstructure/property relationships of carburized steels described in Chapter 2, "Gas Carburizing," in this book.

Hard Chromium Plating

Hard chromium plating is produced by electrodeposition from a solution containing chromic acid (CrO_3) and a catalytic anion in proper proportion. The metal so produced is extremely hard and corrosion resistant. The process is used for applications where excellent wear and/or corrosion resistance is required. This includes products such as piston rings, shock absorbers, struts, brake pistons, engine valve stems, cylinder liners, and hydraulic rods.

Other applications are for aircraft landing gears, textile and gravure rolls, plastic rolls, and dies and molds. The rebuilding of mismachined or worn parts makes up large segments of the industry. One specialized application is a thin chromium layer used as a lacquer adhesive layer in the manufacture of "tin" cans.

Hard chromium plating is also known as *industrial, functional,* or *engineering chromium plating.* It differs from decorative chromium plating in the following ways:

- Hard chromium deposits are intended primarily to increase the service life of functional parts by providing a surface with a low coefficient of friction that resists galling, abrasive and lubricated wear, and corrosion. Another major purpose is to restore dimensions of undersized parts.
- Hard chromium normally is deposited to thicknesses ranging from 2.5 to 500 μm (0.1 to 20 mils) and, for certain applications, to considerably greater thicknesses, whereas decorative coatings seldom exceed 1.3 μm (0.05 mil).
- With certain exceptions, hard chromium is applied directly to the base metal; decorative chromium is applied over undercoats of nickel or of copper and nickel.

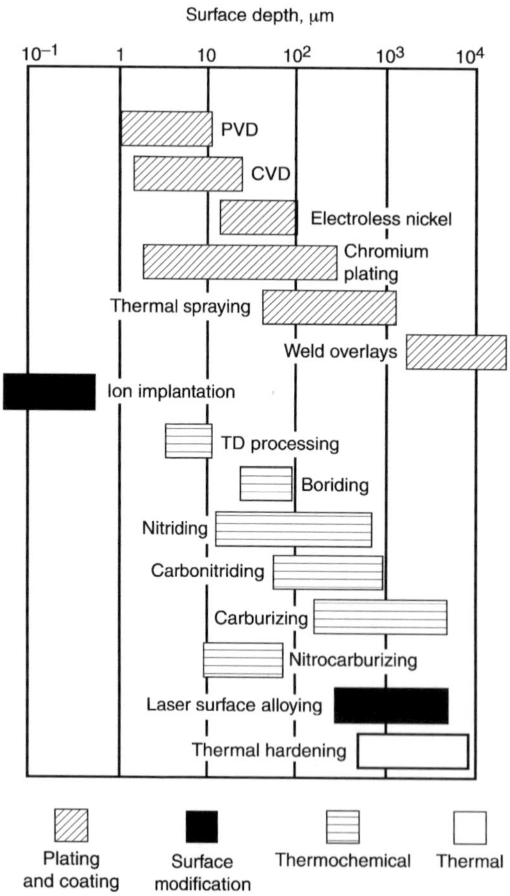

Fig. 1 Classification and typical surface depths of various surface-engineering treatments. Adapted from Ref 1

Hardness

The hardness of chromium electrodeposits is a function of the type of chemistry selected and the plating conditions. In general, chromium plated in the bright range is optimally hard. Typically, bright chromium deposits from conventional plating solutions have hardness values of 850 to 950 HV; those from mixed-catalyst solutions have values of 900 to 1000 HV. Those from fluoride-free chemistries have values of 950 to 1100 HV or higher.

Principal Uses

The major uses of hard chromium plating are for wear-resistance applications, improvement of tool performance and tool life, and part salvage. Table 1 lists parts to which hard chromium plate is applied and representative data regarding plate thickness and plating times. Plating times can be reduced by using high-efficiency or mixed-catalyst solutions.

Wear Resistance. Extensive performance data indicate the effectiveness of chromium plate in reducing the wear of piston rings caused by scuffing and abrasion. The average life of a chromium-plated ring is approximately five times that of an unplated ring made of the same base metal. Piston rings for most engines have a chromium plate thickness of 100 to 200 μm (4 to 8 mils) on the bearing face, although thicknesses up to 250 μm (10 mils) are specified for some heavy-duty engines.

In the automotive industry, hard chromium is also applied to shock absorber rods and struts to increase their resistance to wear and corrosion. Valve stems are plated with a flash coating (about 2.5 μm, or 0.1 mil) to reduce wear. Hydraulic shafts for all kinds of equipment are plated with 20 to 30 μm (0.8 to 1.2 mils) of hard chromium to increase service life.

The low friction coefficient and good wear properties of chromium have been attributed to a self-healing Cr_2O_3 film that forms on the surface. In general, hard chromium has a lower wear rate than either electroplated or electroless nickel, which are the two competing materials. This effect is illustrated in Fig. 2 and 3.

Tooling Applications. Various types of tools are plated with chromium to minimize wear, prevent seizing and galling, reduce friction, and/or prevent or minimize corrosion. Steel dies for molding of plastics are usually plated with chromium, especially when vinyl or other corrosive plastic materials are to be molded. Plating thicknesses of 2.5 to 125 μm (0.1 to 5 mils) usually are recommended for preventing wear in parts sticking in molds and for reducing frequency of polishing when plastics that attack steel are being molded. Chromium-plated dies should not be used when plastics containing fire-retardant chlorides are molded.

The service life of plug gages and other types of gages may be prolonged by hard chromium plating. Most gage manufacturers provide chromium-plated gages. Records in one plant indicate that plug gages made from hardened O1 tool steel wore 0.0025 mm (0.0001 in.) after gaging 5000 cast iron parts. Hard chromium plating of these gages allowed the gaging of 40,000 parts per 0.0025 mm (0.0001 in.) of wear.

Worn gages can be salvaged by being built up with hard chromium plate. Also, chromium plate provides steel gages with good protection against rusting in normal exposure and handling. Chromium plating is not recommended, however, for gages that are subjected to impact at exposed edges during operation.

Deep drawing tools often are plated with chromium, in thicknesses up to 100 μm (4 mils), for improvement of tool performance and/or building up of worn areas. The life of draw rings and punches may be prolonged by plating. In addition, plating reduces frictional force on punches and facilitates removal of workpieces from punches in instances where sticking is encountered with plain steel surfaces. If deep drawing tools are chromium plated, the base metal should be harder than 50 HRC. Steel dies used for drawing bars and tubes are often plated with relatively heavy thicknesses (up to 250 μm, or 10 mils) of chromium to minimize die wear, reduce friction, and prevent seizing and galling.

The service life of cutting tools is often extended by chromium plate, in thicknesses

Table 1 Typical thicknesses and plating times for selected applications of hard chromium plating

Part	Base metal	Thickness of plate		Plating time(a)
		μm	mils	
Computer printer type	Carbon steel	25	1	60 min
Face seals	Steel or copper	75–180	3–7	10 h
Aircraft engine parts	Nickel-based alloys, high-strength steel	75–180	3–7	10 h
Plastic molds	Tool steel	5–13	0.2–0.5	30 min
Textile guides	Steel	5–100	0.2–4	20–240 min
Piston rings	Steel or cast iron	150–255	6–10	8 h
Balls for ball valves	Brass or steel	7.5–13	0.3–0.5	20 min
Micrometers	Steel	7.5–13	0.3–0.5	20 min
Golf ball molds	Brass or steel	7.5–25	0.3–1	20–60 min
Lock cases	Brass	5–7.5	0.2–0.3	20 min
Cylinder	Cast iron	255	10	300 min
Bushing	1018 carburized, 56 HRC	25	1	45 min
Crankshafts	Steel	255–3800	10–150	. . .
Cutting tools	Tool steel	1.3	0.05	5 min
Forming and drawing dies	Steel	25	1	60 min
Gage	Steel	125	5	150 min
Gun barrels, 30 caliber(b)	Steel	25	1	40 min
Hydraulic cylinder	1045 steel	13	0.5	40 min
Pin	Steel	13	0.5	30 min
Pin	1045 steel, 60 HRC	125	5	40 min
Plug gage	1040 steel, 55 HRC	125	5	150 min
Relief-valve plunger	1113 steel, soft	100	4	60 min
Ring gage	Steel	205	8	240 min
Rolls	Steel	13–255	0.5–10	20–300 min

(a) Times shown are for conventional plating solutions; plating times for the proprietary fluoride-free solution are half of those shown.
(b) M-16 rifle, barrel and chamber

ranging from less than 2.5 to 13 μm (0.1 to 0.5 mil). Taps and reamers are examples of tools on which chromium plate has proved advantageous. In one case, a flash plate on taps used to thread cold-worked 1010 steel improved tap life from 250 (for unplated taps) to 6000 parts per tap. The poor tool life of the unplated taps was caused by buildup of metal on the cutting edges. Hard chromium plating is not recommended for cold extrusion tools for severe applications where extreme heat and pressure are generated, because the plate is likely to crack and spall and may be incompatible with phosphate-soap lubricants.

Part Salvage. Hard chromium plating is sometimes used for restoring mismachined or worn surfaces. Since 1970, the use of this process for part salvage has been frequently replaced by thermal spraying and plasma coatings, which can be applied more quickly. The fact that a chromium deposit can significantly reduce fatigue strength must be considered in determining whether chromium plating can be safely used.

Other Applications. Hard chromium plate is applied to printing plates and stereotypes, especially to those intended for long runs, because compared to other materials or coatings used for this application, it wipes cleaner, provides sharper reproduction, and increases the length of press runs. It is used on press rams because of its excellent resistance to corrosion, seizing, galling, and other forms of wear.

Corrosion Properties

In addition to their excellent wear properties, hard chromium electrodeposits also exhibit resistance to corrosion in many harsh environments. Factors influencing corrosion properties include coating stresses and microcracks, coating hardness, and coating thickness.

Stress and Microcracks. The tensile stress in most electroplated chromium deposits increases until microcracks are formed (Ref 3). The microcracks decrease the stress in the deposit as the thickness of the deposit increases. Stress is inversely proportional to the number of microcracks. The number of microcracks is more important in controlling stress than the type of bath chemistry. Crack-free deposits are highly stressed.

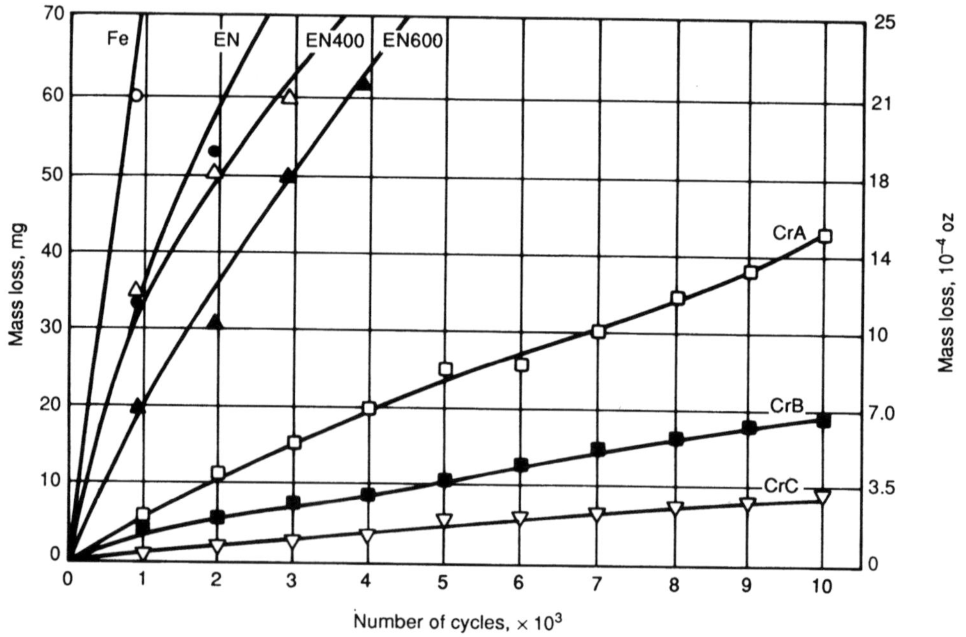

Fig. 2 Effect of number of cycles on mass loss in the Taber abrasion test for uncoated steel substrate (Fe), three chromium deposits (CrA, CrB, CrC), and three electroless nickel deposits: as-plated nickel (EN), heat treated at 400 °C (750 °F) (EN400), and heat treated at 600 °C (1110 °F) (EN600). Source: Ref 2

Microcracks are present in most electroplated hard chromium deposits. Figure 4 shows a typical microcrack structure. The density of the microcracks in chromium deposits varies from 0 to more than 120 cracks/mm (3000 cracks/in.), depending on bath chemistry, current density, and temperature. The number of microcracks increases with the concentration of the catalyst in the plating bath. The depth of a microcrack is less than about 8 μm (0.3 mil) on a deposit that is 130 μm (5 mils) thick with crack counts of about 80 cracks/mm (2000 cracks/in.).

Because chromium protects substrates by forming a barrier, the coatings must be thicker than the microcracks to provide good corrosion resistance. Thin coatings may not form microcracks and can offer as much corrosion resistance as thicker coatings (see the subsequent section "Coating Thickness"). Chromium electrodeposits that are about 25 μm (1 mil) thick with crack counts of about 40 cracks/mm (1000 cracks/in.) are as resistant to corrosion as deposits with crack counts of about 10 cracks/mm (250 cracks/in.). Deposits with very low crack counts have deeper microcracks than deposits with higher crack counts. Therefore, highly microcracked deposits are as resistant to corrosion as sparsely microcracked deposits.

Microcracks are not as detrimental to corrosion resistance as might be expected. There are

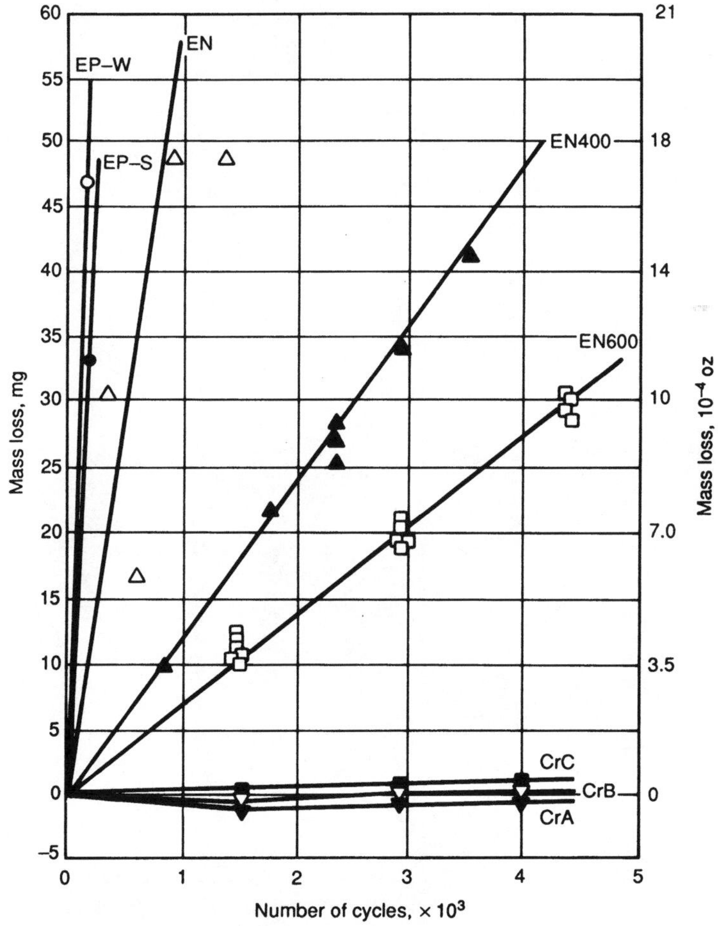

Fig. 3 Effect of number of cycles on mass loss of plated pin versus steel blocks in a Falex test for the three chromium deposits (CrA, CrB, CrC) and the three electroless nickel deposits (EN, EN400, EN600) shown in Fig. 2. Effects on two electroplated nickels from a sulfamate solution (EP-S) and a Watts solution (EP-W) are shown. Source: Ref 2

two reasons for this. First, the microcracks are not voids but are areas with a structure and composition different from those of the bulk. Second, because the microcracks are very narrow (about 0.1 μm wide) and because water does not wet chromium, the water does not readily enter the microcracks.

Microcrack-free thick chromium deposits can be plated from baths at low current densities and high temperatures. These microcrack-free deposits provide better corrosion protection than microcracked chromium. However, these deposits are highly stressed and are not as hard as microcracked chromium. Crack-free deposits can be used when corrosion protection is the only requirement for the deposit. Postplating grinding or cutting may cause pickout (chromium fracturing from chromium) in highly stressed deposits. The conditions under which some crack-free coatings are deposited results in a deposition efficiency that is lower than that normally observed for the plating bath.

Additional Deposit Properties Influencing Corrosion. Hardness is related to microcracking, which is related to corrosion. Chromium coatings have hardnesses between 850 and 1050 HK (100 gf load). Microcrack-free deposits can have hardnesses as low as 600 or 300 HK. According to one study, as deposit hardness increases or crystal size decreases, the rate of attack by sulfuric acid (H_2SO_4), hydrochloric acid (HCl), and CrO_3 decreases (Ref 4).

Coating Thickness. Figure 5 shows that the corrosion resistance of hard chromium-plated steel in salt spray undergoes a maximum and a minimum and then increases with the chromium thickness (Ref 5). Figure 5 also shows the average of two panels in a salt spray exposure. Maximum corrosion resistance occurred at a chromium thickness of about 5 μm (0.2 mil). As the thickness increased above 5 μm (0.2 mil), microcracking occurred and corrosion resistance decreased. When the chromium thickness increased to about 10 μm (0.4 mil), the initial cracks were covered by more chromium, there were fewer corrosion paths to the substrate, and the corrosion resistance of the deposit increased. These deposits were plated from a conventional bath containing 250 g/L of CrO_3 and 2.5 g/L of sulfate (SO_4^{2-}) at 31 A/dm^2 (2 A/in.^{2}); no temperature was specified.

Figure 6 shows additional data on corrosion resistance and chromium thickness. The electrodeposits were prepared from a conventional bath containing 295 g/L of CrO_3 and 3 g/L of H_2SO_4. Data are given for two plating conditions: 30 °C (85 °F) at 20 A/dm^2 (1.3 A/in.^{2}) and 60 °C (140 °F) at 43 A/dm^2 (2.8 A/in.^{2}). The first condition produced cold chromium that was crack-free and soft. The second condition

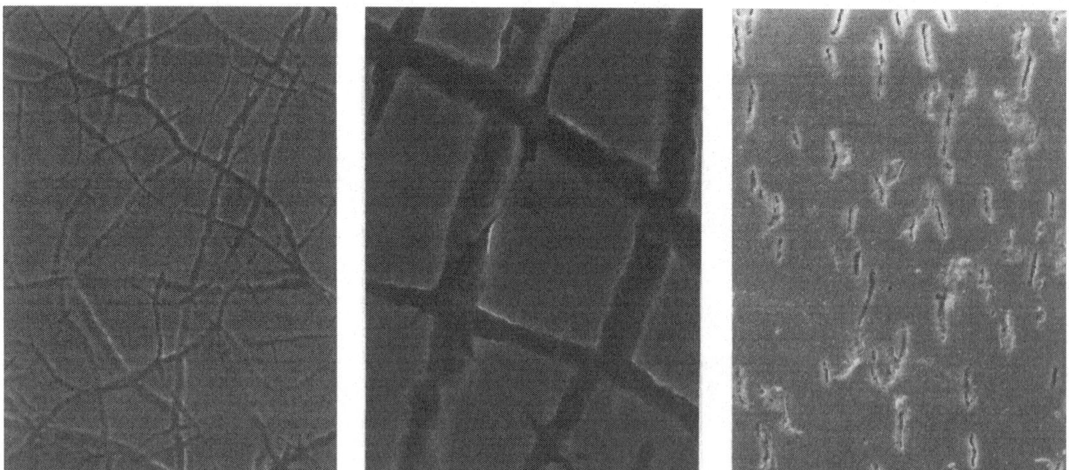

(a) (b) (c)

Fig. 4 Photomicrographs of chromium deposits (plated in a high-efficiency etch-free bath) after etching. (a) and (b) Deposit plated at 78 A/dm^2 (5 A/in.^{2}) and at 55 °C (130 °F). (a) 400×. (b) 1700×. (c) Cross section of a chromium deposit plated at 93 A/dm^2 (6 A/in.^{2}) and at 58 °C (135 °F). The specimen was polished before etching. 650×. Both deposits contain 80 microcracks/mm (2000 microcracks/in.).

produced conventional microcracked hard chromium. The cold chromium deposit showed excellent corrosion resistance at thicknesses of 4.8, 9.1, and 12.4 μm (0.2, 0.36, and 0.49 mil), but the corrosion resistance was very poor at a thickness of 15.5 μm (0.6 mil). The 15.5 μm (0.6 mil) coating was not porous, and no reason was given for its poor corrosion resistance. The high stress in the coating and the poor adhesion of cold chromium may have resulted in coating failure. The thinner (<15 μm, or 0.59 mil) cold chromium coatings performed much better than the conventional chromium deposits.

The conventional deposits shown in Fig. 5 do not exhibit the trend observed in Fig. 6. The thickest deposit (18.8 μm, or 0.74 mil) did show a slight decrease in corrosion resistance relative to the 14.4 μm (0.57 mil) deposit. Statistically, however, the corrosion resistance was the same for the two coating thicknesses.

Corrosion Resistance in Specific Environments. Hard chromium deposits exhibit excellent resistance to corrosion in the atmosphere and a wide range of chemical environments. Electroplated chromium for atmospheric-corrosion applications should be between 20 and 30 μm (0.8 and 1 mil) thick. For corrosion resistance in chemical exposures, electroplated chromium should be 50 to 75 μm (2 to 3 mils) thick.

Corrosion of hard chromium deposits usually begins at microcracks or intersections of microcracks (Ref 3). After moderate acid attack, the corrosion reveals the microcrack pattern. Attack will continue on all of the chromium, and the microcrack pattern will no longer be visible. The corrosion of chromium in sodium chloride (NaCl) solutions will produce mounds of corrosion products. When these mounds are removed, concentric rings define the attacked area. The center and outside area are unattacked.

Chromium-plated steel with and without diffusion treatment at 1000 °C (1830 °F) resists corrosion by sodium polysulfides (Na_2S_4, Na_2S_5)

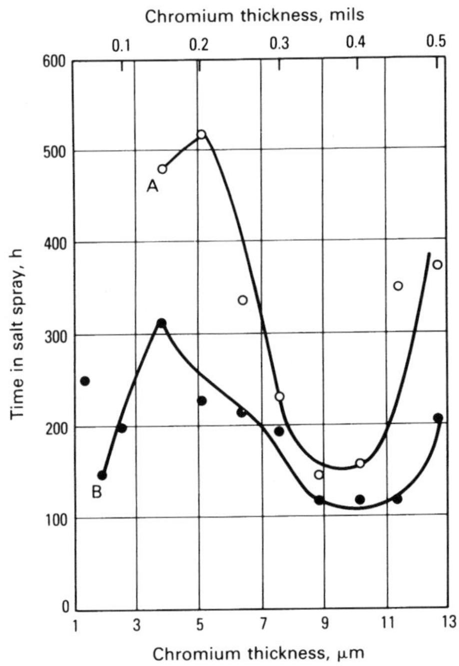

Fig. 5 Chromium corrosion in salt spray versus thickness of deposit. Curve A shows time to general rust; curve B is for time to initial corrosion. Parts were plated in a conventional bath (250 g/L CrO_3 and 2.5 g/L SO_4^{2-} at 31 A/dm², or 2 A/in.²). Source: Ref 5

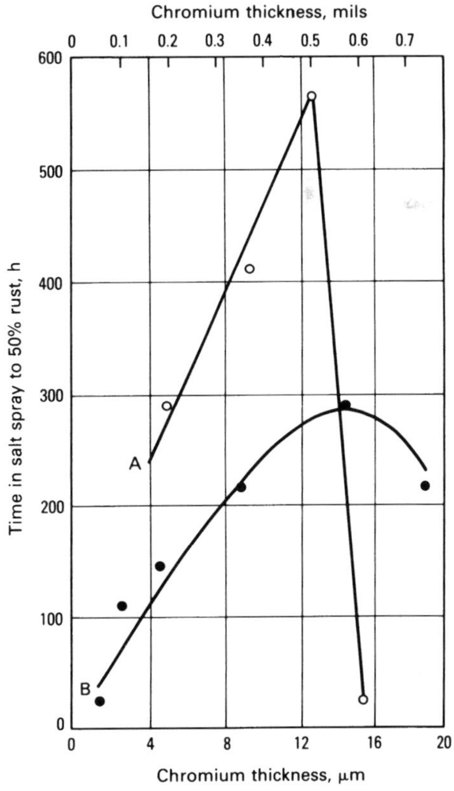

Fig. 6 Chromium corrosion in salt spray versus thickness of deposit. The electrodeposits were plated from a conventional bath (295 g/L CrO_3 and 3.0 g/L SO_4^{2-}) at two conditions: 20 A/dm² (1.3 A/in.²) at 30 °C (85 °F) (curve A) and 43 A/dm² (2.8 A/in.²) at 60 °C (140 °F) (curve B). Source: Ref 6

and sulfur at temperatures to 440 °C (825 °F). In a 12 month static test, chromium-diffused samples (plating thickness: 50 to 200 μm, or 2 to 8 mils) performed better than the as-plated samples. Electroplated chromium is attacked at 58 °C (135 °F) in formic acid (HCOOH), hydrobromic acid (HBr), HCl, perchloric acid ($HClO_4$), H_2SO_4, and trichloroacetic acid (CCl_3COOH) and is attacked at 12 °C (55 °F) in hydrofluoric acid (HF). Hot (58 °C, or 135 °F) solutions of ferric chloride ($FeCl_3$), mercuric chloride ($HgCl_2$), and stannous chloride ($SnCl_2$) attack electroplated chromium more severely than most other salt solutions. Detailed data on the corrosion of hard chromium in various media can be found in the article "Corrosion of Electroplated Hard Chromium Plating" in *Corrosion,* Volume 13 of *ASM Handbook.*

Chromium Replacement

Because the use and emission of hexavalent chromium originating from chromium plating baths has come under increased scrutiny by various regulatory bodies due to adverse health and environmental effects, there has been a tremendous amount of work carried out in recent years to determine suitable replacements for electroplated chromium. Potential replacements include electroless nickel coatings described in the following section and various high-velocity oxyfuel thermal spray coatings such as the WC-Co, Co-Mo-Cr, WC-Co-Cr, and CrC-Ni-Cr alloy systems.

Electroless Nickel Plating (Ref 7)

Electroless nickel plating is used to deposit nickel without the use of an electric current. The coating is deposited by an autocatalytic chemical reduction of nickel ions by hydrophosphite, aminoborane, or borohydride compounds. Details of the process can be found in the article "Electroless Nickel Plating" in Volume 5 of *ASM Handbook.* When sodium hypophosphite is the reducing agent, the deposit generally contains between 3 and 11% phosphorus. The boron contents of electroless nickel range from 0.2 to 4 wt% and from 4 to 7 wt% when the reducing agents are an aminoborane and sodium borohydride, respectively. These coatings are deposited on carbon and alloy steels and 300 and 400 series stainless steels.

Electroless nickel is an engineering coating normally used because of excellent corrosion and wear resistance. Electroless nickel coatings are also applied to molds and dies to improve lubricity and part release. Because of these properties, electroless nickel coatings have found many applications, including those in petroleum, chemicals, plastics, optics, printing, mining, aerospace, nuclear, automotive, electronics, computers, textiles, paper, and food machinery. Some advantages and limitations of electroless nickel coatings include the following:

Advantages

- Good resistance to corrosion and wear
- Excellent uniformity
- Solderability and brazeability
- Low labor costs

Limitations

- Higher chemical cost than electroplating
- Brittleness
- Poor welding characteristics due to contamination of nickel plate with nickel-phosphorus deposits
- Need to copper-strike the plate alloys containing significant amounts of lead, tin, cadmium, and zinc before electroless nickel can be applied
- Slower plating rate, as compared to the rates of electrolytic methods

Electroless Nickel-Phosphorus Coatings

Hypophosphite-reduced electroless nickel is an unusual engineering material because of both its method of application and its unique properties. As applied, nickel-phosphorus coatings are amorphous, uniform, hard, relatively brittle, lubricious, easily solderable, and highly corrosion resistant. They can be precipitation hardened to very high levels through the use of low-temperature treatments, producing wear resistance equal to that of commercial hard chromium coatings. This combination of properties makes the coating well suited for many severe applications and often allows it to be used in place of more expensive or less readily available alloys. Table 2 provides a summary of the properties of nickel-phosphorus as well as nickel-boron electroless deposits.

Hardness and Wear Resistance. As deposited, the microhardness of electroless nickel coatings is about 500 to 600 HV_{100}, which is approximately equal to 48 to 52 HRC and equivalent to many hardened alloy steels. Heat treatment causes these alloys to age harden and can produce hardness values as high as 1100 HV_{100}, equal to most commercial hard chromium coatings. Figure 7 shows the effect of different 1 h heat treatments on the hardness of electroless nickel containing 10.5% P.

Because of their high hardness, electroless nickel coatings have excellent resistance to wear and abrasion, both in the as-deposited and hardened conditions. Taber Abraser Index values for electroless nickel and for electrodeposited nickel and chromium are summarized in Table 3. Additional wear data are presented in Fig. 2 and 3.

Corrosion Resistance. Electroless nickel is a barrier coating, protecting the substrate by sealing it off from the environment, rather than using sacrificial action. Therefore, the deposit must be free of pores and defects. Because of its amorphous nature and passivity, the corrosion resistance of the coating is excellent and, in many environments, superior to that of pure nickel or chromium alloys. Amorphous alloys have better resistance to attack than equivalent polycrystalline materials, because of their freedom from grain or phase boundaries and because of the glassy films that form on and passivate their surfaces. Some examples of the corrosion experienced in different environments are shown in Table 4. The resistance to attack in neutral and acidic environments is increased as the phosphorus content is increased in the deposit. The reverse is true in alkaline corrosive

Table 2 Physical and mechanical properties of electroless nickel-phosphorus and nickel-boron deposits

Properties are for coatings in the as-deposited condition, unless noted.

Property	Electroless nickel-phosphorus(a)	Electroless nickel-boron(b)
Density, g/cm^3 (lb/in.^{3})	7.75 (2.8)	8.25 (2.98)
Melting point, °C (°F)	890 (1630)	1080 (1980)
Electrical resistivity, μΩ · cm	90	89
Thermal conductivity, W/m · K (cal/cm · s · °C)	4 (0.01)	. . .
Coefficient of thermal expansion (22–100 °C, or 72–212 °F), μm/m · °C (μin./in. · °F)	12 (6.7)	12.6 (7.1)
Magnetic properties	Nonmagnetic	Very weakly ferromagnetic
Internal stress, MPa (ksi)	Nil	110 (16)
Tensile strength	700 (100)	110 (16)
Ductility, % elongation	1.0	0.2
Modulus of elasticity, GPa (10^6 psi)	200 (29)	120 (17)
As-deposited hardness, HV_{100}	500	700
Heat-treated hardness, 400 °C (750 °F) for 1 h, HV_{100}	1100	1200
Coefficient of friction vs. steel, lubricated	0.13	0.12
Wear resistance, as-deposited, Taber mg/1000 cycles	18	9
Wear resistance, heat treated 400 °C (750 °F) for 1 h, Taber mg/1000 cycles	9	3

(a) Hypophosphite-reduced electroless nickel containing approximately 10.5% P. (b) Borohydride-reduced electroless nickel containing approximately 5% B

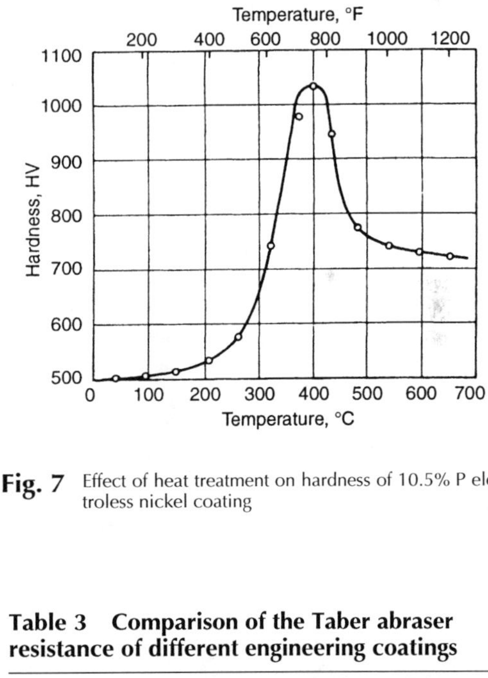

Fig. 7 Effect of heat treatment on hardness of 10.5% P electroless nickel coating

Table 3 Comparison of the Taber abraser resistance of different engineering coatings

Coating	Heat treatment for 1 h °C	Heat treatment for 1 h °F	Taber wear index(a), mg/1000 cycles
Watts nickel	None	None	25
Electroless Ni-P(b)	None	None	17
	300	570	10
	500	930	6
	650	1200	4
Electroless Ni-B(c)	None	None	9
	400	750	3
Hard chromium	None	None	2

(a) CS-10 abraser wheels, 1000 g load, determined as average weight loss per 1000 cycles for total test of 6000 cycles. (b) Hypophosphite-reduced electroless nickel containing approximately 9% P. (c) Borohydride-reduced electroless nickel containing approximately 5% B environments. The effect of phosphorus content on the corrosion rate of electroless coatings is shown in Tables 5 and 6. Where corrosion resistance is required, hardened (heat treated) coatings should not be used.

Electroless Nickel-Boron Coatings

The properties of deposits from borohydride-reduced or aminoborane-reduced baths are similar to those of electroless nickel-phosphorus alloys with a few exceptions. The hardness of nickel-boron alloys is very high, and these alloys can be heat treated to levels equal to or greater than that of hard chromium. Nickel-boron coatings have outstanding resistance to wear and abrasion. These coatings, however, are not completely amorphous and have reduced resistance to corrosive environments; furthermore, they are much more costly than nickel-phosphorus coatings. The physical and mechanical properties of borohydride-reduced electroless nickel are summarized in Table 2.

Hardness and Wear Resistance. The principal advantage of electroless nickel-boron is its high hardness and superior wear resistance. In the as-deposited condition, microhardness values of 650 to 750 HV_{100} are typical for borohydride-reduced and aminoborane-reduced coatings. After 1 h heat treatments at 350 to 400 °C (660 to 750 °F), hardness values of 1200 HV_{100} can be produced.

The wear resistance of electroless nickel-boron is exceptional and after heat treatment equals or exceeds that of hard chromium coatings. Typical Taber wear test results for a 5% B coating are shown in Tables 2 and 3.

Corrosion Resistance. In general, the corrosion resistance of electroless nickel-boron coatings is less than that of high-phosphorus

Table 4 Corrosion of electroless nickel coatings in various environments

			Corrosion rate			
	Temperature		Electroless nickel-phosphorus(a)		Electroless nickel-boron(b)	
Environment	°C	°F	μm/yr	mil/yr	μm/yr	mil/yr
Acetic acid, glacial	20	68	0.8	0.03	84	3.3
Acetone	20	68	0.08	0.003	Nil	Nil
Aluminum sulfate, 27%	20	68	5	0.2	...	...
Ammonia, 25%	20	68	16	0.6	40	1.6
Ammonia nitrate, 20%	20	68	15	0.6	(c)	(c)
Ammonium sulfate, saturated	20	68	3	0.1	3.5	0.14
Benzene	20	68	Nil	Nil	Nil	Nil
Brine, 3.5% salt, CO_2 saturated	95	205	5	0.2	...	...
Brine, 3.5% salt, H_2S saturated	95	205	Nil	Nil	...	...
Calcium chloride, 42%	20	68	0.2	0.008	...	...
Carbon tetrachloride	20	68	Nil	Nil	Nil	Nil
Citric acid, saturated	20	68	7	0.3	42	1.7
Cupric chloride, 5%	20	68	25	1	...	...
Ethylene glycol	20	68	0.6	0.02	0.2	0.008
Ferric chloride, 1%	20	68	200	8	...	...
Formic acid, 88%	20	68	13	0.5	90	3.5
Hydrochloric acid, 5%	20	68	24	0.9	...	...
Hydrochloric acid, 2%	20	68	27	1.1	...	...
Lactic acid, 85%	20	68	1	0.04	...	...
Lead acetate, 36%	20	68	0.2	0.008	...	...
Nitric acid, 1%	20	68	25	2	...	...
Oxalic acid, 10%	20	68	3	0.1	...	...
Phenol, 90%	20	68	0.2	0.008	Nil	Nil
Phosphoric acid, 85%	20	68	3	0.1	(c)	(c)
Potassium hydroxide, 50%	20	68	Nil	Nil	Nil	Nil
Sodium carbonate, saturated	20	68	1	0.04	Nil	Nil
Sodium hydroxide, 45%	20	68	Nil	Nil	Nil	Nil
Sodium hydroxide, 50%	95	205	0.2	0.008	...	...
Sodium sulfate, 10%	20	68	0.8	0.03	11	0.4
Sulfuric acid, 65%	20	68	9	0.4	...	...
Water, acid mine, 3.3 pH	20	68	7	0.3	...	...
Water, distilled, N_2 deaerated	100	212	Nil	Nil	Nil	Nil
Water, distilled, O_2 saturated	95	205	Nil	Nil	Nil	Nil
Water, sea (3.5% salt)	95	205	Nil	Nil	...	...

(a) Hypophosphite-reduced electroless nickel containing approximately 10.5% P. (b) Borohydride-reduced electroless nickel containing approximately 5% B. (c) Very rapid. Specimen dissolved during test.

alloys. That is illustrated by Table 4, which compares the attack experience by hypophosphite-reduced and borohydride-reduced coatings in different media. In environments that cause little corrosion of nickel-phosphorus, such as alkalis and solvents, electroless nickel-boron is also very resistant. In environments, however, that cause moderate attack of nickel-phosphorus, such as acids and ammonia solutions, nickel-boron coatings can be severely corroded. In strongly oxidizing media, of course, neither coating is satisfactory.

Ternary Electroless Nickel Alloy Coatings

Ternary alloy coatings are used to provide higher performance in specific properties over conventional nickel-phosphorus and nickel-boron coating systems. By incorporating a third element in significant levels, the basic structure and physical properties of the coating can be altered. As shown in Table 7, ternary alloy systems include Ni-P-Mo, Ni-Cu-P, Ni-Co-P, Ni-Fe-P, Ni-Re-P, Ni-W-P, Ni-Tl-B, and Ni-Sn-B.

Electroless Nickel Composite Coatings

Composites are one of the most recently developed types of electroless nickel coatings. These cermet deposits consist of small particles of intermetallic compounds, fluorocarbons, or diamonds dispersed in an electroless nickel-phosphorus matrix. These coatings have a high apparent hardness and superior wear and abrasion resistance.

Chemistry. Most composite coatings are applied from proprietary baths. Typically, they consist of 20 to 30 vol% of particles entrapped in an electroless nickel containing 4 to 11% P. Most commonly silicon carbide, diamond particles, fluorinated carbon powders, and polytetrafluoroethylene (PTFE) are used, although calcium fluoride is also occasionally codeposited. The particles are carefully sized and are normally 1 to 3 μm in diameter for silicon carbide

Table 5 Comparison of the corrosion rates of electroless nickel-phosphorus coatings in chemical process environments with other commonly used materials

Corrodent	Corrosion rate(a), μm/yr					
	Nickel 200 (UNS N02200)	Ni-P Coatings(b)			Mild steel	Type 316 stainless steel (UNS S31600)
		LP	MP	HP		
Thionyl chloride	7.0	900.0	1.8	2.5	200.0	5.1
Orthochlorobenzyl chloride (crude)	12.7	3.8	7.4	7.1	NA(c)	25.0
Orthochlorobenzyl chloride	12.7	5.3	13.5	9.4	NA(c)	2.5
Phosphoric acid	10.0	900.0	193.0	19.3	1270.0	2.5
Phosphorus oxychloride	10.0	28.4	1.5	2.5	100.0	18.8
Benzotrichloride	5.1	2.5	5.6	6.1	9.0	5.1
Benzoyl chloride	5.1	1.0	0.8	0.5	8.6	5.1

(a) 60 days exposure at 40 ± 2 °C (105 ± 4 °F). (b) LP, low-phosphorus (1 to 4%) coating; MP, medium-phosphorus (5 to 8%) coating; HP, high-phosphorus (9 to 12%) coating. (c) NA, no data available

Table 6 Comparison of the corrosion rates of electroless nickel-phosphorus coatings in caustic solutions with other commonly used materials

Corrodent	Corrosion rate(a), μm/yr					
	Nickel 200 (UNS N02200)	Ni-P Coatings(b)			Mild steel	Type 316 stainless steel (UNS S31600)
		LP	MP	HP		
45% NaOH + 5% NaCl at 40 ± 2 °C (105 ± 5 °F)	2.5	0.3	0.3	0.8	35.6	6.4
45% NaOH + 5% NaCl at 140 ± 2 °C (285 ± 5 °F)	80.0	5.3	11.9	F(c)	NA(d)	27.9
35% NaOH at 93 ± 2 °C (200 ± 5 °F)	5.1	5.3	17.8	13.2	94.0	52.0
50% NaOH at 93 ± 2 °C (200 ± 5 °F)	5.1	6.1	4.8	533.4	83.8	. . .
73% NaOH at 120 ± 2 °C (250 ± 5 °F)	5.1	2.3	7.4	F(c)	1448.0	332.7

(a) 100 days exposure temperature indicated. (b) LP, low-phosphorus (1 to 4%) coating; MP, medium-phosphorus (5 to 8%) coating; HP, high-phosphorus (9 to 12%) coating. (c) F, failed. (d) NA, no data available and diamonds and 0.35 μm for PTFE. The baths used for composite plating are conventional sodium hypophosphite-reduced electroless nickel solutions, with the desired particles suspended in them. These baths, however, are heavily stabilized to overcome or inhibit the very high surface area produced by the particles. The baths otherwise are operated normally, and the nickel-phosphorus matrix is produced by the traditional hypophosphite reduction of nickel. The particles are merely caught or trapped in the coating as it forms. Their bond to the coatings is purely mechanical.

Hardness and Wear. The primary use for electroless nickel composite coating is for applications requiring maximum resistance to wear and abrasion. The hardnesses of diamond and silicon carbide are 10,000 and 4,500 HV, respectively. In addition, the coatings are normally heat treated to provide maximum hardness (1000 to 1100 HV_{100}) of the electroless nickel matrix. The resulting apparent surface hardness of the composite is 1300 HV_{100} or more.

The wear surface of a composite coating consists of very hard mounds separated by lower areas of hard electroless nickel. During wear, the mating surface usually rides on the particles and slides over the matrix. Thus, the wear characteristics of these coatings approach that of the particle material. Typical wear test results for a silicon carbide composite coating are shown in Table 8.

Corrosion Resistance. In general, the corrosion resistance of composite coatings is significantly less than that of other electroless nickel coatings. The electroless nickel matrix contains large amounts of codeposited inhibitor, which reduces the passivity and corrosion resistance of the alloy. Also, heat treated coatings are less protective than are as-applied coatings, both because of the conversion of the amorphous deposit to crystalline nickel and Ni_3P and because of cracking of the coating. With composites, this problem is amplified because of the presence of the diamond or intermetallic particles. The mixture of phosphides, nickel, and particles creates a very strong galvanic couple accelerating attack. For applications requiring good corrosion resistance, electroless nickel composite coatings are not normally used.

Hardfacing

Hardfacing can be broadly defined as the application of a wear-resistant material, in depth, to the vulnerable (or worn) surfaces of a component by a weld overlay or thermal spray process. This section deals with weld overlay

Table 7 Electroless ternary nickel alloy plating systems

Alloy	Hardness, HK100	Environments in which plating has demonstrated corrosion resistance	Significant properties and applications
Hypophosphite-reduced alloys			
Nickel-phosphorus-molybdenum (5–9% P, 0.5–1% Mo, bal Ni)	550–650	Alkali, brine, caustics, weak acid solutions	Pitting corrosion protection
Nickel-copper-phosphorus (4–8% P, 1–3% Cu, bal Ni)	430–520	Alkali, brine, caustic solutions	Nonmagnetic, conductive, high modulus
Nickel-cobalt-phosphorus (15–40% Co, 3–8% P, bal Ni)	...	...	High-coercivity coating for use in magnetic memory applications
Nickel-iron-phosphorus (1–4% Fe, 2–4% P, bal Ni)	...	...	Magnetic applications in electronics
Nickel-rhenium-phosphorus (1–45% Re, 3–8% P, bal Ni)	...	...	High melting point (1700 °C, or 3090 °F); high-temperature wear resistance
Nickel-tungsten-phosphorus (4–8% P, 1–4% W, bal Ni)	550–620	...	High melting point (1550 °C, or 2820 °F); high-temperature wear resistance
Boron-reduced alloys			
Nickel-thallium-boron (3–5% Tl, 3–5% B, bal Ni)	650–850	...	Wear applications requiring resistance to galling, fretting, and erosion; coating is self-lubricating in contact with ferrous materials
Nickel-tin-boron (3–5% B, 1–3% Sn, bal Ni)	650–850	...	Wear applications requiring resistance to galling, fretting, and erosion; this coating is also self-lubricating

processing/materials; the thermal spray process/materials are covered in the following section of this chapter.

Hardfacing coatings offer unique advantages over other coating systems in that the overlay/substrate weld provides a metallurgical bond that is not susceptible to spallation and can easily be applied free of porosity or other defects. Welded deposits on surfacing alloys can be applied in thicknesses greater than those for most other techniques, typically in the range of 3 to 10 mm (0.12 to 0.4 in.) as shown in Fig. 1. Most welding processes are used for application of hardfacing alloy coatings, and on-site deposition can be more easily carried out, particularly for repair purposes. Weld overlays are very versatile because a large number of commercially available alloys can be selected to provide protection from a wide range of environmental degradation mechanisms.

Hardfacing applications for wear control vary widely, ranging from very severe abrasive wear service, such as rock crushing and pulverizing (metal-to-earth applications/wear) to applications to minimize metal-to-metal wear, such as control valves where a few thousandths of an inch of wear is intolerable. Hardfacing is used for controlling abrasive wear, such as that encountered by mill hammers, digging tools, extrusion screws, cutting shears, parts of earthmoving equipment, ball mills, and crusher parts. It is also used to control the wear of unlubricated or poorly lubricated metal-to-metal sliding contacts such as control valves, undercarriage parts of tractors and shovels, and high-performance bearings.

Hardfacing is also used to control combinations of wear and corrosion, as encountered by mud seats, plows, knives in the food processing industry, and valves and pumps handling corrosive liquids or slurries. In most instances, parts are made of either plain carbon steel or stainless steel, materials that do not provide desirable wear on their own.

In addition, hardfacing alloys are applied to critical wear areas of original equipment or during reclamation of parts. These alloys, which are referred to as buildup alloys, are not designed to resist wear but to return a worn part back to, or near, its original dimensions and/or to provide adequate support for subsequent layers of more wear-resistant hardfacing alloys.

Hardfacing materials include a wide variety of alloys, carbides, and combinations of these materials. Conventional hardfacing materials are normally classified as steels or low-alloy ferrous materials, high-chromium white irons or high-alloy ferrous materials, carbides, nickel-base alloys, or cobalt-base alloys. A few copper-base alloys are sometimes used for hardfacing applications, but for the most part, hardfacing alloys are either iron-, nickel-, or cobalt-base.

Table 8 Comparison of the Taber abraser resistance of silicon carbide composite coatings with other engineering materials

Material	Hardness	Taber wear index, mg/11,000 cycles
400-C stainless steel	57 HRC	5.6
A2 tool steel	60–62 HRC	5.0
Electroless nickel (hardened)	900–1000 HV	3.7
Hard chromium	1000–1100 HV	3.0
Tungsten carbide	1300 HV	2.0
Electroless nickel and silicon carbide composite	1300 HV	0.18–0.22

Note: Taber wear index determined for an average of three 5000 cycle runs with 100 g load and CS17 abrasive test wheels

Applicable Welding Processes

A number of welding processes are available for applying protective weld overlays, and many welding parameters must be considered when attempting to optimize a particular process for a given application. The weld process characteristics are summarized for comparison purposes in Table 9. The processes can be grouped as torch processes, arc welding processes, and high-energy-beam techniques. The torch process, oxyacetylene welding (OAW), is the oldest and simplest hardfacing process and involves simply eating the substrate with the flame and then melting the filler rod to get the hardfacing to melt. High-energy-beam techniques use laser beam welding (LBW) or electron beam welding (EBW) to alloy the surface by adding alloy powders to the weld pool.

In arc welding, the heat is generated by an arc between an electrode and the workpiece. Arc welding processes can be grouped into nonconsumable electrode processes and consumable electrode processes. Nonconsumable electrode processes, gas tungsten arc welding (GTAW) and plasma arc welding (PAW), both involve a tungsten electrode and the introduction of the filler metal (in the form of rod or wire in GTAW and powder in PAW). The arc melts the filler metal to form a molten pool that is protected from the atmosphere by an inert gas shield. In plasma arc welding, an additional inert gas flows through a constricted electric arc in the welding torch to form the plasma. In general, for consumable electrode processes, the arc is maintained between the consumable electrode and the workpiece. In shielded metal arc welding (SMAW), the electrode consists of a core wire surrounded by a flux covering that on melting forms a liquid slag and gas to protect the molten metal pool. In flux cored arc welding (FCAW), the flux is contained in the core of the metallic tubular electrode, whereas in gas metal arc welding (GMAW), the consumable wire electrode and substrate metal is protected from the atmosphere by a gas fed axially with the wire through the welding gun nozzle. In submerged arc welding (SAW), the arc, which is submerged beneath a covering of flux, dispenses from a hopper and melts the electrode, the surface of the workpiece, and some of the flux that protects the molten pool from oxidation. Electroslag welding (ESW) uses equipment similar to SAW for strip cladding.

Iron-Base Hardfacing Alloys

Iron-base hardfacing alloys are more widely used than cobalt- and/or nickel-base hardfacing alloys and constitute the largest volume use of hardfacing alloys. Iron-base hardfacing alloys offer low cost and a broad range of desirable properties. Most equipment that undergoes severe wear, such as crushing and grinding equipment and earthmoving equipment, is usually very large and rugged. Parts subjected to wear usually require downtime for repair. For this reason, there is a general temptation to hardface them with the lowest cost and most readily available materials. As a result, literally hundreds of iron-base hardfacing alloys are in use today.

Due to the great number of alloys involved, iron-base hardfacing alloys are best classified by their suitability for different types of wear and their general microstructures rather than by chemical composition. Most iron-base hardfacing alloys can be divided into the following classes:

- Pearlitic steels
- Austenitic (manganese) steels
- Martensitic steels
- High-alloy irons
- Austenitic stainless steel

Pearlitic steels are essentially low-alloy steels with minor adjustments in composition to achieve weldability. These alloys contain low carbon (<0.2% C) and low amounts of other alloying elements (for example, up to 2% Cr), resulting in a pearlitic structure. Pearlitic steels are useful as buildup overlays, primarily to rebuild carbon steel or low-alloy steel machinery parts back to size. Typically, this group of alloys has high impact resistance and low hardness (in the range of 25 to 37 HRC), as well as excellent weldability. Table 10 lists the composition and properties of a typical low-alloy

Table 9 Weld surfacing processes

Process	Approximate deposit thickness (min), mm	Deposition rate, kg/h	Dilution single layer, %	Typical uses
Oxyacetylene (OAW)	1.5	≤1	1–5	Small area deposits on light sections
Powder weld (PW)	0.1	0.2–1		Small area deposits on light sections
Shielded metal arc (SMAW)	3	1–4	15–30	Multilayers on heavier sections
Gas tungsten arc (GTAW)	1.5	≤2	5–10	High-quality low-dilution work
Plasma transferred arc (PAW)	2	≤10	2–10	High-quality lowest-dilution work
Gas metal arc (GMAW)	2	3–6	10–30	Faster than SMAW, no stub-end loss; positional work possible
Flux cored arc (FCAW)	2	3–6	15–30	Similar to GMAW. Mainly for iron-base alloys for high abrasion resistance
Submerged arc (SAW)				
Wire	3	10–30	15–30	Heavy section work; higher-quality deposits than FCAW
Strip	4	10–40	10–25	Corrosion-resistant cladding of large areas
Bulk	...	...	...	Similar to SAW wire but other alloys possible
Electroslag (ESW)	4	15–35	5–20	High-quality deposits at higher deposit rates than SAW. Limited alloy range

Source: Ref 8

pearlitic steel (EFe1) used for buildup applications.

Austenitic (manganese) steels are modeled after Hadfield steels. Most commercially available alloys in this category can be broadly subdivided into low-chromium and high-chromium alloys.

Low-chromium alloys usually contain up to 4% Cr and 12 to 15% Mn and some nickel or molybdenum (Table 10). Low-chromium austenitic manganese steels generally are used to build up manganese steel machinery parts subjected to high impact (impact crusher or shovel lips).

High-chromium austenitic steels, which may normally contain 12 to 17% Cr in addition to approximately 15% Mn, are used for the buildup and joining of austenitic manganese steels, as well as carbon and low-alloy steels. In addition, the as-deposited hardness of high-chromium steels is higher (~24 HRC) than that of low-chromium steels (~18 HRC). Compositions and properties of austenitic manganese steel hardfacing alloys used for buildup applications are given in Table 10.

Martensitic steels are designed to form martensite on normal air cooling of the weld deposit. As a result, these steels are often termed *self-hardening* or *air-hardening,* and they resemble tool steels with hardnesses in the range of 45 to 60 HRC. The carbon content of the martensitic steels ranges up to 0.7%. Other elements such as molybdenum, tungsten, nickel, and chromium (up to 12%) are added to increase hardenability and strength and to promote martensite formation. Manganese and silicon usually are added to aid weldability. Table 10 lists compositions and properties of martensitic steels.

The major hardfacing applications for martensitic steels include unlubricated metal-to-metal rolling or sliding parts. The impact resistance of martensitic steels is inferior to that of pearlitic or austenitic alloys, but there is a compensating increase in hardness and resistance to abrasive wear.

In hostile environments, a higher chromium content is beneficial. American Welding Society (AWS) ER420 and modified versions containing nickel, molybdenum, and niobium (or vanadium) are therefore the natural choice when high temperatures and mildly corrosive environments are encountered. For applications using steel mill hot-work rolls (which demand considerable hot hardness, resistance to oxidation, and resistance to thermal fatigue), both ER420 and EFe3 have been found suitable.

The high-chromium irons encompass a wide range of compositions in which chromium may vary between approximately 6 and 35 wt%, and carbon may vary from approximately 2 to 6 wt%. Other possible alloying additions include molybdenum, manganese, and silicon. Table 11 lists typical compositions.

The most important microstructural feature in the high-chromium irons, at least from a wear standpoint, is an M_7C_3 carbide, which forms in abundance during solidification and contains chromium, iron, and (if present) molybdenum. The matrix around these carbide particles can be austenitic, pearlitic, or martensitic.

With regard to industrial applications of the high-chromium abrasion alloys, the low-carbon

Table 10 Compositions, hardness, and abrasion data for a pearlitic low-alloy steel (EFe1) and austenitic manganese steels (EFeMn-C and EFeMn-Cr) used for buildup overlays and air-hardening martensitic steels used for metal-to-metal wear applications

	Composition, wt %								Abrasion, volume loss			
									Low-stress(a)		High-stress(b)	
Alloy	Fe	Cr	C	Si	Mn	Mo	Ni	Hardness, HRC	mm³	in.³ × 10⁻³	mm³	in.³ × 10⁻³
Buildup weld overlay												
EFel(c)	bal	2	0.1	1.0	1	1.5	...	37	88	5.4	49	3.0
EFeMn-C(c)	bal	4	0.8	1.3	14	...	4	18	65	4.0	57	3.5
EFeMn-Cr(c)	bal	15	0.5	1.3	15	2.0	1	24	93	5.7	46	2.8
Metal-to-metal weld overlay												
EFe2(c)	bal	3	0.2	1.0	1	1.0	1	48	54	3.3	66	4.0
EFe3(c)	bal	6	0.7	1.0	1	1.0	...	59	60	3.7	68	4.1
ER420(d)	bal	12	0.3	1.0	2	...	...	45	84	5.1	62	3.8

(a) Dry sand/rubber wheel test (ASTM G 65, Procedure B): load 13.6 kg (30 lb); 200 rev. (b) Slurry/steel wheel test (ASTM B 611, modified): load 22.7 kg (50 lb); 250 rev. (c) Two-layer shielded metal arc deposit process. (d) Two-layer submerged arc welding deposit process

(2 to 3% C) hypoeutectic materials are usually selected for situations involving moderate abrasion and impact, whereas the higher-carbon (4 to 6% C) hypereutectic alloys are used in applications involving severe abrasion and little or no impact. The hardness range for high-chromium irons is typically 52 to 62 HRC.

Austenitic stainless steel hardfacing alloys have been developed, which can be considered alternative materials to the cobalt-base alloys described below. These exhibit similar antigalling characteristics and possess equal resistance to cavitation erosion. Some of these have been developed for general use, and others have been developed for specific applications. For example, the cobalt-free composition in Table 12 has been used for hardfacing nuclear plant valves. (Co^{60} produced from cobalt hardfacing wear particles poses a radiation hazard.) Cobalt-containing austenitic grades (Table 12) have been used for the repair of the cavitation damage of hydraulic turbines in hydroelectric power plants.

Nonferrous Hardfacing Alloys

Nonferrous hardfacing alloys are used either for high resistance to specific types of wear (other than abrasion) or for wear resistance (including abrasion) in environments that are too corrosive or beyond the service temperature of ferrous alloys. The cobalt-base alloys and bronzes are particularly resistant to galling and to those wear processes involving microfatigue as the degradation mechanism (such as cavitation erosion). The cobalt-base alloys also possess high resistance to deformation at temperatures in excess of 750 °C (1380 °F). Table 13 lists typical compositions of the nonferrous hardfacing alloys.

Cobalt-Base/Carbide-Type Alloys. The chief difference between the various cobalt-base/carbide-type alloys is in carbon content (hence, carbide volume fraction, room-temperature hardness, and level of abrasion resistance). Chromium-rich M_7C_3 is the predominant carbide in these alloys, although tungsten-rich M_6C is evident in those alloys having a high tungsten content, and chromium-rich $M_{23}C_6$ is common in the low-carbon alloys (see Table 13). Hardness values for cobalt-base/carbide-type alloys range from approximately 40 to 55 HRC. As shown in Fig. 8, the abrasion resistance of these alloys (ERCoCr-A, -B, -C, and -E) increases with increasing carbon content.

Laves Phase Alloy Compositions. Laves phase is a type of topologically close-packed intermetallic compound. Historically, metallurgists have avoided the presence of Laves phase in most alloys due to its detrimental effect on mechanical properties. However, in the 1960s, the usefulness of Laves phase in resisting metal-to-metal wear was discovered; subsequently, alloys containing Laves phase have become commercially available.

As shown in Table 13, there are two Laves phase cobalt-base alloys available for hardfacing applications (Co-29Mo-9Cr-2.5Si and Co-

Table 11 Composition of high-chromium white irons used for metal-to-earth abrasion applications

Alloy	Composition, wt %							
	Cr	C	Si	Mn	Mo	Ni	B	Fe
ERFeCr-A3	11	2.6	1.3	1.8	1.5	...	...	bal
ERFeCr-A4(Mod)	29	3.5	1.1	0.9	...	2.6	0.7	bal
ERFeCr-A2	28	4.3	0.8	1.7	1.4	...	...	bal

Table 12 Nominal compositions of austenitic stainless steel hardfacing alloys

Alloy	Composition, %							
	C	Cr	Mn	Si	Ni	Mo	Co	N	Fe
NOREM(a)	0.7–1.3	24–26	5–12	2.5–5	5–9	1.7–2.3	...	0.05–0.24	bal
IRECA(b)	0.2	17–19	9–10	2–3	...	...	9–10	0.2	bal

(a) Developed by the Electric Power Research Institute. (b) Developed by Hydro-Québec. IRECA denotes Improved REsistance to CAvitation.

29Mo-18Cr-3.5Si). Both of these alloys contain at least 50 vol% of Laves phase. Laves phase has a hardness value between 1000 and 1200 DPH, which is less than that of carbides. Consequently, the Laves phase-containing alloys are less abrasive to mating materials than carbide-containing alloys in metal-to-metal wear situations.

Hardness values for cobalt-base Laves phase alloys range from 54 to 64 HRC. The nickel-base Laves phase alloy (Ni-33Mo-16Cr-3.5Si in Table 13) has a nominal macrohardness of 45 HRC. Abrasion data for Laves phase alloys are given in Fig. 8. Although these alloys have been deposited by weld techniques, they are more frequently applied by thermal spray methods.

Nickel-base/boride-type alloy compositions represent a progression in terms of iron, chromium, boron, and carbon contents (Table

Table 13 Composition of selected nonferrous hardfacing alloys

	Composition, wt %										
Alloy	Fe	Cr	Mo	W	Si	C	B	Al	Cu	Co	Ni
Cobalt-base/carbide type											
ERCoCr-A	...	28	...	5	...	1.2	...	...	...	bal	...
ERCoCr-B	...	29	...	8	...	1.5	...	...	...	bal	...
ERCoCr-C	...	31	...	13	...	2.5	...	...	...	bal	...
ERCoCr-E	...	27	6	...	...	0.2	...	...	...	bal	...
Cobalt- and nickel-base/Laves type											
T-400	...	9	29	...	2.5	...	...	...	...	bal	...
T-700	...	16	33	...	3.5	...	...	...	...	...	bal
T-800	...	18	29	...	3.5	...	...	...	...	bal	...
Nickel-base/boride type											
Alloy 40	1.5	7.5	...	...	3.5	0.3	1.5	...	...	...	bal
ERNiCr-B	3	11	...	...	4	0.5	2.5	...	...	...	bal
ERNiCr-C	4	16	...	...	4	0.7	3.5	...	...	...	bal
Aluminum bronze type											
ECuAl-B	4	...	...	...	1.0	...	...	9	bal	...	...
ECuAl-D	4	...	...	...	...	...	...	13.5	bal	...	...

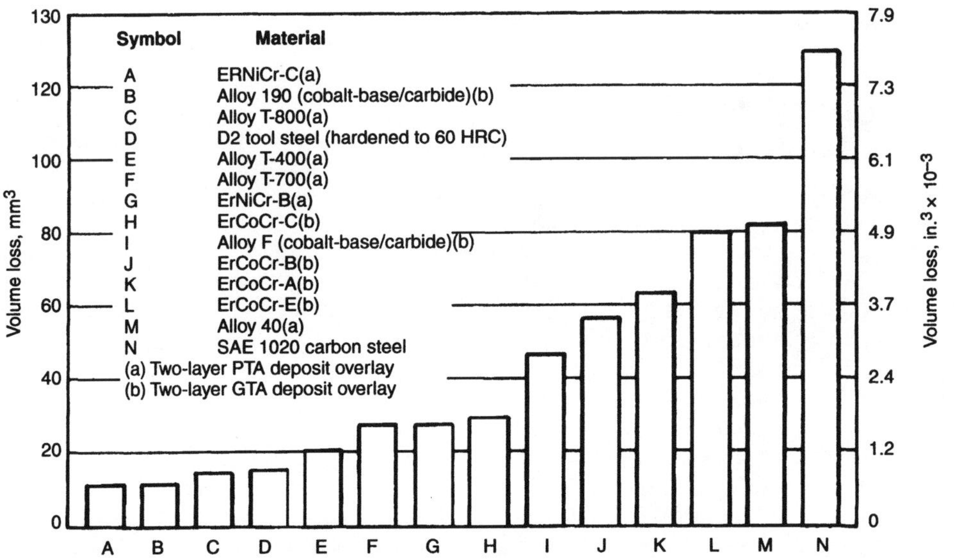

Fig. 8 Comparison of nonferrous hardfacing alloys to tool steel and carbon steel reference materials using ASTM G 65 low-stress, abrasion test. G 65 test parameters: procedure B; room temperature; 13.6 kg (30 lbf) load; quartz grain sand diameter of 212 to 300 μm; 2000 rev at 200 rev/min; 390 g/min (0.86 lb/min) feed rate

13). Iron content is largely incidental, allowing the use of ferrocompounds during manufacture. Together with nickel, the other three elements determine the level and type of hard phase within the structure upon solidification, with boron being the primary hard-phase forming element. Hardness values for boride-containing nickel alloys range from 51 to 57 HRC.

Because of the boride and carbide dispersions within their microstructures, the nickel-base/boride-type alloys exhibit excellent resistance to abrasion (ERNiCr-C, ERNiCr-B, and Alloy 40 in Fig. 8). Low-stress abrasion resistance generally increases with boron and carbon contents, hence the hard-phase volume fraction for these materials. Although their performance is not as good under self-mated sliding conditions as the cobalt-base materials, the nickel-base/boride-type alloys possess moderate resistance to galling. Of the nonferrous materials, the nickel-base/boride-type alloys are the least resistant to corrosion. This is attributed to the lack of chromium in the matrix.

Bronze Type. The two bronze compositions given in Table 13 fall into the aluminum bronze category. Silicon and other types of bronze are also available in welding consumable form for hardfacing.

Components typically protected using the aluminum bronzes include gears, cams, and cold drawing/forming dies. These alloys are not recommended for elevated-temperature use because their mechanical properties considerably decrease at temperatures >200 °C (>390 °F).

Tungsten Carbide Hardfacing Materials

In contrast to the other weld overlay materials, the tungsten carbide composites do not rely on the formation of suitable hard phases during weld pool solidification. Instead, these overlay materials rely on the transfer of tungsten carbide particles from the welding consumable (carbides are inserted in a steel tube) to the overlay. It is important, therefore, to limit the heat input of the welding process in order to prevent melting of the tungsten carbide particles. If the tungsten carbide particles melt, they mix with iron to form much softer iron-tungsten carbides, thus reducing abrasion resistance. For this reason, oxyacetylene deposits usually exhibit higher abrasion resistance than arc-welded tungsten carbide overlays.

An advantage of the tungsten carbide composites is that the size of the hard particles in the overlay can be controlled. This is important because abrasion resistance is dependent on the size relationship between microstructural features (such as carbides) and the abrading particles. If the abrading particles are large in comparison to the microstructural particles, then, after a running-in period (during which the softer matrix material at the surface is worn down), the abrading particles ride over the hard microstructural outcrops. Conversely, if the abrading particles are small in comparison to the microstructural particles, the opportunity exists for wear of the matrix around the microstructural particles. Eventually, these may drop out, having played only a small role in resisting abrasion.

Several tungsten carbide composites are available in a variety of tubular product forms. Popular compositions are 38, 50, 55, and 60 wt% tungsten carbide, with the carbon steel tube making up the balance. For each composition, several carbide size ranges are available. As an example, for the 60% WC oxyacetylene welding consumable, four mesh size ranges are available:

AWS designation	Mesh size range
RWC-12/20	12–20
RWC-20/30	20–30
RWC-30/40	30–40
RWC-40/120	40–120

The same composition is also available in flux-coated form for SMAW and as a continuous wire (with an internal flux) for open arc welding.

Tungsten carbide composites generally possess very high resistance to abrasion and very low impact strength. Performance in a given situation depends on (a) carbide volume fraction; (b) size relationship between the carbides and the abrasive medium; and (c) welding technique applied. Important factors are the distribution of carbides in the overlay (because the particles tend to sink, turbulence in the molten weld pool is an advantage) and the amount of carbide dissolution and reprecipitation in the steel matrix during welding. Impact strength generally decreases with increasing carbide volume fraction.

The tungsten carbide composites have been used to solve a wide variety of industrial sliding and drilling abrasion problems. Table 14 gives abrasion data. For extremely hostile environments, some nonferrous tungsten carbide products (cobalt- and nickel-base products in the form of bare cast rods) are available. Also, several alternative composite materials, utilizing other carbides (for example, vanadium, titanium, or niobium), are available that have the advantage of creating a more homogeneous deposit because of their lower densities.

Thermal Spraying

Thermal spraying comprises a group of processes in which divided molten metallic or nonmetallic material is sprayed onto a prepared substrate to form a coating. The sprayed material is originally in the form of wire, rod, or powder. As the coating materials are fed through the spray unit, they are heated to a molten or plastic state and propelled by a stream of compressed gas onto the substrate. As the particles strike the surface, they flatten and form thin platelets that conform and adhere to the irregularities of the prepared surface and to each other. They cool and accumulate, particle by particle, into a lamellar, castlike structure. In general, the substrate temperature can be kept below approximately 200 °C (400 °F), eliminating metallurgical change of the substrate material. The spray gun generates the necessary heat for melting through combustion of gases, an electric arc, or a plasma.

The deposited structures of thermal spray coatings differ from those of the same material in the wrought form because of the incremental nature of the coating buildup and because the coating composition is often affected by reaction with the process gases and the surrounding atmosphere while the materials are in the molten state. For example, where air or oxygen is used as the process gas, oxides of the material applied may be formed and become part of the coating. The as-applied structures of all thermal spray coatings are similar in their lamellar nature; the variations in structure depend on the particular thermal spray process used, the processing parameters and techniques employed, and the material applied. Figure 9 illustrates the microstructure that results from the thermal spray process. As shown in this figure, the molten particles spread out and deform (splatter) as they impact the substrate, at first locking onto irregularities on the roughened surface and, then interlocking with each other.

The bond between the sprayed coating and the substrate is generally mechanical. Proper surface preparation of the substrate before spraying is often the most critical influence on the bond strength of the coating.

Advantages and Disadvantages

Advantages. A major advantage of the thermal spray processes is the extremely wide variety of materials that can be used to make a coating. Virtually any material that melts without decomposing can be used. A second major advantage is the ability of most of the thermal spray processes to apply a coating to a substrate without significantly heating it. Thus, materials with very high melting points can be applied to finally machined, fully heat treated parts without changing the properties of the part and without thermal distortion of the part. A third advantage is the ability, in most cases, to strip and recoat worn or damaged coatings without changing the properties or dimensions of the part.

Disadvantages. A major disadvantage is the line-of-sight nature of these deposition processes. They can coat only what the torch or gun can "see." Of course, there are also size limitations prohibiting the coating of small, deep cavities into which a torch or gun will not fit.

Applications

The applications of thermal spray coatings are extremely varied, but the largest categories of use are to enhance the wear and/or corrosion resistance of a surface. Other applications include their use for dimensional restoration, as thermal barriers, as thermal conductors, as electrical conductors or resistors, for electromagnetic shielding, and to enhance or retard radiation. They are used in virtually every industry, including aerospace, agricultural implements, automotive, primary metals, mining, paper, oil and gas production, chemicals and plastics, and biomedical.

Although thermal spray coatings provide the solution to many mechanical, electrical, and

Table 14 Abrasion data for tungsten carbide composites

Material		Abrasion, volume loss			
		Low-stress(a)		High-stress(b)	
Carbide, wt %	Mesh size	mm^3	$in.^3 \times 10^{-3}$	mm^3	$in.^3 \times 10^{-3}$
60	20–30	7.3	0.45	28.7	1.75
61	100–250	10.6	0.65	24.4	1.49

(a) Dry sand/rubber wheel test (ASTM G 65, Procedure B): load 13.6 kg (30 lb); 2000 rev. (b) Slurry/steel wheel test (ASTM B 611, modified): load 22.7 kg (50 lb); 250 rev corrosion-resistance problems involving metal parts and assemblies, there are certain applications where such coatings should not be used. Before a thermal spray coating is specified, its suitability can usually be determined according to these criteria:

- No strength is imparted to the base material by a sprayed deposit. The component to be sprayed must, in its prepared form, be able to withstand any mechanical loading that will be experienced in service. (In a few applications, some strength can be added by a thermal spray coating; however, such uses are unusual and should be carefully tested.)
- If the area on a part to be sprayed or any section of the total area will be subjected to shear loading in service, the part is not a suitable candidate for thermal spraying. Gear teeth, splines, and threads are examples.
- Point loading with line contact on a sprayed metal deposit will eventually spread the deposit, causing detachment. If the deposit is on a moving component with such loading, it will fail rapidly. For example, needle and roller bearing seats, where the bearing elements are in direct contact with the sprayed deposit, may not be good thermal spray candidates.
- If a steel component to be treated has been nitrided or carburized, non-high-velocity thermal spraying processes are not recommended unless the case has been removed.

Thermal Spray Processes

Thermal spray processes can be classified into two categories, arc processes and gas combustion processes, depending on the means of achieving the heat for melting the consumable material during the spraying operation. Thermal spray processes can also be classified as high- and low-energy processes.

In the lower-energy electric arc (wire arc) spray process, heating and melting occur when two electrically opposed charged wires, comprising the spray material, are fed together to produce a controlled arc at the intersection. The

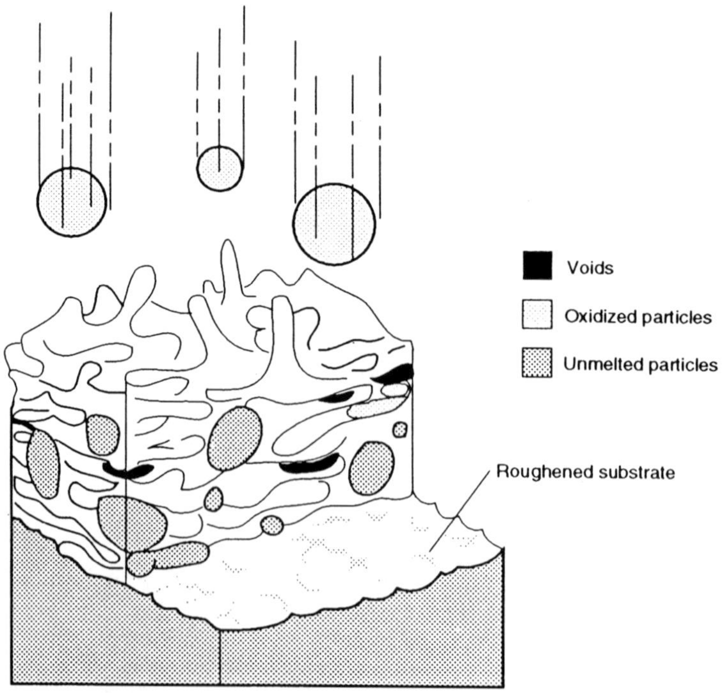

Fig. 9 Schematic showing the buildup of a thermal spray coating. Molten particles spread out and deform (splatter) as they strike the target, at first locking onto irregularities on the substrate, then interlocking with each other. Voids can occur if the growing deposit traps air. Particles overheated in the flame become oxidized. Unmelted particles may simply be embedded in the accumulating deposit.

molten material on the wire tips is atomized and propelled onto the substrate by a stream of gas (usually air) from a high-pressure gas jet. The highest spray rates are obtained with this process, allowing for cost-effective spraying of aluminum and zinc for the marine industry. In the higher-energy plasma arc spray process, injected gas is heated in an electric arc and converted into a high-temperature plasma that propels the coating powder onto the substrate at very high velocities. This process can take place in air with air plasma spraying (APS), or in a vacuum with vacuum plasma spray (VPS) or low-pressure plasma spraying (LPPS).

For gas combustion processes, the lower-energy flame spray process uses oxyfuel combustible gas as a heat source to melt the coating material, which may be in the form of rod, wire, or powder. In the higher-energy, high-velocity oxyfuel combustion spray (HVOF) technique, internal combustion of oxygen and fuel gas occurs to produce a high-velocity plume capable of accelerating powders at supersonic speeds and lower temperatures than the plasma processes. Continuous combustion occurs in most commercial processes, whereas the proprietary detonation gun (D-gun) process uses a spark discharge to propel powder in a repeated operating cycle to produce a continuous deposit.

In the lower-energy processes, electric arc (wire arc) spray and flame spray processes, adhesion to the substrate is predominantly mechanical and is dependent on the workpiece being perfectly clean and suitably rough. Some porosity is always present in these coatings, which may present problems in both corrosion and erosion. The higher-energy processes—APS, VPS, LPPS, and HVOF processes—were developed to reduce porosity and improve adhesion to the substrate. In addition, these processes are capable of spraying materials with higher melting points, thus widening the range of applications to include high-temperature coatings and thermal and mechanical shock-resistant coatings. With these higher-energy processes, bond strengths are higher because of the possible breakup of any oxide films present on the particles or the workpiece surface, allowing for some diffusion bonding to take place. Typical operating features of the various thermal spray processes are listed in Table 15.

Comparing Thermal Spray and Weld Overlay Coatings

The principal characteristics of thermal spraying processes that distinguish them from weld-deposited coatings are as follows:

- The strength of the bond between coating and substrate covers a wide range, depending on the materials and process used. It can vary from a relatively low strength to figures approaching those of welded bonds if the process involves high-temperature diffusion between coating particles and substrate.
- Thermal spraying can apply coating materials to substrates that are unsuited to welding because of their composition or tendency to distort. This feature offers the designer scope to use materials with desirable properties that would not be possible by other means.
- Sprayed deposits can be applied in thinner layers than welded coatings, but thick coatings are possible under certain circumstances.
- Almost all material compositions may be deposited (provided at least one constituent has a stable liquid phase)—metals, ceramics, carbides, polymers, or any combination.
- Most processes are cold, compared with welding, and there is no dilution or metallurgical degradation of the substrate.

Table 15 Operating characteristics of thermal spray processes

Process	Gas temperature, °C	Particle, velocity, m/s	Adhesion, MPa	Oxide content, %	Porosity, %	Spray rate, kg/h	Relative cost, low = 1	Typical deposit thickness, mm
Flame	3,000	40	8	10–15	10–15	2–6	1	0.1–15
Arc wire	NA	100	12	10–20	10	12	2	0.1 to > 50
High-velocity-oxyfuel (JetKote)	3,000	800	>70	1–5	1–2	2–4	3	0.1 to > 2
Detonation gun	4,000	800	>70	1–5	1–2	0.5	NA	0.05–0.3
Air plasma spray	12,000	200–400	4 to >70	1–3	1–5	4–9	4	0.1–1
Vacuum plasma spray	12,000	400–600	>70	ppm	<0.5	4–9	5	0.1–1

NA, not applicable. Source: Ref 8

- Thermal spraying processes are all-positional and most can be operated in air, so they possess great flexibility.

Tables 16 and 17 compare thermal spraying and welding applications and process requirements.

Thermal Spray Coating Properties

The mechanical properties of thermal spray coatings are not well documented, with the exception of their hardness and bond strength. The sensitivity of the properties of the coatings to specific deposition parameters makes universal cataloging of properties by simple chemical composition and general process (e.g., WC-12Co by plasma spray) virtually meaningless. The situation is even more complex because the properties of coatings on test specimens may differ somewhat from those on parts because of differences in geometry and thermal conditions. Nonetheless, coatings made by competent suppliers using adequate quality control will be quite reproducible, and therefore the measurement of various mechanical properties of these standardized coatings may be very useful in the selection of coatings for specific applications. Properties that may be of value include the modulus of rupture, modulus of elasticity, and strain-to-fracture in addition to

Table 16 Applications of thermal spraying and welding processes

Application	Thermal spraying	Welding
Base metal	Identification of alloy by generic type is required. Almost any alloy can be sprayed.	Precise identification of alloy is required. In steels, composition must be known. Some alloys are difficult or impossible to weld.
Operating environment of finished component	Must be known before coating material selection can be made	Need not be known. It is usually sufficient to match the properties of the base metal.
Restoring dimensions	Practical for buildup from 0.25 mm (0.010 in.) to 2.5 mm (0.100 in.) and sometimes greater	Practical, but comparatively costly, particularly on alloys requiring postweld heat treatment
Restoring strength	Not useful	Good to excellent results
Precision-dimensioned	Excellent results—no distortion	Distortion is a serious problem. It is frequently difficult to predict whether distortion will be within acceptable limits.
Rotating or oscillating machinery	Usually excellent. No induced stresses in base metal	Weld stresses may create problems.
Limited fatigue	Excellent results if stress risers are not introduced by machining and if shot peening is done prior to spraying	Not recommended
Antiwear or antifriction surface	Usually excellent	Fair
Galvanic anticorrosion coating	Excellent	Impractical
Corrosion-resistant coating	Excellent	Excellent but expensive

Source: R.B. Alexander & Associates, Huntington Woods, MI

Table 17 Process requirements in thermal spraying and welding

Application	Thermal spraying	Welding
Preheating	Always used to remove moisture, otherwise temperature is held as low as possible. Usual preheat range is 95–150 °C (200–300 °F)	Treatment varies from chilling to heating up to 425 °C (800 °F) depending on composition of base metal.
Auxiliary equipment operation	Work frequently done on a lathe with the spray gun mounted on tool post and the lathe used to machine to plan size	None
Postheating	None required, except for one process variation that fuses the coating after application	Frequently used for dimensional stability, stress relieving, and tempering depending on composition of base metal, geometry, and end use of the part
Restoring to plan dimensions	Special machining and grinding techniques used. Finish machining is sometimes unnecessary after light plasma spray antiwear coatings.	Conventional machining and grinding

Source: R.B. Alexander & Associates, Huntington Woods, MI

hardness. Examples of some of these are given in Table 18.

Any measurement or use of mechanical properties must take into account the anisotropic nature of the coating microstructure and hence its properties (i.e., the coating properties are different parallel to the surface than perpendicular to the surface because of the lamellar nature of the microstructure). Most mechanical properties are measured parallel to the surface, in part because it is easier to produce test specimens in this plane because the coatings are typically thin. Unfortunately, the major load in service is usually perpendicular to the surface. This does not, however, make measurements in the plane of the coating useless. It is frequently important to know, for example, how much strain can be imposed on a coating due to extension or deflection of the part without cracking the coating. Cracks in a coating may not only affect its performance, but also initiate cracks and fatigue failures in the part.

Wear Resistance. One of the most important uses of thermal spray coatings is for wear resistance. They are used to resist virtually all forms of wear, including abrasive, erosive, and adhesive, in virtually every type of industry. The materials used range from soft metals to hard metal alloys to carbide-based cermets to oxides (Table 19). Generally, the wear resistance of the coatings increases with their density and cohesive strength, so the higher-velocity coatings such as HVOF and particularly detonation gun coatings provide the greatest wear resistance for a given composition.

A variety of laboratory tests have been developed to rank thermal spray coatings and compare them with other materials. Examples of abrasive and erosive wear data are shown in Table 20 and 21. It should be kept in mind that laboratory tests can seldom duplicate service conditions. Therefore these tests should only be used to help select candidate coatings for evaluation in service. Only rarely, with good baseline data, can any precise prediction of wear life in service be made from laboratory data.

Chemical and Physical Vapor Deposition Processing (Ref 9)

Two quite different processing approaches are used to deposit the thin coatings or layers of ceramic compounds that provide dramatic improvements in wear resistance and life to hardened tool steel substrates. Physical vapor deposition (PVD) physically generates and deposits atoms or molecules on substrates in high-vacuum environments. The atom flux that impinges on a substrate may be generated by a number of techniques, as described below. Chemical vapor deposition (CVD) exposes substrates to gaseous reactants that chemically react to produce the desired surface-coating products.

The PVD processes are accomplished in high vacuums, and the coating atoms travel relatively large distances without collisions. As a result, the PVD processes are line-of-sight processes. The CVD processes are accomplished at atmospheric pressures and therefore have better throwing power, or the ability to uniformly coat complex shapes, than do the PVD processes. Physical vapor deposition processes for titanium nitride (TiN) coatings are typically

Table 18 Mechanical properties of representative plasma, detonation, and high-velocity combustion coatings

Parameter	Type of coating					
	Tungsten-carbide-cobalt				Alumina	
Nominal composition, wt%	W-7Co-4C	W-9Co-5C	W-11Co-4C	W-14Co-4C	Al_2O_3	Al_2O_3
Thermal spray process	Detonation gun	High-velocity combustion	Plasma	Detonation gun	Detonation gun	Plasma
Rupture modulus, 103 psi(a)	72	...	30	120	22	17
Elastic modulus, 106 psi(a)	23	...	11	25	14	7.9
Hardness, kg/mm2, HV300	1300	1125	850	1075	>1000	>700
Bond strength, 103 psi(b)	>10,000(c)	>10,000(c)	>6500	>10,000	>10,000(c)	>6500

(a) Compression of freestanding rings of coatings. (b) ASTM C 633–89, "Standard Test Method for Adhesion or Cohesive Strength of Flame-Sprayed Coatings," ASTM, 1989. (c) Epoxy failure. Source: Publication 1G191, National Association of Corrosion Engineers

Table 19 Thermal spray coatings for friction and wear applications

Type of wear	Coating material	Coating process(a)	Applications
Adhesive wear	Soft bearing coatings:		
	Aluminum bronze	OFW, EAW, OFP, PA, HVOF	Babbitt bearings, hydraulic press sleeves, thrust bearing shoes, piston guides, compressor crosshead slippers
	Tobin bronze	OFW, EAW	
	Babbitt	OFW, EAW, OFP	
	Tin	OFW, EAW, OFP	
	Hard bearing coatings:		
	Mo/Ni-Cr-B-Si blend	PA	Bumper crankshafts for punch press, sugar cane grinding roll journals, antigalling sleeves, rudder bearings, impeller shatts, pinion gear journals, piston rings (internal combustion); fuel pump rotors
	Molybdenum	OFW, EAW, PA	
	High-carbon steel	OFW, EAW	
	Alumina/titania	OFP, PA	
	Tungsten carbide	OFP, PA, HVOF	
	Co-Mo-Cr-Si	PA, HVOF	
	Fe-Mo-C	PA	
Abrasive wear	Aluminum oxide	PA	Slush-pump piston rods, polish rod liners, and sucker rod couplings (oil industry); concrete mixer screw conveyors; grinding hammers (tobacco industry); core mandrels (dry-cell batteries); buffing and polishing fixtures; fuel-rod mandrels
	Chromium oxide	PA	
	Tungsten carbide	PA, HVOF	
	Chromium carbide	PA, HVOF	
	Ni-Cr-B-SiC/WC (fused)	OFP, HVOF	
	Ni-Cr-B-SiC (fused)	OFP, HVOF	
	Ni-Cr-B-SiC (unfused)	HVOF	
Surface fatigue wear			
Fretting: Intended motion applications	Molybdenum	OFW, PA	Servomotor shafts, lathe and grinder dead centers, cam followers, rocker arms, piston rings (internal combustion), cylinder liners
	Mo/Ni-Cr-B-SiC	PA	
	Co-Mo-Cr-Si	PA, HVOF	
Fretting: Small-amplitude oscillatory displacement applications			
Low temperature (<540 °C, or 1000 °F)	Aluminum bronze	OFW, EAW, PA, HVOF	Aircraft flap tracks (air-frame component); expansion joints and midspan supports (jet engine components)
	Cu-Ni-In	PA, HVOF	
	Cu-Ni	PA, HVOF	
High temperature (>540 °C, or 1000 °F)	Co-Cr-Ni-W	PA, HVOF	Compressor air seals, compressor stators, fan duct segments and stiffeners (all jet engine components)
	Chromium carbide	PA, HVOF	
Erosion	Chromium carbide	PA, HVOF	Exhaust fans, hydroelectric valves, cyclone dust collectors, dump valve plugs and seats, exhaust valve seats
	Tungsten carbide	PA, HVOF	
	WC/Ni-Cr-B-SiC (fused)	OFP, HVOF	
	WC/Ni-Cr-B-SiC (unfused)	OFP, HVOF	
	Chromium oxide	PA	
Cavitation	Ni-Cr-B-SiC-Al-Mo	PA	Wear rings (hydraulic turbines), water turbine buckets, water turbine nozzles, diesel engine cylinder liners, pumps
	Ni-Al/Ni-Cr-B-SiC	PA	
	Type 316 stainless steel	PA	
	Ni-Cr-B-SiC (fused)	OFP, HVOF	
	Ni-Cr-B-SiC (unfused)	HVOF	
	Aluminum bronze	PA, HVOF	
	Cu-Ni	PA, HVOF	

(a) OFW, oxyfuel wire spray; EAW, electric arc wire spray; OFP, oxyfuel powder spray; PA, plasma arc spray; HVOF, high-velocity oxyfuel powder spray applied at substrate temperatures of about 500 °C (930 °F), whereas substrate temperatures in CVD processes are higher, typically about 1000 °C (1830 °F). Thus, tool steels coated by PVD processes need not be heat treated subsequent to deposition, whereas tool steels coated by CVD processes must be hardened after coating. Despite reheating, the hardening treatments applied to CVD TiN-coated D2 steel have been found to maintain homogeneous, defect-free coatings and residual surface compressive stresses of about –1000 MPa (–145 ksi).

PVD Processing. Coating atoms in PVD processes may be generated by evaporation, sputtering, or ion plating in vacuum environments. When gases such as nitrogen, methane, or oxygen are introduced into the vacuum chambers, the metal atoms react with the gas atoms to form nitrides, carbides, or oxides, and the PVD processes are referred to as *reactive processes.* Evaporation is accomplished by heating source materials in high vacuums (10^{-6} kPa, or 7.5×10^{-6} torr, or better) to cause the thermal evaporation of atoms or molecules that travel through the vacuum and deposit on a substrate surface. Deposition processes based solely on evaporation are being replaced by more efficient sputtering or ion-plating processes using glow discharge plasmas.

Sputtering is a PVD coating process in which atoms are ejected mechanically from a target by the impact of ions or energetic neutral atoms. Figure 10 shows schematically the mechanism of sputtering in a simple diode system. The chamber is initially evacuated, then backfilled with argon gas, and the target is made cathodic or negative by the application of a direct current (dc) potential between –500 and –5000 V. A low-pressure glow discharge plasma is produced around the target cathode, creating positively charged argon ions that are accelerated to the target. The momentum transfer due to the impact of the argon ions is sufficient to eject target atoms that travel to the substrate and other parts of the chamber. The mechanical transfer of atoms by sputtering is more readily controlled than transfer of atoms by thermal evaporation, and the sputtered atoms have higher energies.

Simple diode sputtering systems have relatively low rates of deposition. Thus, improved sputtering systems, with magnetic fields applied at the targets, have been developed. The resulting process is referred to as magnetron sputtering, shown schematically in Fig. 11. The magnetic fields trap secondary electrons generated by the target and greatly increase ionization in the cathode plasma. Thus, more argon ions strike the target, and sputtering and deposition rates are significantly increased relative to diode sputtering. A more recent modification of magnetron sputtering is unbalanced magnetron sputtering. In this process (in contrast to conventional magnetron sputtering, where the magnetic field is closely restricted to the target),

Table 20 Abrasive wear data for selected thermal spray coatings

Material	Type	Wear rate, mm³/1000 rev
Carballoy 883	Sintered	1.2
WC-Co	Detonation gun	0.8
WC-Co	Plasma spray	16.0
WC-Co	Super D-Gun	0.7
WC-Co	High-velocity oxyfuel	0.9

Note: ASTM G 65 dry sand/rubber wheel test, 50/70 mesh Ottawa silica, 200 rpm, 30 lb load, 3000-revolution test duration

Table 21 Erosive wear data for selected thermal spray coatings

Material	Type	Wear rate, μm/g
Carballoy 883	Sintered	0.04
WC-Co	Detonation gun	1.3
WC-Co	Plasma spray	4.6
AISI 1018 steel	Wrought	21

Note: Silica-based erosion test; particle size, 15 mm; particle velocity, 139 m/s; particle flow, 5.5 g/min, ASTM Recommended Practice G 75

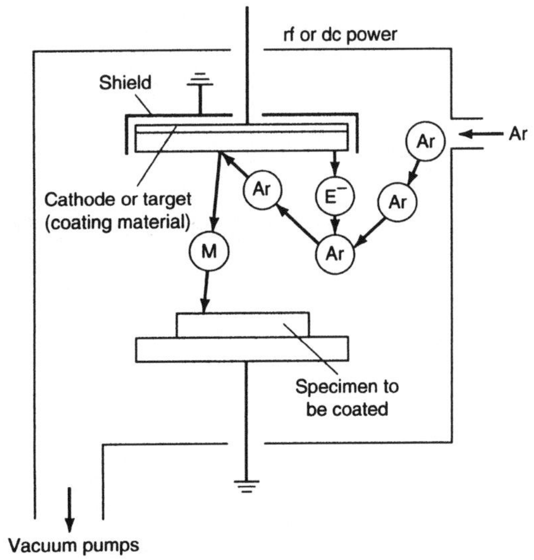

Fig. 10 Schematic of sputtering mechanisms

magnets are arranged to create a plasma that extends between the target and the substrate, with attendant benefits of ion bombardment at the substrate. As a result, unbalanced magnetron sputtering is capable of producing fully dense coatings on large, complex surfaces.

Ion plating, also referred to as *plasma-assisted PVD* or *evaporative-source PVD,* is a plasma-assisted deposition process in which the coating atoms are generated by thermal evaporation from an appropriate source. The sources may be electrically heated wire, arc, electron beam, or hollow cathode designs. The source is made the anode, and the substrate becomes the cathode by the application of a dc or radiofrequency (rf) voltage ranging between –500 and –5000 V. In the resulting substrate, cathode glow discharge, atoms and ions, are accelerated at high energies to the substrate coating. The bombardment of the substrate by the high-energy particles produces dense coatings with excellent adhesion, and the cathode glow discharge enhances substrate coverage. The diode ion plating systems have been further improved by designs that enhance ionization with ion currents that can be controlled independently of the bias voltage between the evaporative source and the substrate. These modified PVD system designs are referred to as *triode ion plating systems.*

CVD Processing. Chemical vapor deposition produces surface coatings by chemical reactions at the surfaces of heated substrates, as shown schematically in Fig. 12. Gaseous reactants are introduced into a reactor, which may be of cold-wall or hot-wall design, then chemically react at the surface of a heated substrate, deposit a solid coating, and create gaseous reaction products that are exhausted from the reactor. General equations for CVD carbide and nitride ceramic coatings deposited on tool steels are of the form:

$$MCl_x + H_2 + 0.5N_2 = MN + xHCl \quad \text{(Eq 1)}$$

$$MCl_x + CH_4 = MC + xHCl \quad \text{(Eq 2)}$$

Specific reactions for TiN and TiC coatings include:

$$TiCl_4 + CH_4 = TiC + 4HCl \quad \text{(Eq 3)}$$

$$TiCl_4 + {}^1/_2N_2 + 2H_2 = TiN + 4HCl \quad \text{(Eq 4)}$$

$$TiCl_4 + NH_3 + {}^1/_2H_2 = TiN + 4HCl \quad \text{(Eq 5)}$$

Often, for improved adhesion, TiN coatings on tool steels are combined with very thin undercoatings of titanium carbide or titanium carbonitride. The deposition temperatures for the CVD TiC and TiN coating reactions are relatively high, around 1000 °C (1830 °F). However, CVD deposition temperatures can be lowered if the CVD reactions are carried out in an environment of glow discharge plasmas maintained at the substrate/vapor interface. These processes are referred to as *plasma-assisted chemical vapor deposition* (PACVD) processes, and such techniques can lower substrate deposition temperatures to the range between 500 and 600 °C (930 and 1110 °F). The CVD and PACVD processes have also been used to deposit diamond and diamondlike coatings on substrates from gaseous mixtures of hydrogen and hydrocarbons, and these coatings show promise for improved performance in cutting tool applications.

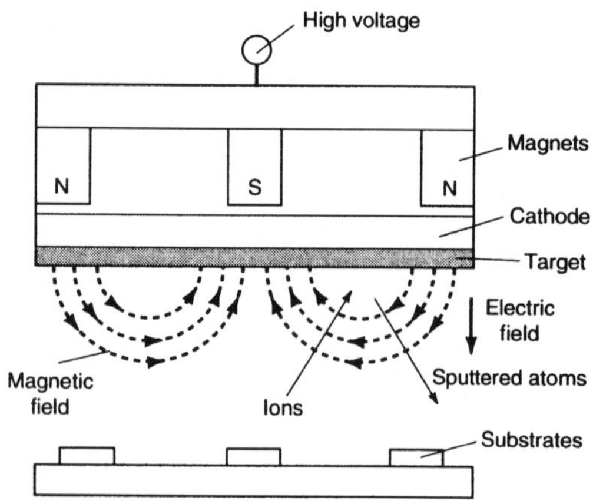

Fig. 11 Schematic of magnetron sputtering

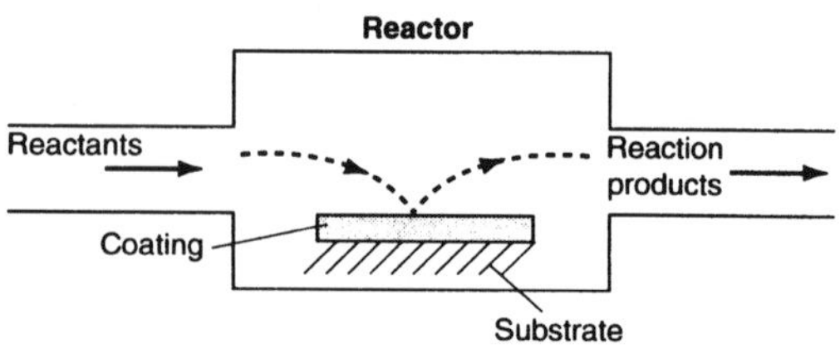

Fig. 12 Schematic of CVD deposition processes in a cold-wall reactor

PVD and CVD Coating Performance. Examples of the improvements in tool life attainable by PVD TiN coating of cutting tools are shown in Table 22. Matthews (Ref 11) has reviewed many of the commercial PVD processes used to apply TiN coatings and also has documented the dramatic improvements in performance that are possible with application of the coatings. Coatings produced by the various processes range in thickness from less than 1 μm (0.04 mil) to 6 μm (0.24 mil) and give the tools a uniform gold color. Deposition temperatures are 500 °C (930 °F) or lower. The hardness of the coatings for TiN is typically around 2500 HV but is a function of coating composition. Coatings applied at low substrate temperatures, which produce dense microstructures, develop high compressive residual stresses, and under certain conditions, stresses high enough to cause plastic deformation and cracking of the coating may develop.

As stated earlier, due to the high coating temperature, almost all CVD coated steel parts must be heat treated after coating to restore core hardness. This is because it is difficult to cool many grades of tool steels quickly enough from the high temperature of the CVD coating process to obtain core hardness. If heat treated in an air furnace, TiC or TiN will oxidize. Therefore, CVD-coated parts must be heat treated in a protective atmosphere or vacuum after coating. In certain cases, heat treating after CVD coating can introduce distortion or a dimensional change. The recent development of the high-pressure gas quenching vacuum furnace (6 bar or greater) with convection gas heating has allowed for less distortion and much better size control. This has increased the number of applications for CVD-coated steel tools (Ref 12).

Ion Implantation (Ref 9)

Ion implantation is a surface-modification process by which surface chemistry and properties are modified by ions forced into workpiece surfaces by very-high-energy beams. Figure 1 indicates that ion implantation represents a unique class of treatments and that the depth of the affected surface zone is quite shallow. The ion beams are produced in a source or gun that ionizes gas molecules by electrons emitted from a source at an energy of about 100 to 200 eV. The ion beam is then focused and extracted from the source by an exit electrode. Figure 13 shows a schematic diagram of a typical ion source. Ion implantation is a line-of-sight process; that is, only areas directly exposed to the beam are implanted. For coverage of areas larger than the ion beam, the specimen must be translated or the ion beam rastered over the specimen surface.

The ion implantation process imparts high strength, high hardness, and residual compressive stresses into substrate surfaces by the lattice damage induced by the ion impact. Point defects, such as vacant lattice sites and ions and

Table 22 Increased tool life attained with PVD coated cutting tools

Cutting tool				Workpieces machined before resharpening	
Type	High-speed tool steel, AISI type	Coating	Workpiece material	Uncoated	Coated
End mill	M7	TiN	1022 steel, 35 HRC	325	1200
End mill	M7	TiN	6061-T6 aluminum alloy	166	1500
End mill	M3	TiN	7075T aluminum alloy	9	53
Gear hob	M2	TiN	8620 steel	40	80
Broach insert	M3	TiN	Type 303 stainless steel	100,000	300,000
Broach	M2	TiN	48% nickel alloy	200	3400
Broach	M2	TiN	Type 410 stainless steel	10,000–12,000	31,000
Pipe tap	M2	TiN	Gray iron	3000	9000
Tap	M2	TiN	1050 steel, 30–33 HRC	60–70	750–800
Form tool	T15	TiC	1045 steel	5000	23,000
Form tool	T15	TiN	Type 303 stainless steel	1840	5890
Cutoff tool	M2	TiC-TiN	Low-carbon steel	150	1000
Drill	M7	TiN	Low-carbon steel	1000	4000
Drill	M7	TiN	Titanium alloy 662 layered with D6AC tool steel, 48–50 HRC	9	86

Source: Ref 10

atoms forced into nonequilibrium interstitial sites, as shown schematically in Fig. 14, account for much of the structural change. The implantation is usually carried out close to room temperature; therefore, the case depth is largely determined by ion trajectories during impact rather than atomic diffusion. However, heat is generated by the ion bombardment, and some short-range diffusion and fine-scale precipitation may take place.

Any kind of ion can be implanted, but nitrogen is commonly implanted in steels to improve near-surface corrosion resistance and tribological properties. As noted above, the case depths produced by ion implantation are very shallow (less than 1 μm), compared to other surface-engineering treatments. The low temperatures of ion implantation result in almost no distortion. Hoyle (Ref 13) reports that ion implantation of nitrogen into M2 steel thread-cutting dies for cast iron and gear-cutting tools results in increased life, but that very little improvement in the life of high-speed steel drills results from ion implantation. Minimal improvements in life may be a result of the very thin case depths of ion-implanted surfaces and the softening of these layers by heat generated in applications such as cutting tools. Ion-implanted forming tools not subjected to significant heating may benefit significantly from enhanced surface properties induced by ion implantation. Several tool steel applications where ion implantation has improved wear resistance have been listed by Hirvonen (Ref 14), and improvements in wear resistance of hardened M2, D2, and 420 steels by nitrogen implantation have been described (Ref 15). Examples of ion implantation in metal forming and cutting applications are listed in Table 23.

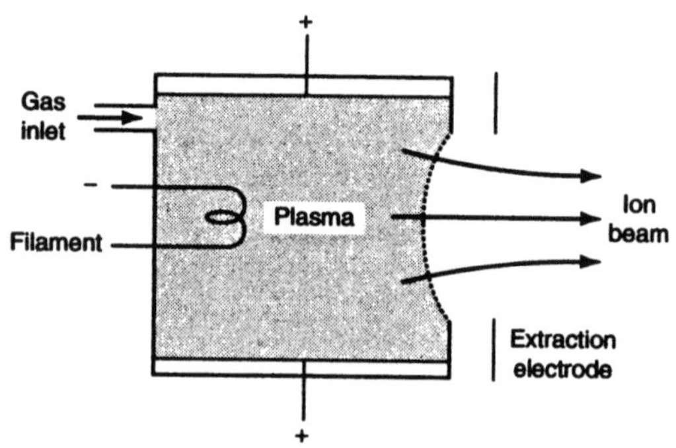

Fig. 13 Schematic of a typical ion source. Shown are an electron-emitting filament, anode, provision for gas input, and the ion extraction system

Laser Surface Processing

Lasers with continuous outputs of 0.5 to 10 kW can be used to modify the metallurgical structure of a surface and to tailor the surface properties without adversely affecting the bulk properties. The surface modification can take the following four forms for steels:

- Laser transformation hardening, in which a surface is heated so that thermal diffusion and solid-state transformations can take place. This process, which is the most commercially successful of the laser surface processing methods, is described in Chapter 10, "Surface Hardening by Applied Energy," in this book.
- Laser surface melting, which results in a refinement of the structure due to the rapid quenching from the melt
- Laser surface alloying, in which alloying elements are added to the melt pool to change the composition of the surface
- Laser cladding (hardfacing), which results in the deposition of a weld overlay coating onto the surface

An excellent review of laser surface processing can be found in Ref 16.

Laser Surface Melting

Microstructures. The most frequently studied steels for laser surface melting are tool steels. Figures 15 and 16 show the dramatic changes in surface microstructure produced by laser surface melting of M42 high-speed tool steel. M42 steel contains nominally 1% C, 8% Co, 1.5% W, 1.1% V, 3.75% Cr, and 9.5% Mo. The wrought microstructure contains a high volume fraction of coarse primary carbides because of the high content of carbide-forming elements. Figure 15 shows the considerable refinement of the microstructure of laser-melted M42 relative to chill-cast M42 and the absence of primary carbides in the laser melt zone. Dissolution of the carbides was a function of traverse speed, and at higher speeds carbides were not dissolved. Figure 16 shows the laser-melted

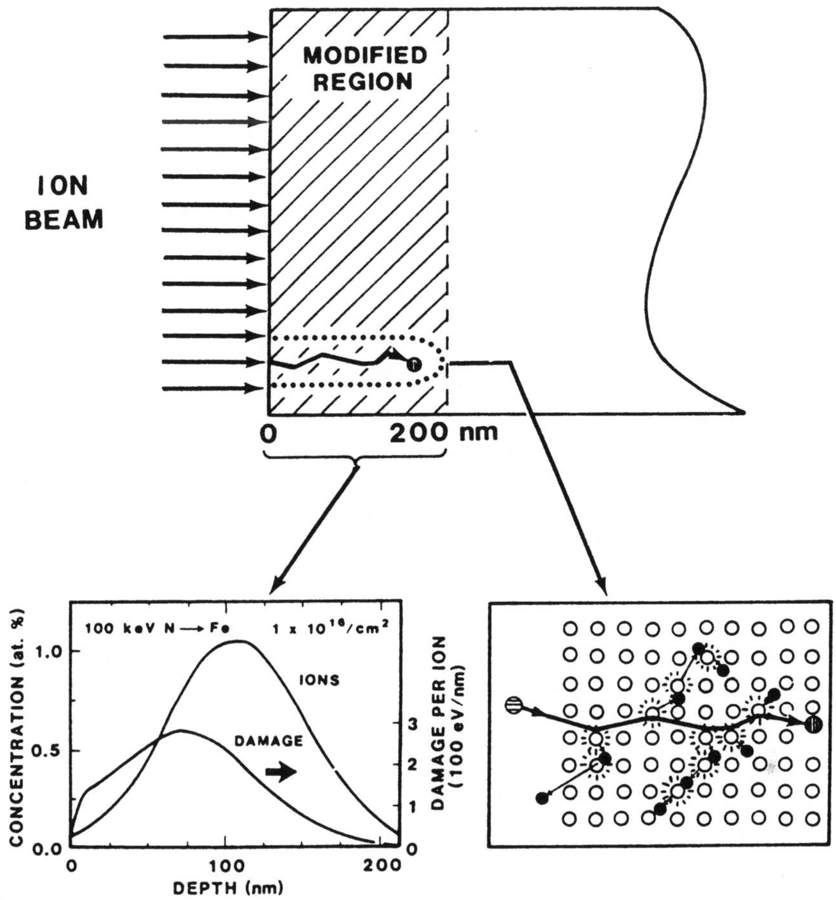

Fig. 14 Schematic of nitrogen atom implantation in iron (top), N and damage profiles (lower left), and defect generation (lower right)

Table 23 Examples of ion implantation in metalforming and cutting applications

Part	Part material	Process	Work material	Ion	Energy, keV	Benefit
Tool inserts	TiN-coated tool steel	Machining	4140	N	80	3 × life
Taps	HSS	Tapping	4140	N	80	3 × life
	HSS	Tapping	4130	N	80	5 × life
	HSS	Tapping	4140	N	50	10 × life
	M35	Tapping	...	N_2	200	4 × life
	M7	Tapping	...	N	100	2 × life
Cutting blade	M2	Cutting	1050	N	100	2 × life
	M2	Cutting	SAE 950	N	100	4 × life
Dies	D2	Forming	321 SS	N	80	2 × life
	M2	Forming	Steel	N	100	2–12 × life
	M2	Forming	1020	N	100	Negligible effect
	D6	Forming	TiO_2 and rubber	N	100	6 × life
Molds	D2	Forming	Polymers	N	50	5 × life
Rollers	H13	Rolling	Steel	N	100	5 × life

Note: HSS, high-speed steel; SS, stainless steel. Source: Ref 10 surface and melting around primary carbides in the matrix below the fine solidification structure of the melt zone. Melting of the carbides is due to a low melting eutectic reaction. The laser-melted surface layer, produced by slow traverse speeds, showed much higher peak hardness after triple tempering than conventionally treated steel does, apparently because of greater solution of alloying elements for subsequent carbide precipitation.

Properties. Hsu and Molian (Ref 18) reported that the tool life of laser-melted M2 steel tool bits was from 200 to 500% higher than if they were conventionally hardened using the catastrophic failure criterion (Fig. 17). For laser-melted M35 steel tool bits, the tool life was from 20 to 125% higher than if the bits were conventionally hardened using the flank wear failure criterion (Fig. 17). High-alloy martensite, fine austenite grain size, and finely dispersed carbides all contributed to high hardness, good toughness, and low coefficient of friction.

Laser Alloying (Ref 16)

Processing. A technique of localized alloy formation is laser surface melting with the simultaneous, controlled addition of alloying elements. These alloying elements diffuse rapidly into the melt pool, and the desired depth of alloying can be obtained in a short period of time. By this means, a desired alloy chemistry and microstructure can be generated on the sample surface; the degree of microstructural refinement will depend on the solidification rate. The surface of a low-cost alloy, such as mild steel, can be selectively alloyed to enhance properties, such as resistance to wear, in such a way that only the locally modified surface possesses properties typical of tribological alloys. This results in substantial cost savings and reduces the dependence on strategic materials. Typical processing parameters for laser alloying are a power density from 10 to 3000 MW/m^2 (6.5 to 1935 kW/in.^{2}) and an interaction time from 0.01 to 1 s. An inert shielding gas is normally used.

One method of alloying is to apply appropriate mixtures of powders on the sample surface, either by spraying the powder mixture suspended in alcohol to form a loosely packed coating, or by coating a slurry suspended in organic binders. The use of metal powders in laser alloying is the least expensive, but with appropriate process modifications, alloys in the form of rods, wires, ribbons, and sheets can also be added. Because of inconsistency in coating

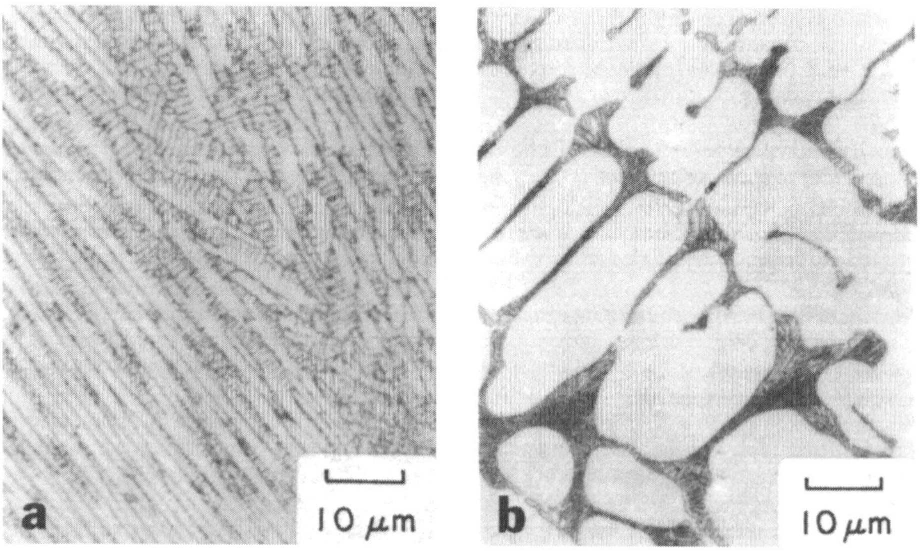

Fig. 15 Effect of laser surface melting on the structure of M42 high-speed steel. (a) Laser-melted dendritic structure. (b) Chill-cast dendritic structure. Source: Ref 17

application and possible loss during processing, the composition of the surface alloy may not reflect that of the applied alloying elements.

Powders that are added in controlled quantities using powder feeders with electronic metering can reduce this problem. Powders usually range from 10 to 50 μm (0.4 to 2 mils); the finer size facilitates dissolution and alloying. Typically, the powder flow rate is from 0.005 to 0.1 cm^3/s (0.0003 to 0.006 $in.^3/s$), and particle velocity is from 1 to 5 m/s (3.3 to 16.4 ft/s). Carrier gas velocity is typically from 3 to 10 m/s (10 to 33 ft/s). Whether preplaced or fed, the powder increases the coupling coefficient of the sur-

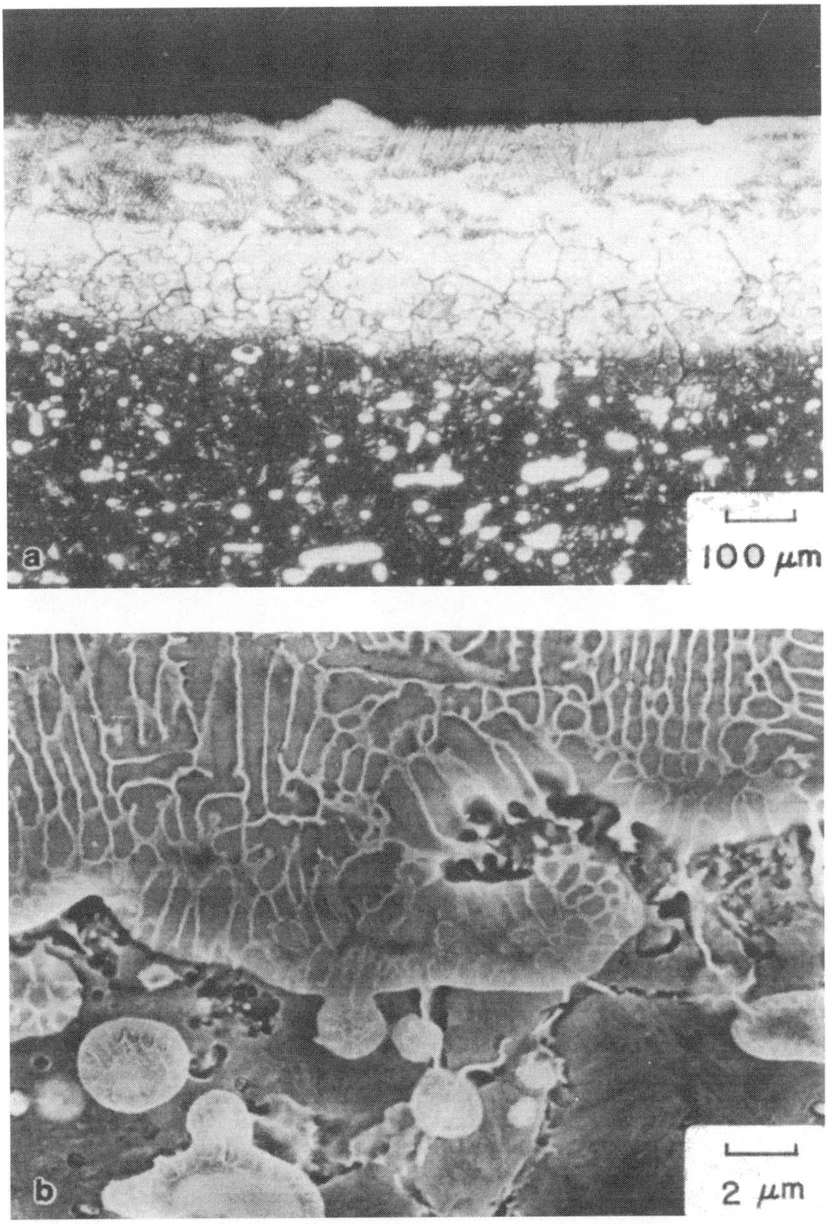

Fig. 16 Laser-melted surface layer on M42 high-speed steel. (a) Lower magnification view of surface cross section. (b) Higher-magnification view showing partial melting of carbides at the melt interface. Source: Ref 17

face, thereby avoiding the need for special absorptive coatings.

During surface alloying, temperature gradients form in the melt pool, which, along with the addition of alloying elements, influences the surface tension. Convection currents are established as a result of the surface tension gradients, and can cause variations in concentration. If the laser beam is oscillated, the melt spreads out over a wider region, and because the beam sweeps the same area several times, a potentially beneficial mixing action can occur. Any alloy concentration gradients that may have formed initially are thus diminished. If the melt pass is made with a very rapid sample translation rate that is greater than 50 mm/s (2 in./s), then inhomogeneities in the microstructure can result, but the high quench rates ensure minimal segregation. The rapid quench also facilitates alloying with hard-to-alloy elements, such as iron, chromium, carbon, and manganese.

Another method of laser alloying is gas reaction, for which a shielding gas of appropriate composition is chosen. For the nitriding of titanium alloys, a dilute (10 to 20%) mixture of N_2 in argon is used as the reactive gas. The extent

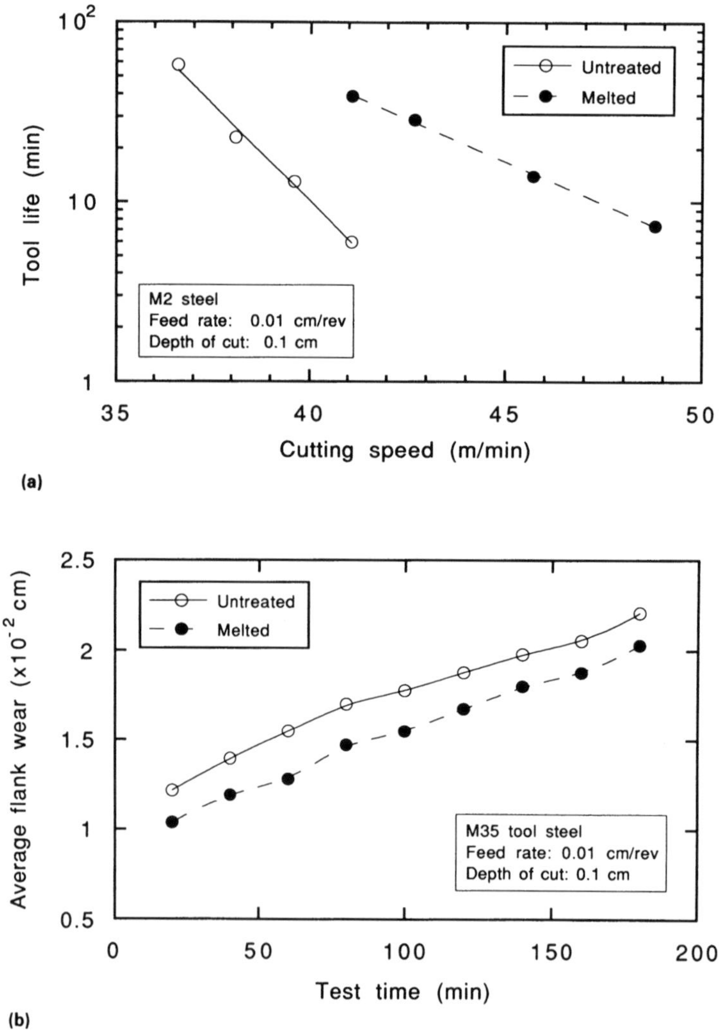

Fig. 17 Tool life of conventionally heat treated and laser-melted tool bits. (a) M2 tool steel. (b) M35 tool steel. Source: Ref 18

of nitriding depends on the partial pressure of N_2 in the atmosphere, which is determined by the law of mass action, and the temperature, which depends on the laser power density.

Applications of Surface Alloying. Examples of laser alloying are carbon steels with chromium, molybdenum, boron, and nickel, and stainless steels with carbon. Laser alloying has been primarily applied to improve corrosion resistance, but wear resistance can also be improved. For example, surface alloying of AISI 1018 steel with carbon and chromium produces stable carbides, such as M_7C_3 and M_3C, in austenitic, pearlitic, or martensitic matrices. The microhardness of these carbides is from 1100 to 1200 HK, and, when uniformly dispersed in a pearlitic or martensitic matrix, results in an improved resistance to abrasive and adhesive wear. With a 16% addition of chromium, the microstructure of laser-alloyed 1018 steel is martensitic, with small islands of ferrite as shown in Fig. 18(a). With additions of 43% Cr and 4.4% C, the microstructure consists of hexagonal-shaped M_7C_3 carbides, as shown in Fig. 18(b).

Laser Cladding (Ref 19)

Laser cladding, also commonly referred to as laser *hardfacing,* differs little in principle from traditional forms of hardfacing; the primary difference is the use of a high-energy laser beam heat source rather than an arc or gas flame. Laser beams offer potential in applying thin overlays or when access to the surface to be hardfaced can be achieved more readily by a laser beam than with an electrode or torch.

Cobalt-, nickel-, and tungsten carbide-base hardfacing alloys are the usual cladding materials used for laser hardfacing. As with conventional hardfacing methods, the materials are used in applications involving metal-to-metal contact, impact, erosion, and abrasion wear resistance. Other laser-cladding materials include titanium carbide, Fe-Cr-Ni-B alloys, aluminum bronzes, and ceramics. Substrates have included carbon and low-alloy steels, stainless steels, and tool steels.

Processing. The hardfacing alloy is melted by a laser beam and allowed to spread freely and freeze over the substrate. The beam also melts a very thin layer of the substrate, which combines with the liquid weld metal to the least extent necessary and solidifies to form a strong metallurgical bond. A good fusion bond can be achieved with a dilution zone that is only 10 to 20 μm thick.

The hardfacing alloy can be in several forms, examples of which are a prealloyed powder that is applied to the sample surface with or without a binder, a self-fluxing powder that is flame-sprayed, a hardfacing alloy that is plasma-

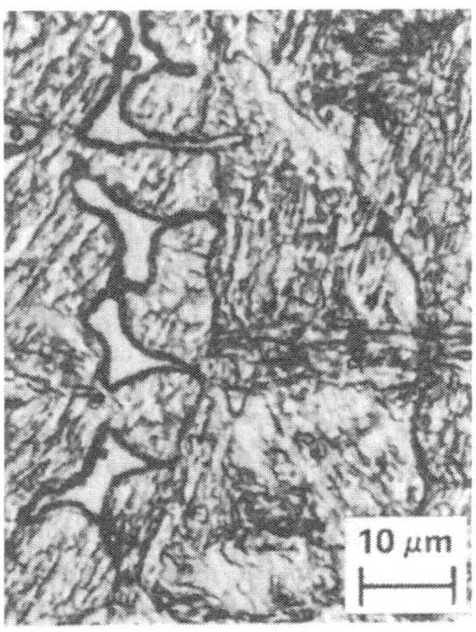

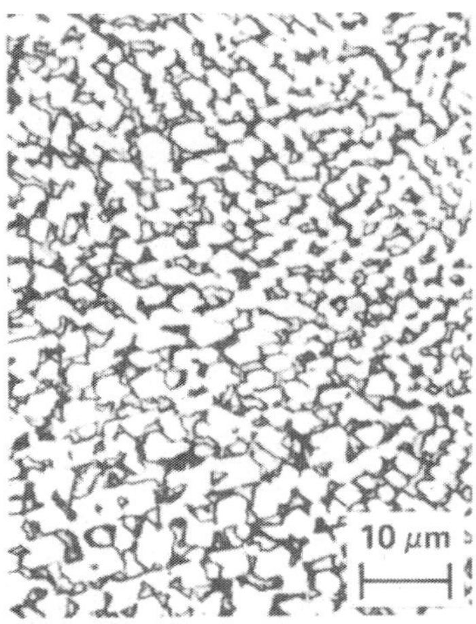

Fig. 18 Microstructures of laser-alloyed 1018 steel. (a) Addition of 16% Cr. (b) Addition of 43% Cr and 4.4% C

sprayed, or a chip that is preplaced. Laser consolidation of these coatings results in densification and smoothening, eliminates channels to the substrate, improves the bonding between coating and substrate, and reduces porosity, all of which contribute to the strength and integrity of the hardfacing layer. Processing parameters for laser cladding are a power density that ranges from 10 to 1000 MW/m^2 and an interaction time from 0.1 to 1 s. The shielding gas could be any of the inert gases or a combination of gases, such as He/Ar and H_2/Ar.

One of the more successful applications of laser cladding involves a proprietary process that utilizes a specially designed powder-feed apparatus. Using such equipment, the powder and an assist gas are fed to the weld area through a ceramic nozzle; a shielding gas of helium and argon surrounds the powder-gas mixture as it leaves the nozzle (Fig. 19). The powder delivery nozzle is positioned in such a manner as to flood the entire melt pool with powder. The feeding angle is generally from 35 to 45° to the horizontal, and the feed tube, which typically has a 3 mm (0.12 in.) diam, is positioned 10 to 12 mm (0.4 to 0.5 in.) from the substrate. Typically, the powder flow rate is from 0.005 to 0.1 cm^3/s (0.0003 to 0.006 $in.^3/s$), particle velocity is from 1 to 2 m/s (3.3 to 6.6 ft/s), and carrier gas velocity is from 3 to 7 m/s (10 to 23 ft/s). Desirable dilutions are from 3 to 8%. Overlay thickness can be varied from 0.15 to 4 mm (0.005 to 0.15 in.). Uniform feeding of the powder ensures uniform surfacing layers. Unlike plasma sprayed coatings, porosity and unmelted powder particles are almost never observed within a laser hardfacing layer.

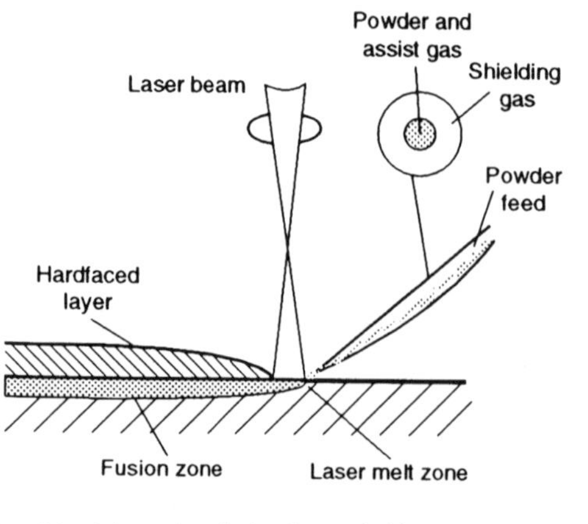

Fig. 19 Schematic of the laser cladding process using dynamic powder feed

Laser-clad deposits are commonly achieved by relatively large doughnut-shape beams or by somewhat focused beams that are oscillated. These ensure a desirable deposit profile. Hardfacing material can also be added in rod, wire, or sheet form, but special procedures are needed because of reflectivity problems.

Thermal stresses in the weld metal can cause harmful cracking, but this can be eliminated by an appropriate preheating practice. Low power densities, large beam diameters, and slow sample translation rates tend to produce crack-free deposits.

REFERENCES

1. G. Krauss, Advanced Surface Modification of Steels, *J. Heat Treat.,* Vol 9, 1992, p 81–89
2. D.T. Gawne and U. Ma, Friction and Wear of Chromium and Nickel Coatings, *Wear,* Vol 129, 1989, p 123
3. A.R. Jones, Corrosion of Electroplated Hard Chromium, *Corrosion,* Vol 13, *ASM Handbook,* ASM International, 1987, p 871–875
4. M. Cymboliste, The Structure and Hardness of Electrochemical Chromium, *J. Electrochem. Soc.,* Vol 73, 1938, p 353–363
5. W.E. Moline, Corrosion Resistance of Chromium Plated and Surface Conditioned 13 Per Cent Chromium Steel, *Mon. Rev. Am. Electroplat. Soc.,* (No. 4), April 1946, p 401–408
6. R. Kausalya and N.V. Parhasaradhy, Chromium Electrodeposits with Improved Corrosion Resistance, *Plating,* Vol 57 (No. 12), 1970, p 1238–1249
7. Nickel Coatings, *ASM Specialty Handbook: Nickel, Cobalt, and Their Alloys,* J.R. Davis, Ed., ASM International, 2000, p 106–123
8. S. Grainger, Ed., *Engineering Coatings: Design and Application,* Abington Publishing, 1989, p 33, 77
9. G. Roberts, G. Krauss, and R. Kennedy, *Tool Steels,* 5th ed., ASM International, 1998, p 309–313
10. J.R. Davis, Surface Engineering of Specialty Steels, *Surface Engineering,* Vol 5,

ASM Handbook, ASM International, 1994, p 762–775
11. A. Matthews, Titanium Nitride PVD Coating Technology, *Surf. Eng.,* Vol 1 (No. 2), 1985, p 93–104
12. M. Podob, CVD and PVD Functional Hard Coatings: Current Market Trends, *Surface Modification Technologies XII,* T.S. Sudarshan, K.A. Khor, and M. Jeandin, Ed., ASM International, 1998, p 15–24
13. G. Hoyle, *High Speed Steels,* Butterworths, London, 1988, p 166–193
14. J.K. Hirvonen, The Industrial Applications of Ion Beam Processes, *Surface Alloying by Ion, Electron, and Laser Beams,* L.E. Rehn, S.T. Picraux, and H. Wiedersich, Ed., American Society for Metals, 1987, p 373–388
15. J.I. Onate, F. Alonso, J.K. Dennis, and S. Hamilton, Microindentation and Tribological Study of Nitrogen Implanted Martensitic Steels, *Surf. Eng.,* Vol 8 (No. 3), 1992, p 199–205
16. K.P. Cooper, Laser Surface Processing, *Friction, Lubrication, and Wear Technology,* Vol 18, *ASM Handbook,* ASM International, 1992, p 861–872
17. T. Bell, I.M. Hancock, and A. Boyce, Laser Surface Treatment of Tool Steels, *Tool Materials for Molds and Dies,* G. Krauss and H. Nordberg, Ed., Colorado School of Mines Press, 1987, p 197–217
18. M. Hsu and P.A. Molian, *Wear,* Vol 127, 1988, p 253
19. J.R. Davis, Hardfacing, Weld Cladding, and Dissimilar Metal Joining, *Welding, Brazing, and Soldering,* Vol 6, *ASM Handbook,* ASM International, 1993, p 789–829

APPENDIX 1

Iron-Carbon Phase Diagram

THE BASIS for the understanding of the heat treatment of steels is the Fe-C phase diagram (Fig. 1). Actually, two diagrams are shown in the figure: the stable iron-graphite diagram (dashed lines) and the metastable Fe-Fe_3C diagram. The stable condition usually takes a very

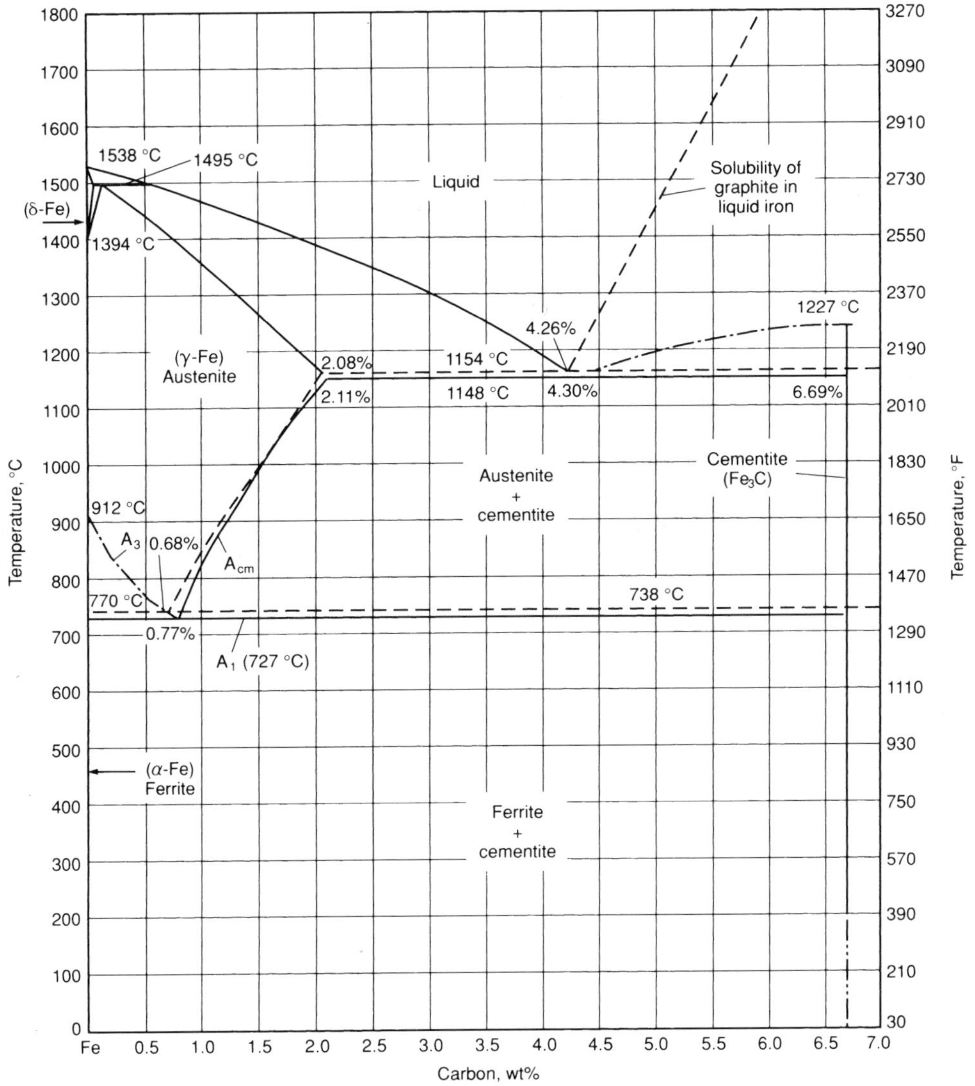

Fig. 1 The Fe-C equilibrium diagram up to 6.67 wt% C. Solid lines indicate Fe-Fe_3C diagram; dashed lines indicate iron-graphite diagram.

Table 1 Important metallurgical phases and microconstituents

Phase (microconstituent)	Crystal structure of phases(a)	Characteristics
Ferrite (α-iron)	bcc	Relatively soft low-temperature phase; stable equilibrium phase
δ-ferrite (δ-iron)	bcc	Isomorphous with α-iron; high-temperature phase; stable equilibrium phase
Austenite (γ-iron)	fcc	Relatively soft medium-temperature phase; stable equilibrium phase
Cementite (Fe_3C)	Complex orthorhombic	Hard metastable phase
Graphite	Hexagonal	Stable equilibrium phase
Pearlite		Metastable microconstituent; lamellar mixture of ferrite and cementite
Martensite	bct (supersaturated solution of carbon in ferrite)	Hard metastable phase; lath morphology when <0.6 wt% C; plate morphology when >1.0 wt% C and mixture of those in between
Bainite	. . .	Hard metastable microconstituent; nonlamellar mixture of ferrite and cementite on an extremely fine scale; upper bainite formed at higher temperatures has a feathery appearance; lower bainite formed at lower temperatures has an acicular appearance. The hardness of bainite increases with decreasing temperature of formation.

(a) bcc, body-centered cubic; fcc, face-centered cubic; bct, body-centered tetragonal

long time to develop, especially in the low-temperature and low-carbon range, and therefore the metastable diagram is of more interest. The Fe-C diagram shows which phases are to be expected at equilibrium (or metastable equilibrium) for different combinations of carbon concentration and temperature.

Important metallurgical phases and microconstituents are summarized in Table 1.

At the low-carbon end of the phase diagram are ferrite (α-iron), which can at most dissolve 0.028 wt% C at 727 °C (1341 °F) and austenite (γ-iron), which can dissolve 2.11 wt% C at 1148 °C (2098 °F). At the carbon-rich side is cementite (Fe_3C). Of less interest, except for highly alloyed steels, is the δ-ferrite existing at the highest temperatures.

Between the single-phase fields are found regions with mixtures of two phases, such as ferrite + cementite, austenite + cementite, and ferrite + austenite. At the highest temperatures, the liquid phase field can be found, and below this are the two-phase fields liquid + austenite, liquid + cementite, and liquid + δ-ferrite. In heat treating of steels, the liquid phase is always avoided.

Some important boundaries at single-phase fields have been given special names. These include:

- A_1, the so-called eutectoid temperature, which is the minimum temperature for austenite
- A_3, the lower-temperature boundary of the austenite region at low carbon contents; that is, the $\gamma/\gamma + \alpha$ boundary
- A_{cm}, the counterpart boundary for high carbon contents; that is, the $\gamma/\gamma + Fe_3C$ boundary

Sometimes the letters c, e, or r are included. Relevant definitions of terms associated with phase transformations of steels are listed in Table 2.

The carbon content at which the minimum austenite temperature is attained is called the

Table 2 Definitions of transformation temperatures in iron and steels

Transformation temperature. The temperature at which a change in phase occurs. The term is sometimes used to denoted the limiting temperature of a transformation range. The following symbols are used for iron and steels.

Ac_{cm}. In hypereutectoid steel, the temperature at which the solution of cementite in austenite is completed during heating.

Ac_1. The temperature at which austenite begins to form during heating, with the c being derived from the French *chauffant*.

Ac_3. The temperature at which transformation of ferrite to austenite is completed during heating.

Ae_{cm}, Ae_1, Ae_3. The temperatures of phase changes at equilibrium.

Ar_{cm}. In hypereutectoid steel, the temperature at which precipitation of cementite starts during cooling, with the r being derived from the French *refroidissant*.

Ar_1. The temperature at which transformation of austenite to ferrite or to ferrite plus cementite is completed during cooling.

Ar_3. The temperature at which austenite begins to transform to ferrite during cooling.

Ar_4. The temperature at which delta ferrite transforms to austenite during cooling.

M_s (or Ar''). The temperature at which transformation of austenite to martensite starts during cooling.

M_f. The temperature at which martensite formation finishes during cooling.

Note all these changes, except the formation of martensite, occur at lower temperatures during cooling than during heating and depend on the rate of change of temperature.

eutectoid carbon content (0.77 wt% C). The ferrite-cementite phase mixture of this composition formed during cooling has a characteristic appearance and is called pearlite. Pearlite can be treated as a microstructural entity or microconstituent. It is an aggregate of alternating ferrite and cementite lamellae that degenerates ("spheroidizes" or "coarsens") into cementite particles dispersed with a ferrite matrix after extended holding at a temperature close to A_1.

APPENDIX 2

Austenitizing Temperatures for Steels

TEMPERATURES RECOMMENDED for austenitizing carbon and low-alloy steels prior to hardening are given in Table 1 (for direct-hardening grades) and Table 2 (for carburized steels). Table 2 is applicable to carburized steels that have been cooled slowly from the carburizing temperature and are to be furnace hardened in a subsequent operation.

For most applications, the rate of heating to the austenitizing temperature is less important than other factors in the hardening process, such as maximum temperature attained throughout the section, temperature uniformity, time at temperature, and rate of cooling. The thermal conductivity of the steel, the nature of the furnace atmosphere (scaling or nonscaling), thickness of section, method of loading (spaced or stacked), and the degree of circulation of the furnace atmosphere all influence the rate of heating of the steel part to the required temperature selected from Tables 1 and 2.

The difference in temperature rise within thick and thin sections of articles of varying cross section is a major problem in practical heat-treating operations. When temperature uniformity is the ultimate objective of the heating cycle, this is more safely attained by slowly heating than by rapidly heating. Furthermore, the maximum temperature in the austenite range should not exceed that required to achieve the necessary extent of solution of carbide. The temperatures listed in Tables 1 and 2 conform with this requirement. When heating with significant cross-section variations, provisions should be made for slower heating to minimize thermal stresses and distortions.

Table 1 Austenitizing temperatures for direct-hardening carbon and alloy steels (SAE)

	Temperature	
Steel	°C	°F
Carbon steels		
1025	855–900	1575–1650
1030	845–870	1550–1600
1035	830–855	1525–1575
1037	830–855	1525–1575
1038(a)	830–855	1525–1575
1039(a)	830–855	1525–1575
1040(a)	830–855	1525–1575
1042	800–845	1475–1550
1043(a)	800–845	1475–1550
1045(a)	800–845	1475–1550
1046(a)	800–845	1475–1550
1050(a)	\800–845	1475–1550
1055	800–845	1475–1550
1060	800–845	1475–1550
1065	800–845	1475–1550
1070	800–845	1475–1550
1074	800–845	1475–1550
1078	790–815	1450–1500
1080	790–815	1450–1500
1084	790–815	1450–1500
1085	790–815	1450–1500
1086	790–815	1450–1500
1090	790–815	1450–1500
1095	790–815(a)	1450–1500(b)
Free-cutting carbon steels		
1137	830–855	1525–1575
1138	815–845	1500–1550
1140	815–845	1500–1550
1141	800–845	1475–1550
1144	800–845	1475–1550
1145	800–845	1475–1550
1146	800–845	1475–1550
1151	800–845	1475–1550
1536	815–845	1500–1550
1541	815–845	1500–1550
1548	815–845	1500–1550
1552	815–845	1500–1550
1566	855–885	1575–1625
Alloy steels		
1330	830–855	1525–1575
1335	815–845	1500–1550
1340	815–845	1500–1550
1345	815–845	1500–1550
3140	815–845	1500–1550
4037	830–855	1525–1575
4042	830–855	1525–1575
4047	815–855	1500–1575
4063	800–845	1475–1550
4130	815–870	1500–1600
4135	845–870	1550–1600
4137	845–870	1550–1600
4140	845–870	1550–1600
4142	845–870	1550–1600
4145	815–845	1500–1550
4147	815–845	1500–1550

(continued)

Table 1 (Continued)

Steel	Temperature °C	Temperature °F	Steel	Temperature °C	Temperature °F
Alloy Steels (continued)					
4150	815–845	1500–1550	51100	775–800(c)	1425–1475(c)
4161	815–845	1500–1550	52100	775–800(c)	1425–1475(c)
4337	815–845	1500–1550	6150	845–885	1550–1625
4340	815–845	1500–1550	81B45	815–855	1500–1575
50B40	815–845	1500–1550	8630	830–870	1525–1600
50B44	815–845	1500–1550	8637	830–855	1525–1575
5046	815–845	1500–1550	8640	830–855	1525–1575
50B46	815–845	1500–1550	8642	815–855	1500–1575
50B50	800–845	1475–1550	8645	815–855	1500–1575
50B60	800–845	1475–1550	86B45	815–855	1500–1575
5130	830–855	1525–1575	8650	815–855	1500–1575
5132	830–855	1525–1575	8655	800–845	1475–1550
5135	815–845	1500–1550	8660	800–845	1475–1550
5140	815–845	1500–1550	8740	830–855	1525–1575
5145	815–845	1500–1550	8742	830–855	1525–1575
5147	800–845	1475–1550	9254	815–900	1500–1650
5150	800–845	1475–1550	9255	815–900	1500–1650
5155	800–845	1475–1550	9260	815–900	1500–1650
5160	800–845	1475–1550	94B30	845–885	1550–1625
51B60	800–845	1475–1550	94B40	845–885	1550–1625
50100	775–800(c)	1425–1475(c)	9840	830–855	1525–1575

(a) Commonly used on parts where induction hardening is employed. All steels from SAE 1030 up may have induction hardening applications. (b) This temperature range may be employed for 1095 steel that is to be quenched in water, brine, or oil. For oil quenching, 1095 steel may alternatively be austenitized in the range 815 to 870 °C (1500 to 1600 °F). (c) This range is recommended for steel that is to be water quenched. For oil quenching, steel should be austenitized in the range 815 to 870 °C (1500 to 1600 °F).

Table 2 Reheating (austenitizing) temperatures for hardening of carburized carbon and alloy steels (SAE)

Carburizing is commonly carried out at 900 to 925 °C (1650 to 1700 °F), slow cooled and reheated to given austenitizing temperature.

Steel	Temperature °C	Temperature °F	Steel	Temperature °C	Temperature °F
Carbon steels			**Alloy steels**		
1010	760–790	1400–1450	3310	790–830	1450–1525
1012	760–790	1400–1450	4320	830–845	1525–1550
1015	760–790	1400–1450	4615	815–845	1500–1550
1016	760–790	1400–1450	4617	815–845	1500–1550
1017	760–790	1400–1450	4620	815–845	1500–1550
1018	760–790	1400–1450	4621	815–845	1500–1550
1019	760–790	1400–1450	4626	815–845	1500–1550
1020	760–790	1400–1450	4718	815–845	1500–1550
1022	760–790	1400–1450	4720	815–845	1500–1550
1513	760–790	1400–1450	4815	800–830	1475–1525
1518	760–790	1400–1450	4817	800–830	1475–1525
1522	760–790	1400–1450	4820	800–830	1475–1525
1524	760–790	1400–1450	8115	845–870	1550–1600
1525	760–790	1400–1450	8615	845–870	1550–1600
1526	760–790	1400–1450	8617	845–870	1550–1600
1527	760–790	1400–1450	8620	845–870	1550–1600
Free-cutting carbon steels			8622	845–870	1550–1600
			8625	845–870	1550–1600
1109	760–790	1400–1450	8627	845–870	1550–1600
1115	760–790	1400–1450	8720	845–870	1550–1600
1117	760–790	1400–1450	8822	845–870	1550–1600
1118	760–790	1400–1450	9310	790–830	1450–1525

APPENDIX 3

Hardness Conversion Tables

Table 1 Approximate equivalent hardness numbers for nonaustenitic steels (Rockwell C hardness range)

For carbon and alloy steels in the annealed, normalized, and quenched-and-tempered conditions.

Rockwell C hardness No., 150 kgf, HRC	Vickers hardness No., HV	Brinell hardness No. 10 mm standard ball, 3000 kgf, HBS	Brinell hardness No. 10 mm carbide ball, 3000 kgf, HBW	Knoop hardness No., 500 gf and over, HK	Rockwell hardness No. A scale, 60 kgf, HRA	Rockwell hardness No. D scale, 100 kgf, HRD	Rockwell superficial hardness No. 15-N scale, 15 kgf, HR-15-N	Rockwell superficial hardness No. 30-N scale, 30 kgf, HR 30-N	Rockwell superficial hardness No. 45-N scale, 45 kgf, HR 45-N	Scleroscope hardness No.	Rockwell C tensile strength (approximate), ksi
68	940	...	...	920	85.6	76.9	93.2	894.4	75.4	97.3	...
67	900	...	...	895	85.0	76.1	92.9	83.6	74.2	95.0	...
66	865	...	...	870	84.5	75.4	92.5	82.8	73.3	92.7	...
65	832	...	(739)	846	83.9	74.5	92.2	81.9	72.0	90.6	...
64	800	...	(722)	822	83.4	73.8	91.8	81.1	71.0	88.5	...
63	772	...	(705)	799	82.8	73.0	91.4	80.1	69.9	86.5	...
62	746	...	(688)	776	82.3	72.2	91.1	79.3	68.8	84.5	...
61	720	...	(670)	754	81.8	71.5	90.7	78.4	67.7	82.6	...
60	697	...	(654)	732	81.2	70.7	90.2	77.5	66.6	80.8	...
59	674	...	(634)	710	80.7	69.9	89.8	76.6	65.5	79.0	351
58	653	...	615	690	80.1	69.2	89.3	75.7	64.3	77.3	338
57	633	...	595	670	79.6	68.5	88.9	74.8	63.2	75.6	325
56	613	...	577	650	79.0	67.7	88.3	73.9	62.0	74.0	313
55	595	...	560	630	68.5	66.9	87.9	73.0	60.9	72.4	301
54	577	...	543	612	78.0	66.1	87.4	72.0	59.8	70.9	292
53	560	...	525	594	77.4	65.4	86.9	71.2	58.6	69.4	283
52	544	(500)	512	576	76.8	64.6	86.4	70.2	57.4	67.9	273
51	528	(487)	496	558	76.3	63.8	85.9	69.4	56.1	66.5	264
50	513	(475)	481	542	75.9	63.1	85.5	68.5	55.0	65.1	255
49	498	(464)	469	526	75.2	62.1	85.0	67.6	53.8	63.7	246
48	484	451	455	510	74.7	61.4	84.5	66.7	52.5	62.4	238
47	471	442	443	495	74.1	60.8	83.9	65.8	51.4	61.1	229
46	458	432	432	480	73.6	60.0	83.5	64.8	50.3	59.8	221
45	446	421	421	466	73.1	59.2	83.0	64.0	49.0	58.5	215
44	434	409	409	452	72.5	58.5	82.5	63.1	47.8	57.3	208
43	423	400	400	438	72.0	57.7	82.0	62.2	46.7	56.1	201
42	412	390	390	426	71.5	56.9	81.5	61.3	45.5	54.9	194
41	402	381	381	414	70.9	56.2	80.9	60.4	44.3	53.7	188
40	392	371	371	402	70.4	55.4	80.4	59.5	43.1	52.6	182
39	382	362	362	391	69.9	54.6	79.9	58.6	41.9	51.5	177
38	372	353	353	380	69.4	53.8	79.4	57.7	40.8	50.4	171
37	363	344	344	370	68.9	53.1	78.8	56.8	39.6	49.3	166
36	354	336	336	360	68.4	52.3	78.3	55.9	38.4	48.2	161
35	345	327	327	351	67.9	51.5	77.7	55.0	37.2	47.1	156
34	336	319	319	342	67.4	50.8	77.2	54.2	36.1	46.1	152
33	327	311	311	334	66.8	50.0	76.6	53.3	34.9	45.1	149
32	318	301	301	326	66.3	49.2	76.1	52.1	33.7	44.1	146
31	310	294	294	318	65.8	48.4	75.6	51.3	32.5	43.1	141
30	302	286	286	311	65.3	47.7	75.0	50.4	31.3	42.2	138
29	294	279	279	304	64.8	47.0	74.5	49.5	30.1	41.3	135
28	286	271	271	297	64.3	46.1	73.9	48.6	28.9	40.4	131
27	279	264	264	290	63.8	45.2	73.3	47.7	27.8	39.5	128
26	272	258	258	284	63.3	44.6	72.8	46.8	26.7	38.7	125
25	266	253	253	278	62.3	43.8	72.2	45.9	25.5	37.8	123
24	260	247	247	272	62.4	43.1	71.6	45.0	24.3	37.0	119
23	254	243	243	266	62.0	42.1	71.0	44.0	23.1	36.3	117
22	248	237	237	261	61.5	41.6	70.5	43.2	22.0	35.5	115
21	243	231	231	256	61.0	40.9	69.9	42.3	20.7	34.8	112
20	238	226	226	251	60.5	40.1	69.4	41.5	19.6	34.2	110

Note: Values in parentheses are beyond the normal range and are presented for information only. Source: ASTM E 140

Table 2 Approximate equivalent hardness numbers for nonaustenitic steels (Rockwell B hardness range)

For carbon and alloy steels in the annealed, normalized, and quenched-and-tempered conditions

Rockwell B hardness No., 100 kgf, HRB	Vickers hardness No., HV	Brinell hardness No., 3000 kgf, HBS	Knoop hardness No., 500 gf and over, HK	Rockwell A hardness No., 60 kgf, HRA	Rockwell F hardness No., 60 kgf, HRF	Rockwell superficial hardness No.			Tensile strength (approximate), ksi	Rockwell B hardness No., 100 kgf, HRB
						15-T scale, 15 kgf, HR 15-T	30-T scale, 30 kgf, HR 30-T	45-T scale, 45 kgf, HR 45-T		
100	240	240	251	61.5	...	93.1	83.1	72.9	116	100
99	234	234	246	60.9	...	92.8	82.5	71.9	114	99
98	228	228	241	60.2	...	92.5	81.8	70.9	109	98
97	222	222	236	59.5	...	92.1	81.1	69.9	104	97
96	216	216	231	58.9	...	91.8	80.4	68.9	102	96
95	210	210	226	58.3	...	91.5	79.8	67.9	100	95
49	205	205	221	57.6	...	91.2	79.1	66.9	98	94
93	200	200	216	57.0	...	90.8	78.4	65.9	94	93
92	195	195	211	56.4	...	90.5	77.8	64.8	92	92
91	190	190	206	55.8	...	90.2	77.1	63.8	90	91
90	185	185	201	55.2	...	89.9	76.4	62.8	89	90
89	180	180	196	54.6	...	89.5	75.8	61.8	88	89
88	176	176	192	54.0	...	89.2	75.1	60.8	86	88
87	172	172	188	53.4	...	88.9	74.4	59.8	84	87
86	169	169	184	52.8	...	88.6	73.8	58.8	83	86
85	165	165	180	52.3	...	88.2	73.1	57.8	82	85
84	162	162	176	51.7	...	87.9	72.4	56.8	81	84
83	159	159	173	51.1	...	87.6	71.8	55.8	80	83
82	156	156	170	50.6	...	87.3	71.1	54.8	77	82
81	153	153	167	50.0	...	86.9	70.4	53.8	73	81
80	150	150	164	49.5	...	86.6	69.7	52.8	72	80
79	147	147	161	48.9	...	86.3	69.1	51.8	70	79
78	144	144	158	48.4	...	86.0	68.4	50.8	69	78
77	141	141	155	47.9	...	85.6	67.7	49.8	68	77
76	139	139	152	47.3	...	85.3	67.1	48.8	67	76
75	137	137	150	46.8	99.6	85.0	66.4	47.8	66	75
74	135	135	147	46.3	99.1	84.7	65.7	46.8	65	74
73	132	132	145	45.8	98.5	84.3	65.1	45.8	64	73
72	130	130	143	45.3	98.0	84.0	64.4	44.8	63	72
71	127	127	141	44.8	97.4	83.7	63.7	43.8	62	71
70	125	125	139	44.3	96.8	83.4	63.1	42.8	61	70
69	123	123	137	43.8	96.2	83.0	62.4	41.8	60	69
68	121	121	135	43.3	95.6	82.7	61.7	40.8	59	68
67	119	119	133	42.8	95.1	82.4	61.0	39.8	58	67
66	117	117	131	42.3	94.5	82.1	60.4	38.7	57	66
65	116	116	129	41.8	93.9	81.8	59.7	37.7	56	65
64	114	114	127	41.4	93.4	81.4	59.0	36.7	...	64
63	112	112	125	40.9	92.8	81.1	58.4	35.7	...	63
62	110	110	124	40.4	92.2	80.8	57.7	34.7	...	62
61	108	108	122	40.0	91.7	80.5	57.0	33.7	...	61
60	107	107	120	39.5	91.1	80.1	56.4	32.7	...	60
59	106	106	118	39.0	90.5	79.8	55.7	31.7	...	59
58	104	104	117	38.6	90.0	79.5	55.0	30.7	...	58
57	103	103	115	38.1	89.4	79.2	54.4	29.7	...	57
56	101	101	114	37.7	88.8	78.8	53.7	28.7	...	56
55	100	100	112	37.2	88.2	78.5	50.0	27.7	...	55
54	...	...	111	36.8	87.7	78.2	52.4	26.7	...	54
53	...	...	110	36.3	87.1	77.9	51.7	25.7	...	53
52	...	...	109	35.9	86.5	77.5	51.0	24.7	...	52
51	...	...	108	35.5	86.0	77.2	50.3	23.7	...	51
50	...	...	107	35.0	85.4	76.9	49.7	22.7	...	50
49	...	...	106	34.6	84.8	76.6	49.0	21.7	...	49
48	...	...	105	34.1	84.3	76.2	48.3	20.7	...	48
47	...	...	104	33.7	83.7	75.9	47.7	19.7	...	47
46	...	...	103	33.3	83.1	75.6	47.0	18.7	...	46
45	...	...	102	32.9	82.6	75.3	46.3	17.7	...	45
44	...	...	101	32.4	82.0	74.9	45.7	16.7	...	44
43	...	...	100	32.0	81.4	74.6	45.0	15.7	...	43
42	...	...	99	31.6	80.8	74.3	44.3	14.7	...	42
41	...	...	98	31.2	80.3	74.0	43.7	13.6	...	41
40	...	...	97	30.7	79.7	73.6	43.0	12.6	...	40
39	...	...	96	30.3	79.1	73.3	42.3	11.6	...	39
38	...	...	95	29.9	78.6	73.0	41.0	10.6	...	38
37	...	...	94	29.5	78.0	72.7	41.0	9.6	...	37
36	...	...	93	29.1	77.4	72.3	40.3	8.6	...	36

(continued)

Table 2 (continued)

Rockwell B hardness No., 100 kgf, HRB	Vickers hardness No., HV	Brinell hardness No., 3000 kgf, HBS	Knoop hardness No., 500 gf and over, HK	Rockwell A hardness No., 60 kgf, HRA	Rockwell F hardness No., 60 kgf, HRF	Rockwell superficial hardness No. 15-T scale, 15 kgf, HR 15-T	30-T scale, 30 kgf, HR 30-T	45-T scale, 45 kgf, HR 45-T	Tensile strength (approximate), ksi	Rockwell B hardness No., 100 kgf, HRB
35	...	...	92	28.7	76.9	72.0	39.6	7.6	...	35
34	...	...	91	28.2	76.3	71.7	39.0	6.6	...	34
33	...	...	90	27.8	75.7	71.4	38.3	5.6	...	33
32	...	...	89	27.4	75.2	71.0	37.6	4.6	...	32
31	...	...	88	27.0	74.6	70.7	37.0	3.6	...	31
30	...	...	87	26.6	74.0	70.4	36.3	2.6	...	30

Source: ASTM E 140

Subject Index

D

E

F

G

H

I

M

N

Q

R

T

U

V

W

X

Y

Z

Alloy Index

This alloy index is divided into four categories: stainless steels, specific types; steels, AISI-SAE, specific types; steels, proprietary classifications, specific types; and tool steels, specific types.

Stainless steels, specific types

Steels, AISI-SAE, specific types

Steels, proprietary classifications, specific types

Tool steels, specific types